Coral reef aquatic ecosystem

Lake ecosystem

Tropical forest biome

Dona Hutchins

Deciduous forest biome

Thomas E. Hemmerly

Environmental Science

Environmental Science
An Introduction

G. Tyler Miller, Jr.
St. Andrews Presbyterian College

Wadsworth Publishing Company
Belmont, California
A Division of Wadsworth, Inc.

Science Editor: Jack Carey

Editorial Associate: Ruth Singer

Production Editor: Gary Mcdonald

Managing Designers: Cynthia Bassett, Merle Sanderson

Print Buyer: Karen Hunt

Copy Editor: Brenda Griffing

Illustrators: Darwen and Vally Hennings, John and
          Judith Waller

Cover and Endpaper Design: Merle Sanderson

Cover Photograph: Jim Brandenberg/Woodfin Camp &
          Associates. The photo shows a young dog
          fox on the Alaskan tundra.

Books in the Wadsworth Biology Series

Biology: The Unity and Diversity of Life, 3rd, Starr and Taggart

Energy and Environment: The Four Energy Crises, 2nd, Miller

Fishes: A Field and Laboratory Manual on Their Structure, Identification, and Natural History, Cailliet, et al.

Living in the Environment, 4th, Miller

Oceanography: An Introduction, 3rd, Ingmanson and Wallace

Biology books under the editorship of William A. Jensen, University of California, Berkeley

Biology: The Foundations, 2nd, Wolfe

Biology of the Cell, 2nd, Wolfe

Cell Ultrastructure, Wolfe

Introduction to Cell Biology, Wolfe

Botany, 2nd, Jensen and Salisbury

Plant Physiology, 3rd, Salisbury and Ross

Plant Physiology Laboratory Manual, Ross

Plants: An Evolutionary Survey, 2nd, Scagel et al.

Printed in the United States of America

1 2 3 4 5 6 7 8 9 10—90 89 88 87 86

ISBN 0-534-05352-1

Library of Congress Cataloging in Publication Data

Miller, G. Tyler (George Tyler), 1931–
    Environmental science, an introduction.

    Bibliography: p.
    Includes index.
    1. Ecology.   2. Human ecology.   3. Natural
resources.   4. Pollution.   I. Title.
QH541.M515   1986   304.2   85-10883
ISBN 0-534-05352-1

The environmental crisis is an outward manifestation of a crisis of mind and spirit. There could be no greater misconception of its meaning than to believe it to be concerned only with endangered wildlife, human-made ugliness, and pollution. These are part of it, but more importantly, the crisis is concerned with the kind of creatures we are and what we must become in order to survive.

Lynton K. Caldwell

This book is dedicated to Mother Earth, who sustains us and all other creatures, and to my life partner, spouse, and best friend, Peggy Sue O'Neal, who understands and attempts to live by the message of this book and who has helped me better understand and appreciate the beauty and complexity of nature.

v

Brief Contents

Part One **Humans and Nature: An Overview** 1

1 Population, Resources, and Pollution 2
2 Human Impact on the Earth 12

Part Two **Some Concepts of Ecology** 19

3 Some Matter and Energy Laws 20
4 Ecosystem Structure: What Is an Ecosystem? 27
5 Ecosystem Function: How Do Ecosystems Work? 44
6 Changes in Ecosystems: What Can Happen to Ecosystems? 57

Part Three **Population** 69

7 Human Population Dynamics 70
8 Human Population Control 81

Part Four **Resources** 91

9 Soil Resources 92
10 Water Resources 106
11 Food Resources and World Hunger 121
12 Land Resources: Wilderness, Parks, Forests, and Rangelands 138
13 Wildlife Resources 151
14 Urban Land Use and Land-Use Planning 164
15 Nonrenewable Mineral Resources 178
16 Energy Resources: Types, Use, and Concepts 191

17 Nonrenewable Energy Resources: Fossil Fuels and Geothermal Energy 208
18 Nonrenewable Energy Resources: Nuclear Energy 222
19 Renewable Energy Resources 239

Part Five **Pollution** 263

20 The Environment and Human Health: Disease, Food Additives, and Noise 264
21 Air Pollution 283
22 Water Pollution 307
23 Pesticides and Pest Control 328
24 Solid Waste and Hazardous Wastes 346

Epilogue Achieving a Sustainable Earth Society 361

Appendixes 363

1 Periodicals 363
2 Units of Measurement 365
3 Major U.S. Environmental Legislation 366
4 Ways to Save Water 367
5 Ways to Save Energy 368
6 National Ambient Air Quality Standards for the United States in 1985 370
7 Federal Drinking Water Standards for the United States in 1985 371

Readings 372

Glossary 379

Index 391

Detailed Contents

Part One   Humans and Nature: An Overview   1

Chapter 1   Population, Resources, and Pollution   2

1-1   A Crisis of Interlocking Problems   2

1-2   Population   2

1-3   Natural Resources   3

1-4   Pollution   6

1-5   Relationships Among Pollution, Population, Resource Use, and Technology   9

1-6   Some Hopeful Trends   9

Chapter 2   Human Impact on the Earth   12

2-1   Hunter-Gatherer Societies   12

2-2   Agricultural Societies   13

2-3   Industrial Societies   14

2-4   Brief History of Natural Resource Conservation and Environmental Protection in the United States   14

Part Two   Some Concepts of Ecology   19

Chapter 3   Some Matter and Energy Laws   20

3-1   Law of Conservation of Matter: Everything Must Go Somewhere   20

3-2   First Law of Energy: You Can't Get Something for Nothing   20

3-3   Second Law of Energy: You Can't Break Even   22

3-4   Matter and Energy Laws and Environmental Problems   24

Chapter 4   Ecosystem Structure: What Is an Ecosystem?   27

4-1   The Ecosphere: Our Life-Support System   27

4-2   The Sun: Source of Energy for Life on Earth   27

4-3   Ecosystems and Ecosystem Structure   28

4-4   Ecosystem Components   31

4-5   Global Patterns of Climate   36

4-6   Limiting Factors in Ecosystems   38

4-7   Biomes: A Look at Major Land Ecosystems   40

Chapter 5   Ecosystem Function: How Do Ecosystems Work?   44

5-1   Energy Flow and Chemical Cycling   44

5-2   Energy Flow in Ecosystems: Food Chains and Food Webs   45

5-3   Chemical Cycling in Ecyosystems: Carbon, Oxygen, Nitrogen, Phosphorus, and Water Cycles   48

5-4   The Niche Concept   52

5-5   Species Interactions in Ecosystems   54

Chapter 6   Changes in Ecosystems: What Can Happen to Ecosystems?   57

6-1   Ecological Succession   57

6-2   Stability in Living Systems   59

6-3   What Can Go Wrong in an Ecosystem?   63

6-4   Humans and Ecosystems   66

Part Three   Population   69

Chapter 7   Human Population Dynamics   70

7-1   Major Factors Affecting Population Change   70

7-2   Birth Rate and Death Rate   70

7-3   Migration   72

7-4   Fertility Rate   74

7-5   Age Structure   76

7-6   Average Marriage Age   79

7-7   The Geography of Population: The Rich–Poor Gap   79

Chapter 8   Human Population Control   81

8-1   Factors Affecting Maximum Population Size   81

8-2   Should Population Growth Be Controlled?   82

8-3 Methods for Controlling Human Population Growth 84

8-4 Efforts at Human Population Control: India and China 88

Part Four Resources 91

Chapter 9 Soil Resources 92

9-1 Components, Formation, and Physical Properties of Soils 92

9-2 Soil Profiles 95

9-3 Soil Erosion 96

9-4 Soil Conservation 100

Chapter 10 Water Resources 106

10-1 Water's Unique Properties 106

10-2 Worldwide Supply, Renewal, Distribution, and Use 107

10-3 Water Supply Problems 109

10-4 Building Dams and Water Diversion Projects 112

10-5 Groundwater Use, Desalination, Towing Icebergs, and Controlling the Weather 116

10-6 Water Conservation 118

Chapter 11 Food Resources and World Hunger 121

11-1 Food Supply, Population Growth, and World Food Problems 121

11-2 Human Nutrition, World Hunger, and Malnutrition 122

11-3 World Agricultural Systems 126

11-4 Simplifying Affluent Diets and Wasting Less Food and Energy 128

11-5 Unconventional, Fortified, and Fabricated Foods 130

11-6 Catching More Fish and Fish Farming 131

11-7 Cultivating More Land 132

11-8 Increasing Crop Yields 135

Chapter 12 Land Resources: Wilderness, Parks, Forests, and Rangelands 138

12-1 Land Use in the United States 138

12-2 Wilderness 140

12-3 Parks 141

12-4 Importance and Management of Forests 142

12-5 Status of World and U.S. Forests 145

12-6 Rangelands 147

Chapter 13 Wildlife Resources 151

13-1 Importance of Wildlife and Wildlife Resources 152

13-2 How Species Become Endangered and Extinct 152

13-3 Wildlife Protection 158

13-4 Wildlife Management 161

13-5 Fisheries Management 162

Chapter 14 Urban Land Use and Land-Use Planning 164

14-1 Urban Growth 164

14-2 The Urban Environment 168

14-3 Urban Transportation 170

14-4 Urban and Nonurban Land-Use Planning and Control 172

14-5 Coping with Urban Problems 175

Chapter 15 Nonrenewable Mineral Resources 178

15-1 Abundance, Mining, and Processing of Mineral Resources 178

15-2 Are We Running Out of Mineral Resources? 182

15-3 Key Resources: The World Situation 186

15-4 Key Resources: The U.S. Situation 188

Chapter 16 Energy Resources: Types, Use, and Concepts 191

16-1 Types and End Uses of Energy Resources 191

16-2 Brief History of Energy Use 192

16-3 How Long Can the Fossil Fuel Era Last? 196

16-4 Energy Concepts: Energy Quality, Energy Efficiency, and Net Useful Energy 197

16-5 Evaluating Energy Resources 205

Chapter 17 Nonrenewable Energy Resources: Fossil Fuels and Geothermal Energy 208

17-1 Conventional and Unconventional Oil 208

17-2 Conventional and Unconventional Natural Gas 211

17-3 Coal 212

17-4 Geothermal Energy 217

Chapter 18 Nonrenewable Energy Resources: Nuclear Energy 222

18-1 Isotopes and Radiation 223

18-2 Conventional Nuclear Fission   226

18-3 Breeder Nuclear Fission   235

18-4 Nuclear Fusion   236

Chapter 19   Renewable Energy Resources   239

19-1 Direct Solar Energy for Producing Heat and Electricity   239

19-2 Indirect Solar Energy from Falling Water and Ocean Waves   246

19-3 Indirect Solar Energy from Thermal Gradients in Oceans and Solar Ponds   248

19-4 Indirect Solar Energy from Wind   250

19-5 Indirect Solar Energy from Biomass   252

19-6 Tidal Power and Hydrogen Fuel   255

19-7 Energy Conservation: Improving Energy Efficiency   256

19-8 Developing an Energy Strategy for the United States   258

Part Five   Pollution   263

Chapter 20   The Environment and Human Health: Disease, Food Additives, and Noise   264

20-1 Types of Disease   264

20-2 Infectious Diseases: Malaria, Schistosomiasis, and Cholera   267

20-3 Chronic Noninfectious Diseases: Cancer   270

20-4 Food Additives   275

20-5 Consumer Protection From Food Additives   277

20-6 Noise Pollution   280

Chapter 21   Air Pollution   283

21-1 Types and Sources of Air Pollution   283

21-2 Industrial and Photochemical Smog   285

21-3 Effects of Air Pollution on Property, Plants, Animals, and Human Health   290

21-4 Effects of Air Pollution on Ecosystems: Acid Fog and Acid Deposition   294

21-5 Effects of Air Pollution on the Ozone Layer and Global Climate   296

21-6 Principles of Pollution Control   299

21-7 Control of Sulfur Dioxide and Particulate Matter in Industrial Smog   301

21-8 Control of Photochemical Smog: The Automobile Problem   303

Chapter 22   Water Pollution   307

22-1 Types, Sources, and Effects of Water Pollution   307

22-2 Drinking Water Quality   309

22-3 Surface Water Pollution of Rivers   310

22-4 Surface Water Pollution of Lakes   313

22-5 Groundwater Pollution   316

22-6 Ocean and Estuarine Zone Pollution   318

22-7 Approaches to Water Pollution Control   323

22-8 Water Pollution Control Laws in the United States   325

Chapter 23   Pesticides and Pest Control   328

23-1 Chemicals Against Pests: A Brief History   329

23-2 Major Types and Properties of Insecticides and Herbicides   330

23-3 The Case for Insecticides and Herbicides   331

23-4 The Case Against Insecticides and Herbicides   332

23-5 Pesticide Bans in the United States   337

23-6 Alternative Methods of Insect Control   339

Chapter 24   Solid Waste and Hazardous Wastes   346

24-1 Solid Waste Production in the United States   346

24-2 Disposal of Urban Solid Waste: Dump, Bury, or Burn?   346

24-3 Resource Recovery from Solid Waste   348

24-4 Types, Sources, and Effects of Hazardous Wastes   350

24-5 Some Examples of Hazardous Wastes: Dioxins, PCBs, and Toxic Metals   352

24-6 Control and Management of Hazardous Wastes   355

Epilogue   Achieving a Sustainable Earth Society   361

Appendixes   363

  1 Periodicals   363

  2 Units of Measurement   365

  3 Major U.S. Environmental
       Legislation   366

  4 Ways to Save Water   367

  5 Ways to Save Energy   368

  6 National Ambient Air Quality
       Standards for the United States in
       1985   370

  7 Federal Drinking Water Standards
       for the United States in 1985   371

Readings   372

Glossary   379

Index   391

Preface

An Introductory Course in Environmental Science—The purposes of this book are **(1)** to cover the diverse materials of an introductory course on environmental studies or environmental science in an accurate, balanced, and interesting way without the use of mathematics, **(2)** to enable both teacher and student to use the material in a flexible manner, and **(3)** to use basic ecological concepts to highlight environmental problems and to indicate possible ways to deal with them.

This book is one of a pair of related textbooks designed for different introductory courses on environmental concepts and problems. *Living in the Environment* (4th ed., Wadsworth, 1985), a longer version of this book, includes additional topics and details and especially a fuller discussion of environmental economics, politics, and ethics.

Flexibility—To provide flexibility, this book is divided into five major parts:

- Humans and Nature: An Overview
- Some Concepts of Ecology
- Population
- Resources
- Pollution

Once Part Two, containing four short chapters on ecological concepts (see brief table of contents), has been discussed, the remainder of the text can be covered in almost any order that meets the needs of each individual instructor.

Other Major Features—This textbook **(1)** emphasizes the use of fundamental ecological concepts (Chapters 3–6) to illustrate the relationships of environmental problems and their possible solutions, **(2)** provides balanced discussions of the opposing views of major environmental issues, **(3)** is based on an extensive review of the literature (from the thousands of references used, key readings for each chapter are listed at the end of the text), **(4)** is based on extensive manuscript review by experts and instructors who have used one or more of four editions of *Living in the Environment* from which this book is derived, and **(5)** offers a realistic but hopeful view that shows how much has been accomplished since 1965 (when the public was made aware of many environmental problems), as well as how much more needs to be accomplished.

As you and your students deal with the crucial and exciting issues discussed in this book, I hope you will take the time to correct errors and suggest improvements for future editions. Please send such information to me, care of Jack Carey, Wadsworth Publishing Company, 10 Davis Drive, Belmont, CA 94002.

Supplementary Materials—Dr. Robert Janiskee at the University of South Carolina has written an excellent student Study Guide and Instructor's Manual for use with this text. In addition, overhead transparencies of some of the major illustrations are available from the publisher.

Acknowledgments—I wish to thank the many students and teachers who responded so favorably to the first four editions and offered suggestions, including the idea that this shorter volume be developed. I am also deeply indebted to the numerous reviewers who pointed out errors and suggested many important improvements. Any errors and deficiencies remaining are mine, not theirs.

It has also been a pleasure to work with many of the talented people at Wadsworth Publishing Company. I am particularly indebted to Gary Mcdonald for his outstanding job as production editor, to Cynthia Bassett and Merle Sanderson as designers, to Brenda Griffing for her superb and most helpful copyediting, and to Darwen and Vally Hennings and John and Judy Waller for their outstanding art work. Above all I wish to thank Jack Carey, science editor at Wadsworth, for his superb reviewing system and for his help and friendship.

G. Tyler Miller, Jr.

Reviewers

Barbara J. Abraham Hampton College
James R. Anderson U. S. Geological Survey
Kenneth B. Armitage University of Kansas
Virgil R. Baker Arizona State University
Ian O. Barbour Carleton College
Albert J. Beck California State University, Chico
Jeff Bland University of Puget Sound
Georg Borgstrom Michigan State University
Arthur C. Borror University of New Hampshire
Leon Bouvier Population Reference Bureau
Michael F. Brewer Resources for the Future, Inc.
Patrick E. Brunelle Contra Costa College
Terrence J. Burgess Irvine Valley College
Lynton K. Caldwell Indiana University
Faith Thompson Campbell Natural Resources
Defense Council, Inc.
E. Ray Canterbery Florida State University
Ted J. Case University of San Diego
Ann Causey Auburn University
Richard A. Cellarius The Evergreen State
University
William U. Chandler Worldwatch Institute
R. F. Christman University of North Carolina,
Chapel Hill
Preston Cloud University of California, Santa
Barbara
Bernard C. Cohen University of Pittsburgh
Richard A. Cooley University of California, Santa
Cruz
John D. Cunningham Keene State College
Herman E. Daly Louisiana State College
R. F. Dasmann University of California, Santa
Cruz
Kingsley Davis University of California, Berkeley
Thomas R. Detwyler University of Michigan
W. T. Edmonson University of Washington
Thomas Eisner Cornell University
David E. Fairbrothers Rutgers University
Paul P. Feeney Cornell University
Nancy Field Bellevue Community College
Allan Fitzsimmons University of Kentucky
George L. Fouke St. Andrews Presbyterian
College
Lowell L. Getz University of Illinois at Urbana-
Champaign
Eville Gorham University of Minnesota
Katherine B. Gregg West Virginia Wesleyan
College

Paul Grogger University of Colorado
J. L. Guernsey Indiana State University
Ralph Guzman University of California,
Santa Cruz
Raymond E. Hampton Central Michigan
University
Ted L. Hanes California State University,
Fullerton
John P. Harley Eastern Kentucky University
Harry S. Hass San Jose City College
Arthur N. Haupt Population Reference Bureau
Denis A. Hayes Environmental consultant
David L. Hicks Whitworth College
Eric Hirst Oak Ridge National Laboratory
C. S. Holling University of British Columbia
Donald Holtgrieve California State University,
Hayward
Marilyn Houck Pennsylvania State University
Richard D. Houk Winthrop College
Donald Huisingh North Carolina State
University
Marlene K. Hutt IBM
David R. Inglis University of Massachusetts
Robert Janiskee University of South Carolina
David I. Johnson Michigan State University
Agnes Kadar Nassau Community College
Nathan Keyfitz Harvard University
Edward J. Kormondy University of Southern
Maine
Judith Kunofsky Sierra Club
Steve Ladochy University of Winnipeg
William S. Lindsay Monterey Peninsula College
Valerie A. Liston University of Minnesota
Dennis Livingston Rensselaer Polytechnic
Institute
James P. Lodge Air Pollution consultant
Ruth Logan Santa Monica City College
T. Lovering University of California, Santa
Barbara
Amory B. Lovins Energy consultant
L. Hunter Lovins Energy consultant
David Lynn
Melvin G. Marcus Arizona State University
Stuart A. Marks St. Andrews Presbyterian
College
Gordon E. Matzke Oregon State University
W. Parker Mauldin The Rockefeller Foundation
Vincent E. McKelvey U. S. Geological Survey

A. Steven Messenger Northern Illinois University
L. John Meyers Middlesex Community College
Ralph Morris Brock University, St. Catherines, Canada
William W. Murdoch University of California, Santa Barbara
Brian C. Myres Cypress College
C. A. Neale Illinois State University
Jan Newhouse University of Hawaii, Manoa
John E. Oliver Indiana State University
Charles F. Park Stanford University
Robert A. Pedigo Callaway Gardens
Harry Perry Library of Congress
Rodney Peterson Colorado State University
David Pimentel Cornell University
Robert B. Platt Emory University
Grace L. Powell University of Akron
James H. Price Oklahoma College
Marian E. Reeve Merritt College
Carl H. Reidel University of Vermont
Roger Revelle California State University, San Diego
Robert A. Richardson University of Wisconsin
William Van B. Robertson, M.D. School of Medicine, Stanford University
C. Lee Rockett Bowling Green State University
Richard G. Rose West Valley College
David Satterthwaite I.E.E.D., London
Frank R. Schiavo San Jose State University
Stephen Schneider National Center for Atmospheric Research
Clarence A. Schoenfeld University of Wisconsin, Madison
Henry A. Schroeder Dartmouth Medical School
Lauren A. Schroeder Youngstown State University

George Sessions Sierra College
Paul Shepard Pitzer College and the Claremont Graduate School
Kenneth Shiovitz
H. F. Siewart Ball State University
Robert L. Smith West Virginia University
Howard M. Smolkin U.S. Environmental Protection Agency
Patricia M. Sparks Glassboro State College
John E. Stanley University of Virginia
Mel Stanley California State Polytechnic University, Pomona
William L. Thomas California State University, Hayward
Tinco E. A. van Hylckama Texas Tech University
Donald E. Van Meter Ball State University
John D. Vitek Oklahoma State University
L. C. Walker Stephen F. Austin State University
Thomas D. Warner South Dakota State University
Kenneth E. F. Watt University of California, Davis
Alvin M. Weinberg Institute of Energy Analysis, Oak Ridge Associated Universities
Brian Weiss
Raymond White San Francisco City College
Charles G. Wilber Colorado State University
John C. Williams College of San Mateo
Ray Williams Whittier College
Samuel J. Williamson New York University
Ted L. Willrich Oregon State University
George M. Woodwell Brookhaven National Laboratory
Robert Yoerg, M. D. Belmont Hills Hospital
Hideo Yanenaka San Francisco State University

Environmental Science

PART ONE

Humans and Nature: An Overview

It is only in the most recent, and brief, period of their tenure that human beings have developed in sufficient numbers, and acquired enough power, to become one of the most potentially dangerous organisms that the planet has ever hosted.

John McHale

Human despair or default can reach a point where even the most stirring visions lose their regenerating powers. This point, some will say, has already been reached. Not true. It will be reached only when human beings are no longer capable of calling out to one another, when the words in their poetry break up before their eyes, when their faces are frozen toward their young, and when they fail to make pictures in the mind out of clouds racing across the sky. So long as we can do these things, we are capable of indignation about the things we should be indignant about and we can shape our society in a way that does justice to our hopes.

Norman Cousins

1

Population, Resources, and Pollution

We travel together, passengers on a little spaceship, dependent on its vulnerable resources of air, water, and soil . . . preserved from annihilation only by the care, the work, and the love we give our fragile craft.

Adlai E. Stevenson

1-1 A Crisis of Interlocking Problems

Today the world is at a critical turning point. The prospect for humanity is both brighter and darker than at any time in history. Prophets of doom warn that the earth's life-support systems are being destroyed, and technological optimists promise a life of abundance for everyone. We spend billions to transport a handful of humans to the moon, only to learn the importance of protecting the diversity of life on the beautiful blue planet that is our home. We use modern medicine and sanitation to lower death rates from disease, only to be faced with a population explosion. We feed more people than ever before, yet millions die each year from lack of food or from diseases brought on or made worse by too little food.

As more people use the earth's resources, increasing stress is placed on the forests, grasslands, and croplands, and on the air, water, and soil that support all life. Tropical forests are cleared to provide lumber and fuelwood and land for growing crops and grazing livestock, but this also threatens thousands of plant and animal species with extinction. Some experts say we are running out of certain fuel and mineral resources; others say we will never run out. We hear of successes in cleaning up rivers, lakes, and the air in some parts of the world, but we are bombarded with stories about new pollution threats such as leaking hazardous waste dumps, acid rain, and potentially harmful changes in the global climate due to the carbon dioxide added to the atmosphere when fossil fuels are burned and forests are cleared without adequate replanting.

The problems associated with increasing population, increasing use of resources, and pollution are all interrelated. The primary aims of this book are to describe major environmental problems, present ecological concepts that connect them, and use these concepts to evaluate the opportunities we have to deal with these problems in coming decades. Let us begin with a brief overview of the related problems of population growth, resource use, and pollution. In later chapters we will look at these problems and proposed solutions in greater depth.

1-2 Population

The J-Shaped Curve of Population Growth You are used to quantities increasing at an *arithmetic rate*—that is, growing in the sequence 1, 2, 3, 4, 5, 6, and so on. The population size of the earth, however, is increasing at an *exponential or geometric rate*—that is, it is growing by doubling: 1, 2, 4, 8, 16, 32, and so on.

Exponential growth can be illustrated by folding a page of this book. The page is about 0.1 millimeter (about 1/254 inch) thick, so after one fold its thickness would be doubled, after 12 doublings the page would be about 410 millimeters (1.34 feet) high, and after 20 doublings about 105 meters (340 feet)—still a relatively unspectacular change. However, after the 35th fold, its height would equal the distance from New York to Los Angeles. After 42 doublings the mound of paper would reach from the earth to the moon, 386,400 kilometers (240,000 miles) away. Slightly past the 51st doubling the pile would reach the sun, 149 million kilometers (93 million miles) from the earth's surface!

When such **exponential** or **geometric growth** is plotted on a graph the result is an *exponential curve*, or *J-shaped curve*, as shown for the human population in Figure 1-1. Notice that it took 2 million to 5 million years to add the first billion people; 80 years to add the second billion; 30 years

---

To the student: At the end of the book is a list of readings that served as major references for each chapter; this material can be a source of further information.

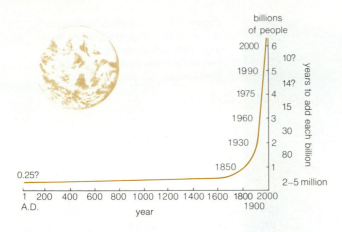

billions
of people

year		
2000	6	
		10?
1990	5	
		14?
1975	4	
		15
1960	3	
		30
1930	2	
		80
1850	1	
0.25?		2–5 million

years to add each billion

1 200 400 600 800 1000 1200 1400 1600 1800 2000
A.D.   1900
year

Figure 1-1 J-shaped curve of the world's population growth. Projections assume that the 1984 growth rate of 1.7 percent will gradually drop to 1.5 percent.

to add the third billion; and only 15 years to add the fourth billion. At present growth rates the fifth billion will be added in the 12 years between 1975 and 1987, and the sixth billion will be added only 11 years later, by 1998.

The average number of live births on this planet is now about 249 babies per minute, or approximately 358,000 per day, while the average number of deaths is only 101 persons per minute, or 146,000 per day. In other words, there are about 2.5 times more births than deaths each day. Population growth for the entire planet over a given period is determined by the difference between births and deaths:

population
increase = births − deaths
= 358,000 people − 146,000 people
/day /day
= **212,000 people/day**

This adds 1.48 million people each week and 77 million people each year to the 4.8 billion passengers already on "spaceship earth." At this rate it takes less than 5 days to replace a number of people equal to all Americans killed in all U.S. wars, less than 12 months to replace the more than 75 million people killed in the world's largest disaster (the bubonic plague epidemic of the fourteenth century), and about 13 months to replace the 86 million soldiers and civilians who died in all wars fought in this century.

All these new passengers must be fed, clothed, and housed. Each will use some resources and will add to global pollution. While some of this population growth is taking place in the *more developed countries* (MDCs) such as the United States and the Soviet Union, most is taking place in the *less developed countries* (LDCs) such as China and India. Cur-

rently, 76 percent of the people in the world live in the LDCs, which have only 20 percent of the world's wealth. As a result, the United Nations estimates that at least half the adults on this planet are illiterate; one-fifth of the people are hungry or malnourished; one-sixth have inadequate housing (Figure 1-2); one out of every four lacks clean water; and one out of three does not have access to adequate sewage disposal and effective medical service.

The World Health Organization (WHO) estimates that while you ate dinner today at least 1,600 people died of starvation, malnutrition, or diseases resulting from or worsened by these conditions. By this time tomorrow, 38,000 will have died from starvation or starvation-related diseases; by next week, 269,000; by next year, about 14 million. Half are children under the age of 5. Because this mass starvation of about 14 million people a year is spread throughout much of the world instead of being confined to one country or region, it is not even classified as a famine by most officials.

Population growth, however, is not our only problem. We are also faced with the environmental problems of increasing resource use and pollution, both related to population growth.

1-3 Natural Resources

Types of Natural Resources A **resource** or **natural resource** is any form of matter or energy that is obtained from the physical environment to meet human needs. *Whether something is considered to be a resource depends on technology, economics, and cultural beliefs.* Most resources are created by human ingenuity. Oil was a useless fluid until humans learned how to locate it, extract it from the ground, separate it by distillation into various components, and use it as gasoline, home heating oil, road tar, and so on. Similarly, coal and uranium fuels were once useless rocks.

Resources can be classified as *renewable* or *nonrenewable* (Figure 1-3). A **renewable resource** is one that either comes from an essentially inexhaustible source (such as solar energy) or can be renewed and replenished relatively rapidly by natural or artificial processes if managed wisely. Examples include food crops, animals, grasslands, forests, and other living things, as well as fresh air, fresh water, and fertile soil. The maximum rate at which a renewable resource can be used without impairing or damaging its ability to be renewed is called its **maximum sustained yield.** If this yield is exceeded, a potentially renewable resource becomes a nonrenewable resource. For example, if a species

UNICEF/Bernard Wolff

Figure 1-2 One-sixth of the people in the world do not have adequate housing. Lean-to shelters like these are homes for many families in Dacca, Bangladesh.

of fish is harvested faster than it can reproduce itself, the species may be threatened with extinction.

A **nonrenewable resource** either is not replaced by natural processes or has a rate of replacement that is slower than the rate of use. Natural geological processes taking place over millions of years have created varying deposits of such metallic and nonmetallic minerals (Figure 1-3). Once such deposits have been mined, they are not replaced fast enough to be useful. The easily available and highly concentrated supplies of nonrenewable minerals are normally depleted first. Then it is necessary to look harder and dig deeper to find the remaining deposits, which usually contain lower concentrations of the desired mineral. This normally costs more, although improvements in resource location and mining technology sometimes reduce costs. Higher costs can stimulate a search for new deposits or make the mining and processing of lower grade deposits more feasible. However, if the cost of finding, extracting, and concentrating a given material becomes too high, the resource will no longer be useful even though some supplies remain.

Sometimes a *substitute* or *replacement* for a re-

source that is scarce or too expensive is discovered. For example, much of the steel used in automobile production is being replaced with aluminum and plastic to produce lighter cars and thus conserve gasoline. Although some resource economists argue that we can use human ingenuity to find a substitute for any nonrenewable resource, this is not always the case. Some materials have unique properties, such that they cannot be replaced; the would-be replacements are inferior, too costly, or otherwise unsatisfactory. For example, nothing now known can replace steel and concrete in skyscrapers, nuclear power plants, and dams. In other cases the proposed substitutes are themselves fairly scarce. Such is the case with molybdenum, the main substitute for tungsten in making hard alloy steels for use in high-speed cutting tools and filaments in electric light bulbs.

Recycling and reuse are other ways of stretching the supplies of some nonrenewable minerals. Nonrenewable resources that can be recycled or reused include *metallic minerals* from which metals such as copper, aluminum, and iron can be extracted. In most LDCs recycling and reuse are necessary for survival. In MDCs, however, economic incentives (such as tax breaks and government

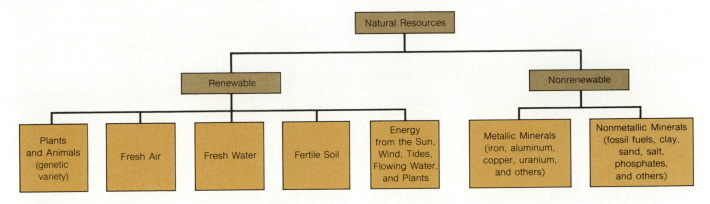

Figure 1-3 Major types of natural resources.

price controls) often encourage the use of virgin resources instead of promoting recycling and reuse.

Examples of nonrenewable resources that cannot be recycled or reused include nonmetallic mineral energy resources such as fossil fuels (coal, oil, and natural gas). **Fossil fuels** are buried deposits of decayed plants and animals that have been converted to organic matter by heat and pressure in the earth's crust over hundreds of millions of years under climatic and geological conditions that no longer exist. We live in a relatively brief period of human history called the *fossil fuel era* (Figure 1-4), in which these deposits are being rapidly depleted to provide us with about 84 percent of the energy we use. *Once a fossil fuel resource has been burned, it is, for all practical purposes, gone forever as a useful source of energy because renewal takes hundreds of millions of years.*

Are We Running Out of Natural Resources? Optimists Versus Pessimists Increasing population causes a rise in resource use, but a rise in the standard of living creates an even greater demand for renewable and nonrenewable natural resources. As income rises, people buy, use, and throw away more resources. Thus, affluent nations have gone around the bend on a J-shaped curve of increasing resource use. For example, *the Western affluent nations, Japan, and the Soviet Union account for only about one-fourth of the world's population but use 80 percent of its natural resources. The United States alone, with about 5 percent of the world's population, produces about 21 percent of all goods and services, uses about 30 percent of all processed natural resources, and produces at least one-third of the world's pollution.*

Natural resource use by affluent nations is expected to rise sharply in coming decades. At the same time, the LDCs of the world hope to become more affluent, further increasing resource use. The Nobel laureate economist Wassily Leontief projects that for even moderate economic growth to occur between 1975 and 2020, production of food and of common minerals must increase fourfold and fivefold, respectively.

This J-shaped curve of increasing resource use raises the question of how long the earth's renewable and nonrenewable resources will last. Great controversy surrounds this question, represented by two distinctly opposing schools of thought. One group called *neo-Malthusians* (or gloom-and-doom pessimists by opponents) believes that if present trends continue, the world will be more crowded and more polluted, heading for economic ruin, increased political instability, and threat of global nuclear war. They cite the following reasons: **(1)** the maximum sustained yield of many of the world's renewable resources may be exceeded through overfishing, destruction of habitat for wildlife, overgrazing, deforestation, overpopulation, and pollution; **(2)** there will be shortages of affordable supplies of nonrenewable fossil fuels (especially oil and possibly natural gas) and selected nonrenewable minerals important to economic well-being; and **(3)** the use of some renewable and nonrenewable resources may be limited by the environmental side effects of more and more people using more and more resources, even if supplies are adequate. The term *neo-Malthusians* reflects belief in an updated and expanded version of the hypothesis proposed by Thomas Robert Malthus in 1803, namely, that human population size tends to outrun food production until poor health and death from starvation and disease restore the balance.

Solutions to these problems suggested by neo-Malthusians usually involve recycling, reuse, resource conservation, reducing average per capita consumption (primarily by eliminating wasteful use of matter and energy resources), increased pollution control, and slowing world population growth.

The opposing group, called *cornucopians* (or

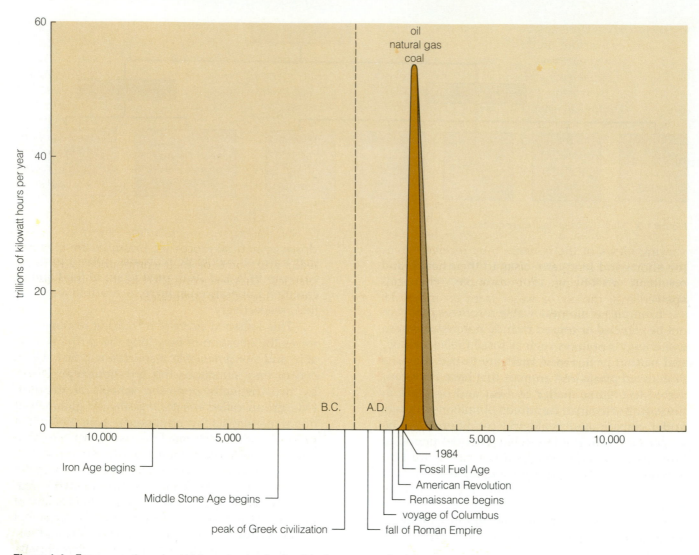

Figure 1-4   Energy use throughout history, showing the fossil fuel era we now live in.

technological optimists by opponents), believes that if present trends continue we will never run out of needed renewable and nonrenewable resources, and that by 2000 continued economic growth and technological advances based on human ingenuity will have produced a less crowded, less polluted world in which most people will be healthier, will live longer, and will have greater material wealth. This group believes that renewable and nonrenewable resources will never be depleted because **(1)** scarcity will cause prices to rise, which will encourage conservation of renewable resources, enable lower grade deposits of nonrenewable resources to be mined, and stimulate searches for new deposits and substitutes; and **(2)** human ingenuity can always develop new forms of technology to mine increasingly lower grade deposits of nonrenewable resources and find acceptable substitutes for any scarce renewable or nonrenewable resources. Some cornucopians also believe that efforts to reduce population growth

are either not necessary because population will be brought down naturally by economic growth in the LDCs, or undesirable because people represent the world's most valuable resource.

The arguments between these two opposing groups have been going on for several decades and will undoubtedly continue. Much of this book is devoted to analyzing this complex debate to help you evaluate each position. As we examine major environmental problems and their possible solutions, we should be guided by Alfred North Whitehead's motto: "Seek simplicity and distrust it." Or as H. L. Mencken put it: "For every problem there is a solution—simple, neat, and wrong."

1-4   Pollution

What Is Pollution?   **Pollution** can be defined as an *undesirable change* in the physical, chemical, or biological characteristics of the air, water, or land that can affect health, survival, or activities of humans or other organisms. Note that pollution does not have to cause physical harm. Pollutants such as noise and heat may cause injury but more often cause psychological distress. Forms of aesthetic pollution such as unpleasant sights and foul odors and sights offend the senses. Some forms of pollution may merely interfere with human activities. For example, a lake may be considered polluted if it cannot be used for boating activities.

It is difficult to specify what is and is not pollution because people often differ in what they consider to be an undesirable change based on varying views of the benefits they receive versus the short- and long-term risks to their health and economic well-being. For example, chemicals spewed into the air or water from an industrial plant might be harmful to humans and other organisms living nearby. However, if the installation of expensive pollution controls were required, the plant might be forced to shut down. Workers who would lose their jobs might feel that the risks to them from contaminated air and water are not as great as the benefits of employment. Such *risk-benefit analysis* enters into all environmental decisions. The nature of tragedy, as the philosopher Hegel pointed out, is the conflict not between right and wrong but between right and right.

Types of Pollution From a biological viewpoint, there are two major types of pollutant: degradable and nondegradable. A **degradable pollutant** can be decomposed, removed, or consumed and thus reduced to acceptable levels either by natural processes or by human-engineered systems (such as sewage treatment plants), as long as these processes and systems are not disrupted by receiving more potential pollutants than they can handle. There are two classes of degradable pollutants: rapidly degradable (nonpersistent) and slowly degradable (persistent). *Rapidly degradable or nonpersistent pollutants*, such as human sewage and animal and crop wastes, normally can be biodegraded quickly by decomposing organisms such as bacteria and fungi, if the target water, air, or soil system is not overloaded. For example, a rapidly flowing river can normally cleanse itself of human sewage if it does not receive too much raw sewage from a large city or a number of small cities or farms. Thus, control of rapidly degradable pollutants involves ensuring that the natural systems receiving them are not overloaded.

Slowly degradable or persistent pollutants remain in the environment for a long time before being broken down or reduced to harmless levels. They

include some radioactive materials and synthetic compounds, such as dichlorodiphenyltrichloroethane (DDT), polychlorinated biphenyls (PCBs), and plastics that are designed to resist decomposition by heat, light, chemicals, and decomposer organisms. Control of slowly degradable pollutants involves (1) preventing them from reaching the environment either by banning their use or by finding a safe way to store them, (2) learning how to speed up their conversion or degradation to harmless materials, or (3) controlling the amount released to the environment so they do not build up to harmful levels.

Nondegradable pollutants are not broken down by natural processes. Examples are mercury, lead, and some of their compounds, and some plastics. They are controlled in the same ways as slowly degradable pollutants.

Determining Harmful Levels of Pollutants Determining the amount of a particular pollutant that can cause a harmful or undesirable effect in humans or other organisms is a difficult scientific problem. The amount of a chemical or pollutant in a given volume of air, water, or other medium is called its **concentration**. Concentrations are often expressed as parts per million or parts per billion but are increasingly being reported as milligrams per liter (mg/L), milligrams per cubic meter (mg/m^3), micrograms per liter (μg/L), and micrograms per cubic meter (μg/m^3). As discussed in Appendix 2, one microgram (1 μg) is one millionth of a gram.

Parts per million (ppm) is the number of parts of a chemical or pollutant found in 1 million parts of a particular gas, liquid, or solid mixture. It is equal to the number of milligrams of a substance in a 1-liter volume of air, water, or other medium (mg/L). **Parts per billion (ppb)** is the number of parts of a chemical or pollutant found in a billion parts of a particular gas, liquid, or solid mixture. It is equal to the number of micrograms of a substance in a 1-liter volume of air, water, or other medium (μg/L).

One part per million and one part per billion may seem very small, but for some organisms and with some pollutants they represent dangerous levels. Some chemicals, called **nonthreshold pollutants**, are harmful to a particular organism in any concentration (Figure 1-5). Examples include mercury, lead, cadmium, and some of their compounds. Other substances, called **threshold pollutants**, are harmful only above given concentrations, or *threshold levels* (Figure 1-5). For these latter pollutants (DDT and arsenic are examples), the concentration can increase with no effect until the threshold is crossed, at which point a harmful

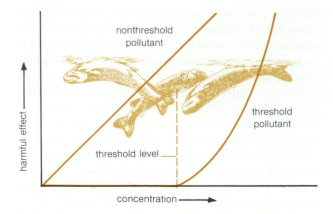

Figure 1-5 Effects of nonthreshold and threshold pollutants.

or fatal effect is triggered—like the straw that broke the camel's back.

One method of indicating the average effect of a pollutant on members of the same species is to determine the concentration required for 50 percent of the population to show some type of nonlethal or lethal response. The amount of exposure to a toxic chemical that results in the death of half of the exposed population is called the **LD-50** (for lethal dose—50 percent).

To complicate matters further, pollutants can have both acute and chronic effects. An *acute effect*, such as a burn, illness, or death, occurs shortly after exposure, often in response to fairly high concentrations of a pollutant. A *chronic effect* is one that takes place over a long period of time, often due to continued exposure to low concentrations of a pollutant. For example, people exposed to a large dose of radiation may die within a few days. However, people receiving the same total dose in small amounts over a long period may develop various types of cancer or may transmit genetic defects to their children or grandchildren.

During a lifetime an individual is exposed to many different types and concentrations of potentially harmful pollutants. Thus, the scientific evidence linking a particular harmful effect to a particular pollutant is usually statistical or circumstantial—as is most scientific evidence. For example, so far no one has been able to show what specific chemicals in cigarette smoke cause lung cancer; however, smoking and lung cancer are linked by an overwhelming amount of statistical evidence from more than 30,000 studies.

Certain pollutants acting together can cause a harmful effect greater than the sum of the individual effects when each substance is acting alone. This is called *synergism*. For example, asbestos workers who smoke have a much higher chance of getting lung cancer than those who do not because of an apparent **synergistic effect** between tobacco and asbestos. Testing all the possible synergistic interactions among the thousands of possible pollutants in the environment would be prohibitively expensive and time-consuming, even for one type of organism.

Sources of Pollution Polluting substances enter the environment naturally (volcanoes) or through human activities (burning coal), as shown in Table 1-1. For example, we are exposed to small amounts of radiation from cosmic rays entering the atmosphere and from radioactive minerals in the earth's crust. Nevertheless, we cannot assume that no risk is attached to such increases in the radiation we are exposed to from other sources such as nuclear weapons tests, nuclear power plant accidents, and unnecessary X rays and other medical tests involving radiation. In addition, most natural pollution is not concentrated in a particular area and is normally diluted or degraded to harmless levels. In

Table 1-1 Pollutants Generated by Natural and Human Activities	
Type of Pollutant	Text Discussion
Class 1: Almost Completely Generated by Human Activities	
DDT, PCBs, and other chlorinated hydrocarbon compounds	Chapters 23 and 24
Lead in the air (from burning leaded gasoline)	Section 24-5
Solid wastes and litter	Chapter 24
Class 2: Primarily Generated by Human Activities	
Radioactive wastes	Section 18-2
Oil in the oceans	Section 22-6
Sewage (animal and plant wastes)	Chapter 22
Phosphates in aquatic systems	Section 22-4
Waste heat in rivers, lakes, and oceans	Section 22-4
Photochemical smog in the air (from burning gasoline)	Sections 21-2 and 21-8
Sulfur dioxide in the air (from burning coal and oil)	Sections 21-2 and 21-7
Noise	Section 20-6
Class 3: Primarily Generated by Natural Sources	
Hydrocarbons in the air	Chapter 21
Carbon monoxide and carbon dioxide in the air	Chapter 21
Solid particles in the air	Chapter 21
Mercury in the ocean	Section 24-5

contrast, the most serious human pollution problems occur in or near urban and industrial areas, where large amounts of pollutants are concentrated in relatively small volumes of air, water, and soil. Also, many pollutants from human activities are synthetic chemicals that are not decomposed by natural processes.

Most analysts agree that the biggest overall threat to the environment and to humans and all other forms of life is global nuclear war. *There already exist enough nuclear weapons to kill 35 times the number of people now living, and an average of 10 more nuclear weapons are made each day.* Concern over nuclear power plant safety is also an important issue (Section 18-2). But by mid-1985 there were 91 operating nuclear power plants in the United States and about 36,000 nuclear weapons, each of which poses a greater potential hazard than a serious nuclear reactor accident. As former presidential science adviser George Kistiakowsky said, "When you get emotional about nuclear plants and don't care about nuclear war, it's worrying about a pimple on your cheek when you have a case of cancer." The environmental effects of nuclear war are discussed in Section 6-3.

1-5 Relationships Among Pollution, Population, Resource Use, and Technology

A crude model has been proposed to estimate the pollution, or harmful environmental impact, caused by people and their consumption activities. In this model, the total pollution in a given area depends on the product of four factors: the number of people, the amount of resources each person uses, the pollution resulting from each unit of resource used, and pollution control per unit of resource used.

$$\begin{matrix} \text{pollution or} \\ \text{environmental} \\ \text{impact in} \\ \text{a given area} \end{matrix} = \begin{matrix} \text{population} \\ \text{size} \end{matrix} \times \begin{matrix} \text{resource} \\ \text{use per} \\ \text{person} \end{matrix}$$

$$\times \begin{matrix} \text{pollution per} \\ \text{unit of} \\ \text{resource used} \end{matrix} \times \begin{matrix} \text{pollution control} \\ \text{per unit of} \\ \text{resource used} \end{matrix}$$

From this model, we see that without adequate pollution control a small number of people using matter and energy resources at a high rate can produce more pollution than a much larger number of people using such resources at a low rate. But the situation is more complicated. The use of some re-

sources creates more pollution than the use of other types. For example, a throwaway aluminum can wastes more resources and creates more pollution than a returnable glass bottle, because making the can requires about three times as much energy as making the bottle. In other words, *pollution also depends on the type of technology used*.

The use of a particular technology, however, is not always harmful. Since World War II, technologies have been introduced that provide important environmental and resource supply benefits. These benefits include: **(1)** substitutes for scarce natural resources, such as rubber, **(2)** improved efficiency and reduced waste in the use of resources such as wood, mercury, and coal, **(3)** development of processes to control and clean up many forms of pollution, and **(4)** the substitution of less harmful products for those previously used. For example, in the early 1900s the major insecticide was lead arsenate—a substance much more toxic and persistent than DDT and most modern insecticides. Decades ago the major red food coloring in the United States was lead chromate—an environmental horror compared with the recently banned red dye no. 2 (Chapter 20).

Our problem and challenge are not to eliminate technology, but to decide how to use it more carefully and humanely. As Stuart Chase reminds us, "To condemn technology *in toto* is to forget gardens made green by desalinization of seawater, while to idealize technology is to forget Hiroshima."

1-6 Some Hopeful Trends

Is the achievement over the next few decades of a less crowded, less polluted world in which most people are healthier and live longer a hopeless, idealistic goal? Fortunately, the answer to this important question is no. Present undesirable trends do not necessarily indicate where we are heading. As René Dubos reminds us, "Trend is not destiny." *We can say no!*

There are growing signs that we can make such a transition. Today in the MDCs there is sophisticated awareness of the global problems of population, pollution, and resource depletion, and this knowledge is spreading rapidly to the LDCs. During the short period between 1965 and today, most U.S. citizens became aware of and concerned about the environment. April 22, 1970, was the first Earth Day in the United States, and 20 million people took to the streets to demand better environmental quality. At that time, polls showed that Americans considered reducing pollution to be the second most important problem. More recent polls showed that 70 percent of those queried indicated greater concern about the environment in 1984 than in the past, and 65 percent indicated they would favor further protection of the environment even if it cost them 10 percent more in taxes and cost of living.

Even more important, this awareness has been translated into action. Today there are more than 4,000 organizations worldwide devoted to environmental issues. Most MDCs have passed laws designed to protect the air, water, land, and wildlife. For example, since 1970 more than 80 federal laws have been enacted in the United States to protect the air, water, land, wildlife, and public health (see Appendix 3), and billions of dollars have been spent on pollution control. By 1985, environmental protection agencies had been established in 151 of the world's 176 nations—including 112 LDCs, compared to only 11 LDCs in 1972.

Since 1965, 70 U.S. rivers, lakes, and streams have been cleaned up. About 3,600 of the nation's 4,000 major industrial water polluters are meeting federal water pollution cleanup deadlines. In 20 major U.S. cities the air is measurably cleaner than it was before the passage of the Clean Air Act of 1970, and about 90 percent of major U.S. factories are in compliance with federal air pollution regulations. Government and industry in the United States are spending $50 billion a year—$209 a year for each American—to reduce pollution.

Other industrialized countries have also made significant progress in pollution control. Smog in London has decreased sharply since 1952, and the Thames River is returning to life. Japan, once regarded as the most polluted country in the world, has dramatically reduced air pollution in most of its major cities and has upgraded the quality of its waters since passing antipollution laws in 1967. The Japanese environment is still highly degraded, however, partly because the country's small size relative to its large population means that most of the people live in crowded cities where pollution levels are concentrated.

On the energy front, as energy prices have risen, we are learning that energy conservation is our cheapest and least environmentally harmful energy option. Some people are driving less; more thermostats have been turned down; more homes have been insulated; and a small but growing number of people are using solar energy to heat their water and homes. However, there is still a long way to go, especially in the United States. Despite higher gasoline prices, carpools and vanpools are not used widely, and the nation lacks extensive and efficient railroad and mass transit systems like those found in Japan and in most European MDCs.

With regard to population, the good news is that between 1965 and 1985 the annual growth of the world's population slowed from 1.99 to 1.7 percent. By 1985, a dozen European countries had reached or were close to zero population growth (ZPG), where the annual number of births equals the number of deaths. Despite these hopeful trends in many *more developed countries*, by 2000 the world's population is projected to be 6.1 billion, assuming that the annual growth rate has dropped to 1.5 percent by then (Figure 1-1). About 90 percent of this growth is expected to take place in the *less developed countries*, where already the lives and dignity of at least 800 million people—one-sixth of humanity—are degraded by disease, malnutrition, illiteracy, exploitation, and fear.

Even with this projected drop in the world's population growth rate, 92 million people are expected to be added in 2000, compared with 77 million in 1985. Based on present trends, United Nations population experts project that the earliest year that the world could reach ZPG is 2040 (with 8 billion people). Their more likely projection is the year 2110 (with a population of 10.5 billion).

The rate of population growth in the United States has slowed significantly, with the average number of live children born per woman decreasing from 3.8 in 1957 to 1.8 in 1985. If these low rates can be maintained, especially through 1987, the United States could have a stable population by 2020, and perhaps as early as 2010, depending primarily on the annual addition of legal and illegal immigrants.

The amazing thing is not the lack of progress in dealing with environmental problems in many parts of the world but that so much has been done in a mere two decades. Nevertheless, we should not get carried away with optimism. Environmentalists must constantly struggle to see that existing enviromental laws (Appendix 3) are enforced and to prevent them from being weakened. At the same time, many new and serious problems such as hazardous wastes (Chapter 24) and acid deposition (Section 21-4) have been identified.

The promising developments since 1965 indicate that many people are beginning to ask the right questions. What are our responsibilities toward our fellow humans and other forms of plant and animal life on this planet? How close are we to overloading the earth's life-support systems? How are we as individuals willing to modify our own life-styles to reduce resource waste and pollution?

What is the use of a house if you don't have a decent planet to put it on?

Henry David Thoreau

Discussion Topics

1. Debate the following resolution: High levels of resource use by the United States are necessary because this means **(a)** purchases of raw materials from poor nations and **(b)** economic growth in the United States, to finance aid to LDCs.

2. Debate the following resolution: The world will never run out of resources because technological innovations either will produce substitutes or will allow use of lower grades of scarce resources.

3. Should economic growth in the United States and in the world be limited? Why or why not? Is all economic growth bad? Which types, if any, do you believe should be limited? Which types, if any, should be encouraged?

4. Is the world overpopulated? Why or why not? Is the United States overpopulated? Why or why not?

5. You have been appointed to a technology assessment board. What drawbacks and advantages would you list for the following: **(a)** intrauterine devices (IUDs), **(b)** snowmobiles, **(c)** sink garbage disposal units, **(d)** trash compactors, **(e)** pocket transistor radios, **(f)** television sets, **(g)** electric cars, **(h)** computers, **(i)** abortion pills, **(j)** effective sex stimulants, **(k)** drugs that would retard the aging process, **(l)** drugs that induce euphoria but are physiologically and psychologically harmless, **(m)** electrical or chemical methods that would stimulate the brain to remove anxiety, fear, and unhappiness, and **(n)** genetic engineering (manipulation of human genes)? In each case, would you recommend that the technology be introduced?

2

Human Impact on the Earth

We found our house—the planet—with drinkable, potable water, with good soil to grow food, with clean air to breathe. We at least must leave it in as good a shape as we found it, if not better.

Rev. Jesse Jackson

The J-shaped curves of increasing population, resource use, and pollution (Chapter 1) are merely symptoms of the fundamental cultural change from humans as hunter-gatherers to humans as shepherds and tillers of the soil to humans in industrial society, as discussed in this chapter.

2-1   Hunter-Gatherer Societies

People have always had an impact on the environment. Throughout most of human history, however, this impact was fairly small and localized, because about 90 percent of all humans who ever lived have been hunter-gathers. They were few in numbers, and their main energy source was their own muscle power used to hunt, fish, and gather edible plants, tubers, and roots. Today fewer than 1 percent of the earth's inhabitants live by hunting and gathering.

Early hunter-gatherers survived and multiplied primarily by means of three major cultural adaptations, all the product of intelligence: **(1)** the use of *tools* for hunting, gathering, and preparing food and making protective clothing, **(2)** learning to live in an often-hostile environment through effective *social organization* and *cooperation* with other human beings, and **(3)** the use of *language* to increase the efficiency of cooperation and to pass on knowledge of survival experiences.

Our hunter-gatherer ancestors cooperated by living in small bands or tribes, clusters of several families typically consisting of no more than 50 persons. The size of each band was limited by the availability of food. If a group became too large, it split up. Sometimes these widely scattered bands had no permanent base, traveling around their ter-

ritory to find the plants and animals they needed to exist. Hunter-gatherers' material possessions consisted mostly of simple tools such as sharpened sticks, scrapers, and crude hunting weapons. Much of their knowledge could be described as ecological—how to find water in a desert, and how to locate plant and animal species useful as food. Studies of Bushmen, Pygmies, and other hunter-gatherer cultures that exist today have shown the uncertainty of success in hunting wild game; this meant that often most of the food of primitive people was provided by women and children, who collected plants, fruits, eggs, mollusks, reptiles, and insects.

Early hunter-gatherers had to struggle to stay alive. But research among the few remaining hunter-gatherer societies shows that they may hunt for only a week out of each month. These "primitive" peoples may have less stress and anxiety and enjoy a more diverse diet than most "modern" people. Although malnutrition and starvation in early hunter-gatherer societies were rare, infant mortality was high, primarily from infectious diseases. This factor, coupled with infanticide (killing the young) and geronticide (killing the old), led to an average life expectancy of around 30 years and kept population size in balance with food resources.

Gradually members of such groups developed improved hunting weapons and learned to cooperate with members of other groups to hunt herds of reindeer, the European bison, and other big game. They also discovered methods of mass killing. Fire was frequently used to flush game from thickets toward hunters lying in wait, or to stampede animals over cliffs. Toward the end of the Pleistocene period (several ice ages that occurred between 3 million and 10,000 years ago), about 70 percent of the large North American mammals such as the mammoth, the mastodon, and the ground sloth became extinct. Although climatic change was a major cause of these extinctions, some scholars believe that the process was hastened by overhunting.

With the use of advanced weapons and fire, hunter-gatherer societies made some significant changes in their environment. Because of their small numbers, however, their impact was insignificant even on a regional scale. They were examples of humans *in* nature, who learned to survive by understanding and cooperating with nature.

2-2 Agricultural Societies

One of the most significant changes in human history developed about 10,000 to 12,000 years ago. People began the long process of learning how to herd game instead of hunting it and to grow selected wild plants close to home instead of gathering them over a large area. Over several thousand years the importance of hunting and gathering declined as more and more people became shepherds and farmers.

This change may have started with the domestication of dogs, found as pets and scavengers around campsites. Wild sheep were probably domesticated next, followed by wild goats, pigs, cattle, and horses. The impact of these early shepherds on the land was often more extensive than that of hunters and gatherers. They burned and cleared forests, producing grasslands and tropical savannas containing annual plants that provided food for their flocks of grazing animals. Some of these grasslands were in turn degraded by overgrazing. As a result, grasslands were destroyed and soils were eroded over large areas of the Mediterranean region. Today nature is still imposing drastic penalties for overgrazing.

The first type of plant cultivation, called *horticulture* ("hoe culture"), began when women found that they could quite easily grow some of their favorite food plants by digging a hole with a stick (a primitive hoe) and placing roots and tubers in the ground. People also learned how to plant and grow *seed crops* such as wheat, barley, rice, peas, lentils, corn, and potatoes. We have domesticated very few new major seed crops in the past 2,000 years, although genetically improved strains of these earlier crops have been developed (Section 11-8).

Another method that was developed to grow food plants is called *slash-and-burn* or *shifting* cultivation. It was important in prehistoric and medieval Europe and was used by some of the early colonists in eastern North America. It is still practiced in tropical areas of Africa, South America, and Southeast Asia by an estimated 150 million to 200 million people. A small patch of forest is cleared and the dried vegetation is burned. The

ash left after the burning provides inorganic fertilizer for the soil, and crops are grown in the forest opening until the plant nutrients in the ash are exhausted—typically after 2 to 5 years. The farmer and his family then move on and clear another patch, leaving the recently used cropland fallow. Wild plants repopulate the original cleared area, making nutrients available again for growing crops. Shifting cultivation works well in the tropical forest environment, provided human population density and industrial activities remain low enough to ensure that abandoned areas are not replanted for 10 to 20 years.

True *agriculture* (as opposed to horticulture) began with the invention of the plow, pulled by domesticated animals or women. This lessened dependence on human muscle power as the prime source of energy for growing food. As people learned to cultivate plants efficiently, they had not only a constant food supply but a regular food *surplus*. This surplus had three important effects: **(1)** without the threat of starvation, populations began to increase; **(2)** people cleared more and more land and began to control and shape the surface of the earth to suit their requirements; and **(3)** urbanization began—villages, towns, and eventually cities slowly formed as people developed specialties other than farming. In hunter-gatherer societies the entire adult population (and sometimes the children) were involved in either hunting or gathering food. In today's LDCs approximately two-thirds of the adult population participates in food production. Today in the United States less than 5 percent of the population is involved directly with agriculture, and in some countries in western Europe the figure is less than 2 percent.

As the number of farmers increased and spread out over much of the earth, they created an environmental impact far exceeding that of the hunter-gatherers. Forests and grasslands were replaced with large areas, typically planted with a single food crop such as wheat. Poor management of many of the cleared areas allowed vital topsoil to wash away and pollute streams, rivers, and lakes with silt (Chapters 9 and 22). Land-clearing activities also destroyed and altered the habitats of plant and animal species, endangering their existence and in some cases causing or hastening their extinction (Chapter 13). Irrigation without proper drainage led to the accumulation of salts in topsoil, diminishing soil fertility. Pests that were controlled naturally by the diverse array of species in forests spread much more rapidly in areas planted with one or only a few crops. Pesticides were used to protect food crops, but this led to a new series of problems that threatened wildlife, polluted the air and soil, and in some cases increased the number and size of pest populations (Chapter 23).

The development of agriculture thus brought about a fundamental modification in humanity's relationship with the environment, as more and more people began shifting from hunter-gatherers *in* nature to shepherds, farmers, and urban dwellers *against* nature. But even more fundamental changes were to follow.

2-3 Industrial Societies

Humans have learned how to find and use more and more energy in their attempts to change and control the environment. Early societies had to rely on their own muscle power to survive. Agricultural societies eventually learned to used draft animals and later wind and water power to help them exert more control over the land and their food supplies. During the eighteenth century, however, industrial societies made a gigantic leap by discovering how to unlock on a larger scale the chemical energy stored in fossil fuels such as coal, oil, and natural gas (Chapters 3, 16, and 17).

The gradual rise of industrial societies, fueled by these new sources of energy, has allowed the creation of useful products and has raised the standard of living of people throughout the world. At the same time it has intensified existing environmental problems and created a series of new ones. By learning to put some of the earth's chemical resources together in new ways, industries have produced metal alloys, plastics, agricultural pesticides and fertilizers, and medicines. But pollution (from DDT, lead, mercury, PCBs, solid wastes, radioactive wastes, and a host of other chemicals discussed throughout this book) has also increased.

Increased mining to provide industries with raw materials has disrupted more and more of the earth's surface and has threatened plant and animal species. By decreasing the need for most people to engage in agriculture, industrialization has caused massive shifts of population from rural to urban areas—creating a new array of social, political, economic, and environmental problems (Chapter 14).

The benefits from the Industrial Revolution are great. Very few people would propose that we abandon the technological achievements of the past two centuries. Increasingly, however, our time, energy, money, and new forms of technology must be used to correct the ill effects of earlier advances. We are learning that in many cases the more we attempt to control nature, the less control we have.

2-4 Brief History of Natural Resource Conservation and Environmental Protection in the United States

Frontier Versus Conservation Mentality When Europeans first settled North America, the entire continent was primitive wilderness, relatively untouched by the Native American peoples who lived there. To the landless colonists and pioneers, the vast American wilderness was a hostile country to be conquered, opened up, owned, cleared, and used. Their lack of concern for the long-term conservation of America's natural resources was reinforced by the belief that the wilderness was too vast to be destroyed.

By the mid-1800s the government owned 80 percent of the total land area of the United States, mostly as a result of the Louisiana Purchase, the Oregon Compromise, the purchase of Alaska, the Mexican Cession, and other acquisitions that ignored the rights of the aboriginal inhabitants. The policy of the government was to dispose of these lands as rapidly as possible to encourage settlement and development of the nation and thus strengthen it against its enemies. By 1900 more than half this publicly owned land had been sold and given to railroads, timber companies, homesteaders, mining companies, state and local governments, and land-grant colleges.

Early Conservation Efforts (1830–1910) In the first half of the nineteenth century, naturalist writers such as Ralph Waldo Emerson and Henry David Thoreau set forth the idea that true progress comes from achieving a harmonious relationship with nature rather than through exploitation. Around 1880 when the nation's frontier was considered to be closed, George Catlin, John Muir, Frederick Law Olmstead, Charles W. Eliot, Stephen Mather, and others began arguing that America's land resources were finite and were being exploited at an alarming rate through overgrazing, overcutting, and general misuse. They proposed that part of the land owned by the government be withdrawn from public use and preserved in the form of national parks.

In the 1860s George Perkins Marsh, a congressman from Vermont, questioned the idea of the inexhaustibility of natural resources and showed how the rise and fall of past civilizations was linked to their use and misuse of nature. Many of his ideas for federal management of publicly owned forests and watersheds were put into effect in the early 1900s by President Theodore Roosevelt, an ardent conservationist. Marsh's ideas were extended later by Paul B. Sears in 1935 and by Fairfield Osborn in 1948, who tried to alert Americans to some of the environmental problems we face today.

In 1872 President Ulysses S. Grant signed into law an act designating the 2 million acres of the newly discovered Yosemite Forest in northeastern Wyoming for preservation as Yellowstone National Park—the world's first national park. This action marked the first phase of the government's effort to keep and manage public lands rather than to give them away or exploit them.

In the late 1800s Gifford Pinchot, Franklin Hough, Theodore Roosevelt, and other conservationists began efforts to halt the overcutting of publicly owned forestlands. In 1891 Congress designated Yellowstone Timberland Reserve as the first federal forest reserve and authorized the president to set aside additional areas of federal land. Such national forest reserves were to ensure the availability of adequate timber in the future and to protect the watersheds of the nation's rivers. By 1897 presidents Benjamin Harrison and Grover Cleveland had set aside 28 forest reserves, mostly in the West. Powerful political foes—especially westerners accustomed to using these public lands as they pleased—called these actions undemocratic and un-American.

Effective protection of the national forest reserves did not exist, however, until 1905 when Theodore Roosevelt became president. He transferred administration of the reserves from the Department of the Interior, which had a reputation for lax enforcement, to the Department of Agriculture. In 1905 Congress created the U.S. Forest Service to manage and protect the forest reserves, and President Roosevelt appointed Gifford Pinchot as its first chief. Pinchot pioneered efforts to manage these renewable forest resources according to the principles of *sustained yield* and *multiple use*—policies that prevail today.

Sustained yield has been achieved when the rate at which timber is removed by cutting, pests, disease, and fire does not exceed the rate at which it is being naturally replenished by new growth. Under **multiple use**, forests are used for a variety of purposes, including timbering, recreation, grazing, wildlife preservation, and water conservation. When there were competing claims on such land resources, Pinchot proposed that the land be reserved for its "highest use." In some areas the highest use was timber cutting; in others it was watershed protection and preservation of natural beauty and wildlife. In 1907 Congress introduced the name "National Forests," implying that these lands could be managed on a multiple-use and sustained-yield basis rather than not being used at all.

Preservation Versus Scientific Conservation (1911–1932) After 1910 the conservation movement split into two schools of thought, one emphasizing preservation and the other scientific conservation. Preservationists proposed that large tracts of public lands be set aside as wilderness, protected from all forms of development, and thus *preserved* in their natural state for future generations. By contrast, scientific conservationists believed that public land resources should be managed and used wisely in a manner that would *conserve* them for future generations.

The scientific conservationists were led by Roosevelt, Pinchot, Powell, Charles Van Hise, and others who advocated the use of public land resources on a sustained-yield, multiple-use basis. Preservationists were led by California woodsman and nature writer John Muir, who founded the Sierra Club in 1890. After Muir's death in 1914, forester Aldo Leopold became a leader calling for preservation. According to his ecological ethic for land use, the role of *Homo sapiens* is that of member, citizen, and protector of the environment—not its conqueror. Another ardent and effective supporter of wilderness preservation was Robert Marshall, an officer in the U.S. Forest Service, who founded the Wilderness Society in 1935. Leopold's land ethic has been continued in more recent years by David Brower, former head of the Sierra Club and founder of Friends of the Earth, and by Ernest Swift, Stewart L. Udall, and others.

In 1912 Congress created the U.S. National Park System, which by 1916 included 16 national parks and 21 national monuments, most of them in the western states. The enabling legislation enacted in 1916 declared that national parks are set aside to conserve scenery, wildlife, and natural and historic objects for the use, observation, health, and pleasure of the people and are to be maintained in a manner that leaves them unimpaired for future generations. The same act created the National Park Service within the Department of the Interior to manage the park system. To carry out these often-conflicting goals, the Park Service's first director, Stephen Mather, recruited a corps of professional park rangers. Mather also began the practice of contracting with private business concessionaires to provide food, lodging, and other services in the parks.

Expanding Federal Role in Land Management (1933–1969) Between 1933 and 1969 the federal government's role in managing the nation's natural resources greatly expanded, especially during the 1930s as President Franklin D. Roosevelt attempted to get the country out of the Great Depression. Shortly after taking office he established the Civilian Conservation Corps (CCC). The CCC hired 2 million unemployed people between the ages of 18 and 25 to work in several thousand camps throughout the country to plant trees, develop parks, improve waterways, provide flood control, develop rangeland, reclaim land, control soil erosion, and protect wildlife. The efforts of the CCC to control soil erosion led to the creation of the Soil Conservation Service under the Department of Agriculture in 1935. Effective soil conservation efforts owe much to the leadership of Hugh H. Bennett, long-time chief of the Soil Conservation Service.

For many decades public lands, especially in the arid West, had been heavily overgrazed because of the combination of ranchers' ignorance and greed, and periodic drought. The Taylor Grazing Act of 1934 placed 80 million acres of public land outside the national forests and parks into grazing districts to be managed jointly by the Grazing Service established within the Department of the Interior and committees of local ranchers. From the start, however, ranchers resented government interference with their long-established use of public land. Since 1934 they have led repeated efforts to have these lands removed from government ownership and turned over to private cattle and sheep interests. Until 1976, western congressional delegations kept the Grazing Service (which in 1946 became the Bureau of Land Management or BLM) so poorly funded and staffed, and without enforcement authority, that many ranchers and miners continued to misuse these lands—a practice that according to environmentalists is not yet completely eradicated.

In 1960 conflicts between competing uses of publicly owned land led to the passage of the Multiple-Use–Sustained-Yield Act. This legislation required the Forest Service and the BLM to attempt to balance outdoor recreation, timber, rangeland, watershed, fish and wildlife habitats, mineral extraction, and other uses of publicly owned land to provide the optimal value to the nation as a whole. Today conflicts continue over what constitutes balanced multiple use of lands managed by these two agencies and what use should have the highest priority when there are competing claims.

In 1964 Congress created the National Wilderness System, in which undeveloped tracts of federally owned lands are to be set aside and retained in their natural state unless and until Congress later decides they are needed for the national good. In this act, **wilderness** is defined as areas where the earth and its community of life have not been seriously disturbed by humans and humans are only temporary visitors. Wilderness areas are open for fishing, hiking, camping, canoeing, and,

in some cases, hunting and horseback riding. Roads, timber harvesting, mining, drilling, commercial activities, toilet facilities, human-made structures, motor vehicles, power saws, and other motors are prohibited. Most areas included in the National Wilderness System are undisturbed, but some contain a few abandoned roads, farms, and buildings. Grazing is permitted only where lands were leased before the law was passed. The act allows exploration and identification of mineral, energy, and other resources in wilderness areas as long as such activities do not involve the use of motorized vehicles and other motors.

The U.S. concept of setting aside land for protection in the form of national parks, wilderness areas, and so on has been adopted by more than 100 other nations. By 1983 about 3 percent of the world's land surface had been set aside for protection from development, overgrazing, and overcutting. The amount of protection actually provided for these designated lands, however, varies widely throughout the world. Despite the warnings of writers such as Marsh, Sears, and Osborn, it was not until 1962, when Rachel Carson published *Silent Spring*, that the American public began embracing the notion of environmental protection.

The Environmental Decade (1970–1980) Until the 1970s the role of the federal government in environmental regulation was generally limited to the management of publicly owned land, minerals, timber, and waterways. This situation changed dramatically between 1970 and 1980, now known as the *environmental decade*, primarily through extensive media coverage and the writings of such biologists as Paul Erhlich, Barry Commoner, and Garrett Hardin, who helped the general public became aware of the interlocking relationships between population growth, resource use, and pollution (Section 1-5).

During the 1970s two dozen separate pieces of legislation (Appendix 3) were passed to protect the air, water, land, and wildlife. In 1970 President Richard Nixon used administrative reorganization to create the Environmental Protection Agency (EPA) to determine environmental standards and see that federal environmental laws are enforced. William D. Ruckelshaus was appointed as its first director.

At the same time citizen-supported environmental organizations such as the Sierra Club, the Wilderness Society, the National Wildlife Federation, Friends of the Earth, the Environmental Defense Fund (EDF), and the Natural Resources Defense Council (NRDC) lobbied for better protection and management of public lands. They also began taking the government to court to secure enforcement of environmental laws. In addition, private organizations such as the Nature Conservancy and the Audubon Society accelerated their efforts to buy and protect unique areas of land threatened by development.

Continuing Controversy: The 1980s The Federal Land Policy and Management Act of 1976 gave the BLM its first real authority to manage its lands. Western ranchers, farmers, miners, off-road motorized vehicle users, and others who had been been doing pretty much as they pleased on BLM lands discovered that this was no longer possible. In the late 1970s western ranchers who had been paying low fees for grazing rights that encouraged overgrazing launched a political campaign that came to be known as the sagebrush rebellion. Its goal, like earlier attempts since the 1930s, was to remove most western public lands, including the national forests, from public ownership, turn them over to the states, and then move them into private ownership or private control. Six western states, led by Nevada, laid claim to federal lands in court, and some western congressional representatives introduced legislation to give public lands to the states. So far the court suits and legislative efforts inspired by the sagebrush rebellion have failed.

In 1981 Ronald Reagan, a declared sagebrush rebel and advocate of less federal government control, became president. He replaced most of the Carter administration's key appointive environmental and consumer advocates in the Environmental Protection Agency and the Department of the Interior. Environmentalists saw the hard-won environmental protection legislation and policies of the 1970s threatened by appointments of people they considered to be opposed to many of their goals and by the administrative actions and legislative proposals that followed.

Between 1982 and 1984 several environmental groups studied the Reagan environmental and land-use policies in detail. Some of their major charges were that the administration **(1)** appointed key personnel in the Environmental Protection Agency and the Department of Interior who came mostly from industries or legal firms that opposed existing federal environmental and land-use legislation and policies, **(2)** made it difficult to enforce existing environmental laws by encouraging drastic budget and staff cuts in environmental programs, **(3)** greatly increased energy and mineral development and cutting of timber by private enterprise on public lands, often at giveaway prices, **(4)** returned much of the management decision making on the use of public rangelands to ranch-

ers, and **(5)** increased the federal budget for the development of nuclear power while drastically cutting the budget for energy conservation and the development of renewable energy from the sun and wind.

The Reagan administration argued that **(1)** it is normal for a newly elected president to appoint key personnel who wish to see the president's policies carried out; **(2)** drastic budget and staff cuts were necessary to reduce waste and to help decrease the mounting national debt; **(3)** increased energy and mineral exploration and sale of timber on federal lands are necessary to encourage private enterprise, stimulate economic growth, and improve national security by ensuring that the nation will have sufficient resources; **(4)** because federal bureaucrats are inept managers, it is better to return many management decisions regarding use of public rangelands to ranchers; and **(5)** the federal government should continue to subsidize nuclear power because it is a safe and proven technology that can provide the nation with much-needed electricity in the future; the budget for energy conservation and alternative renewable energy sources should be decreased so that these emerging technologies can be developed by private enterprise under free market competition.

From this brief summary we see that the history of the use of publicly owned natural resources in the United States has been one of continuing controversy over how much land should be owned by the federal government and how this land should be managed for use by the public. It seems clear that these controversies, along with those over how much tax money should be used to protect the air, water, and land from environmental abuse and how much environmental protection efforts should be carried out by the federal government, will intensify in the future.

A continent ages quickly once we come.
Ernest Hemingway

Discussion Topics

1. It is sometimes argued by those wishing to avoid dealing with environmental problems that "people have always polluted and despoiled this planet, so why all the fuss over ecology and pollution? We've survived so far." Identify the kernel of truth in this position and then discuss its serious deficiencies.

2. Do you think that it is better for public land to be owned by the federal government or by state governments? Why?

3. Argue for or against: **(a)** selling national forests and publicly owned rangelands to private interests, **(b)** selling national forests and publicly owned rangelands to state and local governments, **(c)** eliminating future purchases of new land by the federal government for use as wilderness, national parks, national forests, and national wildlife ref-uges. In each case relate your decision to your own lifestyle and consumption habits.

4. Do you agree or disagree with the major environmental and land-use polices of the Reagan administration discussed in Section 2-4? In each case, defend your position.

PART TWO

Some Concepts of Ecology

USDI/National Park Service/George Grant

Some Environmental Principles

1. *Everything must go somewhere; or, there is no away. (Law of conservation of matter)*
2. *You can't get something for nothing; or, there is no free lunch. (First law of energy, or law of conservation of energy)*
3. *You can't even break even; or, if you think things are mixed up now, just wait. (Second law of energy)*
4. *Everything is connected to everything else, but how?*
5. *A thing is right when it tends to preserve the integrity, stability, and beauty of the biotic community. It is wrong when it tends otherwise.*
6. *Natural systems can take a lot of stress and abuse, but there are limits.*
7. *In nature you can never do just one thing, so always expect the unexpected; or, there are side effects, often unpredictable, to everything we do.*

3

Some Matter and Energy Laws

The laws of thermodynamics control the rise and fall of political systems, the freedom or bondage of nations, the movements of commerce and industry, the origins of wealth and poverty, and the general physical welfare of the human race.

Frederick Soddy, Nobel laureate, chemistry

Look at a beautiful flower, drink some water, eat some food, or pick up this book. The two things that connect these activities and other aspects of life on earth are matter and energy. **Matter,** or anything that has mass and occupies space, is the stuff you and all other things are made of. **Energy** is a more elusive concept. Formally it is defined as the ability or capacity to do work or produce change by pushing or pulling some form of matter. Energy is what you and all living things use to move matter around and to change it from one form to another. Energy is used to grow your food, to keep you alive, to move you from one place to another, and to warm and cool the buildings in which you work and live. The uses and transformations of matter and energy are governed by certain scientific laws, which, unlike the laws people enact, cannot be broken. In this chapter we begin our study of ecological concepts with a look at one fundamental law of matter and two equally important laws of energy. These laws will be used throughout the book to help you understand many environmental problems and to aid you in evaluating proposed solutions.

3-1 Law of Conservation of Matter: Everything Must Go Somewhere

We talk about consuming or using up material resources, but actually we don't consume any matter. We only borrow some of the earth's resources for a while—taking materials from the earth, carrying them to another part of the globe, processing them, using them, and then discarding, reusing, or recycling them. In the process of using matter we may change it to another form, but in every case we neither create nor destroy any measurable amount of matter. This circumstance is expressed in the **law of conservation of matter:** In any physical or chemical change, matter is neither created nor destroyed but merely changed from one form to another.

This law tells us that there is no "away." *Everything we think we have thrown away is still here with us, in one form or another.* We can collect dust and soot from the smokestacks of industrial plants, but these solid wastes must then go somewhere. We can collect garbage and remove solid sludge from sewage, but these substances must either be burned (perhaps causing air pollution), dumped into rivers, lakes, and oceans (perhaps causing water pollution), or deposited on the land (perhaps causing soil pollution and water pollution).

Although we can certainly make the environment cleaner, the law of conservation of matter says that we will always be faced with pollution of some sort. This means that we must *trade off* one form of pollution for another. This tradeoff process involves making controversial scientific, political, economic, and ethical judgments about what is a dangerous pollution level, to what degree a pollutant must be controlled, and how much we are willing to pay to reduce the amount of a pollutant to a harmless level. Now let's look at energy and the two energy laws to learn more about what we can and cannot do on this planet.

3-2 First Law of Energy: You Can't Get Something for Nothing

Types of Energy You encounter energy in many forms: mechanical, chemical (food or fuel), electrical, nuclear, heat, and radiant (such as light). Scientists usually classify most forms of energy as either potential energy or kinetic energy (Figure 3-1). **Kinetic energy** is the energy that matter has because of its motion and mass. A moving car, a falling rock, a speeding bullet, and the flow of electrons or charged particles (electrical energy) are all examples.

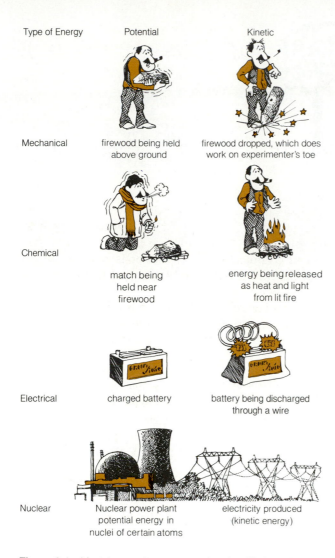

Type of Energy	Potential	Kinetic
Mechanical	firewood being held above ground	firewood dropped, which does work on experimenter's toe
Chemical	match being held near firewood	energy being released as heat and light from lit fire
Electrical	charged battery	battery being discharged through a wire
Nuclear	Nuclear power plant potential energy in nuclei of certain atoms	electricity produced (kinetic energy)

Figure 3-1 Most forms of energy can be classified as either potential energy or kinetic energy.

The energy stored by an object as a result of its position or the position of its parts is called **potential energy**. A rock held in your hand, a bowl of cereal, a stick of dynamite, and a tank of gasoline are all examples. The rock has stored or potential energy that can be released and converted to kinetic energy (in the form of mechanical energy and heat) if it is dropped. **Chemical energy** is the potential energy stored in the chemical bonds that hold the atoms and ions (electrically charged atoms) found in chemicals together. Examples are the energy stored in fuel, explosives, and the carbohydrates, proteins, and fats found in our food.

Doing work involves changing energy from one form to another. When you rub your hands together rapidly, they get warm because the mechanical energy of rubbing is transformed into heat. When you lift this book, chemical energy stored in chemicals obtained from your digested food is converted into the mechanical energy used to move your arm and the book upward and into heat given off by your body.

In an automobile engine, the chemical energy stored in gasoline is converted into mechanical energy that propels the car and into waste heat. A battery converts chemical energy into electrical energy and into heat. In an electric power plant, chemical energy from fossil fuels or nuclear energy from nuclear fuel is converted into a combination of mechanical energy and heat. The mechanical energy is used to spin a turbine that converts the mechanical energy into electrical energy and more heat. When this electrical energy passes through the filament wires in an ordinary light bulb, it is converted into light and still more heat. Note that in all these energy transformations some useful energy is lost as unusable, low-temperature heat that flows into the environment.

First Energy Law What energy changes occur when you drop a rock? Because of its position, the rock in your hand has a higher potential energy than the same rock at rest on the ground. Has energy been lost or used up in this process? At first glance it seems so. But according to the **law of conservation of energy**, also known as the **first law of thermodynamics**, in any ordinary physical or chemical process energy is neither created nor destroyed but merely changed from one form to another. The energy lost by a *system* or collection of matter under study (in this instance, the rock) must equal the energy gained by the *surroundings* or *environment* (in this instance, air molecules and soil particles moved by the impact of the rock). This energy law holds for all systems, living and nonliving.

Let's consider what really happens. As the rock drops, its potential energy is changed into kinetic energy—both its own and that of the air through which it passes. The friction created when the rock drops through the air causes the gaseous molecules in the air to move faster, so their average temperature rises. This means that some of the rock's original potential energy has been transferred to the air as heat. When the rock hits the ground, more of its mechanical energy is transferred to particles of soil. The energy lost by the rock (system) is exactly equal to the energy gained by its surroundings. In studying hundreds of thousands of mechanical processes (such as the rock falling) and chemical processes (such as the burning of a fuel), scientists have found that no detectable amount of energy is either created or destroyed. *Energy input always equals energy output.*

Although most of us know this first energy law, we sometimes forget that regarding energy

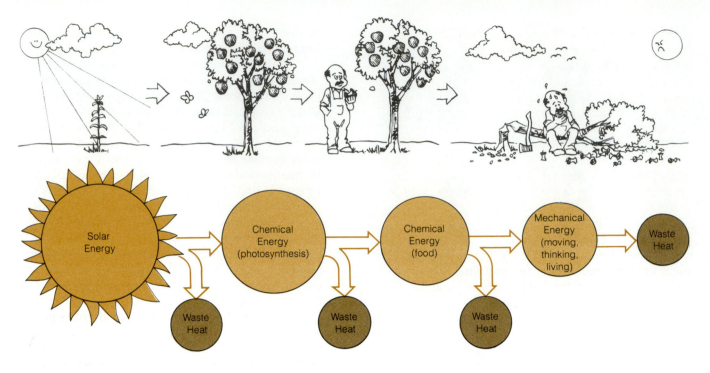

Figure 3-2 The second energy law. When energy is changed from one form to another, some of the initial input of energy is always degraded to low-quality heat, which is added to the environment.

quantity, it means that we can't get something for nothing. In the words of environmentalist Barry Commoner, "There is no free lunch." For example, we often hear that we have huge amounts of energy available from the world's deposits of oil, coal, natural gas, and nuclear fuels (such as uranium). The first law of thermodynamics tells us that we really have much less useful energy available than these estimates indicate because *it takes energy to get this energy*. We must use large amounts of energy to find, remove, and process these fuels. The only energy that really counts is the *net useful energy* or *useful energy yield* available for use after subtracting the energy needed to make this energy available from the total energy available in the resource (Section 16-4).

3-3 Second Law of Energy: You Can't Break Even

Second Energy Law and Energy Quality Because according to the first law energy can neither be created nor destroyed, you might think that there will always be enough energy. Yet when you fill a car's tank with gasoline and drive around, something is lost. If it isn't energy, what is it? The *second law of energy*, also known as the *second law of thermodynamics*, provides the answer to this question.

Energy varies in its *quality* or ability to do use-

ful work. For useful work to occur, energy must move or flow from a level of high-quality (more concentrated) energy to a level of lower quality (less concentrated) energy. The chemical potential energy concentrated in a lump of coal or a tank of gasoline and the concentrated heat energy at a high temperature are forms of high-quality energy. Because they are concentrated, they have the ability to perform useful work in moving or changing matter. In contrast, dispersed or less concentrated heat energy at a low temperature has little remaining ability to perform useful work.

In investigating hundreds of thousands of conversions of heat energy to useful work, scientists have found that some of the energy is always degraded to a more dispersed and less useful form, usually as heat given off at a low temperature to the surroundings. This is a statement of the **second energy law**, another name for the **second law of thermodynamics**: No transfer of heat energy to useful work is 100 percent efficient. Thus, the supply of concentrated, high-quality energy available to the earth is being continually depleted and the supply of low-quality energy is being continually increased so that the total energy remains the same.

Let's consider some examples of the second energy law. When a car is driven, only about 10 percent of the high-quality chemical energy available in the gasoline is converted to mechanical energy used to propel the vehicle. The remaining 90 percent is degraded to low-quality heat that is released into the environment. When electrical energy flows through the filament wires in an ordinary light bulb, it is converted into a mixture of about 5 percent useful radiant energy or light and 95 percent low-quality heat. *Much of modern civilization is built around the internal combustion engine and the incandescent light, which, respectively, waste 90 and 95 percent of their initial energy input*, as discussed further in Section 16-4.

Another example of the degradation of energy is the conversion of solar energy to chemical energy in food. Photosynthesis in plants converts radiant energy (light) from the sun into high-quality chemical energy (stored in the plant in the form of sugar or starch molecules) and low-quality heat energy. If you eat a plant food, such as potatoes, its high-quality chemical energy is transformed within your body to high-quality mechanical energy (used to move your muscles and to perform other life processes) and low-quality heat energy. As shown in Figure 3-2, in each of these energy conversions some of the initial high-quality energy is degraded into low-quality heat that flows into the environment.

According to the first energy law we will never run out of energy, but according to the second law we can run out of high-quality, or useful, energy. *Not only can we not get something for nothing (the first law), we can't even break even in terms of energy quality (the second law).*

The second energy law also tells us that high-grade energy can never be reused. *We can recycle matter but we can never recycle high-quality energy.* Fuels and foods can be used only once to perform useful work. Once a piece of coal or a tank of gasoline has burned, its high-quality potential energy is lost forever. This means that the net useful, or high-quality, energy available from fossil fuels, uranium, or any concentrated energy source is even less than that predicted by the first energy law, as discussed further in Section 16-4.

Second Energy Law and Increasing Disorder   The second energy law can be stated in a number of ways. For example, since energy tends to flow or change spontaneously from a concentrated and ordered form to a more dispersed and disordered form, the *second energy law* also can be stated as follows: Heat always flows spontaneously from hot (high-quality energy) to cold (lower quality energy). You learned this the first time you touched a hot stove. A cold sample of matter such as air has its heat energy dispersed in the random motion of its molecules. This is why heat energy at a low temperature can do little if any useful work.

Figure 3-3   The spontaneous tendency toward increasing disorder or entropy of a system and its surroundings.

Let's look at other spontaneous changes in the world around us. A vase falls to the floor and shatters into a more disordered state. A dye crystal dropped into water spontaneously dissolves, and the spreading of color is evidence that the dye molecules spontaneously tend toward a more dispersed and disordered state throughout the solution. Your desk and room seem spontaneously to become more disordered after a few weeks of benign neglect (Figure 3-3). Smoke from a smokestack and exhaust from an automobile disperse spontaneously to a more random or disordered state in the atmosphere, and pollutants dumped into a river spread spontaneously throughout the water. Indeed, until we discovered that the atmosphere and water systems could be overloaded, we assumed that such spontaneous dilution was the solution to pollution.

These observations all suggest that a *system* of matter spontaneously tends toward increasing randomness or disorder, often called *entropy*. But you may have already thought of some cases that contradict this hypothesis. What about living organisms, with their highly ordered systems of molecules and cells? You are a walking, talking contradiction of the idea that systems tend spontaneously toward increasing disorder or entropy. We must look further.

The way out of our dilemma is to look at changes in disorder or order not only in the system but both in the system *and in its environment or surroundings*. Look at your own body. To form and preserve its highly ordered arrangement of molecules and its organized network of chemical reactions, you must continually obtain high-quality energy and raw materials from your surroundings. This means that disorder or entropy is created in the environment—primarily in the form of low-quality heat. Just think of all the disorder in the

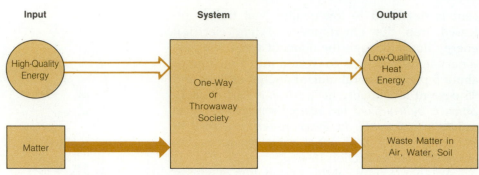

Figure 3-4 The one-way or throwaway society found in most industrialized countries is based on maximizing the rates of energy flow and matter flow. This results in the rapid conversion of the world's mineral and energy resources to trash, pollution, and waste heat. This type of society is sustainable indefinitely only with essentially infinite supplies of mineral and energy resources and an infinite ability of the environment to absorb the resulting heat and matter wastes.

form of heat that is added to the environment to keep you alive. Planting, growing, processing, and cooking foods all require energy inputs that add heat to the environment. The breakdown of the chemicals in food in your body gives off more heat to the environment. Indeed, your body continuously gives off heat equal to that from a 100-watt light bulb—explaining why a closed room full of people gets warm.

Measurements show that the total amount of disorder or entropy, in the form of low-quality heat, added to the environment to keep you alive is much greater than the order maintained in your body. In addition, enormous amounts of disorder go into the environment when concentrated deposits of minerals and fuels are extracted from the earth and burned or dispersed to heat the buildings you use, to transport you, and to make roads, clothes, and shelter.

Thus, *all forms of life are tiny pockets of order maintained by creating a sea of disorder around themselves.* The primary characteristic of modern industrial society is an ever-increasing use or flow of high-quality energy to maintain the order in our bodies and the pockets of order we call civilization. As a result, today's industrialized nations are adding more entropy or disorder to the environment than any society in human history.

In considering the *system and surroundings as a whole*, scientists find that there is always a net increase in disorder with any spontaneous chemical or physical change. Experimental measurements have demonstrated this over and over again. Thus, we must modify our original hypothesis to include the surroundings. *Any system and its surroundings as a whole spontaneously tends toward increasing randomness, disorder, or entropy.* In other words, if you think things are mixed up now, just wait. This is another way of stating the *second energy law*, or *second law of thermodynamics*. No one has ever found a violation of this law. In most apparent violations, the observer has failed to include the greater disorder added to the surroundings when there is an increase in order in the system.

3-4 Matter and Energy Laws and Environmental Problems

As discussed throughout this book, the law of conservation of matter and the first and second laws of energy give us keys for understanding and dealing with environmental problems. The one-way or throwaway society found in most industrial countries is based on using more and more of the earth's matter and energy resources at a faster and faster rate (Figure 3-4). The earth receives a constant flow of energy from the sun, but for all practical purposes little matter enters or leaves the earth. *We have all the matter that we will ever have.*

Some say we should become a *matter-recycling society* so that economic growth can continue indefinitely without depleting material resources. But there is a catch to infinite recycling. *Recycling matter always requires high-quality energy.* However, if a resource is not too widely scattered, recycling often requires less high-quality energy than that needed to find, get, and process virgin ores. In the long run, a recycling society based on indefinitely increasing economic growth must have an essentially inexhaustible and affordable supply of high-quality energy. And high-quality energy, unlike matter, can never be recycled. Although experts disagree on how much usable energy we have, it is clear that supplies of nonrenewable fossil fuels and nuclear fuel resources are finite. Indeed, *affordable* supplies of oil, natural gas, and fuel-grade uranium may last no longer than several decades (Chapters 16, 17, and 18).

"Ah," you say, "but don't we have an essentially infinite supply of solar energy flowing into the earth?" Sunlight reaching the earth is high-quality energy, but the quantity reaching a particular area of the earth's surface each minute or hour is low and is nonexistent at night. With a proper heat storage system, using solar energy to provide hot water and to heat a house to moderate temperatures makes sense (Chapters 16 and 19). However, using solar energy to provide the high

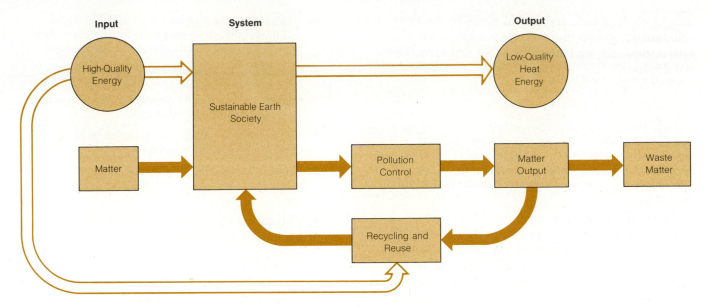

Figure 3-5 A sustainable earth society is based on energy flow and matter recycling. It is characterized by reusing and recycling renewable matter resources, not using renewable matter resources faster than they are replenished by natural process, conserving energy (since it cannot be recycled), increasing pollution control, and deliberately lowering the rate at which matter and energy resources are used, so that the environment is not overloaded and resources are not depleted.

temperatures needed to melt metals or to produce electricity in a solar power plant may not be cost effective. In these cases, solar energy must be collected and concentrated to provide the necessary high temperatures. This requires large amounts of money and high-quality energy to mine, process, and to transport the resources needed to make solar collectors, focusing mirrors, pipes, and other materials.

One way to lessen this problem involves the development and widespread use of *solar photovoltaic cells*, which convert sunlight directly to electricity in one simple, nonpolluting step. If present research increases the efficiency of such cells and decreases their cost, we could be covering entire roofs of houses and buildings with rolls of these cells to provide all the electricity we need (Section 19-1). Such a development could, in a fairly short time, make most large, centralized electric power plants in the world obsolete. Mass production and transportation of solar cells would require energy and matter resources. But most of the matter would come from silicon, one of the most abundant chemical elements on earth.

Even if some breakthrough provides us with an essentially infinite supply of affordable energy, the *second energy law tells us that as we use more and more energy to transform matter into products and then recycle these products, the disorder in the environment will increase.* Thus, the more we try to order, or "conquer," the earth, the greater the disorder we put into the environment. We will always attempt to some extent to order the environment for our benefit, but the second energy law helps us understand that we should act with ecological wisdom, care, and restraint.

Why do some people think we can avoid the effects of the second energy law? Part of the problem is that many people have never heard of the second law of thermodynamics, let alone understood its significance. In addition, this law has a cumulative rather than an individual effect. You accept the law of gravity because it limits you and everyone else on a personal level. By the same token, your automatic entropy-increasing activities seem small and insignificant. But the cumulative impact of the entropic, or disorder-producing, activities of billions of individuals trying to convert more and more of the world's resources to trash and low-quality heat as fast as possible eventually can have a devastating impact on the life-support systems that sustain us all. *The second energy law tells us that, like it or not, we are dependent on each other and on the other parts of nature for our survival.*

This may seem like a rather gloomy situation, but it need not be. The second energy law, along with the first energy law and the law of conservation of matter, tell us what we *cannot* do. But more importantly, these laws tell us what we *can* do. They show us that one way out is to shift to a *sustainable earth society* (Figure 3-5), based on reducing the rate of using matter and high-quality energy so that local, regional, and global limits of the environment to absorb entropy are not exceeded and vital renewable and nonrenewable resources are not depleted.

The law that entropy increases—the second law of thermodynamics—holds, I think, the supreme position among laws of nature. . . . If your theory is found to be against the second law of thermodynamics, I can give you no hope; there is nothing to do but collapse in deepest humiliation.

Arthur S. Eddington

Discussion Topics

1. Explain why we don't really consume anything and why there is no such thing as a throwaway society.

2. A tree grows and increases its mass. Explain why this isn't a violation of the law of conservation of matter.

3. List six different types of energy that you have used today and classify each as kinetic or potential.

4. According to the first law of energy, the world will never run out of energy. Why, therefore, has there been so much talk about an energy crisis?

5. Use the first and second energy laws to explain why the usable supply of energy from fossil and nuclear fuels is usually considerably less than that given by most official estimates.

6. Use the second energy law to explain why a barrel of oil can be used only once as a fuel.

7. Criticize the statement "Any spontaneous process results in an increase in the disorder of the system."

8. Criticize the statement "Life is an ordering process, and since it goes against the natural tendency for increasing disorder, it breaks the second law of thermodynamics."

9. a. Use the law of conservation of matter to explain why a matter-recycling society will sooner or later be necessary.

   b. Use the second energy law to explain why there should be more emphasis on reusing than on recycling matter.

   c. Use the second energy law to explain why energy can never be recycled.

Ecosystem Structure: What Is an Ecosystem?

If we love our children, we must love the earth with tender care and pass it on, diverse and beautiful, so that on a warm spring day 10,000 years hence they can feel peace in a sea of grass, can watch a bee visit a flower, can hear a sandpiper call in the sky, and can find joy in being alive.

Hugh H. Iltis

What plants and animals live in a forest? How do they get the matter and energy needed to stay alive? How do these plants and animals interact with one another and with their physical environment? What changes will this dynamic system of life undergo with the passing of time?

Ecology is the branch of science that attempts to answer such questions. In 1866 German biologist Ernst Haeckel coined the term *ecology* from two Greek words: *oikos*, meaning "house" or "place to live," and *logos*, meaning "study of." Literally, then, "ecology" is a study of organisms in their home. In more formal terms, **ecology** is the study of the structure and function of nature, or the study of the relationships among living organisms and of the totality of physical and biological factors making up their environment.

Ecologists call a *self-sustaining* collection of living organisms and their environment such as those in a forest an *ecological system* or *ecosystem*—a term introduced by English botanist A. G. Tansley in 1935. The remainder of Part Two is devoted to a study of ecosystems. This chapter considers ecosystem structure and type, Chapter 5 ecosystem function, or what happens within them, and Chapter 6 some of the changes that can occur in ecosystems as a result of natural events and human activities.

4-1   The Ecosphere: Our Life-Support System

The earth can be divided into three intersecting regions: **(1)** the **atmosphere**—a region of gases and particulate matter extending above the earth; **(2)**
the **hydrosphere**—a region that includes all the earth's liquid water (oceans, smaller bodies of water, and underground aquifers), frozen water (polar ice caps, floating ice, and frozen upper layers of soil known as permafrost), and the small amounts of water vapor found in the earth's atmosphere; **(3)** the **lithosphere**—a region of soil and rock consisting of the earth's crust, a mantle of partially molten rock beneath this crust, and the earth's inner core of molten rock called magma.

All life exists within a thin spherical shell of air, water, and soil known as the **ecosphere** or **biosphere** (Figure 4-1). It is found within the atmosphere, the hydrosphere, and the lithosphere. If the earth were an apple, the ecosphere would be no thicker than the apple's skin. Everything in this skin of life is interconnected: Air helps purify water and keeps plants and animals alive, water keeps plants and animals alive, plants keep animals alive and help renew the air and soil, and the soil keeps plants and many animals alive and helps purify water.

The ecosystems that make up the ecosphere also help to **(1)** moderate the weather, **(2)** recycle vital chemicals needed by plants and animals, **(3)** dispose of our wastes, **(4)** control more than 95 percent of all potential crop pests and causes of human disease, and **(5)** maintain a gigantic genetic pool, which we use to develop new food crop strains and medicines. The ecosphere, then, is a remarkably effective and enduring system—and endure it must, or life will become extinct. *The goal of ecology is to find out just how everything in the ecosphere is related.* Let's begin this study of interrelationships by learning how solar energy sustains all life in the ecosphere.

4-2   The Sun: Source of Energy for Life on Earth

The source of the radiant energy that sustains essentially all life on earth is the sun. It warms the earth and provides energy for photosynthesis in green plants. These plants, in turn, synthesize the car-

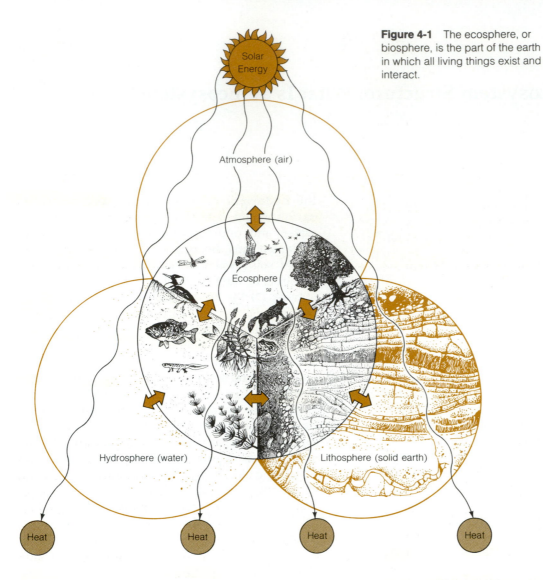

Figure 4-1 The ecosphere, or biosphere, is the part of the earth in which all living things exist and interact.

bon compounds that keep them alive and serve as food for almost all other organisms. Solar energy also powers the water cycle (Section 5-3), which purifies and removes salt from ocean water to provide the fresh water on which land life depends.

The sun's energy comes to us as *radiant energy*, or **electromagnetic radiation** (Figure 4-2), traveling through space as waves of oscillating electric and magnetic fields. These electromagnetic waves travel at a speed of 300,000 kilometers (186,000 miles) per second. At this speed the light striking your eyes made the 150-million-kilometer (93-million-mile) trip from the surface of the sun to the earth in about 8 minutes. Only about one two-billionth (0.000000002) of the sun's total radiated energy is intercepted by the earth, a minute target in the vastness of space.

The visible light rays we call sunlight are only a tiny part of the wide range, or spectrum, of energies given off by the sun. The energies in this **electromagnetic spectrum** range from high-energy cosmic rays to low-energy radio waves (Figure 4-2).

Each type of energy can be treated as an electromagnetic wave with a different **wavelength**, the **distance between the crest of one wave and the next**. High-energy electromagnetic waves have short wavelengths (⌇⌇⌇⌇), whereas low-energy electromagnetic waves have long wavelengths (⌇‿⌇‿). The higher energy, shorter wavelength rays—cosmic rays, gamma rays, X rays, and most ultraviolet rays—are harmful to the genetic material known as deoxyribonucleic acid (DNA). Fortunately, most of this harmful electromagnetic radiation is screened out by ozone and water vapor in the earth's atmosphere (Section 21-5). Without this screening, most life on earth could not exist as it does today.

4-3 Ecosystems and Ecosystem Structure

Levels of Organization of Matter Looking at earth from space, we see mostly a blue sphere with

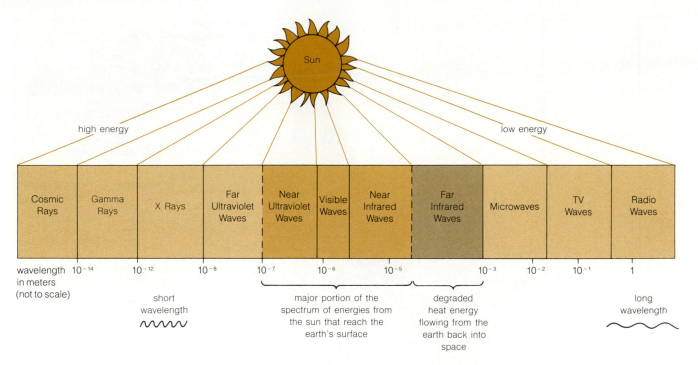

Figure 4-2 The electromagnetic spectrum. The sun radiates a wide range of energies with different wavelengths. Since much of this incoming radiation is either reflected or absorbed by the earth's atmosphere, mostly moderate-to low-energy radiation actually reaches the earth's surface.

irregular green, red, and white patches on its surface. As we zoom closer, these colorful patches appear as deserts, forests, grasslands, mountains, seas, lakes, oceans, farmlands, and cities. Each zone is different, having its own characteristic set of organisms and climatic conditions. Moving in closer, we can pick out a variety of *organisms* or living things. If these plant and animal organisms were greatly magnified, we would see that they are made up of *cells*—groups of atoms and molecules interacting in an organized way to exhibit what we call life.

The *molecules* or *compounds* such as water and proteins found in these cells are chemical combinations of more fundamental building blocks of all matter called *atoms*. All matter on earth is made up of various combinations of atoms of only 108 chemical elements. For convenience, each element is given a symbol: H for hydrogen, O for oxygen, N for nitrogen, P for phosphorus, Cl for chlorine, Na for sodium, and so on. Atoms in turn are made up of even smaller *subatomic particles* such as electrons, protons, and neutrons. All matter, in fact, can be viewed as being organized in identifiable patterns, or *levels of organization*, ranging in complexity from subatomic particles to far galaxies (Figure 4-3).

The Realm of Ecology   As Figure 4-3 shows, ecology is primarily concerned with interactions among only five levels of organization of matter:

organisms, populations, natural communities, ecosystems, and the *ecosphere* or *biosphere.* All organisms of a given kind constitute a **species**. Each species represents a particular array of hereditary material called a *gene pool,* which is distinct from the gene pools of other species. Every member of a species is potentially capable of breeding with other members of the same species but normally not with members of other species.

Every species in nature is composed of smaller units, known as **populations**, or groups of individual organisms (such as squirrels or oak trees) of the same species that interbreed and occupy given areas at given times. Each organism and population in a natural community has a **habitat**, the place where it lives. Habitats vary widely in size from an entire forest to the intestine of a termite.

The populations of plant and animal species living and interacting in a given area at a given time are called a **natural community**. In many natural communities one or two organisms dominate. For example, in an oak–hickory forest natural community, oak and hickory trees are the *dominant* species. Other kinds of trees are less frequent. A natural community of living things that interact with each other and with chemicals and physical factors such as solar energy, temperature, light, wind, and water currents in a way that perpetuates the community is called an **ecosystem**.

The ecosystem concept was developed to facilitate the study of any patch of the earth, of any

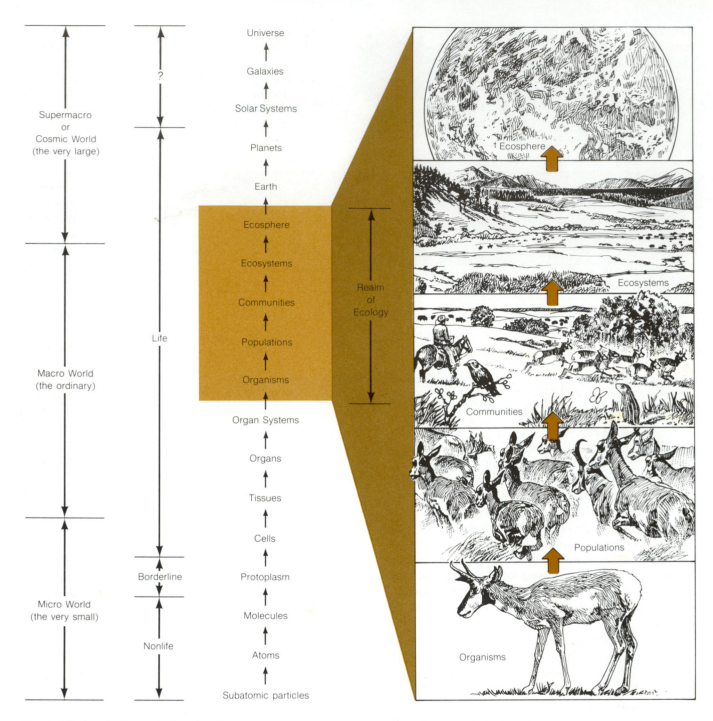

Figure 4-3 Levels of organization of matter.

convenient size, to determine interactions. The boundaries drawn around ecosystems are arbitrary, selected for convenience in studying each system. An ecosystem can be a planet, a tropical rain forest, an ocean, a fallen log, a puddle of water, or a culture of bacteria in a petri dish. The variation in ecosystems is essentially infinite. However, because certain plant and animal species are often found together, it is useful to classify ecosystems according to their similarities in structure.

There are two major classes of *aquatic ecosys-*

tems based on salinity: **(1)** *marine and estuarine ecosystems* such as oceans (see inside front cover), seas, estuaries (where fresh water from rivers and streams mixes with seawater—see inside back cover), and inland bodies of brackish or saline water; and **(2)** *freshwater ecosystems* consisting of inland bodies of standing fresh water such as lakes (see inside front cover) and ponds and inland bodies of flowing water such as springs, creeks, and rivers.

On land, the large major ecosystems—called

biomes—are forests, grasslands, and deserts (see map inside front cover). Each **biome** is a large terrestrial ecosystem with a distinct group of plants and animals, as described in more detail in Section 4-7. All the various ecosystems on the planet, along with their interactions, make up the largest unit, or planetary ecosystem, called the *ecosphere* or *biosphere* shown earlier (Figure 4-1).

The ecosystem approach, however, is not the only way of studying ecology. Some ecologists prefer to approach the subject from the standpoint of biological evolution, which usually involves populations. Because the evolutionary approach requires a fairly detailed and technical background in biology, this book basically relies on the ecosystem approach.

Ecosystem Structure   For convenience, scientists divide an ecosystem into two major components: the **abiotic** or nonliving components, and the **biotic** or living components. The *abiotic* parts include (1) an outside energy source (usually the sun), (2) various other physical factors such as wind and heat, which determine the climate and weather of the ecosystem, and (3) all the chemicals found in the soil, air, and water that are essential nutrients for life and are obtained by breaking down the matter in dead plants and animals. The type, quantity, and variation of various physical and chemical abiotic factors in a given area determine the types of plants and animals that can exist in a particular ecosystem.

The *biotic* or living portion of an ecosystem consists of *producers*, *consumers*, and *decomposers*. *Producers* are green plants and some bacteria that use certain wavelengths of solar energy (Figure 4-2) in the photosynthetic production of organic food substances such as glucose from carbon dioxide gas and water obtained from the environment. Animals called *primary consumers* or *herbivores* (for example, deer and grasshoppers) feed directly on green plants. Other animals called *secondary or higher order consumers*, or *carnivores* (for example, lions and snakes), feed indirectly on green plants by consuming the tissue of herbivores. *Decomposers* or *microconsumers* such as some bacteria, fungi, and protozoans, rot, decompose, or otherwise break down organic wastes from live organisms and tissue from dead plants and animals into simpler substances that are returned to the environment for use as nutrients by other living organisms.

Chemical Cycling and Energy Flow in an Ecosystem   The major components of an ecosystem (Figure 4-4) are related by the processes of chemical cycling within the ecosystem and the one-way energy flow through the ecosystem. Notice that the chemicals (represented by solid arrows in Figure 4-4) that serve as nutrients for living organisms in an ecosystem are *cycled* from the abiotic environment to producers, to consumers, to decomposers, and then back to the abiotic environment for reuse. At the same time solar energy flows one way into the ecosystem and is used by green plants to produce organic food substances such as glucose. In accordance with the second law of thermodynamics (Section 3-3), much of this input of high-quality solar energy is degraded to low-quality heat that flows back into the environment. This one-way energy flow through an ecosystem is represented by the open arrows in Figure 4-4. These processes of *chemical cycling and one-way energy flow* allow the groupings of producers, consumers, and decomposers found in a particular ecosystem to perpetuate themselves, as discussed in more detail in the next chapter.

Any self-sustaining ecosystem must have a mix of producers, consumers, and decomposers. However, as Figure 4-4 shows, it is not necessary for an ecosystem to have primary, secondary, and higher order consumers for the vital processes of chemical cycling and energy flow to take place. Figures 4-5 and 4-6, respectively, show greatly simplified portions of the structures of ecosystems found in a field and in a freshwater pond.

4-4   Ecosystem Components

Abiotic Components   The major abiotic components of an ecosystem are *physical factors* such as light, temperature, wind, and water currents, and *chemicals* that serve as nutrients for the producers, consumers, and decomposers found in an ecosystem. The chemicals found in ecosystems throughout the world are either elements or compounds.

Elements are chemicals such as iron (Fe), sodium (Na), carbon (C), nitrogen (N), and oxygen (O) whose distinctly different atoms serve as the basic building blocks of all matter. Elements exist as *atoms*, which can combine with other atoms to form *molecules*. For example, two atoms of hydrogen (H) can combine to form a hydrogen molecule with the shorthand *chemical formula* H_2, (read as *H-two*). Note that the number of atoms of each kind in a molecule is shown by placing a numerical subscript to the right of the symbol for each element. When no subscript appears, one atom is understood. Similarly, two nitrogen atoms can combine to form a nitrogen molecule represented by the for-

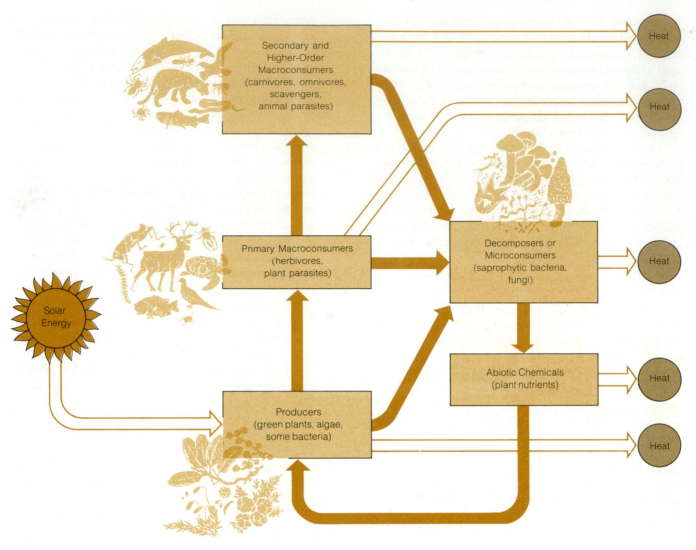

Figure 4-4 The basic components of an ecosystem. Solid arrows represent the cyclical movement of chemicals through the system; open arrows indicate one-way energy flows.

mula N_2 (read as *N-two*). This molecular form of the element nitrogen is the gas that makes up about 78 percent of the atmosphere. Oxygen gas, which makes up almost 20 percent of the atmosphere, consists of oxygen molecules (O_2) formed by the combination of two atoms of oxygen. Gaseous oxygen also exists as molecules of ozone, with the formula O_3, formed when three atoms of oxygen combine; it is found primarily in the *ozone layer*, located in the upper region of the atmosphere known as the stratosphere.

Each atom of an element is made up of a tiny center or *nucleus*, consisting of one or more positively charged *protons (p)* and uncharged *neutrons (n)*, and one or more negatively charged *electrons (e)* whizzing around outside the nucleus. For uncharged atoms, the number of positively charged protons in the nucleus equals the number of negatively charged electrons outside the nucleus, and

this distinguishes one element from another. For example, an atom of the simplest element, hydrogen (H), has one proton in its nucleus and one electron outside its nucleus, whereas sodium (Na) has 11 protons in its nucleus and 11 electrons outside its nucleus.

Atoms or groups of atoms of many of the elements can lose or gain one or more negatively charged electrons outside their nuclei to form positively or negatively charged **ions**. Positive ions, such as sodium ion (Na^+), calcium ion (Ca^{2+}), and ammonium ion (NH_4^+), are formed when an atom or group of atoms loses one or more of its electrons. Negative ions, such as chloride ion (Cl^-), nitrate ion (NO_3^-), and phosphate ion (PO_4^{3-}), are formed when an atom or group of atoms gains one or more electrons.

Most matter in the world exists as **compounds**—substances composed of atoms or ions of two or more different elements held together in a fixed ratio by one of two major types of chemical bonds called *covalent bonds* and *ionic bonds*. The type of bond depends on the type of structural units making up a compound. The basic structural units of compounds are either *molecules*, held together by covalent bonds, or *formula units* of ions with opposite electrical charges, held together by ionic bonds. Water, for example, is a compound composed of H_2O (read as H-two-O) molecules, each consisting of two hydrogen atoms bonded covalently to an oxygen atom.

Because opposite electrical charges attract each another, the oppositely charged ions of different elements can attract one another to form compounds made up of formula units held together by ionic bonds between ions rather than molecules held together by covalent bonds between atoms. Thus, sodium chloride, the main ingredient in table salt, is composed of extremely large numbers of individual formula units of oppositely charged sodium ions (Na^+) and chloride ions (Cl^-) represented by the formula Na^+Cl^- or more often simply by $NaCl$ (where the electrical charges are not shown).

Compounds are represented in shorthand form by *chemical formulas* in which the number of atoms (except one) of each type of element present is shown by placing a numerical subscript to the right of the chemical symbol for each element. For example, the chemical formula for carbon dioxide is CO_2 (read as *C-O-two*); that for glucose, a sugar, is $C_6H_{12}O_6$ (read as *C-six-H-twelve-O-six*); and that for ammonium nitrate, an ingredient in some commercial fertilizers, is NH_4NO_3 (read as *N-H-four-N-O-three*).

Compounds are usually classified as *organic* or *inorganic*. Hydrocarbon compounds containing only atoms of the elements carbon (C) and hydrogen (H), and other compounds derived from hydrocarbons that also contain atoms of one or more elements such as oxygen (O), nitrogen (N), sulfur (S), phosphorus (P), and chlorine (Cl), are called **organic compounds**. Examples include methane (CH_4), the major component of natural gas, proteins, carbohydrates such as glucose ($C_6H_{12}O_6$) and sucrose or table sugar ($C_{12}H_{22}O_{11}$), lipids (fats), vitamins, and complex molecules such as DNA. Organic compounds are major components of the tissues of living and dead organisms.

Compounds not classified as organic compounds are called **inorganic compounds**. Examples include water (H_2O), ammonia (NH_3), nitrogen gas (N_2), carbon dioxide gas (CO_2), sulfur dioxide (SO_2), an air pollutant produced by volcanoes and by the burning of coal and oil containing sulfur impurities, and sodium chloride ($NaCl$).

Some inorganic compounds are found in the tissues of living and dead organisms; others exist apart from organisms. The critical inorganic and organic chemicals found in the air, water, and soil must be continually recycled through the ecosphere.

Just as words can be combined to make sentences, elements and compounds (the "words" of chemistry) can be combined by *chemical reactions* (the "sentences" of chemistry). A chemical reaction is represented in shorthand form by a *chemical equation* using the chemical formulas for the elements and compounds involved. Formulas of the original starting chemicals, called *reactants*, are placed to the left and formulas of the new chemicals produced, called *products*, are placed to the right. Each different reactant and product is separated by a plus sign (+), and the series of reactants and the product or products are separated by an arrow, which stands for "produces" or "yields." For example, the combustion (burning) reaction of carbon (C)—the main ingredient in coal—with oxygen gas (O_2) in the atmosphere to produce carbon dioxide gas (CO_2) can be represented by the chemical equation:

$$\text{reactants} \longrightarrow \text{products}$$
$$C + O_2 \longrightarrow CO_2$$

The law of conservation of matter (Section 3-1) tells us that no atoms (matter) can be created or destroyed in a chemical reaction. Instead, a chemical reaction represents a rearrangement of existing atoms in different forms. Notice that in the chemical equation for the combustion of carbon there is one carbon atom and two oxygen atoms on each side of the equation. A chemical equation that contains the same number of atoms of each element on each side in accordance with the law of conservation of matter is said to be *balanced*.

However, the simple listing of atoms in the elements or compounds involved in a chemical reaction, followed by the formulas for one molecule or formula unit of each substance, sometimes gives us an unbalanced equation. For example, the chemical equation

$$H_2 + O_2 \rightarrow H_2O \qquad (\textit{unbalanced})$$

is not balanced because the number of oxygen atoms on the left-hand side does not equal the number on the right. To balance this equation, we need to have the same number of atoms of oxygen on both sides. This can be achieved by multiplying the number of *hydrogen* atoms on each side by 2:

$$2H_2 + O_2 \rightarrow 2H_2O \qquad (\textit{balanced})$$

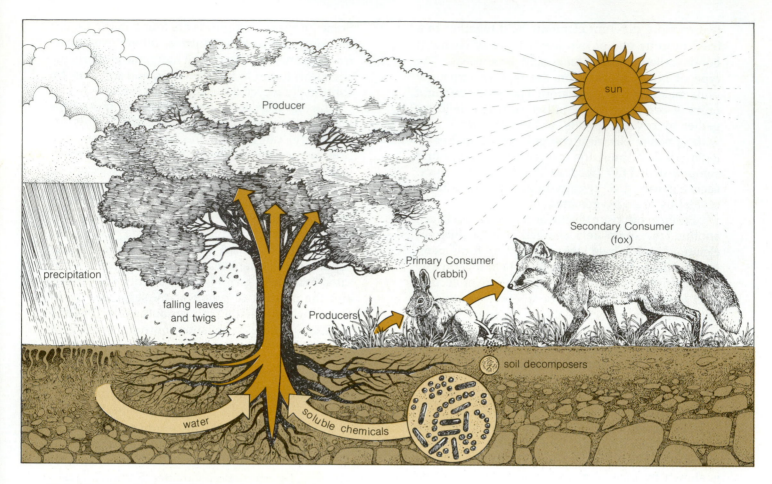

Figure 4-5   Greatly simplified version of the structure of an ecosystem in a field.

Now there are four atoms of hydrogen and two atoms of oxygen on each side of the equation, so that the law of conservation of matter is not violated.

Biotic Components   The major biotic components of an ecosystem are producers, consumers, and decomposers. **Producers** are green plants and some types of bacteria that use either solar energy (green plants) or chemical energy (some bacteria) to convert simple inorganic molecules such as carbon dioxide gas and water obtained from the environment to form complex organic compounds such as glucose ($C_6H_{12}O_6$) and other compounds necessary to life. Producer plants contain chlorophyll and often other pigments that can absorb certain wavelengths of solar energy; they use this energy to combine carbon dioxide and water in the photosynthetic formation of glucose and oxygen gas, as described in more detail in Section 5-3.

Producer plants range in size from tiny, floating phytoplankton such as algae and diatoms in aquatic ecosystems to large trees in terrestrial ecosystems. Because they use either solar energy or

chemical energy to synthesize the organic nutrients they need to stay alive, producers are also called *autotrophs* ("self-feeders").

It is important to note that *not all plants are producers*. Nonproducer plants that lack chlorophyll or any other pigment capable of absorbing solar energy usually are whitish and cannot carry out photosynthesis. Examples include flowering plants such as the Indian pipe and fungi such as mushrooms and molds. Most nonproducer plants are decomposers.

Organisms that feed either directly or indirectly on producers are called **consumers**. Because these organisms cannot manufacture the organic nutrients they need to stay alive, they are also known as *heterotrophs* ("other-feeders"). An organism that feeds directly on a producer is called a primary consumer or **herbivore** ("plant eater"). Examples are shown in Figures 4-5 and 4-6.

An organism that feeds on a primary consumer or herbivore is called a secondary consumer or **carnivore** ("meat eater"). For example, in Figure 4-5 the rabbit that feeds on green plants is a primary consumer and the fox that feeds on the rabbit is a secondary consumer. Higher levels of consum-

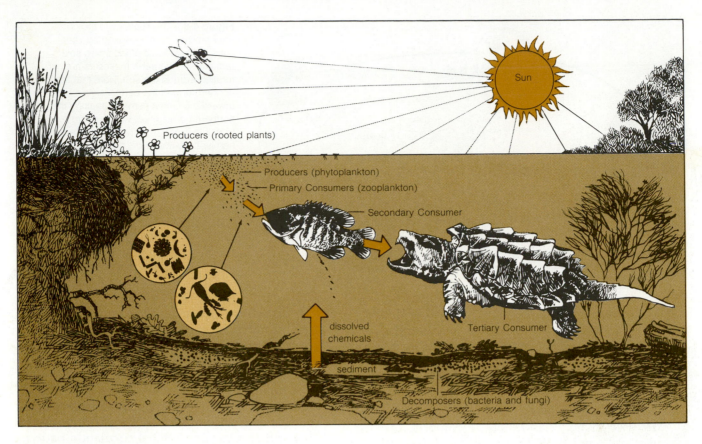

Figure 4-6 Greatly simplified version of the structure of a freshwater pond ecosystem.

ers also exist. For example, in Figure 4-6 the fish that feeds on zooplankton (primary consumers) is a secondary consumer and the turtle that feeds on this fish is a tertiary consumer. Organisms such as pigs, rats, cockroaches, and humans that can eat either plants or other animals at the same or different times are called **omnivores (generalists)**.

Another important type of consumer is the *parasite*. Instead of devouring their prey, parasites slowly feed on the exterior or interior of an organism called the *host* in what is called a *host–parasite relationship*. Parasites such as tapeworms that live *inside* the host are called <u>endo</u>parasites. Others, such as ticks, fleas, leeches, lice, and the sea lamprey (Figure 4-7), attach themselves to the outside of an organism and are known as <u>ecto</u>parasites. Bacteria and other microorganisms that cause plant and animal diseases are specialized endoparasites. Some flowering plants such as mistletoe are parasitic on other plants. Parasites are discussed further in Section 5-5.

Another class of primary, secondary, and higher order consumers are known as *detritus feeders*. These organisms feed on organic material called **detritus**, consisting of dead plant material, the remains of dead animal matter, and fecal matter. Detritus feeders include vultures, earthworms,

termites, ants, beetles, and crayfish. Some animals can act as both regular consumers and detritus consumers. For example, a goat eating grass is a primary consumer, but when it eats fallen leaves the goat is a detritus consumer. Some flowering plants, such as the Indian pipe, obtain their energy from detritus rather than from photosynthesis and are called *saprophytes* (dead feeders).

Most of the dead matter in ecosystems (espe-

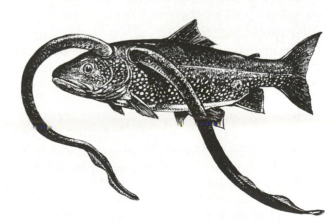

Figure 4-7 The sea lamprey is an ectoparasite that attaches itself to the body of a lake trout or other fish, opens a hole in the skin, and sucks blood and other body fluids from the host organism.

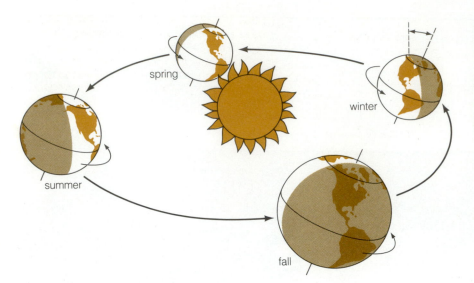

Figure 4-8 The seasons in the northern hemisphere are caused by variations in the amount of incoming solar radiation as the earth makes its annual rotation around the sun. Note that the northern end of the earth's axis is tilted toward the sun in summer, making the northern hemisphere warmer, and away from it in winter, making the northern hemisphere cooler.

cially dead leaves and wood) rots, decays, or decomposes instead of being eaten by detritus feeders. Organisms that obtain nutrients by breaking down into simpler substances the complex organic molecules in the remains of dead animals and plants or the waste products of living organisms are called **decomposers.**

The two major classes of decomposer organisms are *fungi* and certain types of *bacteria* (microscopic, single-celled organisms). Molds, mushrooms, coral fungi, and puffballs are all fungi. Beneath the visible part of a fungus such as a mushroom is a network of tiny threadlike filaments called *mycelia*. These filaments penetrate into detritus and secrete specialized chemicals called *enzymes*. Each enzyme accelerates the breakdown of a certain type of dead material into simpler organic nutrients, which are then absorbed by the cells of the fungus. Certain bacteria can break down dead material into simpler organic nutrients in the same way. These bacteria and fungi in turn are fed on by other organisms such as protozoans, mites, worms, and insects living in the soil or water.

4-5   Global Patterns of Climate

Weather and Climate   **Weather** is the day-to-day variation in atmospheric conditions such as temperature, moisture (including precipitation and humidity), sunshine (solar electromagnetic radiation), and wind. Where the atmosphere thins to nothing, as on the moon or in space, there is no weather. **Climate** is the generalized weather at a given place on earth over a fairly long period. It involves seasonal and annual averages, totals, and occasional extremes of the day-to-day weather pattern for an area. Climate then is the weather you expect to occur at a particular time of the year in your hometown, whereas weather is the actual atmospheric conditions there on a particular day.

Global Air Circulation Patterns   Heat from the sun and evaporated moisture are distributed over the earth as a result of global circulation patterns of atmospheric air masses. Three major factors affecting the pattern of this global air circulation are **(1)** the uneven heating of the equatorial and polar regions of the earth, which creates the driving force for atmospheric circulation, **(2)** the rotation of the earth around its axis, which causes deflection of air masses moving from the equator to the poles and from the poles back to the equator, and **(3)** unequal distribution of landmasses, oceans, ocean currents, mountains, and other geological features over the earth's surface.

An *air mass* is a vast body of air in which the conditions of temperature and moisture are much the same at all points in a horizontal direction. A warm air mass tends to rise and a cold air mass tends to sink. Air in the earth's atmosphere is heated more at the equator, where the sun is almost directly overhead, than at the poles, where the sun strikes the earth at an angle. Because of this unequal heating, warm equatorial air tends to rise and spread out northward and southward, carrying heat from the equator toward the poles. At the poles the warm air cools, sinks, and moves back toward the equator.

In addition, because of the earth's annual rotation around the sun, the sun is higher in the sky in summer (July for the northern hemisphere and January for the southern hemisphere) than in winter (January for the northern hemisphere and July for the southern hemisphere). Such annual varia-

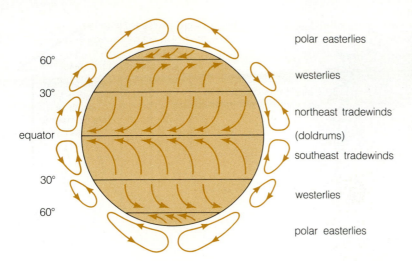

60°

30°

equator

30°

60°

polar easterlies

westerlies

northeast tradewinds

(doldrums)

southeast tradewinds

westerlies

polar easterlies

Figure 4-9 The earth's daily rotation on its axis deflects the general movement of warm air from the equator to the poles and back, to the right in the northern hemisphere and to the left in the southern hemisphere. This twisting motion causes the airflow in each hemisphere to break up into three separate belts of prevailing winds.

tions in the duration and intensity of sunlight lead to seasonal variations in the different hemispheres and at the poles (Figure 4-8).

The general tendency for large air masses to move from the equator to the poles and back is modified by the twisting force associated with the earth's rotation on its axis (Figure 4-8). This force deflects airflow in the northern hemisphere to the right and in the southern hemisphere to the left (Figure 4-9). This distortion of the earth's general air circulation causes the single air movement pattern that would exist in each hemisphere on a non-rotating earth to break up into three separate belts of moving air or *prevailing ground winds*: the polar easterlies, the westerlies, and the tradewinds. The equatorial calm is known as the doldrums. These three major belts of prevailing winds in each hemisphere contribute to the distribution of heat and moisture around the planet, which leads to differences in climate in different parts of the world. The general global circulation pattern of air masses also influences the distribution of precipitation over the earth's surface.

The amount of water vapor in the air is called its *humidity*. The amount of water vapor the air is holding at a particular temperature expressed as a percentage of the amount it could hold at that temperature is its *relative humidity*. When air with a given amount of water vapor cools, its relative humidity rises; when the same air is warmed, its relative humidity drops. Thus, warm air can hold more water vapor than cold air, and this is why the humidity tends to rise in warmer months.

When an air mass rises, it cools, which causes its relative humidity to increase. Once the relative humidity of the rising air mass has reached 100 percent, any further decrease in temperature causes tiny water droplets or ice crystals to condense on particles of dust in the atmosphere to form *clouds*. As these droplets and ice crystals in

clouds are moved about in turbulent air, they collide and coalesce to form larger droplets and crystals. Eventually they can become big enough to be pulled by gravitational attraction toward the earth's surface in the form of *precipitation* such as rain, sleet, snow, or hail. Thus, almost all clouds and forms of precipitation are caused by the cooling of an air mass as it rises. Conversely, when an air mass sinks, its temperature rises and its relative humidity can decrease until release of its moisture as rain or snow becomes impossible.

The climate and the weather of a particular area are affected by the distribution of land and water over the earth's surface because an air mass tends to take on the temperature and moisture characteristics of the surface over which it moves. In general, land surfaces are heated rapidly by the sun. Because this heat does not penetrate deeply, land surfaces also cool rapidly. This means that interior land areas not near a large body of water usually have great differences in daily high and low temperatures.

Water, however, warms up slowly, holds a much larger quantity of heat than the same volume of land surface, and cools slowly. As a result, the surface layer of air over the oceans is cooler in summer and warmer in winter than that over the continents. Because warm air rises, a net inflow of cool ocean air moves onto the continents in summer, and in winter there is a net outflow of cool air from the continents onto the oceans. Land and sea breezes result from the land being colder than the water at night and early morning but warmer later in the day.

Albedo, Emissivity, and Climate Another factor affecting climate is the amount of incoming solar energy reaching the earth's surface (Figure 4-10). About 30 percent of the incoming solar radiation is immediately reflected back to space by clouds,

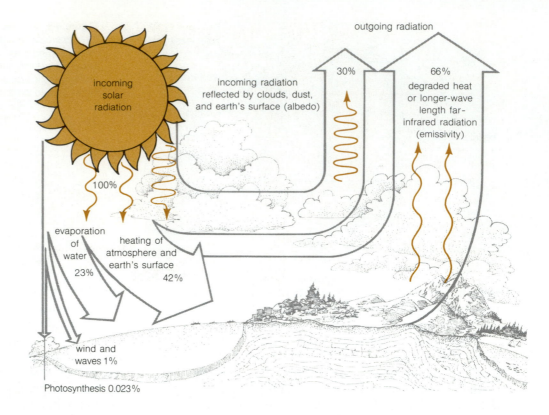

Figure 4-10 The flow of energy to and from the earth.

outgoing radiation

incoming
solar
radiation

incoming radiation
reflected by clouds, dust,
and earth's surface (albedo)

30%

66%
degraded heat
or longer-wave
length far-
infrared radiation
(emissivity)

100%

evaporation
of
water
23%

heating of
atmosphere and
earth's surface
42%

wind and
waves 1%

Photosynthesis 0.023%

chemicals and particulate matter in the air, and the earth's surface. This reflectivity of the earth and its atmosphere is call planetary **albedo**. The remaining 70 percent of the incoming radiation is absorbed by the earth's atmosphere, lithosphere, hydrosphere, and ecosphere.

About 42 percent of the incoming solar energy heats the land and warms the atmosphere. Another 23 percent is used to evaporate water and cycle it through the ecosphere. A tiny fraction (1 percent) of the incoming solar energy generates winds, which in turn move the clouds and form waves in the ocean. An even smaller fraction, 0.023 percent, is captured by green plants and converted by photosynthesis to carbohydrates, proteins, and other molecules essential for life.

Most of the 70 percent of the solar energy that is not reflected away is degraded into longer wavelength heat or far infrared radiation, in accordance with the second law of thermodynamics (Section 3-3), and flows back into space (Figure 4-10). The total amount of energy returning to space is called the **emissivity** of the earth. Emissivity is affected by the presence of chemicals such as water, carbon dioxide, ozone, and solid particles of matter in the atmosphere. These molecules act as gatekeepers, allowing some of this heat to flow back into space and absorbing and reradiating some of it back toward the earth's surface.

There is concern that human activities can affect global climate patterns. For example, increases in the average levels of carbon dioxide in the earth's atmosphere due primarily to the burning of fossil fuels may lead to a so-called *greenhouse effect*, whereby far infrared radiation (Figure 4-2) that otherwise would escape into space is trapped and raises the average temperature of the atmosphere. On the other hand, it has been theorized that dust from volcanic explosions and particulate matter emitted into the atmosphere from human activities such as land clearing and smokestack, chimney, and automobile emissions could reflect significant amounts of sunlight away from the upper atmosphere and cause global cooling. These possible effects of human activities on global climate patterns are discussed in Section 21-5.

4-6 Limiting Factors in Ecosystems

The Limiting Factor Principle   Why is one area of the earth a desert, another a forest, and another a grassland? Why are there different types of desert, grassland, and forest, and what determines the variations of plant and animal life found in each type?

You know that the speed of traffic moving along a crowded one-lane, one-way street is determined by the slowest moving car. Similarly, the overall structure of an ecosystem can often be determined by a single *limiting factor* such as precipitation or average annual rainfall. Thus, the

Figure 4-11 axis labels:

average rainfall (centimeters per year) — 0, 10, 20, 30, 40, 50, 60, 70, 80, 90, 100, 110, 120

Northern coniferous forest

Deciduous forest

Tropical rain forest

Polar grassland (tundra)

Temperate grassland

Tropical grassland (savannah)

Cool desert

Temperate desert

Tropical desert

cool   temperate   hot

average temperature

Figure 4-11 The type of desert, grassland, or forest ecosystem found in a particular area is determined primarily by the two abiotic factors, average precipitation and average temperature, that together determine the major climate type of the area.

structure of an ecosystem is often determined by the **limiting factor principle**: The single factor that is most deficient in an ecosystem is the one that determines the presence or absence of particular plant and animal species.

Precipitation and Temperature as Limiting Factors *Precipitation is the limiting factor that determines whether the biomes found on most of the world's land areas are desert, grassland, or forest.* For example, regardless of the average conditions of temperature and sunlight and the concentrations of certain minerals in the soil, the result will still be a desert if the average amount of precipitation is less than 25 centimeters (10 inches) per year. With so little rainfall, desert areas typically have sparse vegeta-

tion with large areas of bare ground between plants. Similarly, areas with 75 centimeters (30 inches) or more of rainfall a year are generally dominated by a forest ecosystem unless the vegetation is cut down by humans. Regions that have about 25 to 75 centimeters (10 to 30 inches) of precipitation a year are typically grasslands because there is not enough moisture to support large stands of trees.

Average precipitation by itself does not determine the particular type of desert, grassland, or forest found in a given area. Average precipitation and temperature acting together determine the climate of an area. The combined action of these two abiotic factors also determines the type of desert, grassland, or forest biome found in a particular area. For example, Figure 4-11 shows that a humid tropical climate with hot temperatures and lots of rainfall produces a *tropical rain forest*, with a wide variety of broad-leaved evergreen trees (species that keep their leaves all year). *Temperate forests* are found in areas having temperate climates and ample precipitation. These forests are dominated by a few deciduous tree species, such as oak and hickory, that drop their leaves and become dormant during winter when temperatures sometimes drop below freezing. In very cold areas with ample precipitation are *northern coniferous forests*, dominated by a few species of coniferous (cone-bearing), needle-leaved evergreen trees such as spruce and fir.

Figure 4-11 also shows that temperature variations combined with very little precipitation lead to development of tropical, temperate, and cold deserts. Similarly, temperature variations in areas with moderate rainfall lead to tropical grasslands (savanna), temperate grasslands, and polar grasslands (tundra). Each of these major types of biome contains many variations and many smaller ecosystems. The map inside the front cover shows the distribution of the major types of forest, grassland, and desert biomes throughout the world. These various biomes are described in more detail in the next section.

Other Limiting Factors   Light is often the limiting factor that determines the trees, bushes, small plants, mosses, fungi, and bacteria found in a forest. The tops of the trees receive most of the light and their leaves provide shade for other species that require less light to grow. Only a fraction of the full light reaches the bushes, still less reaches the small plants, and even less reaches the mosses. In dense pine and fir forests so little light penetrates that the forest floor contains only an evergreen layer of certain mosses.

In some cases the *soil* that forms a thin layer over much of the land surface of the earth can act as a limiting factor. *Soil* is a complex mixture of small pieces of inorganic rock, gravel, and minerals, and organic compounds, living organisms, air, and water. Soils are formed by the combined action of water, wind, chemical activity, and living organisms on the parent rock. Climate, topography, time, and various forms of life that inhabit the soil determine some of its characteristics, as discussed in Chapter 9.

4-7   Biomes: A Look at Major Land Ecosystems

Tropical Biomes   In "winterless" tropical climates with high temperatures, different types of biome

are found depending on the average amount and seasonality of precipitation. These include tropical rain forests, seasonal forests, thorn scrubs, thorn woodlands, and savannas.

Tropical rain forests are found near the equator (see inside front cover). Although such a forest has more different kinds of organism than any other biome, there are fewer individuals of any one species in a given area. This biological diversity is the result of this biome's almost unchanging climate: high but not excessive annual mean temperature of about 28°C (82°F), high humidity, and heavy, almost daily rainfall.

These forests consist of several layers of plant and animal life. The top layer is dominated by tall trees with slender trunks, reaching heights of 30 to 60 meters (100 to 200 feet). Trees in a rain forest normally have broad, flat leaves that provide a large surface area, which helps absorb heat from the sun and cool the trees by evaporation of rainwater. These leathery, evergreen leaves form a canopy, making the forest floor dark and very humid. With so little sunlight, the ground is relatively free of vegetation except along river banks and clearing edges. Decomposition is rapid, and everything that falls to the ground is quickly carried off, consumed, or decomposed by beetles, termites, ants, and other often unusually large insects.

Below the upper canopy, thick, woody vines called *lianas* hang from branches that are covered with thousands of species of ferns, strange pineapplelike plants called bromeliads, and *epiphytes* or air plants (such as orchids). Lacking underground roots, epiphytes stay alive by getting minerals from falling leaves and animal wastes and by trapping water in their flowers or leaves. These airborne pools of water are miniature ecosystems containing entire communities of insects, spiders, and even tiny frogs.

The animal life in tropical forests is so varied that more different kinds of organism may be found living in a single tree than in an entire forest to the north. Though relatively few animals are found on the forest floor, many species of huge colorful butterflies and exotic birds fly around the forest canopy. They are joined by monkeys swinging and hopping about as well as lemurs, snakes, frogs, and other animal species.

Humans have steadily been clearing tropical rain forests to get lumber and to plant crops. If clearing continues at present rates, within 50 years few of these biomes may be left (Section 12-5).

North and south of the equator are the *tropical deciduous* or *seasonal forests* where temperatures are fairly high and most of the precipitation occurs during a rainy (monsoon) season (see inside front cover). Here canopy heights are lower than those in tropical rain forests and there are more deciduous trees that lose their leaves during the dry season.

In tropical climates less humid than those of tropical seasonal forests are *tropical thorn scrub* and *thorn woodland* biomes. There is enough precipitation to support small thorny plants in thorn scrub biomes and a few medium-sized trees in thorn woodland biomes, but not enough to support forests. Grass or dense shrubs grow below the trees.

The *tropical savanna* biome appears in areas having high temperatures and long dry seasons without enough rainfall to support forests (see inside back cover). It is typically covered with grass and scattered groups of small trees and shrubs. During the dry season large areas of savanna often burn, hindering the development of forests. Undisturbed tropical savannas are usually populated by large herds of grazing hoofed mammals such as wildebeest, gazelle, zebra, and antelope in Africa, and kangaroos and wallabies in Australia. The great herds of such grazing animals and their predators are disappearing rapidly except in protected areas because of ranching, farming, hunting, and other human activities (Section 13-2).

Desert In areas too dry to support grasslands, where evaporation exceeds rainfall, you will find the deserts that make up more than one-third of the world's land surface. There are *cold deserts*, with cold winters (there may even be snow on the ground) and hot summers, like those found in Oregon, Utah, and Nevada, and *warm deserts*, with warm to hot temperatures throughout the year. Typical warm deserts are found in Arizona, New Mexico, California, Texas, northern Mexico, Africa (the Sahara), and Saudi Arabia (see inside back cover).

Though a few hot and very dry deserts like the Sahara consist of endless stretches of barren sand dunes, most deserts consist of widely scattered thorny bushes and shrubs, occasional cacti (especially in the western hemisphere), and some small flowers that quickly carpet the desert floor after a brief rain. These desert plants use a number of strategies to get and conserve water. There is so little water available that desert plants cannot cool themselves by evaporation. Thus, if desert plants had leaves they would heat up during the day, be cooked, and die. Cacti and other succulent plants also have extensive shallow root systems so they can take up water rapidly during the brief rains and store it in their fleshy stems. Other desert plants like mesquite bushes get groundwater with taproots as long as 30 meters (100 feet).

Desert animals include a number of rodents (such as the kangaroo rat), lizards, toads, snakes and other reptiles, owls, eagles, vultures, many small birds, and numerous insects. Most of these creatures are small and avoid the heat by coming out only at night, especially toward dusk and dawn. Some desert animals get water from the morning dew. Others, such as the kangaroo rat, never drink at all and instead obtain water from their food. The scales of reptiles and the exterior covering of insects prevent evaporation of water.

Temperate Biomes The world's temperate regions on the average experience moderate temperatures and are found north and south of the tropics and their adjacent deserts. Major types of temperate biome are temperate rain forests, evergreen forests, deciduous forests, woodlands, shrublands, and grasslands (see inside front and back covers).

Deciduous trees in temperate forest biomes with fairly mild climates most of the year have broad, flat leaves to help control their heat input and output. To prevent excessive heat loss during winter, these trees survive by losing their leaves and becoming dormant until spring. Coniferous tree species such as spruce and fir found in the much colder northern coniferous forests have thick clusters of thin, needle-shaped leaves. These leaves lose their heat less rapidly than broad leaves because they have a smaller surface area in relation to their mass. These trees also keep their leaves all year, to take advantage of brief warm, sunny spells.

Temperate rain forests are found in cool climates in areas near the sea with abundant rainfall and summer cloudiness or fog. Along the Pacific Coast of North America temperate rain forests range from the mixed coastal coniferous forests of Washington, dominated by Douglas fir and several species of pine, to the majestic coastal redwood forests of northern California.

Temperate evergreen forests cover large areas where poor soils and numerous droughts and forest fires favor needle-leaved conifer or broad-leaved evergreen tree species over deciduous tree species. In the western United States large stands of these forests are dominated by spruce, fir, and ponderosa pine.

A somewhat cooler and drier climate produces *temperate deciduous forests* (inside front cover) with abundant (but not excessive) precipitation, moderate temperatures that gradually change with the seasons, and a long growing season (4 to 6 months). As temperatures drop in autumn, leaves exhibit a rainbow of colors before falling to the ground. Dominant tree species vary among deciduous forests, depending largely on average local precipitation and temperature.

The dominant herbivore of most deciduous forests in the eastern United States is the whitetail deer. Because its predators (such as the wolf) have largely been eliminated or driven out, the whitetail population can grow out of control, leading to destruction of vegetation and mass starvation of the deer.

Predator birds, such as owls, hawks, and eagles, play an important ecological role by keeping down populations of mice and other small rodents that otherwise would destroy vegetation. As winter approaches, many of the bird species in these forests migrate south, and many of the mammal species go into hibernation.

Temperate woodlands are found in areas with climates too dry to support forests but with enough precipitation to support a mixture of grasses and some trees. Such woodlands consisting of scattered oak trees spread across rolling grassland are found in central California and are largely used to grow wheat or barley or as rangeland to feed livestock.

When a temperate area has a so-called Mediterranean climate—mild, damp winters and hot summers with little or no rain—a biome known as *temperate shrubland* (or *chaparral* in Mexico and parts of the southwestern United States, Chile, western Australia, and the Mediterranean region) occurs (see inside back cover). The leathery evergreen leaves of many of the shrubby plants found in this biome have waxy coatings that reduce evaporation and help them survive long rainless periods. They also contain volatile and flammable compounds that, when exposed to dry summers and lightning or human activities, lead to periodic fires. These fires play an important ecological role in clearing out dead material and old growth and allowing various plant species that have become adapted to fire to reproduce. But they also destroy homes of people who have built in chaparral areas such as Marin County, California, to take advantage of the mild year-round climate.

Temperate grasslands are found where rainfall is great enough to keep deserts from forming yet so low and erratic that cycles of drought and fire prevent forests from growing (inside front and back covers). These biomes have carpets of high and low grasses, mixed in some areas with small bushes and shrubs and even a few widely dispersed trees. Here the winters are cold with snow covering the ground at times, and summers are hot and dry. The winds blow almost continuously, and there are few natural windbreaks to blunt their force. The soils of the tall-grass prairie are among the richest in the world, explaining why most of the grasslands in Canada and the United States have been selected for vast plantings of corn and wheat.

At one time the grasslands of the Great Plains of North America were dominated by large herbivores, such as the bison, pronghorn antelope, elk, and wild horses. These grazing animals were preyed on by wolves, coyotes, cougars, grizzly bears, and Plains Indians. Today these herds have been replaced by domesticated cattle, sheep, and goats. Because they are confined, these animals often overgraze the land, especially in years of severe drought. The exposed soil is quickly blown away and the productivity of the grassland is destroyed, as discussed in more detail in Section 9-3.

Undisturbed temperate grasslands also contain numerous species of small herbivores, such as the prairie dog, ground squirrel, pocket gopher, and jackrabbit. The prairie dog's elaborate tunnels and burrows help aerate the soil, and its droppings enrich the soil. Settlers on the North American prairie, however, considered the species a nuisance, and it has been largely eliminated and replaced by the ground squirrel.

Taiga (Boreal Forests) Moving north toward the arctic we encounter the *taiga* (from a Russian word meaning "primeval forest") or *boreal forest* biome stretching across the northern portions of North America, Europe, and Asia (see inside front and back covers). In this biome winters are long and extremely cold, and much of the precipitation falls as snow.

Species diversity is low, and these forests are dominated by only a few species of needle-leafed, coniferous evergreen trees such as spruce and fir, which can survive the extreme winter cold. Lakes, ponds, and bogs are also often found. The carpet of needles and leaf litter on the forest floor decomposes slowly because of the cold.

The major large herbivores of these coniferous forests are moose, mule deer, caribou (which migrate down from the tundra during the fall), and elk. Because of large-scale killing by farmers and ranchers, major predators such as the timber wolf, which once inhabited most North America taiga, are now found primarily in Canada and Alaska. When wolf populations are killed off or driven away, the increased populations of moose, caribou, and deer they once controlled can devastate taiga vegetation.

Tundra: Arctic and Alpine Grasslands Only a few hardy plant and animal species can survive the harsh climate of bitter cold and low precipitation of the *tundra*, an icy, treeless grassland found between the tree line and the Arctic Circle (*arctic tundra*) and just above the timberline on mountaintops (*alpine tundra*) (see inside front and back covers). In many areas the deeper layers of soil remain frozen as *permafrost*. Even if trees could survive the cold air, the permafrost prevents them from putting down roots deep enough to grow. Summers last only a few weeks, just long enough to thaw a thin veneer of soil above the permafrost. During this short thaw, the normally frozen plain turns into a quagmire because the permafrost retards drainage. Humans and other animals entering this biome during the thaw are attacked by hordes of mosquitoes, deerflies, and blackflies.

The tundra landscape is covered with a mat of low-growing lichens, mosses, grasses, sedges, dwarf woody shrubs, and (in alpine tundra) some small shrubs and mountain wildflowers. Most of the tundra's permanent animal residents are creatures that burrow under the snow, such as lemmings—small, furry herbivores whose population rises rapidly and then crashes about every 4 years. Large herbivores, such as caribou, reindeer, and musk ox, slowly migrate south during the fall. During the short summer, large numbers of migrating birds, especially waterfowl, nest in the tundra to feed on the swarms of insects.

Because the ground is so cold most of the year, decomposition occurs very slowly. Combined with the shallow soil and slow growth rate of plants, this means that tundra takes a long time to recover when it is destroyed or disrupted. This is why environmentalists are concerned about the long-term ecological effects of running oil and gas pipelines through this type of biome.

In this chapter we have seen how various physical and chemical factors can influence the distribution of plants and animals in an ecosystem. With this background in ecosystem structure, we are ready to learn more about what goes on in an ecosystem as its plant and animal species interact with one another and with their abiotic environment.

We sang the songs that carried in their melodies all the sounds of nature—the running waters, the sighing of winds, and the calls of the animals. Teach these to your children, that they may come to love nature as we love it.

Grand Council Fire of American Indians

Discussion Topics

1. Would you rather be exposed to electromagnetic radiation with a short or a long wavelength? Explain why people who spend a lot of time acquiring suntans may get skin cancer.

2. How can an area have bad weather and a good climate?

3. **a.** How would you set up a self-sustaining aquarium for tropical fish?

   b. Suppose you have a balanced aquarium sealed with a transparent glass top. Can life continue in the aquarium indefinitely as long as the sun shines regularly on it?

   c. Which of the following will probably be the limiting factor: the oxygen supply in the air above the water, the original oxygen supply dissolved in the water, or the supply of nitrogen in the soil at the bottom?

4. A friend cleans out your aquarium and removes all the soil and plants, leaving only the fish and water. What will happen?

5. **a.** A bumper sticker asks "Have you thanked a green plant today?" Give two reasons for thanking a green plant.

   b. Trace back the materials comprising the sticker and see whether the sticker itself represents a sound application of the slogan.

5

Ecosystem Function: How Do Ecosystems Work?

Earth and water, if not too blatantly abused, can be made to produce again and again for the benefit of all. The key is wise stewardship.

Stewart L. Udall

5-1 Energy Flow and Chemical Cycling

To survive, you and every other form of life must have an almost continuous *input* of both energy and matter. Merely receiving energy and matter, however, will not keep you alive. An *output* of degraded energy (heat) and waste matter must also flow from an organism. For the organism to remain alive, the input and output of energy and matter must be in balance. Thus, life depends on the *one-way flow of both matter and energy*.

To be self-sustaining, the ecosphere and any ecosystem *depend on matter cycling*—not matter flow—and *energy flow*, as summarized in Figure 5-1. At the ecosystem and ecosphere levels, life depends on energy flow because, according to the second law of thermodynamics (Section 3-3), *energy quality can never by recycled*. In any ecosystem, high-quality energy enters (usually as sunlight), moves through organisms, and eventually escapes to space as low-quality heat energy.

Life at the ecosystem and ecosphere levels, however, depends on matter cycling, not on one-way matter flow, because according to the law of conservation of matter (Section 3-1) matter can neither be created nor destroyed, only changed from one form to another. While the chemicals essential for life must eventually be cycled completely in the ecosphere, chemical cycling in an ecosystem need not be as complete. Some of this matter flows from one ecosystem to another because the various ecosystems on earth are interconnected.

Thus, *we can generally answer the question "What*

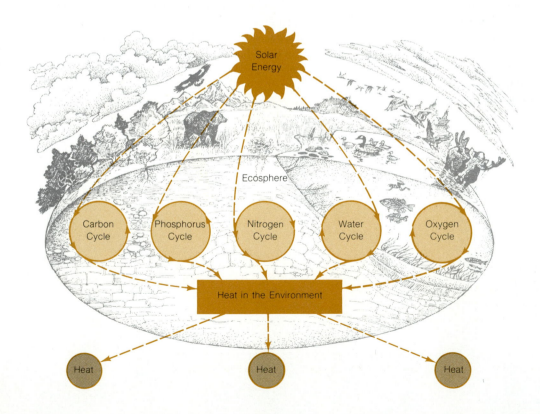

Figure 5-1 Life on earth depends on the cycling of critical chemicals (solid lines) and the one-way flow of energy through the ecosphere (dashed lines).

happens in an ecosystem?" by saying that *energy flows and matter cycles*. These two major ecosystem functions connect the various structural parts of an ecosystem so that life is maintained, as summarized in Figure 5-2. In the remainder of this chapter we examine these two functional processes in ecosystems.

5-2 Energy Flow in Ecosystems: Food Chains and Food Webs

Food Chains   In general, the flow of energy and the cycling of nutrient matter through an ecosystem is the study of what eats or decomposes what. One organism's waste or death is another organism's food. A caterpillar eats a leaf; a robin eats the caterpillar; a hawk eats the robin. When the plant, caterpillar, robin, and hawk die, they are in turn consumed by decomposers (Figure 5-3). A sequence of transfers of nutrients and energy from one organism to another when one organism eats or decomposes another is called a **food chain** (Figure 5-3).

The various feeding levels of producers and consumers in a food chain are called **trophic levels** (from the Greek *trophikos* for "nourishment" or "food"). As shown in Figure 5-3, all producers belong to the first trophic level, all primary consumers, whether feeding on living or dead producers, belong to the second trophic level, and so on.

Ecologists sometimes distinguish between *grazing food chains* and *detritus food chains* (Figure 5-4). In a **grazing food chain**, green plants are eaten by herbivores, which in turn may be eaten by carnivores. In a **detritus food chain**, decomposers consume the organic waste products and dead organic matter (*detritus*) or partially decomposed tissues of other organisms. These two types of food

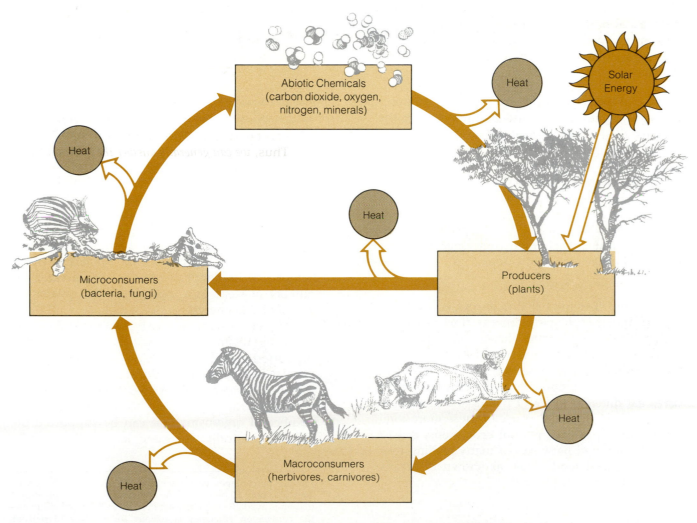

Figure 5-2   Summary of ecosystem structure and function. The major structural components (energy, chemicals, and organisms) of an ecosystem are connected through the functions of energy flow (open arrows) and chemical cycling (solid arrows).

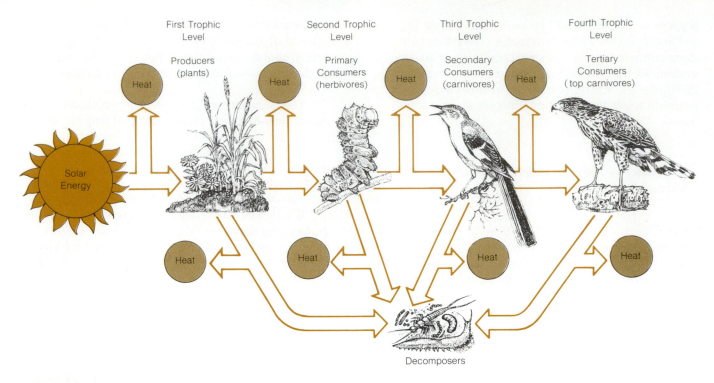

First Trophic Level

Second Trophic Level

Third Trophic Level

Fourth Trophic Level

Producers (plants)

Primary Consumers (herbivores)

Secondary Consumers (carnivores)

Tertiary Consumers (top carnivores)

Solar Energy

Heat

Decomposers

Figure 5-3 A food chain. The arrows show how chemical energy in food flows through various trophic levels, with most of the high-quality chemical energy being degraded to low-quality heat in accordance with the second law of thermodynamics.

chains are interrelated (Figure 5-3) because eventually all organisms die and become part of the detritus food chain. Figure 5-4 also shows that aquatic food chains often have a larger number of trophic levels, or links, than land-based food chains. As omnivores, humans can eat plants and animals. However, *most people on earth function as herbivores in grazing food chains*, getting an average of 89 percent (worldwide) and 64 percent (in the United States) of their food from vegetables, cereals, and fruits.

Food Webs The food chain concept is useful for tracing matter cycling and energy flow in an ecosystem, but simple food chains, such as those shown in Figure 5-4, rarely exist by themselves. Many animals feed on several different types of food at the same trophic level. In addition, omnivores eat different kinds of plants and animals at several trophic levels. Because of these more complex feeding patterns, natural ecosystems consist of interconnected networks of many feeding relationships called **food webs**, as shown in greatly simplified form in Figure 5-5.

Food Chains, Food Webs, and the Second Law of Thermodynamics *Only about 10 percent of the high-*

quality chemical energy available at one trophic level is transferred and stored in usable form in the bodies of the organisms at the next trophic level. This is sometimes called the **ten percent rule**.* The remaining 90 percent of the chemical energy transferred from one trophic level to another is degraded and lost as low-quality heat to the environment according to the second law of thermodynamics.

Figure 5-6 gives a striking picture of this loss of usable energy at each step in a simple food chain. Such a diagram is known as a *pyramid of energy* because of its shape. Note that the greater the number of steps in a food chain, the greater the loss of usable energy.

We get the same picture by looking at the number of organisms of a particular type that can be supported at each trophic level from a given input of solar energy at the producer trophic level. This *pyramid of numbers* in Figure 5-6 shows that in going from one trophic level up to another, the total number of organisms that can be supported decreases drastically. For example, a million phyto-

---

*Actual percentages vary from 2 to 30 percent with species. Typically, only 10 percent of the energy entering the plant population is available to herbivores. For warm-blooded carnivores the conversion efficiency is usually lower than 10 percent, whereas for cold-blooded ones it may be 20 or 30 percent. Thus 10 percent seems to be an appropriate average.

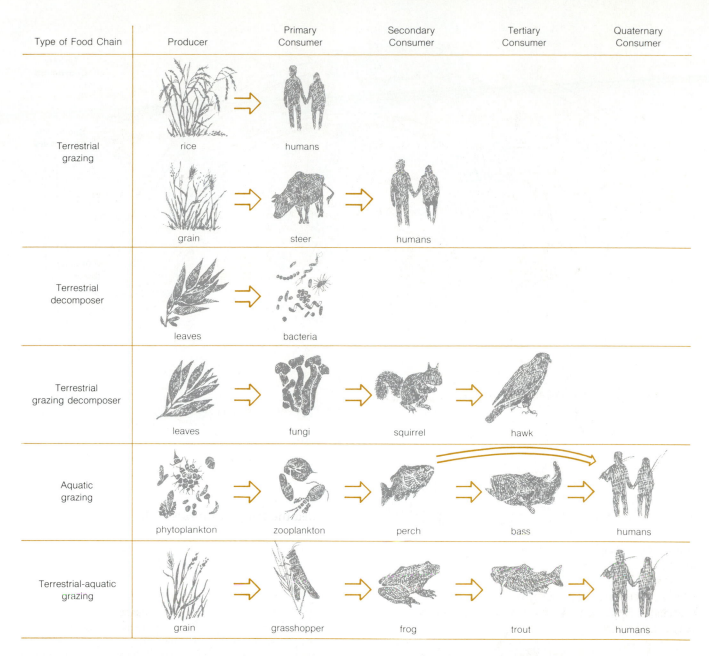

Type of Food Chain	Producer	Primary Consumer	Secondary Consumer	Tertiary Consumer	Quaternary Consumer
Terrestrial grazing	rice	humans			
	grain	steer	humans		
Terrestrial decomposer	leaves	bacteria			
Terrestrial grazing decomposer	leaves	fungi	squirrel	hawk	
Aquatic grazing	phytoplankton	zooplankton	perch	bass	humans
Terrestrial-aquatic grazing	grain	grasshopper	frog	trout	humans

Figure 5-4   Some typical food chains. Upon their death, the plants and consumer animals shown in these simplified chains are broken down by decomposers.

plankton producers in a small pond may support 10,000 zooplankton primary consumers. These in turn may support 100 perch, which might feed one human for a month or so. The pyramid of numbers helps explain why there are more plant-eating rabbits than flesh-eating tigers.

Two important principles emerge from a consideration of the loss of available energy at successively higher trophic levels in the food chains that make up more complex food webs. *First, most life and forms of food begin with sunlight and green plants. Second, the shorter the food chain, the less the loss of usable energy.* This means that a larger population

of humans can be supported if people eat grains directly (for example, rice→human), rather than eating animals that fed on the grains in a longer food chain (such as grain→steer→human). However, a diet based on only one or two plants lacks certain amino acid molecules that link together in different sequences to form the variety of proteins essential for good health, as discussed in Section 11-2.

With this overview of energy flow in ecosystems, we are now ready to look at chemical cycling, the second major functional process occurring in ecosystems.

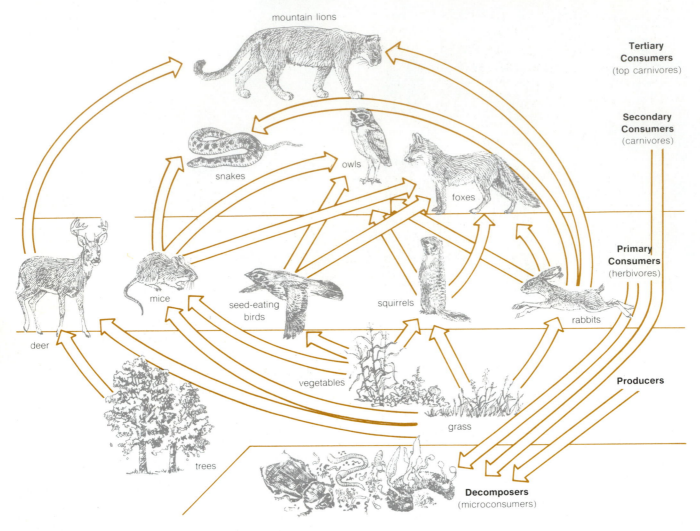

Figure 5-5 Greatly simplified food web for a terrestrial ecosystem.

5-3 Chemical Cycling in Ecosystems: Carbon, Oxygen, Nitrogen, Phosphorus, and Water Cycles

Biogeochemical Cycles In chemical terms, life can almost be summed up in six words: *carbon, oxygen, hydrogen, nitrogen, phosphorus,* and *sulfur*— the half-dozen elements that make up more than 95 percent of the mass of all living organisms. These six, plus a few others required in relatively large quantities, are called **macronutrients**. Iron, manganese, copper, iodine, and other elements needed only in minute quantities are called **micronutrients**.

Because we have a fixed supply of the six macronutrient elements, they must continuously cycle from their reservoirs of air, water, and soil through the food webs of the ecosphere and back again to their reservoirs in **biogeochemical cycles** (*bio-* meaning "living," *-geo-* for water, rocks, and soil, and *-chemical* for the matter that is cycled by

being changed from one chemical to another). This means that one of the oxygen molecules you just inhaled may be one you inhaled 2 years ago, or it may have been one inhaled by your grandmother before you were born, or by Cleopatra, many centuries ago.

There are three types of biogeochemical cycle: gaseous, sedimentary, and hydrologic (water). The **gaseous cycles**, in which the atmosphere is the primary reservoir, include the *carbon, oxygen,* and *nitrogen cycles*. The **sedimentary cycles** move materials from land to sea and back again. They include the *phosphorus, sulfur* (Section 21-2), *calcium, magnesium,* and *potassium cycles*. The *hydrologic cycle* represents the cyclical movement of water from the sea to the atmosphere, to the land and back to the sea again.

In all chemical cycles, both the nature of the cycling process and the rate at which critical chemicals are cycled are important. For example, most water on earth eventually goes through the photosynthesis process in plants, but at a rate esti-

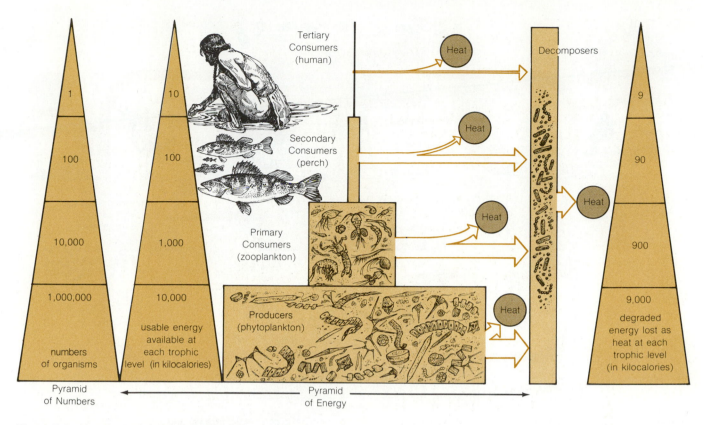

Figure 5-6 Hypothetical pyramids of energy and numbers showing the decrease in usable energy available at each succeeding trophic level in a food chain.

mated to be once every 2 million years. Similarly, the oxygen gas produced by green plants through photosynthesis cycles from the atmosphere and back again about every 2,000 years, and the gaseous carbon dioxide given off by all plants and animals when they break down food molecules by cellular respiration cycles about once every 300 years.

Carbon and Oxygen Cycles Carbon is the basic building block of the large organic molecules necessary for life. The chlorophyll and often other pigments in green plants can absorb solar energy and use it to combine carbon dioxide with water to form organic nutrients such as glucose in the process of **photosynthesis**, summarized as follows:

$$\text{carbon dioxide} + \text{water} + \textbf{solar energy} \rightarrow \text{sugars, such as glucose} + \text{oxygen}$$

$$6CO_2 + 6H_2O + \textbf{solar energy} \rightarrow C_6H_{12}O_6 + 6O_2$$

This ability of green plants to synthesize sugars makes most other forms of life possible.

Producers, consumers, and decomposers transform a portion of the carbon in the food they synthesize or eat back into carbon dioxide and water by the process of *cellular respiration*. This **cellular respiration** process provides the energy plants and animals need to live. For most organisms, it can be summarized as follows:

$$\text{sugars, such as glucose} + \text{oxygen} \rightarrow \text{carbon dioxide} + \text{water} + \textbf{energy}$$

$$C_6H_{12}O_6 + 6O_2 \rightarrow 6CO_2 + 6H_2O + \textbf{energy}$$

Both photosynthesis and cellular respiration consist of a large number (80 to 100) of chemical reactions operating in sequence. However, from the equations just given, we see that the overall reaction for the sequence involved in respiration is the opposite of that for the photosynthesis process.

Thus, photosynthesis and respiration operate together as a closed cycle through which plants produce oxygen needed by animals and absorb the carbon dioxide given off by animals, as shown in greatly simplified form in Figure 5-7. Note that some carbon is tied up in minerals such as fossil fuels and carbonate rock formations (for example,

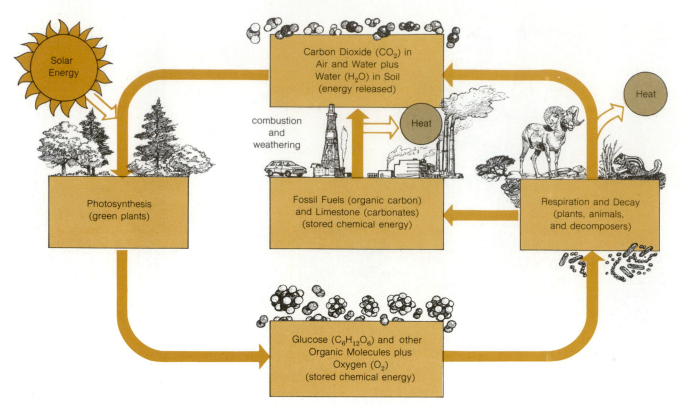

Figure 5-7 Simplified version of the carbon and oxygen cycles, showing chemical cycling (solid arrows) and one-way energy flow (open arrows).

limestone, or $CaCO_3$). This carbon is returned to the cycle as carbon dioxide and water when fossil fuels are burned and when carbonate rock formations slowly weather away.

Nitrogen Cycle Nitrogen is often a factor limiting the growth of plants. Too little nitrogen can also cause malnutrition in humans because many of the body's essential functions require nitrogen-containing molecules, such as proteins, nucleic acids, vitamins, and hormones (Section 11-2).

Although nitrogen gas (N_2) makes up about 78 percent of the earth's atmosphere by volume, it is useless in this form to most plants and animals. Fortunately, in the gaseous nitrogen cycle—shown in *simplified* form in Figure 5-8—nitrogen-fixing bacteria in soil, blue-green algae in water, and symbiotic bacteria in nodules on alfalfa, clover, and other legumes (members of the pea family) convert, or "fix," gaseous nitrogen to solid nitrate salts (containing nitrate, or NO_3^-, ions). A small amount of nitrogen gas is also fixed by lightning. These nitrate salts dissolve easily in soil water and are taken up by plant roots. The plants then convert the nitrates to large nitrogen-containing protein molecules and other organic nitrogen molecules necessary for life.

Some of these nitrogen-containing protein molecules are transferred to plant-eating animals and eventually to other animals that feed on them. When plants and animals die, decomposers break down these large organic nitrogen molecules into ammonia gas (NH_3) and water-soluble salts containing ammonium ions (NH_4^+). Ammonia and ammonium are then converted by other groups of soil bacteria into water-soluble nitrite ions (NO_2^-), nitrogen gas, which returns to the atmosphere, or nitrous oxide (N_2O) gas, which also ends up in the atmosphere (Figure 5-8).

Some plants can absorb the ammonium ions from salts dissolved in soil water and convert them to nitrogen-containing protein molecules. Another group of bacteria can add a third oxygen atom to nitrite ions and convert them to nitrate ions, which can be taken up by plants to begin the cycle again. Some nitrogen is temporarily lost from the cycle when soluble nitrate salts are washed from the soil into rivers and streams and eventually into the oceans (Figure 5-8).

Crop growth can be limited if there is not the right amount of nitrogen in the soil, primarily as nitrate (NO_3^-) and ammonium ions (NH_4^+). During World War I the German chemist Fritz Haber developed the industrial process that bears his name to convert nitrogen gas (N_2) by reacting it with hy-

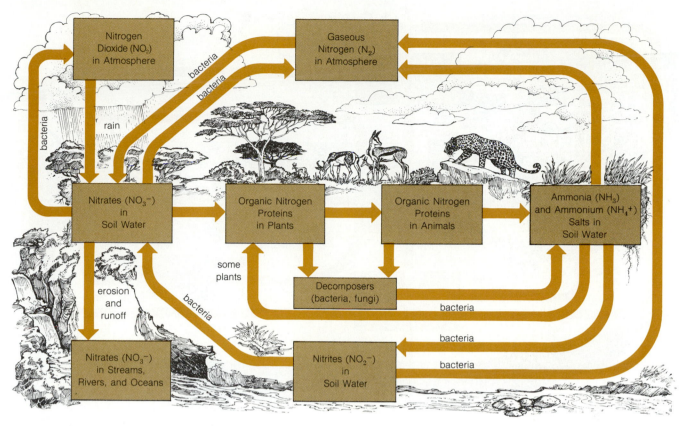

Figure 5-8 Simplified version of the nitrogen cycle (energy flow not shown).

drogen gas (H_2) at high temperatures and pressures to produce ammonia gas (NH_3), which can be converted to NO_3^- compounds used to make gunpowder. The ammonia gas can be converted to ammonium salts and used as commercial fertilizer to improve or restore soil fertility (Section 9-4). Nitrate salts are also mined and used, along with ammonium salts, as commercial fertilizers.

Phosphorus Cycle Phosphorus (mainly in the form of phosphate ions, PO_4^{3-}) is an essential nutrient of both plants and animals. It is also a major constituent of the genetic material coded in DNA molecules, the energy storage compound adenosine triphosphate (ATP), and a component of cell membranes, bones, and teeth. Phosphate, together with nitrates and potassium, is a major ingredient in commercial fertilizers.

The *phosphorus cycle,* shown in greatly simplified form in Figure 5-9, is a sedimentary cycle in which the earth's crust is the major reservoir. Phosphorus moves through this cycle primarily in the form of phosphate ions. The major reservoirs of phosphorus are phosphate rock deposits on land and in shallow ocean sediments. Slowly, through weathering and erosion, the phosphates from these rock deposits are released into the eco-

sphere. Many of these phosphates wash into rivers and eventually to the oceans, where they form insoluble phosphate deposits on the bottom of shallow ocean areas near the coast and in deep ocean sediments.

Fish catches and *guano* (the phosphate-rich waste matter from fish-eating birds, such as pelicans, gannets, and cormorants) return some of this phosphate to the land. These returns, however, are small compared with the larger amounts of phosphate eroding from the land to the oceans each year as a result of natural processes and human activities. Large quantities of phosphate are mined each year in the United States (mostly from shallow ocean deposits off Florida) to produce commercial fertilizers that replace some of the phosphates lost from farmland and lawns, as well as phosphate detergents. Most phosphates in detergents are released to rivers as effluents from urban sewage treatment plants and eventually end up in the oceans.

Some of the phosphates released by chemical weathering of phosphate rocks flow fairly rapidly through plants and animals on their way to the ocean. This occurs when phosphate rock is dissolved in water in the soil. The roots of plants absorb the phosphate ions and, when the plant is eaten, the phosphorus is passed on to animals. It

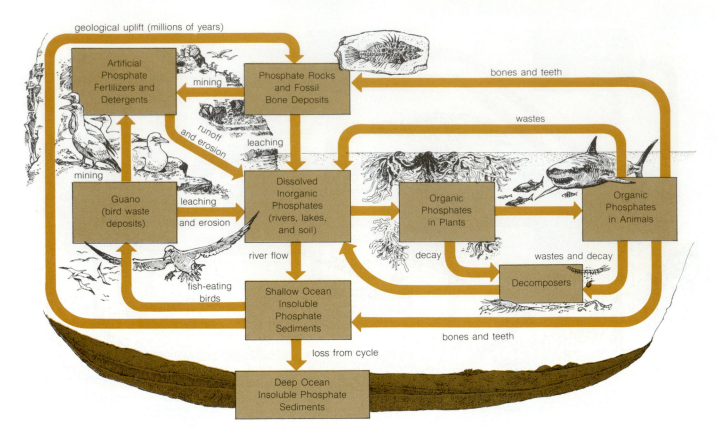

Figure 5-9 Simplified version of the phosphorus cycle (energy flow not shown).

eventually returns to the soil, rivers, and oceans as animal wastes and decay products (Figure 5-9).

Phosphorus, more than any other element, can become the limiting factor for plant growth in a number of ecosystems. In Asia, for example, it is estimated that between 1975 and 2000 the supply of phosphate fertilizer must increase tenfold to meet food demands. Even at this rate of growth, world phosphate supplies are not likely to run out, although local and regional shortages exist and could worsen. These shortages are due primarily to uneven distribution of phosphate deposits and the rising cost of mining, processing, and shipping phosphate fertilizers.

Water or Hydrologic Cycle The **hydrologic or water cycle** (Figure 5-10) is a gigantic water distillation and distribution system. Incoming solar energy evaporates water from the oceans, lakes, rivers, soil, and plants (evaporation and transpiration) into the atmosphere. As warm air masses rise and cool, this water vapor can condense to form tiny droplets or ice crystals that make up clouds. Eventually these droplets or ice crystals can coalesce and precipitate out of the clouds as rain, sleet, or snow falling onto the land or back into oceans, rivers, and lakes. Some of this fresh water

becomes part of glaciers. Some sinks, or percolates, down through the soil and ground to form the groundwater system. When rain falls faster than the water can infiltrate the soil (or cement or other material covering the land), water collects in puddles and ditches and runs off into nearby streams, rivers, and lakes. These streams and rivers carry water back to the oceans, completing the cycle. Water resources provided by this cycle are discussed further in Chapter 10.

5-4 The Niche Concept

Ecological Niches The **ecological niche** (pronounced *nitch*) is a description of all the physical, chemical, and biological factors that a species needs to survive, stay healthy, and reproduce in an ecosystem. It describes what a particular species does in the ecosystem—how it transforms matter and energy, and how it responds to and modifies its physical and biotic environment. To describe a species' niche, we must know what it eats and what eats it, where it leaves its wastes, the ranges of temperature, wind, shade, sunlight, and various chemicals it can tolerate, its effects on other species and on the nonliving parts of its environ-

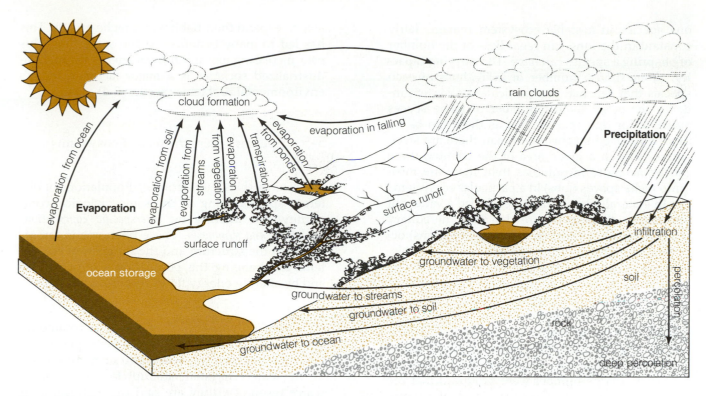

Figure 5-10 Simplified version of the hydrologic cycle. (Source: U.S. Department of Agriculture.)

ment, and what effects other species have on it. Determining the niche of an organism is extremely complex, explaining why we will never know as much as we would like about the niches of most plants and animals.

"Ecological niche" should not be confused with an organism's *habitat*, which is a physical location or locations, such as the height above the ground in a forest canopy or the depth in the mud at the bottom of a pond. A common analogy is that an organism's *habitat* refers to its "address" in an ecosystem and its *ecological niche* is its "occupation" and "life-style." For example, the habitat of a robin includes such areas as woodlands, forests, parks, pasture land, meadows, orchards, gardens, and yards. In contrast, its ecological niche includes such characteristics as using trees for nesting sites and roosts, eating insects, earthworms, and fruit, and dispersing fruit and berry seeds in its droppings.

Competitive Exclusion Principle According to the **competitive exclusion principle**, *no two species in the same ecosystem can occupy exactly the same ecological niche indefinitely*. The more similar the niches of two species, the more they will compete for the same food, shelter, space, and other critical resources. When two species try to occupy the same niche for a relatively long time, one will probably relocate, switch its habits, or become extinct, as discussed in more detail in the next section.

Different species may live together in the same habitat yet have quite different ecological niches. For example, two different species of fish-eating birds, the common cormorant and the shag cormorant, look alike, fish in a similar manner, and live on the same cliffs near the ocean. Careful observations, however, reveal that the species have different ecological niches. The shag cormorant fishes mainly in shallow water for sprats and sand eels and nests on the lower portion of the cliffs. The common cormorant fishes primarily for shrimp and a few fish farther out to sea and lives nearer the top of the cliffs than the shag cormorant.

Different species, however, can occupy similar ecological niches in different ecosystems. Such species are called **ecological equivalents**. Grasslands, for example, are found all over the world. The niche of grassland grazers can be occupied by bison and pronghorn antelope in North America, wild horses and antelope in Eurasia, kangaroo in Australia, and antelope and zebras in Africa. In many regions, a number of these ecologically equivalent herbivores have been replaced by domesticated cattle and sheep.

Niche and Population Size The niche concept helps explain why the population size of the various species in a stable ecosystem remains fairly constant in the long run regardless of the number of offspring a species can have. The **carrying capacity**, or maximum number of individuals of each species that can live in a particular ecosystem indefinitely, is set by the number of niche spaces available for that species. Just as there can be no more employed steelworkers than there are steelworker jobs in a given area of the country, there can be no more mosquitoes than there are mosquito niche spaces (jobs) in a particular ecosystem.

In general, species have developed a variety of approaches to help ensure that their hereditary material (genes passed on to later generations) occupies as many future niche spaces as possible. Ecologist Paul Colinvaux calls the extremes of this continuum of breeding strategies "the small-egg gambit" and "the large-young gambit." The most common breeding strategy is for an individual in a species such as the housefly, mosquito, or dandelion to produce thousands of tiny eggs or seeds. Although most of these offspring die at an early age, such a large number were produced that the chances of a few of the young surviving and reproducing are increased.

On the other hand, instead of having numerous tiny offspring and letting them fend for themselves, other species—such as horses, tigers, white sharks, and people—have a few large offspring and nurse them until they are big and strong. This is the surest method of populating future ecological niches with one's descendants, *provided an individual does not have too many or too few offspring*. If there are too many fairly helpless offspring, the parents may not be able to protect them and feed them adequately. If there are too few offspring and all are killed prematurely (for example, by disease, accident, predation, or starvation), the genetic line ends.

Our Ecological Niche What niche do human beings occupy? Most plants and animals are limited to specific habitats in the ecosphere because they can tolerate only a narrow range of climatic and other environmental conditions. But some generalist species, such as flies, cockroaches, mice, and humans, are very adaptable, can live over much of the planet, and eat a wide range of foods.

Humans occupy a generalist niche. In addition to obtaining energy and nutrients by eating a variety of plants and animals, humans also have learned to utilize solar energy stored for millions of years in deposits of coal, oil, and natural gas. With this fossil fuel energy subsidy as well as other forms of energy (draft animals, biomass, nuclear, hydropower, wind, and geothermal), humans have been able to expand their habitat and niche greatly. This has led to many benefits. At the same time, this ever-increasing use and flow of energy through industrialized societies is a major factor in today's environmental problems (Section 3-4).

5-5 Species Interactions in Ecosystems

Types of Species Interactions Populations of different species living in the same biological community can interact by *interspecific competition, predation, and symbiosis*. These phenomena can affect both the structure and the function of ecosystems by changing the population size of species and by altering energy flows through food webs.

Interspecific Competition As long as commonly used resources are abundant, different species can share them. However, when two or more species in the same ecosystem attempt to use the same scarce resources, they are said to be engaging in **interspecific competition**. The scarce resource may be food, water, a place to live, sunlight, or anything needed for survival.

One competing species may have such an advantage over other species that its population grows and it dominates the use of scarce resources by reproducing more young in a given amount of time, obtaining food or defending itself more effectively, or being able to tolerate a greater range of variance in some limiting factor (Section 4-6). For example, if one shade-intolerant tree species in a patch of forest can multiply and grow fast, it can create a dense overhead canopy that prevents other shade-intolerant species from flourishing on the forest floor.

Populations of some animal species can avoid or reduce competition with more dominant species by moving to another area, switching to a less accessible or less readily digestible food source, or hunting for the same food source at different times of the day or in different places. For example, hawks and owls feed on similar prey. Competition is reduced, however, because hawks hunt during the day and owls hunt at night.

Evolutionary adaptation can also give a species a competitive advantage. Thus insects, with short generation times and numerous offspring, can undergo rapid evolutionary adaptation to a new environmental condition, as discussed in more detail in the next chapter. For example, pesticides have been used to reduce insect populations that compete with humans for food resources. However, through natural selection, the few surviving members of a particular insect species can rapidly breed new populations that are genetically resistant to a particular pesticide (Section 23-4).

Predation The most obvious form of species interaction is **predation**; that is, an organism of one species (the *predator*) captures and feeds on either parts or all of an organism of another species (the *prey*). In most cases a predator species has more than one prey species. Likewise, a single prey species may have several different predators. Eagles, owls, and hawks eat rats, mice, rabbits, and other prey.

We usually think of predation in terms of animal–animal interactions, but it can also involve plant–animal interactions and even plant–plant interactions. A cow can be classified as a predator when it eats grass (the prey), and carnivorous plants, like the Venus flytrap, catch and digest insects.

The ability of predators to find and feed on their prey determines the rate at which energy and various types of matter flow from one trophic level to another (Section 5-3). Some predator species have helpful specialized features and behavior patterns. For example, the remarkable eyesight of hawks, falcons, and eagles allows them to spot moving rodents from great heights. Bats have a kind of sonar system that helps them find prey and avoid obstacles. Wolves can take prey much larger than themselves by hunting in packs. Spiders build complex webs to trap insects.

Prey species have also developed ways to avoid predators. Species that rely on running, swimming, or flying out of reach of predators usually live in places where there aren't many hiding places. Fast-swimming fish and seals are found in open seas, and fast-running hooved animals such as zebra inhabit open areas or savannas. Rabbits run and hide under the cover of vegetation, mice dash into holes to hide, and lizards hide under rocks.

Some species defend themselves by inflating to a larger size (blowfish), giving off an obnoxious odor (skunk), tasting bad to predators (milkweed bug and monarch butterfly), having a protective shell or covering (turtles and armadillos), having brightly colored tails that break off when seized by an attacker (many lizard species), running in herds (wildebeest), having thorns (roses) and spines (cacti and porcupine), stinging (nettle), and producing chemicals harmful or unpalatable to predators (poison oak and poison ivy). Others have natural camouflage abilities: Chameleon lizards can change color, and some cacti look like rocks;

the walkingstick looks like a twig; and edible viceroy butterflies resemble bad-tasting monarch butterflies.

Often sport hunting by humans is justified as means of population control of animals. For example, wolves and other natural predators of a species, such as deer, may be killed off or driven out of an ecosystem. Then human hunters must act as predators to keep the deer population from exploding and destroying much of the vegetation. But sport hunting by humans for the sake of population control can cause a serious problem. Human sport hunters usually don't kill the old, sick, and weak animals. Instead, they normally go after the strongest and healthiest ones—the animals most needed for reproduction. Thus, sport hunting can control a population, but the quality of the remaining population can be lowered.

Symbiosis An interaction in which two different species exist in close physical contact, with one living on or in the other so that one or both species benefit from the association, is called **symbiosis.** This "living together" is an ecological association involving some transfer of energy or other ecological benefit. The three major types of symbiosis are *mutualism*, *commensalism*, and *parasitism*.

Mutualism is a symbiotic relationship between two different species in which both parties benefit from the interaction. For example, butterflies and bees depend on flowers for food in the form of pollen and nectar. In turn, the flowering plants depend on the bees, butterflies, and other pollinating insects to carry the male reproductive cells contained in their pollen grains to the female flowering parts of other flowers of the same species. Another mutualistic relationship exists in the nitrogen cycle (Figure 5-8) between legume plants (such as peas, clover, beans, and alfalfa) and *Rhizobium* bacteria, which live in nodules on the roots of these plants. Large colonies of bacteria in these nodules "fix," or convert, gaseous nitrogen to forms of nitrogen such as nitrate and ammonium ions that can be used as nutrients by the plant and by the bacteria themselves.

In the second type of symbiosis, called **commensalism,** one species benefits from the association while the other is apparently neither helped nor harmed. For example, in tropical and subtropical forests rootless *epiphytes*, living high above the ground on tree trunks and branches, are able to use the tree hosts to get access to sunlight in the relatively dark forest. They also use their own leaves and cupped petals to collect water and minerals that drip down from the tops of trees. The tree is neither harmed nor benefited.

In **parasitism**, the third form of symbiosis, the parasite benefits and the *host* species is harmed. The parasite usually takes its nourishment from the host, either externally from skin, feathers, hair, or scales (*ectoparasitism*), or internally from the cells, tissues, and other parts of the host (*endoparasitism*). Examples of parasites are leeches (which suck blood from their victims), lice, ticks, fleas, bacteria, protozoans, fungi, and worms such as tapeworms, hookworms, and pinworms. Human diseases caused by parasitic bacteria include typhoid fever, tuberculosis, cholera, syphilis, and gonorrhea. Diseases such as amoebic dysentery, African sleeping sickness, and malaria are caused by parasitic protozoans (Section 20-2).

Plants are also plagued by parasites. Parasite fungi cause wheat rust, corn leaf blight, Dutch elm disease, and other destructive plant diseases. By planting large fields of a single species, such as corn, humans allow parasites to wipe out an entire crop. A successful parasite may harm its host, but it does not kill the host. If the host dies, the parasite must find a new host, or die itself.

Parasitism is sometimes classifed as a special type of predation rather than an example of symbiosis. Parasitism, like predation, involves one species feeding on another. But there are significant differences between these two types of species interaction. Parasites are usually smaller than their host, whereas most predators are larger than their prey. Parasites tend to live in, on, or near their host and slowly consume only part of the host. In contrast, predators live apart from their prey and tend to kill and then consume most or all of it.

In this chapter and the preceding one, we have seen that *the essential feature of the living and nonliving parts of an ecosystem is interdependence.* In the next chapter we shall see that this interdependence is a key to understanding how ecosystems can change in response to stresses from natural and human sources.

We cannot command nature except by obeying her.
Sir Francis Bacon

Discussion Topics

1. Explain how the survival of an individual organism depends on energy flow and matter flow, whereas the ecosystem in which it lives is able to survive only by energy flow and matter cycling.

2. Using the second law of thermodynamics, explain why there is such a sharp decrease in high-quality energy along each step of a food chain. Doesn't an energy loss at each step violate the first law of thermodynamics? Explain.

3. Using the second law of thermodynamics, explain why many people in LDCs must exist primarily on a vegetarian diet.

4. Explain why a balanced vegetarian diet is sound ecological practice.

5. Why don't lions hunt mice?

6. Why would you expect to find more rabbits than coyotes on earth, even if humans did not kill and poison large numbers of coyotes?

7. When you throw away your trash, where does it go? Trace it through the carbon, oxygen, and nitrogen cycles.

8. Compare the ecological niches of humans in a small town and in a large city and those of humans in a more developed country and in a less developed country.

9. Discuss the advantages and disadvantages of sport hunting. Do you think sport hunting should be banned? Why or why not? What restrictions, if any, would you put on hunting for sport? Why?

10. Explain how parasitism differs from predation.

6

Changes in Ecosystems: What Can Happen to Ecosystems?

When we try to pick out anything by itself, we find it hitched to everything else in the universe.

John Muir

Ecosystems are dynamic, not static. The natural plant and animal communities found in some ecosystems gradually change their environment in ways that eliminate some species and allow for invasion by others. The plants and animals in ecosystems also undergo change as they attempt to adapt to environmental (natural and human-induced) stresses. Although ecosystems are always changing, they also resist disturbance and possess the ability to restore themselves if the outside disturbance is not too drastic. This ability to adapt and yet sustain themselves is truly a remarkable feature of ecosystems. In this chapter we examine first how ecosytems evolve and change normally without human influence and then some human influences on ecosystems.

6-1   Ecological Succession

Types of Ecological Succession   Tropical rain forest, oak–hickory forest, and coral reef ecosystems develop over many decades or centuries, starting with the colonization of an uninhabited site by a natural community of *pioneer species* (such as lichens and mosses)—hardy plants that can survive intense sunlight, wide temperature swings, and soils poor in nutrients. Typically, such species have short life cycles, must start over each year, and are small and short. Pioneer species are sometimes called *opportunist species* because their strategy is to maximize production of seeds that can be scattered widely. The pioneers are slowly joined, and gradually replaced, by other species to form a new natural community as the ecosystem matures.

Most of the plants in mature ecosystems, such as perennial herbs and trees, are larger, more vigorous, and do not put most of their energy into seed production. Instead, they produce relatively few but large seeds, and most nontropical plants develop big roots for underground storage of some of their energy input, helping them to last the winter and begin their seasonal growth early the next year, with a survival advantage over pioneer species. Species adopting this survival strategy are sometimes called *equilibrium species*.

With time the opportunist species that first invade a patch of barren ground are replaced by natural communities with increasing numbers of other species. This repeated replacement of one kind of natural community of organisms by another over time is called **ecological succession**.

If not severely disrupted by natural disasters or by human activities, most ecosystems attain a stage that is much more stable than those that preceded it. This is sometimes called a **climax ecosystem**, or *climax natural community*. Most climax ecosystems tend to be self-perpetuating and long-lived, as long as climate and other major environmental factors remain essentially the same. Some ecosystems, however, require disturbances such as fire to reach the climax stage of succession.

Ecologists recognize two types of ecological succession: primary and secondary. Successional changes taking place in a soilless area previously devoid of life constitute **primary succession**. Examples of such areas are bare rocks newly exposed by retreating glaciers, cooled volcanic lava, newly exposed sand dunes, and surface-mined areas from which all topsoil has been removed. On such barren surfaces, primary succession from bare rock to a mature forest may take hundreds to thousands of years.

More common is **secondary succession**, that is, successional changes that occur in a previously inhabited area that has been disturbed and set back to an earlier stage of succession. Since soil is present, new vegetation can sprout within a few weeks. Examples include new growth on abandoned farmland, in forests that have been burned or cut, and in heavily polluted streams.

Primary Succession   During the early part of this century, William S. Cooper was able to trace the stages of primary succession from bare rock to a

exposed rocks   lichens and mosses   small herbs and shrubs   heath mat   jack pine, black spruce, and aspen   balsam fir, paper birch, and white spruce climax community

Time ———►

Figure 6-1 Primary ecological succession over several hundred years on a single patch of land on Isle Royale, in northern Lake Superior.

balsam, fir, paper birch, and white spruce climax natural community on Isle Royale in northern Lake Superior (Figure 6-l).

First, retreating glaciers exposed bare rock. Wind, rain, and frost weathered the rock surfaces to form tiny cracks and holes. Water collecting in these depressions slowly dissolved minerals out of the rock's surface. These minerals were able to support hardy pioneer plants, such as lichens and mosses. Gradually these early invaders covered the rock surface, dissolving additional minerals from the rock and, when they died, depositing organic matter. Decomposer organisms moved in to feed on the dead lichens and mosses and were followed by a few small animals such as ants, mites, and spiders. This first combination of plants, animals, and decomposers is called the *pioneer natural community*.

After many years, the pioneer natural community built up enough organic matter in the thin soil to support the roots of small herbs and shrubs such as bluebell, yarrow, bearberry, blueberry, and juniper. These newcomers slowed the loss of moisture and provided food and cover for new plants,

animals, and decomposers. Under these modified conditions, the species in the pioneer natural community species were crowded out and gradually replaced with a different type of natural community.

As the new natural community thrived, it added organic matter to the slowly thickening crust of soil. This led to the next stage of succession, a compact layer of vegetation called a heath mat. This mat, in turn, provided thicker and richer soil, needed for the germination and growth of trees such as jack pine, black spruce, and occasionally aspen. Over several decades these trees increased in height and density, and the plants of the heath mat were crowded out. The shade and other conditions created by these trees allowed the germination and growth of balsam fir, paper birch, white spruce, and other shade-tolerant climax species. These taller climax tree species created a canopy, and most of the earlier shade-intolerant tree species were eliminated because they could not reproduce. After several centuries, what was once bare rock became a mature or climax ecosystem (Figure 6-1).

Labels on figure (right side, top to bottom): canopy, lower canopy trees, understory trees, tall shrub understory

Labels along bottom axis: annual weeds, perennial weeds and grasses, shrubs, young pine forest, mature oak forest

Time ⟶

Figure 6-2 Secondary ecological succession on an abandoned farm field over about 150 years.

Secondary Succession Secondary succession, as shown in Figure 6-2, occurs when land in an oak–hickory forest in the eastern United States is cleared for growing corn and abandoned after harvest. The abandoned field already has a thick layer of soil that serves as a well-developed nutrient pool, so the early stages of primary succession are not necessary. The bare field is quickly covered wth crabgrass in the fall. In the spring horseweed takes over, and during the summer the field is invaded by white aster plants, which shade out the crabgrass within a year. After 2 or 3 years, enough organic matter in the soil has built up to support a perennial grass. As dead plants and other debris accumulate, decomposers build up the soil. This prepares the way for the growth of young pine trees, which can thrive in direct sunlight and open fields. Pine seeds are blown in, and young pine trees invade the grass, grow within 5 or 10 years to the low-shrub stage, and begin to shade out the sun-loving weeds and grasses. Over the next 20 years a pine forest develops. The tall pine trees shade out their own seedlings. In the cool shade beneath the pine forest, seedlings of shade-loving hardwoods such as red gum, red maple, black oak, and hickory begin to grow. These species have long taproots and can obtain moisture unavailable to the shallow-rooted pines. Over about 40 to 120 years the hardwoods, especially oak and hickory, replace the shade-intolerant pines as the latter die out. Other shade-tolerant trees and shrubs, such as dogwood, sourwood, and redbud, fill in the understory below the canopy of oak and hickory trees, resulting in a mature oak–hickory deciduous forest (Figure 6-2).

6-2 Stability in Living Systems

What Is Stability? Organisms, populations, communities, and ecosystems all have some ability to withstand or recover from externally imposed changes or stresses. In other words, they have some degree of *stability*.

Stability in a system implies persistence of structure over time. This stability is maintained, however, only by constant change. You are continually adding and losing matter and energy, but your body maintains a fairly stable structure over your life span. Similarly, in the oak–hickory forest ecosystem in Figure 6-2, some trees will die, and others will take their place. Some species may also disappear, and the numbers of individual species

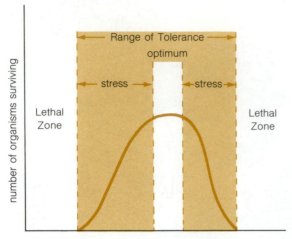

Figure 6-3 Range of tolerance for a large number of organisms of the same species to an environmental factor, such as temperature.

may change. But unless it is cut, burned, or blown down, you will recognize it as a deciduous forest 50 years from now. Let's look at some of the ways individual organisms, populations, and ecosystems maintain their stability by responding to stress.

Organism Responses to Stress Each organism has a particular **range of tolerance** to variations in certain chemical and physical factors making up its abiotic environment (Figure 6-3). There may be too much moisture or not enough, too high a temperature or too low, too much light or not enough, too many minerals dissolved in the soil or too few.

This is summarized in the *limiting factor principle* or *law of tolerance*: The existence, abundance, or distribution of an organism can be determined by whether the levels of one or more limiting factors fall above or below the levels tolerated by the organism. For example, suppose a farmer plants corn in a field containing too little phosphorus (as phosphates). Even if the corn's requirements for water, nitrogen, potassium, and other chemicals are met, the corn will stop growing when it has used up the available phosphorus. In this case, availability of phosphorus is the *limiting factor*.

Organisms of the same species have the same general range of tolerance to various stresses. Individual organisms within a large population, however, may have slightly different tolerance ranges because of genetic variability. It may take a little more heat or a little more of a poisonous chemical to kill one cat or one human than another. This is why Figure 6-3 is plotted for a large number of organisms of the same species rather than for a single organism.

Normally an organism's sensitivity to a particular stress also varies with its physical condition and with its life cycle. Organisms already weakened by fatigue or disease are usually more sensitive to stresses than healthy individuals. For most animals, tolerance levels are much lower in juveniles (where body defense mechanisms may not be fully developed) than in adults.

The ability to adapt slowly to new conditions is a useful protective device, but it can also be dangerous. Many changes in pollution levels, for instance, occur so gradually that we tolerate them. But with each change we come closer to our limit of tolerance. Suddenly, without any warning signals, we cross a threshold that triggers a harmful or even fatal effect. This **threshold effect**—the straw that breaks the camel's back—partly explains why so many ecological problems seem to pop up, even though they have been building or accumulating for a long time.

Often pollutants are diluted to relatively harmless level in the air or water, or degraded to harmless forms by decomposers and other natural processes. However, some synthetic chemicals, such as DDT (Chapter 23), some radioactive materials (Chapter 18), and some mercury and lead compounds (Section 24-5) are neither diluted nor broken down by natural processes. Instead, they can become more concentrated as they go through various food chains and food webs in an ecosystem. As a result, organisms at high trophic levels in these food webs receive large doses of such chemicals even though relatively small amounts are in the air, water, or soil. Figure 6-4 illustrates this phenomenon of **biological magnification** in an estuarine ecosystem adjacent to Long Island Sound.

Biological magnification in a food chain or web depends on three factors: the second law of thermodynamics (Section 3-3), chemicals that are soluble in fat but insoluble in water, and chemicals that are nondegradable or only slowly degradable in the environment. Since energy transfer at each link in a food chain or web is so inefficient (Figure 5-6), a small fish must eat a great deal of plankton, larger fish must eat a great many small fish, and a pelican must eat a great many larger fish to survive. Any substance that is not degraded or excreted as it moves through this chain becomes more concentrated, especially if it dissolves in and remains in fatty tissues of organisms. For example, if each phytoplankton concentrates 1 unit of DDT from the water, a small fish eating thousands of phytoplankton will store thousands of units of DDT in its fatty tissue. If DDT were water soluble, the fish would excrete it at each level, but it is not. Thus, a large fish that eats ten of these smaller fish

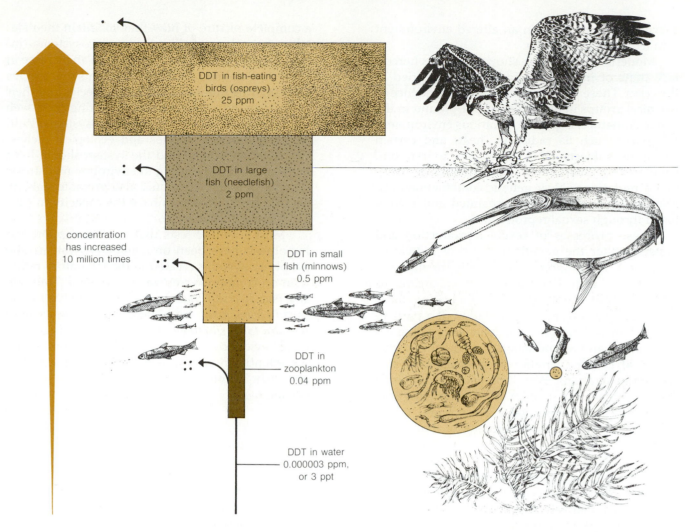

DDT in fish-eating
birds (ospreys)
25 ppm

DDT in large
fish (needlefish)
2 ppm

concentration
has increased
10 million times

DDT in small
fish (minnows)
0.5 ppm

DDT in
zooplankton
0.04 ppm

DDT in water
0.000003 ppm,
or 3 ppt

Figure 6-4 The concentration of DDT in organisms is magnified approximately 10 million times in this food chain found in an estuarine ecosystem adjacent to Long Island Sound. Dots represent DDT, and arrows show small losses through respiration and excretion. Figure is not drawn to scale.

will receive and store tens of thousands of units of DDT. A bird or human that feeds on several large fish can ingest hundreds of thousands of units of DDT.

Organisms also have a number of responses to environmental stress: They may migrate (some birds fly south for the winter), hibernate to wait out the stress period in a less active physiological state (chipmunks, bears, and ground squirrels hibernate in winter), change their appearance or certain body functions to counteract the stress (arctic hares are white during the snowy winter season but shed their white coat and become brown during the spring to blend in with the brown tundra vegetation), or slowly become used to the new conditions.

Population Responses to Stress Populations respond in a number of ways to stress. Death rates may increase or birth rates may decrease, reducing the population size to one that can be supported by available resources. Also, the structure of the population may change. The old, very young, and weak members may die, leaving a population better capable of surviving such a stress as a more severe climate or an increase in predators, parasites, or disease organisms.

Populations are also capable of adapting to environmental changes through *natural selection*. Most organisms in a population produce more offspring than can possibly survive. Because these offspring vary somewhat in their genetic makeup, some are better adapted to survival in a given environment than others. The individuals better suited to their environment can leave more offspring and thus contribute more of their genetic material to future generations. This increase in a population in individuals with new combinations of genetic material that make them more likely to survive and reproduce in an altered environment is called **natural selection**.

Modern research has shown that the hereditary traits of individual organisms are carried in the *genes*. These traits are coded in the sequence of chemical groups in molecules of DNA found in the genes. In nature, exposure to various environmental factors such as radiation, heat, and certain chemicals subjects DNA molecules to random, spontaneous changes in their chemical makeup called **mutations**. New genetic combinations can also arise when genes are separated and recombined through sexual reproduction.

These processes of sexual reproduction and mutation alter the genetic makeup of individual organisms. If a random, new genetic variation detracts from survival and reproduction, organisms with this trait eventually die—eliminating these particular genes from the population. If a particular random variation in genetic makeup enhances the survival and reproduction of an organism, this beneficial genetic variation is passed on to succeeding generations. With time, more and more of the individuals in a population of a particular species possess the beneficial genetic traits, which enhance survival and reproduction of that population in its environment.

Populations of species with short generation times and large numbers of offspring, such as bacteria, insects, and rats, can make adaptive genetic changes in a relatively short time. For example, in only a few years a number of species of mosquitoes have become genetically resistant to DDT and other pesticides, as discussed in Section 23-4. Similarly, species of bacteria can evolve new strains that are genetically resistant to widely used bacteria-killing drugs, such as penicillin.

In marked contrast, our species, like many others, has a long generation time and cannot reproduce a large number of offspring rapidly. For such species, adaptation to an environmental stress by changes in genetic makeup takes hundreds of thousands and in some cases millions of years.

The incredible genetic diversity created within populations of the same species and between different species on this planet through evolution is nature's insurance policy against disaster. Although many more species than currently exist have appeared and disappeared throughout earth's history, every species here today represents stored genetic information that allows the species to adapt to certain changes in environmental conditions.

Ecosystem Stability and Species Diversity Ecosystems are so complex that ecologists do not have a complete picture of how they maintain their stability. In many—perhaps most—ecosystems, increasing species diversity can lead to an increase in ecosystem stability. Intuitively it seems that species diversity (the number of different species and their relative abundances) and the resulting food web complexity should help stabilize ecosystems. With so many different species and ecological niches, risk is widely spread and the system should have more ways of responding to environmental stress. A complex food web should also promote stability because many predators have the capacity to shift to alternate food sources.

Most scientific tests that tend to support this hypothesis have been on aquatic ecosystems and simple ecosystems created in the laboratory rather than on complex terrestrial ecosystems. Thus, *the idea that diversity leads to ecosystem stability may be valid in many cases, but we should be wary of applying this idea to all situations.* However, there is considerable evidence that simplifying an ecosystem by the intentional or accidental removal of a species often has harmful effects, as discussed in Sections 6-3 and 6-4.

6-3 What Can Go Wrong in an Ecosystem?

Effects of Environmental Stress Since ecosystems tend to be self-maintaining and self-repairing, why not drop all our wastes into the environment and let Nature take care of them? By now you realize that there are serious problems with this idea. First, as we have seen, organisms, populations, and ecosystems all have certain limits of tolerance. Second, ecosystems do not have the decomposers and disposal mechanisms for coping with many of the synthetic chemicals produced by humans. Third, even though populations might evolve so that ecosystems could digest these new chemicals and absorb many of today's environmental insults, for most populations—especially the human population—such evolutionary changes would take hundreds of thousands or millions of years.

Table 6-1 summarizes what can happen to organisms, populations, and ecosystems if one or more limits of tolerance are exceeded. The stresses that can cause the changes shown in Table 6-1 may result from natural hazards (earthquakes, volcanic eruptions, hurricanes, drought, floods, fires ignited by lightning) or from human activities (industrialization, warfare, transportation, and agriculture), as discussed in the remainder of this section and throughout much of this book.

Table 6-1 Some Effects of Environmental Stress

Organism Level

Physiological and biochemical changes
Psychological disorders
Behavioral changes
Fewer or no offspring
Genetic defects in offspring (mutagenic effects)
Birth defects (teratogenic effects)
Cancers (carcinogenic effects)
Death

Population Level

Population increase or decrease
Change in age structure (old, young, and weak may die)
Survival of strains genetically resistant to a stress
Loss of genetic diversity and adaptability
Extinction

Community–Ecosystem Level

Disruption of energy flow
   Decrease or increase in solar energy input
   Changes in heat output
   Changes in trophic structure in food chains and food webs

Disruption of chemical cycles
   Depletion of essential nutrients
   Excessive addition of nutrients

Simplification
   Reduction in species diversity
   Reduction or elimination of habitats and filled ecological niches
   Less complex food webs
   Possibility of lowered stability
   Possibility of ecosystem collapse

Disrupting Food Webs with Synthetic Chemicals

Malaria once infected nine out of ten people on the island of North Borneo, now a state of Indonesia. In 1955 the World Health Organization (WHO) began spraying dieldrin (a pesticide similar to DDT) to kill malaria-carrying mosquitoes.* The program was very successful, almost eliminating this dreaded disease. But other things happened. The dieldrin killed other insects besides mosquitoes, including flies and cockroaches inhabiting the houses. The islanders applauded. But then small lizards that also lived in the houses died after gorging themselves on dead insects. Then cats began dying after feeding on the dead lizards. Without cats, rats flourished and began overrunning the villages. Now people were threatened by sylvatic plague carried by the fleas on the rats. The situa-

---

*Most reports of this episode cite DDT as the insecticide used. According to a personal communication from A. J. Beck (at one time a medical zoologist at the Institute for Medical Research in North Borneo), dieldrin, not DDT, was used.

tion was brought under control when WHO had the Royal Air Force parachute cats into Borneo.

On top of everything else, thatched roofs began to fall in. The dieldrin also killed wasps and other insects that fed on a type of caterpillar that either avoided or was not affected by the insecticide. With most of its predators eliminated, the caterpillar population exploded. The larvae munched their way through one of their favorite foods, the leaves used in thatching roofs. In the end, the Borneo episode was a success story in that both malaria and the unexpected side effects of the spraying program were brought under control. But it shows the unpredictable results that may be encountered when we interfere in an ecosystem. More details on the dilemma of using pesticides are found in Chapter 23.

Reducing Variety in Ecosystems by Eliminating and Introducing Species

We tend to divide plants and animals into "good" and "bad" species and to assume that we have a duty to wipe out the villains. Consider the American alligator. Its marsh and swamp habitats are destroyed to make way for agriculture and industry, and the animal itself is hunted for its hide. Between 1950 and 1960, Louisiana lost 90 percent of its alligators; the population in the Florida Everglades was also threatened.

Why care? The alligator is a key factor in the ecological balance of the Everglades—a balance on which much of the urbanizing state of Florida depends for water. Alligators dig deep pools, or "gator holes," which collect water during dry spells and provide a sanctuary for birds and animals that repopulate the area after droughts. The large alligator nesting mounds are nest sites for birds, including the herons and egrets essential to other life cycles. As alligators move from gator holes to nesting mounds, they help keep waterways open. In addition, they preserve game fish balances by consuming large numbers of predator fish, such as gar.

In 1968 the U.S. government placed the American alligator on the endangered species list. Protected from hunters, by 1975 the alligator population had made a comeback in many areas. Indeed, it had come back too far. People began finding alligators in their backyards and swimming pools. Now the American alligator has been removed from the endangered species list in most areas, and limited hunting is allowed in some areas (especially Louisiana) to keep its population in check. More details on threatened and endangered species are given in Chapter 13.

Hazardous Chemical Dumps: The Love Canal Episode "Out of sight, out of mind" does not always apply. Hazardous industrial wastes buried decades ago can bubble to the surface, find their way into groundwater supplies, or end up in back-yards and basements, as residents of a suburb of Niagara Falls, New York, discovered in 1977. Between 1942 and 1953, Hooker Chemicals and Plastics Corporation dumped more than 19 million kilograms (21,800 tons) of chemical wastes (mostly contained in steel drums) into an old canal, known as the Love Canal, and sealed the dump with a clay cap.

In 1953 Hooker Chemicals sold the canal area to the Niagara Falls school board for $1 on the condition that the company would have no future liability for any injury or property damage caused by the dump's contents. The company says that it warned the school board against carrying out any kind of construction at the disposal site. An elementary school and a housing project, eventually containing 949 homes, were built in the Love Canal area. Though residents began complaining to city officials about chemical smells and about chemical burns received by children playing in the canal, these complaints were ignored. In 1977 chemicals from badly corroded barrels filled with hazardous wastes began leaking into storm sewers, gardens, and basements of homes adjacent to the canal.

Informal health surveys conducted by alarmed residents revealed an unusually high incidence of birth defects, miscarriages, assorted cancers, and nerve, respiratory, and kidney disorders among people who lived near the canal. Complaints to local elected and health officials had little immediate effect. Pressure from residents and unfavorable publicity, however, led state officials to make a preliminary health survey and tests. They found that **(1)** women between the ages of 30 and 34 in one area of the canal had a miscarriage rate four times higher than normal, and **(2)** the air, water, and soil of the canal area and basements of nearby houses were contaminated with a wide range of toxic and carcinogenic chemicals.

In 1978, after local health officials failed to act, the state closed the school and relocated the 239 families whose homes were closest to the dump. In 1980, after protests from the outraged 710 families still living nearby, President Carter declared Love Canal a federal disaster area and had these families temporarily relocated. Federal and New York State funds were then provided to buy the homes of those who wanted to move permanently.

Since that time homes within a block and a half of the canal have been torn down and the state has purchased 550 of the homes; 140 families still live in the area. The dump site has been covered with a clay cap and surrounded by a barrier drain system that includes a permanent water treatment facility for removing contaminants. Local officials have pressed the federal government for a clean bill of health so that the state can resell homes it bought from fleeing homeowners and begin rehabilitating the neighborhood. In 1983 the EPA declared that contamination from the toxic dump site was more extensive than originally believed and developed a revised cleanup plan that will probably postpone decisions on the habitability of various zones until 1988 or later.

No definitive study has been made to determine the effects of exposure to these hazardous chemicals on the former Love Canal residents. All studies made so far have been criticized on scientific grounds. Even if the effects of exposure to these chemicals proves to be less harmful than expected, the psychological damage to the evacuated residents is enormous. For the rest of their lives they will wonder whether a disorder will strike and will worry about the possible effects of the chemicals on their children and grandchildren.

Regardless of the final outcome, the Love Canal episode vividly reminds us that we can never really throw anything away. It is more frightening when we realize that the EPA estimates that at least 2,000 other hazardous chemical dump sites in the United States could pose similar problems, as discussed in Chapter 24.

Nuclear War: The Ultimate Pollution In the 1970s some military strategists and politicians began talking about a nuclear attack that would be "survivable" by many people in the United States or the Soviet Union. In 1982, however, chemist Paul Crutzen, director of the Max Planck Institute of Chemistry in West Germany, made some previously overlooked calculations. Crutzen showed that any nation engaging in even a relatively small nuclear war involving detonation of nuclear warheads with a total explosive power equivalent to that from the detonation of 100 megatons of TNT—only 0.8 percent of the combined nuclear arsenals of the United States and the Soviet Union—would be committing suicide. In addition, such a "small" nuclear war would probably kill most people in all the other nations in the northern hemisphere and, very likely, many in the southern hemisphere.

In 1983 Crutzen's calculations were reviewed and expanded by an international group of more than 100 prominent scientists. They agreed that even a limited nuclear attack would inject a huge, dark cloud of soot and smoke over most of the earth that would blot out the sun. This effect has been called the *nuclear winter* because regardless of when the warheads were detonated, most of the earth would experience darkness and below-freezing temperatures for months. Eventually, most humans and other forms of life not killed by the initial blast, fires, and radiation would die of the effects of a nuclear winter.

Let's consider in more detail the effects of thermonuclear war. According to the World Health Organization, a nuclear war involving about one-third of U.S. and Soviet nuclear arsenals—about 5,000 megatons—would probably kill from 43 to 65 percent of the world's population, including about 2.1 billion people in the northern hemisphere (mostly in the United States, Europe, and the Soviet Union) and an additional 1 billion in the southern hemisphere. In the United States at least 165 million people—about 70 percent of the population—would be killed. Another 60 million would be injured, with 30 million of these contracting radiation sickness, 20 million experiencing trauma and burns, and 10 million suffering from a combination of trauma, burns, and radiation sickness.

An estimated 80 percent of all physicians would be killed and 80 percent of the hospital beds would be destroyed, along with most stores of blood plasma, antibiotics, and other drugs. Most of the injured would have no morphine for pain, no facilities for emergency surgery, and no antibiotics to fight infection. There would be little food or water that was not contaminated with radiation or with toxic chemicals released into the atmosphere from the destruction and burning of chemical plants and storage tanks.

Those who escaped death and serious injury would have to stay in shelters under grim conditions for 1 week to 3 months or more, depending on location, to avoid exposure to dangerous radiation levels. Many shelters would be overcrowded, and supplies of food, water, and medicine would be inadequate. Since national electric power and communications systems would be knocked out by the explosions and by the electromagnetic pulses (EMPs) produced during the first few billionths of a second of the attack, shelters would be dark, damp, cold, and isolated. Waste disposal systems would be primitive, and diseases such as cholera and dysentery would run rampant—adding to the panic, stress, and acute psychological shock.

Because of the nuclear winter effect, survivors emerging from fallout shelters would face so many short- and long-term survival problems that they might envy the dead. Within a week or two, the massive amounts of dust and soot injected in the lower atmosphere and into the stratosphere would coalesce into a massive dark cloud that would prevent 96 percent of incoming sunlight from reaching most of the northern hemisphere, particularly the mid-latitude belt encompassing most of the United States, Canada, Europe, the Soviet Union, and Japan. Incoming sunlight would remain at less than 50 percent of normal for nearly 2 months. With most of the sunlight blocked, temperatures would drop well below freezing, whatever the season, and would remain that way for many weeks.

During the resulting nuclear winter of darkness and subfreezing temperatures, most livestock, wild mammals, and cultivated and uncultivated food supplies, at least in the northern hemisphere, would die from the effects of radiation and cold. Food supplies from crops, livestock, wildlife, or fish and other aquatic species would be very scarce for at least a year because in the absence of sunlight photosynthesis by plants on the land and phytoplankton in the sea would cease. Thick ice would cover inland surface waters and plagues of insects—the animal life form best equipped to survive nuclear war—would damage stored food and spread disease. Huge coastal storms created by land–sea temperature gradients would cause further destruction.

Most crops, animals, and freshwater fish not destroyed would be contaminated with radioactive fallout particles. Survivors would be able to drink groundwater supplies only, and only if they had hand-operated pumps or emergency diesel-fueled generators.

Countless populations of plant species and animal wildlife would be extinct or contaminated, reducing for years the feasibility of trying to stay alive by hunting and gathering. Large areas of forest and grasslands would be devastated by blast and by fires that would burn out of control for weeks, perhaps months, adding more soot to the atmosphere. Much of the soil—essential to the reestablishment of both plants and animals—would wash into rivers and eventually, when the rains from winter storms arrived, into the sea. People who went to the shore hoping to subsist on seafood would find most surviving aquatic species contaminated with radioactivity, silt, and runoff from ruptured tanks of industrial liquids and oil pouring out of damaged offshore rigs. In areas where crops could still be grown, farmers would be cut off from supplies of seeds, fertilizer, pesticides, and fuel.

The nuclear explosions would also cause nitrogen and oxygen in the stratosphere to combine and produce large amounts of nitrogen oxides, which would destroy most of the ozone layer found in the stratosphere for 2 to 20 years. This stratospheric layer of ozone gas shields most forms of life on earth from damaging ultraviolet radiation, as discussed in more detail in Section 21-5.

During this period the sharp increase in ultraviolet rays reaching the earth's surface would lead to an increase among unprotected people of skin cancers, lethal sunburns, lethal levels of vitamin D, and blindness and genetic damage. Levels of nitrogen oxides would also be increased 5 to 50 times in the lower atmosphere—causing large increases in acid precipitation (Section 21-4), which would further threaten food crops, trees, and many aquatic and land-based life forms.

Although limited global nuclear war probably would not completely extinguish the human species, technological society as we know it wouldn't survive. Much of humanity would be reduced to small scattered bands, necessarily returning to a species of hunters, gatherers, and simple farmers (Chapter 2).

6-4 Humans and Ecosystems

Simplifying Ecosystems In modifying ecosystems for our own use we simplify them. We bulldoze fields and forests containing thousands of interrelated plant and animal species and cover the land with buildings, highways, or single crops. Modern agriculture deliberately keeps ecosystems in early stages of succession to stimulate the growth of one or only a few plant species (such as corn or wheat).

But such fast-growing, single-crop systems (monocultures) are highly vulnerable. Weeds and a single disease or pest can wipe out an entire crop unless protected with chemicals such as insecticides (insect-killing chemicals) and herbicides (weed-killing chemicals). When quickly breeding insects develop genetic resistance to pesticides, farmers must either use stronger doses or switch to new pesticides. This approach kills other species that prey on the pests, thus simplifying the ecosystem further and allowing pest populations to expand (Section 23-4).

Cultivation is not the only factor that simplifies ecosystems. Ranchers don't want bison or prairie dogs competing with sheep for grass, so these species are eliminated from grasslands, along with wolves, coyotes, eagles, and other predators that occasionally kill sheep. We tend to overfish and overhunt some species to extinction or near extinction, further simplifying ecosystems, as discussed in Chapter 13.

Achieving a Balance Between Simplicity and Diversity There is nothing wrong with simplifying a reasonable number of ecosystems to provide food for the human population. But the price we pay for simplifying, maintaining, and protecting such stripped-down ecosystems includes time, money, and matter and energy resources, as summarized in Table 6-2.

As the human population grows, we may simplify too many of the world's ecosystems to young, productive, but highly vulnerable forms. Simplified human systems depend on the existence of neighboring natural ecosystems. For example, simple farmlands on the plains must be balanced by diverse forests on nearby hills and mountains. These forests hold water and minerals, releasing them slowly to the plains below. If the forests are cut for short-term economic gain, water and soil will wash down the slopes in a torrent instead of a nourishing trickle. Thus, forests are valued not only for their short-term production of timber but for their role in protecting watersheds (Section 12-4).

Some Lessons from Ecology What can we learn from the brief overview of ecological principles presented in the past few chapters? It should be clear that ecology forces us to recognize five major features of all life: *interdependence*, *diversity*, *resilience*, *adaptability*, and *limits*. Its message is not that we should avoid change, but that we should recognize that human-induced changes can have far-reaching and often unpredictable consequences. Ecology is a call for wisdom, care, and restraint as we alter the ecosphere.

Table 6-2 Comparison of Properties of a Natural Ecosystem and a Simplified Human System	
Natural Ecosystem (marsh, grassland, forest)	Simplified Human System (cornfield, factory, house)
Captures, converts, and stores energy from the sun	Consumes energy from fossil or nuclear fuels
Produces oxygen and consumes carbon dioxide	Consumes oxygen and produces carbon dioxide from the burning of fossil fuels
Creates fertile soil	Depletes or covers fertile soil
Stores, purifies, and releases water gradually	Often uses and contaminates water and releases it rapidly
Provides wildlife habitats	Destroys some wildlife habitats
Filters and detoxifies pollutants and waste products free of charge	Produces pollutants and waste materials, much of which must be cleaned up at our expense
Usually capable of self-maintenance and self-renewal	Requires continual maintenance and renewal at great cost

What has gone wrong, probably, is that we have failed to see ourselves as part of a large and indivisible whole. For too long we have based our lives on a primitive feeling that our "God-given" role was to have "dominion over the fish of the sea and over the fowl of the air and over every living thing that moveth upon the earth." We have failed to understand that the earth does not belong to us, but we to the earth.

Rolf Edberg

Discussion Topics

1. Explain how organisms can change local conditions so that they become extinct in a given ecosystem. Could humans do this to themselves?

2. What does it mean to say that modern farming consists of keeping an ecosystem at an early stage of succession? Why is this necessary? What undesirable effects does this have?

3. Someone tells you not to worry about air pollution because the human species through natural selection will develop lungs that can detoxify pollutants. How would you reply?

4. What characteristics must a chemical have before it can be biologically magnified in a food chain or web?

5. What responsibility, if any, do you feel Hooker Chemicals should have for damages and cleanup costs resulting from the leakage of hazardous wastes from the Love Canal toxic waste dump? Defend your position.

6. There are complex arguments between two opposing groups in the United States, one favoring a freeze on the production of nuclear weapons and the other saying that this would put the nation at a strategic disadvantage and increase the risk of nuclear war. Use the library to research the claims of each group and determine your own position. Debate this crucial issue, including in your arguments the implications of the nuclear winter effect.

Population

We need that size of population in which human beings can fulfill their potentialities; in my opinion we are already overpopulated from that point of view, not just in places like India and China and Puerto Rico, but also in the United States and in Western Europe.

George Wald, Nobel laureate, biology

7

Human Population Dynamics

The present extended period of rapid population growth in the world is unique when seen from a long-range perspective; it has never occurred before and is unlikely to occur again.

Jonas and Jonathan Salk

For about 97 percent of the time since *Homo sapiens* appeared, the human population was less than 5 million. World population is now at a record 4.9 billion people and climbing (Figure 1-1). Based on present trends, United Nations population experts estimate that world population size will probably reach 6.1 billion by the year 2000 and 7.8 billion by 2020, not leveling off until 2110, at about 10.5 billion—more than twice the number of people on earth today.

What are the major factors affecting these dramatic changes in the size of the human population? How can the size and growth rate of the human population be controlled? The first question is discussed in this chapter and the second one in Chapter 8.

7-1 Major Factors Affecting Population Change

Five major factors affect the size and growth rate of the human population:

1. *Birth and death rates*. As long as the birth rate is greater than the death rate, population size will grow at a rate that depends on the difference between birth rate and death rate.

2. *Net migration rate.* If more people *immigrate* to than *emigrate* from a particular country, city, or area during a given period, the population of that area will grow at a rate that depends on the difference between the immigration rate and the emigration rate. This factor does not affect world population but does affect the size and rate of growth in various countries, cities,

and areas as people move from one place to another.

3. *Total fertility rate*. World population size can level off only when the average number of children the women in the world have during their reproductive years stays at or below an average replacement level of 2.1 children per woman for a considerable time.

4. *Age structure*. The time necessary for world population to stabilize after average total fertility rates have reached or remain below the replacement level will depend on the number or percentage of persons at each age level in the population. The larger the number and percentage of women in their reproductive years (15 to 44) and in their prereproductive years (under age 15), the longer it takes for population size to stabilize.

5. *Average marriage age or average age when first child is born.* Normally the later the average marriage age, the lower the average number of children a woman has between ages 15 and 44 (total fertility rate), and the sooner population size will stabilize.

In the remainder of this chapter we will look more closely at these five factors.

7-2 Birth Rate and Death Rate

Net Population Change   The difference between the total number of live births and the total number of deaths throughout the world during a given period of time gives the **net population change**.

$$\text{net population change} = \begin{array}{c} \text{number of} \\ \text{live births} \end{array} - \begin{array}{c} \text{number of} \\ \text{deaths} \end{array}$$

If there are more live births than deaths, population will increase. *Today there are about 2.5 births for each death, causing the world's population to increase by 139 people a minute, 8,833 an hour, 212,000 a day, and 77 million a year.*

Table 7-1 The World's 10 Most Populous Nations in 1985, with Projections for 2020

Nation	1985 Population (in millions)	Projected 2020 Population (in millions)	Increase (%)
China	1,042	1,288	24
India	762	1,246	64
Soviet Union	278	364	31
United States	239	297	24
Indonesia	168	293	74
Brazil	138	251	82
Japan	121	127	5
Bangladesh	102	207	103
Pakistan	96	196	98
Nigeria	91	258	184

Sources: Population Reference Bureau, *1985 World Population Data Sheet*. Washington, DC: Population Reference Bureau, Leon F. Bouvier, "Planet Earth 1984–2034: A Demographic Vision," *Population Bulletin*, vol. 39, no. 1, pp. 1–39 (1984).

Table 7-2 1985 Birth Rates, Death Rates, and Infant Mortality Rates in 10 Geographic Regions

Region	Birth Rate	Death Rate	Infant Mortality Rate
World	27	10	81
More developed nations	15	9	18
All less developed nations	31	11	90
Less developed nations, excluding China	36	12	101
Africa	45	16	110
Asia	28	10	87
Europe	13	10	15
Latin America	31	8	62
North America	15	8	10
Oceania	21	8	39

Source: Population Reference Bureau, *1985 World Population Data Sheet*. Washington, DC: Population Reference Bureau.

Words like "million" or "billion" often make little impression on us. But suppose you decide to take 1 second to say hello to each of the 77 million persons added during the past year. Working 24 hours a day, you would need about 2.4 years to greet them, and during that time more than 185 million additional babies would have been born. Table 7-1 shows the world's 10 most populous nations in 1985 with their projected population in 2020. Because six of the most populous nations are in Asia, it is not surprising that Asia is by far the most populous continent, with 2.8 billion persons or 58 percent of the world's population.

Rate of Natural Change Demographers, or population specialists, normally use the **birth rate** and **death rate** (also called *crude birth rate* and *crude death rate*) rather than total births and deaths to describe population change. These rates give the number of births and deaths per 1,000 persons in the population at the midpoint of a given year (July 1), since this should represent the average population for that year. The birth and death rates are calculated as follows:

$$\text{birth rate} = \frac{\text{births per year}}{\text{midyear population}} \times 1,000$$

$$\text{death rate} = \frac{\text{deaths per year}}{\text{midyear population}} \times 1,000$$

Table 7-2 shows the average birth rates, death rates, and infant mortality rates (the number of deaths of infants under 1 year of age per 1,000 live births in a given year) in various geographic areas. Notice that the death rates do not vary significantly between MDCs and LDCs, whereas there is a sharp difference between the birth rates and infant mortality rates between nations of these two types.

The difference between the birth and death rates is the **rate of natural change** (increase or decrease) during a given year.

$$\text{rate of natural change} = \text{birth rate} - \text{death rate}$$

In 1985 the world, with 27 births per 1,000 population and 10 deaths per 1,000 population, had a rate of natural increase of 17 persons per 1,000 population.

Percentage Growth Rates and Doubling Times Sometimes the rate of population growth is expressed as a percentage:

$$\begin{aligned}\text{percentage annual} \atop \text{growth rate} &= \frac{\text{birth rate} - \text{death rate}}{10} \\ &= \frac{27 - 10}{10} = \textbf{1.7\% in 1985}\end{aligned}$$

In the late 1970s, a series of newspaper headlines, such as "Population Time Bomb Fizzles," "Another Non-Crisis," and "Population Growth May Have Turned Historic Corner," falsely implied that world population growth had almost stopped. Actually, the annual percentage *population growth rate* peaked at about 2 percent in 1965 and then began a slow decline, to 1.9 percent by 1970 and to 1.7 percent in 1985. Despite this encouraging slowdown in the *annual population growth rate*, the *annual net population growth* increased from 69 million in 1969 to 77 million in 1985. It is projected that by 2000, the world's population growth rate will have declined to 1.5 percent, but annual net population growth will have increased to 92 million persons a year.

A population growth rate of 1 to 3 percent a year may seem relatively small. However, a population growing at only 1 percent a year will increase by 270 percent in 100 years; one growing by 2 percent a year increases 724 percent in 100 years; and one with a 3 percent annual growth rate increases by 1,922 percent over a century. For example, Nigeria, with a population of 91 million and a 3.1 percent growth rate in 1985, is projected to increase to 258 million by 2020 and eventually to 623 million, more people than now live on the entire continent of Africa.

Another indication of the rate at which a population is growing is called **doubling time**: the time it takes for a population to double in size. The *approximate* doubling time in years can be found by dividing the annual percentage growth rate into 70—using the *rule of 70*.

$$\frac{\text{doubling time}}{\text{(years)}} = \frac{70}{\text{annual percent growth rate}}$$

In 1985 the doubling time for the world's population was about 40 years (70/1.7 = 40). The doubling time for the MDCs in 1985 was 118 years, compared to 34 years for the LDCs including China and 29 years excluding China. Figure 7-1 shows the relationship between annual percent population growth and doubling times and gives the approximate doubling times for the population of selected countries throughout the world. Although doubling time gives a picture of *present* growth rates, it is at best a crude way of estimating *future* population size. This is because it assumes a constant growth rate over decades, whereas growth rates normally change over much shorter periods.

The rapid growth of the world's population over the past 100 years was not the result of a rise in birth rates; rather, it was due largely to a decline in death rates—especially in the LDCs, as shown in Figure 7-2. The interrelated reasons for this general decline in death rates include **(1)** an increase in food supplies because of improved agricultural production, **(2)** better food distribution due to improved transportation, **(3)** better nutrition, **(4)** reduction of diseases associated with crowding—such as tuberculosis—because of better housing, **(5)** improved personal hygiene, including the use of soap, which reduces

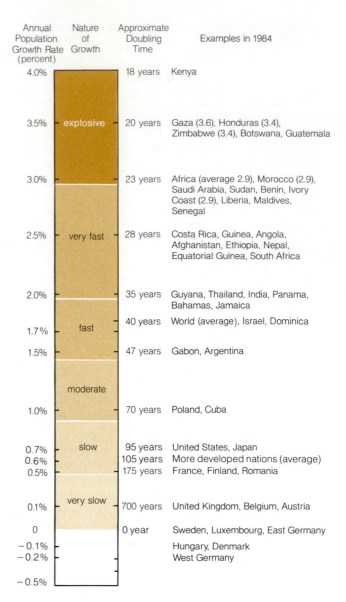

Figure 7-1 Relationship between population growth rate and doubling time (70 divided by annual percentage population growth).

the spread of disease, **(6)** improved sanitation and water supplies, which reduce death rates from plague, cholera, typhus, dysentery, diphtheria, and other fatal diseases, and **(7)** improvements in medical and public health technology through the use of antibiotics, immunization, and insecticides.

7-3 Migration

Net Migration Rate For the world, population growth occurs only when there are more births than deaths. For a county, city, or other area, however, we must also consider *immigration* and *emi-*

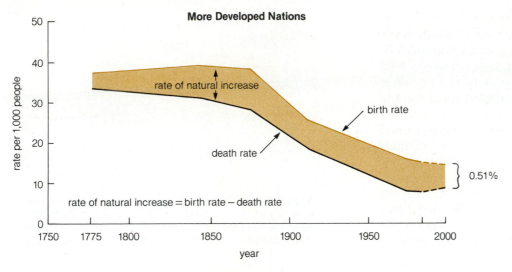

More Developed Nations

rate of natural increase

birth rate

death rate

0.51%

rate of natural increase = birth rate − death rate

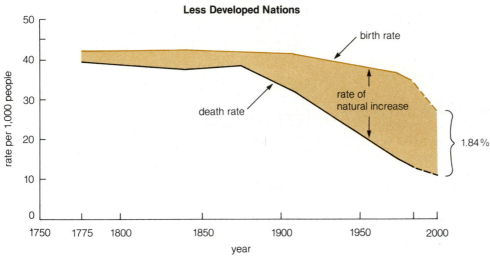

Less Developed Nations

birth rate

rate of natural increase

death rate

1.84%

Figure 7-2 Estimated birth and death rates and rates of natural population increase in more developed and less developed countries between 1775 and 1985 and projected rates (dashed lines) to 2000. (Sources: Population Reference Bureau, Annual *World Population Data Sheets*; United Nations medium population projections.)

gration as people migrate from one place to another. For a given place on earth, the **net migration rate** per year is calculated as follows:

$$\text{net migration rate} = \frac{\dfrac{\text{immigrants}}{\text{per year}} - \dfrac{\text{emigrants}}{\text{per year}}}{\text{midyear population}} \times 1{,}000$$

If more persons immigrate than emigrate, the net migration rate is positive. If more leave than enter, it is negative.

The **rate of population change** per year for a given country, city, or area is the difference between its birth rate and death rate plus its net migration rate.

$$\text{rate of population change} = \frac{\text{birth}}{\text{rate}} - \frac{\text{death}}{\text{rate}} + \frac{\text{net migration}}{\text{rate}}$$

U.S. Migration Rates The United States was founded by immigrants and their children.

Throughout its history the United States has admitted a larger number of immigrants and refugees than any other country in the world. Indeed, this total number is almost twice as large as that received by all other nations combined.

In 1985 an estimated 1.3 million people entered the United States as legal or illegal immigrants and refugees, and about 100,000 people emigrated from the United States. Thus, the estimated net migration rate for the United States in 1985 was

$$\text{net migration rate} = \frac{\dfrac{\text{immigrants}}{\text{per year}} - \dfrac{\text{emigrants}}{\text{per year}}}{\text{midyear population}} \times 1{,}000$$

$$= \frac{1{,}300{,}000 - 100{,}000}{238{,}900{,}000} \times 1{,}000$$

$$= \textbf{5 per 1,000 population}$$

Population Growth in the United States Figure 7-3 shows the variation in birth rates, death rates, and rate of natural increase in population (excluding migration) for the United States between 1900 and 1985. Although the rate of natural population increase has declined since 1947 (except for the slight upturn since 1976), the population continues to grow. This is because the birth rate has remained considerably larger than the death rate. In mid-1985 the U.S. population was about 239 million.

In 1985 the birth rate per 1,000 persons in the United States was 16, the death rate 9, and the net immigration rate (including 1.3 million legal and illegal immigrants and refugees) was 5 per 1,000 persons. Thus, the rate of population change was

$$\frac{\text{rate of population}}{\text{change}} = \frac{(\text{birth rate} - \text{death rate})}{+ \text{ net immigration rate}}$$

$$\text{rate of population change} = (16 - 9) + 5$$
$$= \textbf{12 per 1,000 persons}$$

Including migration, the percent annual growth rate in the United States in 1985 was 1.2 percent, with a doubling time of 58 years compared to 100 years when net migration is not included. Thus *with migration included, the U.S. population grew by an average of 5 people a minute, 321 an hour, 7,700 a day, and 2.8 million a year during 1985.*

7-4 Fertility Rate

Total Fertility Rate To improve their ability to understand and project population changes, demographers use the **total fertility rate (TFR)**, a *projection* of the average number of live children that would be born to each woman if she were to live through her childbearing lifetime (ages 15 to 44), bearing children at the same rate as all other women did in each of those years. The TFR gives us the best idea of the average number of live births per woman for the women who survive to the end of their childbearing period (i.e., age 44).

Logic suggests that two children would replace two parents. The actual average number of children needed for replacement, however, is slightly higher. In MDCs the *replacement level* is 2.1 children per woman and in the LDCs about 2.7. Both these numbers are greater than 2 for two major reasons: **(1)** some female children die before reaching their reproductive years (especially in LDCs), and **(2)** there is a slightly higher percentage of male children born than female children.

In 1985 the average total fertility rate in the

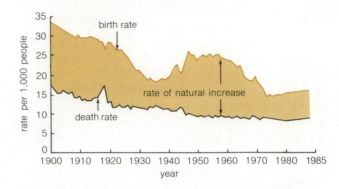

Figure 7-3 Birth and death rates and the rate of natural population increase (migration excluded) in the United States between 1900 and 1985. (Source: Population Reference Bureau, Annual *World Population Data Sheets*.)

Table 7-3 Average 1985 Total Fertility Rates for 10 Geographical Regions	
Region	Average Total Fertility Rate
World	3.7
More developed nations	2.0
All less developed nations	4.2
Less developed nations, excluding China	5.0
Africa	6.3
Asia	3.7
Europe	1.8
Latin America	4.2
North America	1.8
Oceania	2.7

Source: Population Reference Bureau, *1985 World Population Data Sheet*. Washington, DC: Population Reference Bureau.

world was 3.7 children per woman. The average rates were 4.2 in the LDCs including China and 5.0 excluding China, and 2.0 in the MDCs (Table 7-3). This shows clearly why the world has a long way to go before its population size will level off and begin to decline. For example, Bangladesh, with a population of 102 million and a TFR of 6.4 in 1985, could have a population of 207 million by 2020—all squeezed into an area the size of Wisconsin.

Possibilities for World Population Stabilization By 1985 a dozen European nations **(1)** *were close to zero population growth* (Great Britain, Norway, Belgium, Switzerland, East Germany, and Italy), **(2)** *had achieved ZPG* (Austria, Luxembourg, and Sweden), or **(3)** *were experiencing decreases in population size because death rates were higher than birth rates* (Denmark, West Germany, and Hungary). Although these nations represent only 5.2 percent of the world's population, this is a step toward eventual stabilization of world population. If present trends continue, several other more developed European nations should reach ZPG by 1990 and then enter a period of population decrease.

Achieving ZPG in the world, however, is more difficult because the LDCs have a much higher average TFR (Table 7-3). In 1982 the United Nations made three projections of when world population size would reach its peak level (ZPG), followed by a slow decrease in population size. According to the medium projection, *if the global average total fertility rate declines to 2.5 by 2025 and is maintained, world population should stabilize around 10.5 billion about 85 years later (in 2110).* If fertility decreases at a faster pace, with a TFR of 2.5 reached around 2005, world population would stop growing much sooner (in 2040), with a peak size of 8 billion. If a TFR of 2.5 is not reached until 2045, world population would not level off until 2130, with a peak population size of 14.2 billion. Table 7-4 shows the projected stable population size and year of stabilization for different geographical regions using the medium UN projection.

Of course, no one knows whether any of these projections will prove to be accurate. All are based on the assumption that there will be adequate supplies of food, energy, and other natural resources. If such supplies are not adequate or if there is global nuclear war, population size could stabilize or be greatly reduced by a sharp increase in death rates.

Possibilities for U.S. Population Stabilization

Figure 7-4 shows that the total fertility rate in the United States has oscillated wildly. Note that the

Table 7-4 United Nations Medium Projections for Stable Populations in 10 Geographical Regions

Region	Stable Population (in millions)	Projected Year of Stabilization
World	10,529	2110
More developed nations	1,390	2080
Less developed nations	9,139	2110
Africa	2,193	2110
East Asia	1,725	2090
South Asia	4,145	2100
Europe	540	2030
Latin America	1,187	2100
North America	318	2060
Oceania	41	2070

Source: United Nations Fund for Population Activities, *State of World Population: 1982.* New York: United Nations Fund for Population Activities.

peak of the post–World War II baby boom (1947–1964) occurred in 1957, when the average TFR reached 3.7 children per woman. This was followed by a sharp drop to an average of 1.77 children per woman by 1976, with a slight rise to 1.84 between 1976 and 1985 (Figure 7-4)—still considerably below the 2.1 replacement level cited in Section 7-1 (item 3). Since 1980 there has been an increase in the birth rate of women in their late 20s and 30s. This delayed childbearing, however, has caused only a small rise in the overall birth rate and total fertility rate.

Figure 7-4 also shows why birth rates dropped sharply between 1957 and 1976. The average American woman simply had fewer children during her reproductive years. This drop was probably caused by a mixture of factors, including reduction in the number of unwanted and mistimed births because of the widespread use of effective birth control methods, the availability of legal abortions, changed motivation in favor of smaller families, rising costs of raising a family ($145,000 to raise one child born in 1985 to age 18), greater social acceptance of childless couples, and increasing numbers of women working outside the home. For example, the birth rate among working women is only one-third the rate of those not in the labor force.

In the late 1970s some newspapers incorrectly reported that the United States had reached ZPG because the total fertility rate had fallen below the replacement level of 2.1. *However, reaching a total fertility rate at or even below replacement level does not mean that zero growth of the population has been reached.* Zero population growth occurs in a country or region when, over time, the number of births plus immigrants equals the the number of deaths plus emigrants. The United States is a long way from ZPG in spite of the dramatic drop in the TFR. This is due primarily to the age structure of the population, as discussed in the next section.

No one knows whether or how long U.S. total fertility rates will remain below replacement level. The U.S. Bureau of the Census has made several projections for population growth assuming various TFRs, life expectancies, and net immigration rates. According to the medium projection, usually taken to be the best projection, U.S. population could reach 268 million in the year 2000, peak at 309 million by 2050, and then begin to decrease, assuming an average of 1.9 live births per woman and an annual net migration rate of 450,000 persons.

Groups such as The Environmental Fund and Zero Population Growth, however, have questioned some of the assumptions on which the cen-

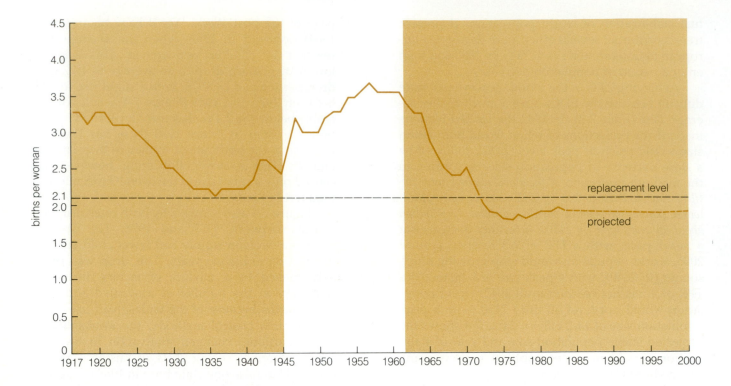

Figure 7-4 Total fertility rate for the United States between 1917 and 1985 and projected rate (dashed line) to 2000; shaded area shows the peak years of the baby boom. (Source: Population Reference Bureau, 1982, *U.S. Population: Where We Are; Where We're Going*; 1982 U.S. Bureau of Census medium population projections.)

sus bureau based its projections. First, they argue that the net migration rate is more likely to be 750,000 to 1,000,000 people per year than 450,000 people per year, as assumed in the medium projection. Second, they question the assumption that the current fertility rate, the lowest in history, will continue. They suggest that this rate may rise for a variety of reasons, some of which reflect important changes in the ethnic and racial composition of the population. The proportion of Hispanic and black women in the population is increasing, and historically these groups have had fertility rates higher than the average. Some analysts project that by 1990 Hispanics will surpass blacks as the biggest minority group in the United States.

7-5 Age Structure

Number of Women of Childbearing and Younger Ages Why will world population keep growing for at least 100 or more years (assuming death rates don't rise), even after replacement TFRs have been reached? Why do some demographers expect the U.S. birth rate to rise between now and 1993, even though the TFR may drop or stay at low levels?

To answer these questions we must consider a fourth factor in population dynamics, namely, the **age structure**, or *age distribution*, of a population—the number or percentage of persons at each age level in a population. A major factor in population growth is the number or percentage of women of childbearing age (especially the number in the prime reproductive years of ages 20 to 29) and the number or percentage worldwide below age 15 who will soon be entering their prime childbearing years. If a large number of women are of or near childbearing age, births can rise even though women on the average have fewer children. Thus, *any population with a large number of people below age 29, and especially below age 15, will have a powerful built-in momentum to maintain population growth.*

In 1985 about 35 percent of the people on this planet were under 15 years of age. The presence of these young people explains why population will continue to grow—especially in LDCs—70 to 100 years after replacement level total fertility rates are reached, unless death rates rise sharply. Table 7-5 shows the percentage of the population under age 15 and over age 64 in various geographic regions. Notice that in the LDCs an average of 39 percent of the population is under 15, whereas in the MDCs this figure is only 23 percent.

Table 7-5 also shows a significant difference between the percentage of people above age 64 in the

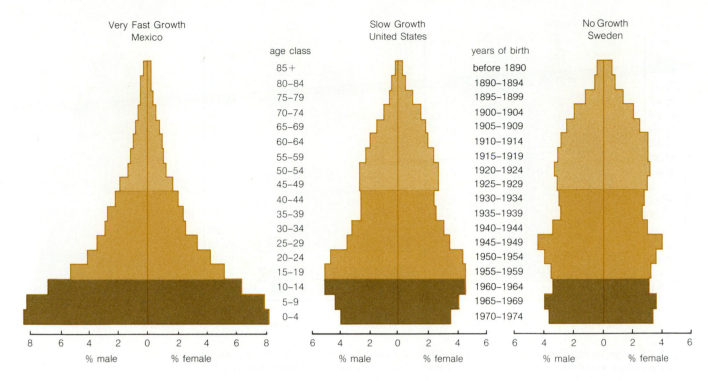

Figure 7-5 Population age structure diagrams for countries with rapid, slow, and zero population growth rates. Dark portions represent preproductive periods (0 to 14), shaded portions represent reproductive years (15 to 44), and clear portions are postproductive years (45 to 85 +). (Source: Population Reference Bureau.)

MDCs and the LDCs. The UN projects that by 2025 there will be almost 1 billion people over 64 who will need to be cared for, with relatively fewer people of working age to care for them. This dramatic shift in population age structure by 2025 is projected because of a combination of fewer children per woman and longer lives. Whereas world birth rates are expected to be halved between 1950 and 2025, average life expectancy is expected to rise from 47 to 70. In the LDCs this could mean more old people left to fend for themselves.

Population age structure also explains why it will probably take at least 50 years for the United States to reach ZPG, even if the present historically low TFRs are maintained at 1.7 to 2.0 births per woman (Figure 7-4). *Even though many U.S. couples are now having smaller families, the number of births could easily rise during the 1980s—not because each woman will have more babies, but because there are more women to have babies.* The 37 million women born during the baby-boom period will affect U.S. population growth through 1993. This year is when women born in 1963, at the end of the baby boom, will reach age 30, thus moving out of their prime reproductive years.

Age Structure Diagrams We can obtain an age structure diagram for the world or for a given country or region by plotting the percentages of males and females in the total population in three age categories: *preproductive* (ages 0 to 14), *reproductive* (ages 15 to 44, with prime reproductive ages 20 to 29), and *postproductive* (ages 45 to 85 +). Figure 7-5 compares the age structure for countries with rapid, slow, and zero population growth rates.

	Population Under Age 15 (%)	Population Over Age 64 (%)
Table 7-5 Percentage of Population Under Age 15 and Over Age 64 for 10 Geographical Regions in 1985		
Region		
World	35	6
More developed nations	23	12
All less developed nations	39	4
Less developed nations, excluding China	41	4
Africa	47	3
Asia	37	4
Europe	22	13
Latin America	38	5
North America	22	12
Oceania	29	8

Source: Population Reference Bureau, *1985 World Population Data Sheet*. Washington, DC: Population Reference Bureau.

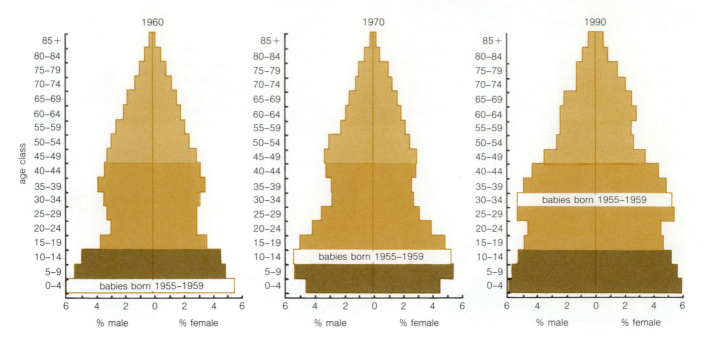

Figure 7-6 Age structure of the U.S. population in 1960, 1970, and 1990 (projected). The population bulge of babies born between 1955 and 1959 will slowly move up. (Source: Population Reference Bureau.)

Making Projections from Age Structure Diagrams Figure 7-6 shows that the baby boom in the United States caused a bulge in the age structure that will move through the prime reproductive ages of 20 to 29 between 1970 and 1993 and through older age groups in later years. This helps explain why the 1960s and 1970s have been called the "generation of youth." Similarly, the period between 1975 and 1990 could be called the "age of young adults," that between 1990 and 2010 the "age of middle-aged adults," and that between 2010 and 2030 the "age of senior citizens."

A diagram like Figure 7-6 can be used to project some of the social and economic changes that may occur in the United States in coming decades. In the 1970s and early 1980s large numbers of baby-boom adults flooded the job market, causing high unemployment rates for teenagers and adults under 29. This situation won't begin to ease until after 1992, when the last of the people born during the baby boom turn 29. For those with jobs, advancement is likely to be much slower than in the present generation, and there will be more competition for upper level jobs. It is estimated that between 1980 and 1990, nearly one-quarter of the baby-boom college graduates will be overeducated for the jobs they will be able to get—as secretaries, store clerks, cab drivers, and factory workers.

In the 1990s the large baby boom generation will be settling into middle age (ages 35 to 64). Many of these adults may face relatively little op-

portunity for professional advancement unless they somehow force those aged 45 to 59 to retire. We should also see more young leaders in government, politics, and private industry.

Assuming that death rates don't rise, the number of people over age 64 will go from 29 million (12 percent of the population) in 1985 to 65 million (21 percent of the population) by 2030. The burden of supporting so many retired baby boomers will fall on the shoulders of the much smaller group of the contemporary work force—people born in the 1960s, 1970s, and 1980s, primarily aged 20 to 65, who traditionally support children and retired people. This increase in retired citizens could put a severe strain on Social Security and Medicare.

The latest "baby-bust" generation, born in the 1970s and early 1980s when total fertility rates were low, should have a much easier time than the preceding baby-boom generation. Few will be competing for education, jobs, and services. Labor shortages should drive up wages for the baby-bust generation, and the United States could end up relaxing current immigration laws to bring in new workers. With a shortage of young adults, the armed forces will be hard pressed to meet recruiting levels, and the draft may be reinstated. Those in this baby-bust group, however, may find it hard to get job promotions as they reach middle age because most upper level positions will be occupied by the baby-boom group.

From these few projections, we see that any

Table 7-6 Major Characteristics of Less Developed and More Developed Countries in 1985

Less Developed Countries	More Developed Countries
High birth rates (25–50 births per 1,000 population, average 31, or 36 excluding China)	Low birth rates (10–20 births per 1,000 population, average 15)
Low to high death rates (9–25 deaths per 1,000 population, average 11, or 12 excluding China)	Low death rates (9–11 deaths per 1,000 population, average 9)
Low to fairly high average life expectancy (average 58 years, or 56 excluding China)	High average life expectancy (average 73 years)
Rapid population growth (average 2.0%, or 2.4% excluding China)	Slow population growth (average 0.6%)
Large fraction of population under 15 years old (average 39%, or 41% excluding China)	Moderate fraction of population under 15 years old (average 23%)
Moderate to high infant mortality rate (40–200 deaths of infants under 1 year old per 1,000 live births, average 90, or 101 excluding China)	Low infant mortality (8–20 deaths of infants under 1 year old per 1,000 live births, average 18)
Moderate to high total fertility rate (average 4.2 children per woman, or 5.0 excluding China)	Low total fertility rate (average 2.0 children per woman)
Low to moderate per capita daily food supply (1,500–2,700 calories per person per day)	High per capita daily food supply (3,100–3,500 calories per person per day)
High illiteracy level (25–75%)	Low illiteracy level (1–4%)
Mainly rural, farming population (68% rural, 32% urban)	Mainly urban, industrial population (72% urban, 28% rural)
Low per capita energy use (average 3 million kilocalories per person per year)	High per capita energy use (average 30 million kilocalories per person per year)
Low to moderate average per capita income (widespread poverty: $90–$3,000 per person per year, average $700, or $880 excluding China)	High average per capita income (widespread affluence: $3,000–$25,600 per person per year, average $9,380)

bulge or indentation in the age structure of a population creates a number of social and economic changes that ripple through a society for decades.

7-6 Average Marriage Age

Another factor affecting population size and rate of change is the *average marriage age of women*, or, more precisely, the *average age at which women give birth to their first child*. Because older brides on the average have fewer children, an increase in average marriage age generally leads to a decrease in the total fertility rate.

Although data are lacking for a number of countries, average marriage ages are around 19 in Africa, 21 in Asia and Latin America, 22 in Oceania, and 23 in Europe. Recent studies have shown that to lower fertility, the average marriage age would have to rise to at least 25.

Since 1955 the average marriage age in the United States has been increasing and in 1984 was 25.4 for men and 23 for women. This narrows the average childbearing period for American women from ages 15 to 44 to ages 23 to 44. Even more important, the prime reproductive period is almost cut in half, from ages 15 to 29 to ages 23 to 29.

7-7 The Geography of Population: The Rich–Poor Gap

The world is polarized into two major groups: one rich and one poor; one literate, the other largely illiterate; one overfed and overweight, one hungry and malnourished; one with a moderate to low rate of population growth, the other with a very high rate. Table 7-6 compares LDCs and MDCs. In this context, the terms *less developed nations, developing nations, poorer nations, Third World nations,* and *South* describe nations with nonindustrial economies and average per capita gross national products (GNP) below $3,000 per year.

The term *rich–poor gap* dramatizes the differences between nations, but it is an oversimplification. That a country is classified as less developed or poor does not always mean that living conditions are hopeless. Many people in LDCs are well fed and live reasonably comfortable lives. At the same time, these regions contain at least 800 million people in dire poverty, characterized by malnutrition, disease, illiteracy, squalor, high infant mortality rates, and short life expectancy. In Africa and Asia, almost half the population exists on an average per capita gross national product of less than $300 per person. Up to 80 percent of this amount is spent on food.

Some members of affluent societies attribute the large families of the poor in LDCs to ignorance. However, from the parents' viewpoint, personal survival depends on having six, seven, or more children—especially boys—to help beg or work in the fields and to help them in their old age. One poor couple in India would have to bear an average of 6.3 children to have a 95 percent chance of keeping one son alive to adulthood. For those living near the edge of survival, having many children may cause problems, but not having enough can cause death at an early age. In the next chapter we will look at possibilities and methods for bringing world population growth under control.

Not to decide is to decide.

Harvey Cox

Discussion Topics

1. Why are falling birth rates not necessarily a reliable indicator of future population growth trends?

2. Why do the deaths of millions of human beings by starvation and famine usually make little impression on many of us, while the deaths of individual human beings by murder, drowning, or being trapped in a coal mine receive nationwide attention and sympathy? Analyze this response in terms of dealing with world population growth.

3. Suppose modern medicine finds cures for cancer and heart disease. What effects would this have on population growth in MDCs and LDCs? On age structure? On social problems?

4. Explain the difference between achieving replacement level and ZPG. Why is the replacement level in LDCs higher than in MDCs?

5. What must happen to the total fertility rate if the United States is to attain ZPG in 40 to 60 years? Why will it take so long?

6. Explain how the U.S. population has the potential to grow rapidly again through 1993.

7. Project what your own life may be like at ages 25, 45, and 65 on the basis of the present age structure of the U.S. population or that of the country in which you live. What changes, if any, do such projections make in your career choice and in your plans for marriage and children?

8. Explain why raising the average first marriage age to 25 or higher is an effective means of reducing population growth rates.

9. Criticize each of the following headlines or statements. Be specific.
   a. "Baby Boom Replaced by Bust—U.S. in Danger of Instant ZPG."
   b. "Birth Rates Falling—Prophets of Doom Wrong Again."

10. Explain why it may be rational for a couple in India to have six or seven children. What possible changes might induce such a couple to think of their behavior as irrational?

11. Do you think the world is more likely to reach the high (14.2 billion), medium (10.5 billion), or low (8 billion) population size projected in 1982 by the United Nations? Explain.

8

Human Population Control

Reproduction is a private act, but it is not a private affair. It has far-reaching social consequences.

Lincoln Day

What factors affect the maximum human population size? Should population growth in the world and in the United States be controlled? What methods are available for controlling the size of the human population? These questions and some possible answers are the subjects of this chapter. Let's begin with a discussion of the factors that can limit the population of any species and then relate these ideas to the human population.

8-1 Factors Affecting Maximum Population Size

J Curves and S Curves The population story for a species usually can be told with two very simple curves—a *J-shaped curve* and an *S-shaped curve*. With unlimited resources and ideal environmental conditions, a species can produce offspring at its maximum rate, called its **biotic potential**. Such growth starts slowly and increases rapidly to produce an exponential or J-shaped curve of population growth (Figure 8-1). Bacteria, insects, mice, and rodents have high biotic potentials; larger species, such as elephants and humans, have relatively low biotic potentials. Since environmental conditions usually are not ideal, a population rarely reproduces at its biotic potential.

The population size of a particular species in a given ecosystem is limited by availability of one or more resources that can act as *limiting factors* (Section 4-6). The maximum population size an ecosystem can support *indefinitely* under a given set of environmental conditions is called the ecosystem's *carrying capacity*. Thus, in an ecosystem with finite resources, a J-shaped population growth curve for a particular species cannot continue forever.

All the limiting factors that act to reduce the growth rate of a population are called the population's **environmental resistance**. As a population encounters environmental resistance, the J-shaped curve of population growth bends away from its steep incline and eventually levels off at a size that typically fluctuates above and below the carrying capacity (Figure 8-l). In other words, environmental resistance opposes the biotic potential of a population and converts a J-shaped curve into a sigmoid, or S-shaped curve.

Some populations (especially rapidly reproducing ones such as insects, bacteria, and algae) may overshoot the carrying capacity and then undergo a rapid decrease in population size, known as a **population crash** (Figure 8-2). Some population crashes involve sharp increases in the death rate; others can involve combinations of a rise in the death rate coupled with migration of large numbers of individuals to other areas. A population crash can also occur when a change in environmental conditions lowers the carrying capacity level for a population.

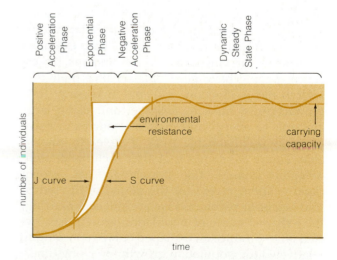

Figure 8-1 The J-shaped curve of population growth is converted to an S-shaped curve when a population encounters environmental resistance caused by one or more limiting factors.

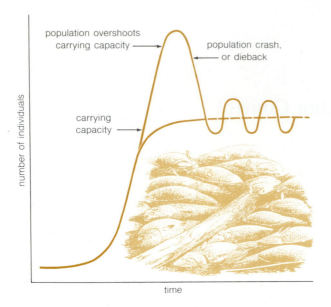

number of individuals

population overshoots carrying capacity →

← carrying capacity

population crash, or dieback

time

Figure 8-2 A population crash, or dieback, can occur either when a population temporarily overshoots its carrying capacity or when a change in environmental conditions lowers the carrying capacity.

Such a population crash occurred as a result of the potato famine in Ireland around 1845. Since the 1700s the Irish people had depended heavily on the potato for a major portion of their diet, but in 1845 a blight (a fungus infection) destroyed the potato crop that fed most of Ireland's 8 million people. Between 1845 and 1850, 1 million people starved to death in Ireland and another million moved to other countries. These emigrants were followed by others, and by 1900 the Irish population was about half the size it had been in 1845.

Projections for the Future The human population has continued to grow in size because human cleverness, technological and social adaptations, and other forms of *cultural evolution* have extended the earth's carrying capacity for humans. In essence, people have been able to alter their ecological niche (Section 5-4) by increasing food production, controlling disease, and using large amounts of energy and matter resources to make normally uninhabitable areas of the earth inhabitable. Figure 8-3 shows that after each major technological change the population has grown rapidly. Then, except for the period of the industrial–scientific revolution we are still experiencing, the J-shaped curve of population growth has leveled off to produce an S-shaped curve representing relatively little population growth for a long time.

The dashed lines in Figure 8-3 show three projections for the human population based on our present understanding. No one, of course, knows

what the present or future *carrying capacity* for the human population is or what the limiting factor or factors might be—food, air, water, pollution, or lack of space.

Some observers believe we have already gone beyond the carrying capacity point at which all the earth's inhabitants can be adequately fed, sheltered, and supported. H. R. Hulett of Stanford University estimates that present world food production could feed only 1.2 billion people (one-fourth of the present world population) based on U.S. dietary standards, and only 600 million (one-eighth of the present world population) at the U.S. rate of energy consumption. Some argue, however, that U.S. rates of food, energy, and resource consumption are wasteful and unnecessarily high.

At the other end of the spectrum, agricultural economist Colin Clark sees a world in which 45 billion people could have a U.S. type of diet by cultivating all arable land, using nuclear power for energy, and mining much of the earth's crust to a depth of 1.6 kilometers (1 mile). If the human diet were based on grain rather than a mixture of grain and meat, he estimates the earth could support 157 billion people. Bernard Gilland labels such a vision a "nightmare to be avoided at all costs" and disputes such estimates of potentially arable land.

A second possibility shown in Figure 8-3 is continued growth past the carrying capacity followed by a dieback of billions—perhaps 50 to 80 percent of the world's population—through a combination of famine, disease, nuclear war, and widespread ecological disruption. Some think that this, unfortunately, may be the most likely path to ZPG.

8-2 Should Population Growth Be Controlled?

World Population Most nations in the world either favor stabilizing their population size or at least hope to reduce their rate of population growth. The problems of peace, poverty, racism, disease, pollution, urbanization, ecosystem simplification, and resource depletion won't be solved by population stabilization. But without it each of these problems is very likely to become much worse.

Economist Julian Simon, however, does not favor stabilizing the world population. He argues that **(1)** people are the world's most vital resource; **(2)** the problems of resource depletion and pollution can be solved by human ingenuity and technology so that the more people we have, the more likely it is that these problems will be solved; and

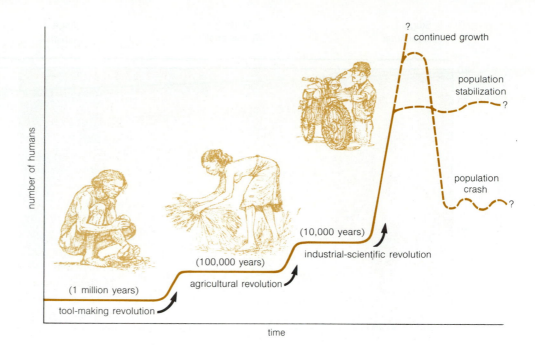

Figure 8-3 So far humans have extended the earth's carrying capacity by technological innovation. After each major technological revolution, the population has grown rapidly and then, except for the ongoing industrial–scientific revolution period, leveled off for a long time. Dashed lines represent different projections for the human population: continued growth, population stabilization, and continued growth, followed by a population crash and population stabilization at a much lower level. All curves are suggestive and are not drawn to scale.

number of humans

continued growth

population stabilization

population crash

(1 million years)

tool-making revolution

(100,000 years)

agricultural revolution

(10,000 years)

industrial-scientific revolution

time

(3) continued economic growth is not threatened but enhanced by population growth because more people generate more production.

Critics counter that most people are being added in LDCs, where education, health, and nutrition levels are so low that continued rapid population growth can condemn hundreds of millions of people to an early death. Technological advances do not necessarily come only from people who are well educated or well paid. But encouraging significant increases in births in the *hope* that in the long run a few of these individuals will solve the world's pollution and resource problems is a very risky proposition. Critics argue that this hardly seems like a humane way to preserve and improve the lives of people who already exist. Instead of blindly encouraging births, they suggest that providing better education, nutrition, health care, and work opportunities for a smaller population has a better chance of stimulating human ingenuity and technological breakthroughs and preventing human misery. They also point out that poor people are not able to buy many, if any, products and thus do not stimulate production. Instead the services they require constitute an economic drain on government and private capital that could be used to stimulate production and technological breakthroughs.

U.S. Population At present the United States has no official population control policy, primarily because population control is a controversial issue. At the 1984 United Nations International Conference on Population the United States stated that its official position is that population growth is neither good nor bad. The United States does provide financial support for national and international family planning and for contraceptive and reproductive research. However, there are attempts by antiabortion groups to cut off these funds, and in 1985 the United States refused to provide funds for any family planning or population control programs involving abortion.

Some observers argue that the United States should increase its population to maintain economic growth and power throughout the world, pointing out that there are populous nations that are not powerful but no powerful nations that are not populous. Other observers disagree. They point out that Japan, with a relatively small population, has become one of the world's leading industrial nations since 1945. It is also argued that continued population growth in the United States is desirable because a ZPG society would be more conservative and less innovative. However, such fears may be unfounded. Sweden has achieved ZPG and is one of the world's more energetic and innovative societies. Furthermore, the more conservative societies tend to be LDCs with a youthful age structure and a mostly rural population. South Africa and Portugal, for example, have young populations, but are more conservative than the older populations of Sweden and West Germany.

Other analysts argue that a good reason for stabilizing population is that adding millions to the U.S. population intensifies many environmental and social problems because increased resource use and pollution are caused by the combination of population growth and increased affluence. An

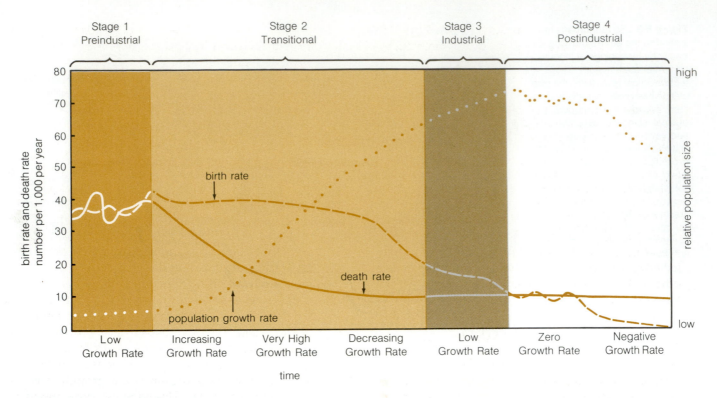

Figure 8-4 A generalized model of the demographic transition.

official goal of stabilizing U.S. population also would set an example for other nations to do likewise.

8-3 Methods for Controlling Human Population Growth

Controlling Birth Rates A government can alter the size and growth rate of its population by encouraging a change in any of the three basic demographic variables: births, deaths, and migration. All governments are presumably committed to reducing death rates as much as possible. In addition, most governments restrict emigration and immigration, so migration is a significant factor in only a few countries, such as the United States, Canada, and Australia. Thus, *controlling the birth rate is the primary way of controlling human population growth in the world.*

There are two general approaches to controlling the birth rate of the human population: *economic development* and *family planning*. Family planning enables people to have no more than the number of children they desire, whereas economic development changes their motivation for a certain desired number of children. Let's look at these two approaches more closely.

Economic Development and the Demographic Transition Between 1929 and 1950, demographers began to examine the birth and death rates in western European nations that became industrialized during the nineteenth century. From this analysis they formed a model of population change and control known as the **demographic transition**, based on the hypothesis that population growth decreases with economic development.

Analyses of industrialized western European nations suggest that the demographic transition has four distinct phases (Figure 8-4). In the first, or *preindustrial stage*, harsh living conditions lead to high birth rates (to compensate for high infant mortality) and high death rates, with the population stable or growing slowly. In the second, or *transitional stage*, as industrialization spreads and living conditions improve, birth rates remain high but death rates fall sharply. This leads to a temporary but prolonged period of rapid population growth. In the third, or *industrial stage*, industrialization is widespread. Birth rates fall and approach death rates, as better educated and more affluent couples (especially in cities) become aware that numerous children hinder them from taking advantage of job opportunities in an expanding economy. Population growth continues but at a slower and perhaps fluctuating rate, depending on economic conditions. According to this hypothesis, the United States, Japan, the Soviet Union, Canada, Australia, New Zealand, and most of the industrialized western European nations are in the third phase.

A fourth *postindustrial stage* takes place when birth rates decline even further to equal death rates (ZPG) and then possibly continue to fall, so that total population size begins slowly to decrease (Figure 8-4). West Germany has entered the fourth stage, if this model is valid. If West Germany maintains its present low fertility rate of 1.3 and does not allow immigration, its population will decrease from 61 million in 1985 to 40 million by 2050. West German leaders are concerned about this projected population decrease and would like to see birth rates rise again. They fear that there will be too few workers to support continued economic growth and to pay taxes to support the increasing fraction of its population over age 64.

According to the demographic transition hypothesis, the LDCs of Africa, Asia, and Latin America should be able to reduce their population growth and become MDCs if they receive enough economic aid and technical assistance from today's MDCs. However, some population analysts argue that many of today's LDCs could become caught in the transitional stage (Figure 8-4), or if death rates rose because of food shortages, they could be forced back to the preindustrial stage of the demographic transition. First, new evidence suggests that improved and expanded family planning programs may bring about a more rapid decline in the birth rate at a lower cost than economic development alone. Second, economic development will be more difficult for today's LDCs than it was for the nations that developed a century ago for at least four reasons:

1. Even with a large and growing population, they do not have enough trained workers for economic development.

2. Most low- and middle-income nations lack the capital and resources needed for development, and the amount of money being given or loaned to LDCs—already struggling under tremendous debt burdens—has been decreasing.

3. If future world economic growth averages closer to 2 percent than 4 percent, as it has in the early 1980s, half the world's population will soon live in countries where population growth could exceed economic growth.

4. Today's LDCs face stiff competition from MDCs and recently modernized LDCs in selling the products on which their growth depends.

Family Planning Basic **family planning** is a purely voluntary approach whereby information is provided and contraceptives are distributed to help couples have the number of children they want, when they want them. Family planning services were first introduced in LDCs in the 1940s and 1950s by private physicians and groups of self-motivated women. Since that time, organizations such as the International Planned Parenthood Federation, the Planned Parenthood Federation of America, the United Nations Fund for Population Activities (UNFPA), the U.S. Agency for International Development (AID), the Ford Foundation, and the World Bank have been helping nations carry out voluntary family planning by providing technical assistance, funding, or both.

Family planning depends primarily on helping couples use various forms of birth control. Table 8-1 lists the typical effectiveness of these methods in the United States. By 1984 the most widely used form of birth control in the world and in the United States was sterilization, and surveys show that its use would increase sharply if it were more readily available worldwide. The second most popular form of birth control was oral contraceptives, followed by intrauterine devices (IUDs), and then condoms. Researchers throughout the world are at work trying to develop new and better methods of fertility control including: (1) a morning-after pill, (2) long-lasting injections, pills, and hormone implants, (3) an antipregnancy vaccine, (4) male contraceptives, (5) a procedure for reversible vasectomies, and (6) a means of influencing brain chemistry to trigger the release of the hormones that suppress ovulation in females and sperm formation in males.

Between 1965 and 1985, family planning was claimed to be a major factor in reducing birth and fertility rates in China, Mexico, and Indonesia, and in some small LDCs, such as Singapore, Hong Kong, Barbados, Taiwan, Cuba, Mauritius, Thailand, Colombia, Costa Rica, South Korea, Fiji, and Jamaica.

However, only moderate to poor results have been claimed in other populous LDCs like India, Brazil, Bangladesh, Pakistan, and Nigeria, and in 79 less populous LDCs—especially in much of Africa and Latin America, where population growth rates are usually very high.

Despite the importance of family planning, its official goal is not to lower the birth rate, but to help couples have the number of children they want, when they want them. Thus, some experts argue that family planning, even coupled with economic development, cannot bring birth and fertility rates down fast enough because surveys show that most couples

Table 8-1 Typical Effectiveness of Birth Control Methods Used in the United States

Method	Typical Effectiveness (%)
Extremely Effective	
Abortion	100
Sterilization	
Tubal ligation and tubal occlusion (women)	99.5
Vasectomy (men)	99.5
Highly Effective	
Oral Contraceptives	98
IUD with slow-release hormones	98
IUD plus spermicide	98
Diaphragm plus spermicide	98
IUD	
Copper T	97
Older loops	95
Condom (good brand) plus spermicide	95
Effective	
Condom (good brand)	90
Diaphragm alone	87
Spermicide (vaginal foam)	85
Vaginal sponge impregnated with spermicide	83
Moderately Effective	
Rhythm method based on daily temperature readings	76
Spermicide (creams, jellies, suppositories)	75
Relatively Ineffective	
Condom (cheap brand)	70
Withdrawal	70
Rhythm method not based on daily temperature readings	Variable but normally below 60
Unreliable	
Douche	40

in LDCs want to have three to four children—well above the 2.5 average fertility rate needed to bring about eventual population stabilization.

Many analysts believe that effective population control requires a combination of economic development with better income distribution among the poor, family planning, and the use of methods that go beyond voluntary family planning. Four of the most widely used of these methods are: **(1)** voluntary abortion, **(2)** increased rights, education, and work opportunities for women, **(3)** economic incentives and disincentives, and **(4)** restricting immigration. Let's look at these approaches in more detail.

Abortion Induced abortion is one of the oldest and most widely used methods of human birth control. The total number of legal and illegal abortions throughout the world each year is estimated at 40 million—about one abortion for every four live births. About half these procedures are performed illegally and represent a leading cause of death among women of childbearing age.

One of the major social trends throughout the world since 1965 has been the legalization of abortion. *By 1983, only 10 percent of the world's population lived in countries where abortion is prohibited without exception.* Nevertheless, hundreds of millions of women are either too poor to afford abortion or live in rural areas where such services are not available. Where legal abortion is not available, illegal abortion is normally widespread. Table 8-2 shows that illegal abortions cause 50 to 100 times more deaths of women per 100,000 cases than legal abortions and that legal abortions are considerably less risky to women than are live births and surgical procedures such as cesarean sections and hysterectomies.

Perhaps half of all illegal abortions are self-induced. Women swallow large and dangerous doses of chemicals sold as home remedies. Should this method fail, these women often resort to backstreet abortionists or worse: Most poor women abort themselves with sharpened sticks. Infection is understandably a major killer. The most dangerous bacterium, *Clostridium welchii*, kills in only 12 hours unless medical treatment is provided. Antibiotics decrease the risk but must be given quickly; most are often not available in LDCs.

In 1973 the U.S. Supreme Court ruled that during the first 3 months (12 weeks) of pregnancy the decision to have an abortion must be left up to a woman. Between 1973 and 1981 the number of legal abortions per year in the United States doubled, increasing from almost 745,000 to more than 1.54 million—almost one-third of all pregnancies. Despite this increase, an estimated 29 percent of U.S. women who wanted legal abortions during 1980 were unable to obtain them, primarily because 80 percent of U.S. counties, mostly in rural areas, are without a clinic or hospital that regularly performs abortions. In addition, since 1977 federal law has banned the use of Medicaid funds to pay for abortions.

Abortion is a highly emotional issue that does not lend itself to compromise or cool debate. Basically, the argument is between those who regard

Table 8-2 Mortality Risks Associated with Various Obstetrical and Gynecological Procedures

Procedure	Deaths per 100,000 Cases	
	United States*	LDC's** (estimated)
Legal abortion	1	4–6
Female sterilization	4	10–100
Live birth delivery	14	250–800
Cesarean section	41	160–220
Hysterectomy	160	300–400
Illegal abortion (non-medical)	50	100–1,000

*Data from U.S. Centers for Disease Control.
**Data from Population Crisis Committee (1982).

abortion as murder and those who believe that each woman has the right to control her own body. According to numerous polls, most Americans are for reproductive freedom: That is, they believe that no one should be forced to have an abortion, and at the same time, no one should be forced to have a baby. Some supporters of reproductive freedom also argue that making abortion a crime imposes the religious or moral views of one group on other women who may not hold the same views. Other individuals, however, view abortion as an act of murder. They argue that the right to life of an unborn child should take precedence over a woman's right to choose whether to terminate a pregnancy.

The issue is complicated by inconclusive and controversial medical, theological, and legal debates about when life begins—at conception, at birth, or at some difficult-to-define point in between. Many scientists say that no precise answer is possible. Some believe that life begins when the brain starts functioning (usually at 8 weeks after conception). Others make the distinction between an embryo and a viable fetus, arguing that until the fetus can survive outside the uterus, it is not a true person. Because of incubators and modern medical techniques, survival of a fetus is now possible after 28 weeks and in some circumstances after 24 weeks.

Changes in Women's Roles Numerous studies have shown that increased female education is one of the strongest factors leading women to have fewer children. This occurs because educated women **(1)** are more likely to be employed outside the home rather than stay home and raise children, **(2)** are more likely to marry at a later age, reducing their prime fertility years, either because they are in school or because they may take more time to find suitable partners, and **(3)** have fewer

infant deaths—a major factor in reducing fertility rates.

This means little, however, unless women are offered the opportunity to become educated and to express their lives in meaningful work and social roles outside the home. In most nations women do not have the same access to education as men. About 60 percent of the world's 1.4 billion illiterates are women, and in most cases males are given preference in education and vocational training. Competition between men and women for jobs in many countries should become even more intense by 2000, when there will be another billion people looking for work.

In the rural areas of LDCs women typically do all the housekeeping and child care and more than half the work associated with growing food, gathering firewood, and hauling water. In most cases they are not paid for this vital and exhausting labor. These women also suffer the most malnutrition because men and children are given primary access to limited food supplies.

About 50 percent of the world's paid work force in 1984 was female, including 69 percent in Sweden, 68 percent in the Soviet Union, 53 percent in the United States, and 1.8 percent in Algeria. In the United States most women working for pay have clerical and service jobs and in 1984 earned an average of 41 percent less than male workers.

Using Economic Incentives and Disincentives An increasing number of countries are using economic incentives, economic disincentives, or both to help reduce fertility. One-time, relatively small payments to individuals who agree to use contraceptives or become sterilized, and payments to doctors and family planning workers for each sterilization they perform and each IUD they insert, have been used in about 20 countries, including China and India—the world's two most populous countries.

Such payments, however, are most likely to attract people who already have all the children they want. Although the payments are not physically coercive, they have been criticized as being psychologically coercive because in some cases the poor might have little choice but to accept them.

China, parts of India, Taiwan, and several other countries have used deferred incentives in the form of old-age pensions and health care, life insurance, education funds, and the like to be paid in the future to people who have succeeded in keeping the family small. Such schemes leave the choice of birth control to the individual and pose less risk of charges of unfair influence.

Economic disincentives or penalties that impose extra taxes and other costs or withhold or reduce such benefits as income tax deductions for children, health care, food allotments, and job preferences are also ways to reduce fertility. China has a comprehensive program of such economic disincentives designed to promote one- or two-child families, as discussed in Section 8-4.

Like economic incentives, economic disincentives can be psychologically coercive for the poor. Programs that withhold food or increase the cost of raising children can also be unjust and inhumane, punishing innocent children because parents plan poorly and have more children than they can afford.

Experience has shown that economic rewards and penalties designed to reduce fertility work best when they (1) nudge rather than push people to have fewer children, (2) reinforce existing customs and trends toward smaller families without departing too far from prevailing norms, (3) do not penalize people who have produced large families before the new programs were established, and (4) increase a poor family's income or land ownership.

Restricting Immigration Controlling population growth in a particular country by limiting immigration is widely used by most countries in the world. Today about 500,000 legal immigrants and refugees enter the United States each year. In addition, an estimated 500,000 illegal immigrants are added each year to the estimated 2 to 6 million illegal immigrants already in the United States. This annual addition of about 1 million refugees and legal and illegal immigrants now accounts for 40 to 50 percent of U.S. population growth each year.

Government officials estimate that illegal immigrants cost the United States $4 billion annually in benefits, lost income taxes, and wages lost by displaced U.S. workers. Another study by the Environmental Defense Fund puts the cost at $14 billion a year. Others point out, however, that many illegal immigrants work in menial jobs that most U.S. workers won't take—mostly because of subminimum wages and exploitation by employers. In times when unemployment is high, however, union officials argue that illegal immigrants take many such jobs away from legal residents looking for any kind of work, especially young people, minorities, and unskilled workers.

Studies have shown that in general the social security, income taxes, and other payments made by illegal immigrants far exceed what they take away in benefits and that only a small fraction use government services such as welfare (1 percent), food stamps (1 percent), unemployment insurance (4 percent), and free medical services (5 percent). On the other hand, refugees are eligible for public services and benefits, and have put a severe strain on locally supported services in cities in Florida and California, where most Asian and Latin American refugees settle.

Present U.S. immigration policy has been criticized because it encourages highly trained and skilled scientists, physicians, and other LDC citizens to immigrate to the United States. This "brain drain" policy causes LDCs to lose not only some of their most talented people but also the scarce capital that has gone into their training. In effect, such a policy is a reverse form of foreign aid, from the LDCs to the United States. However, defenders of present policy point out that some of these professionals (1) are unable to get jobs in their own countries, (2) send money back to relatives, which is a major source of income for these LDC residents, (3) are more likely to be productive and make contributions that can benefit the world as a whole in a place of their own choosing, and (4) create jobs for other U.S. citizens.

In 1984 Congress considered but was unable to pass revisions in immigration laws that would have (1) limited immigration to 425,000 annually, excluding refugees (who would continue to be admitted at levels to be determined by consultation between the president and Congress), (2) granted amnesty to all illegal aliens who entered before a specified date but made them ineligible for food stamps, Medicaid, and other federal benefits for a certain number of years, (3) fined employers who hire illegal aliens and sent repeat offenders to jail, and (4) provided additional funds to increase border surveillance, help catch employers who hire illegal aliens, and deport illegal aliens not granted amnesty.

8-4 Efforts at Human Population Control: India and China

India India started the world's first national family planning program in 1952, when its population was nearly 400 million, with a doubling time of 53 years. In 1985, after 33 years of population control effort, India was the world's second most populous country, with a population of 762 million and a doubling time of 32 years. In 1952 India was adding 5 million persons to its population each year. In 1985 it added 16 million.

India's population is projected to reach 1 billion by 2000, 1.2 billion by 2020, and 1.6 billion before leveling off early in the twenty-second century. In 1985 at least one-third of its population had an annual per capita income less than $70 a year, and the average per capita income was only $260. To add to the problem, nearly half of India's labor force is unemployed or underemployed. Each *week* 100,000 more people enter the job market, and for most of them, jobs do not exist.

Without its long-standing national family planning program, India's numbers would be growing even faster. But the program has yielded disappointing results, primarily because of poor planning, bureaucratic inefficiency, low status of women (despite constitutional guarantees of equality), extreme poverty, and lack of administrative and financial support until 1965—more than a decade after the program began.

But the problem is deeper. For one thing, 77 percent of India's people live in 560,000 rural villages, where birth rates are still close to 40 per thousand. The overwhelming economic and administrative task of delivering contraceptive services and education to the mostly rural population is complicated by an illiteracy rate of about 71 percent, with 80 to 90 percent of the illiterate being rural women. Although for years the government has provided information about the advantages of small families, rural Indian couples have an average of five children. Such couples remain convinced that many children are needed as a source of cheap labor and old-age survival insurance. This belief is reinforced by the fact that almost one-third of all Indian children die before age 5. Population control is also hindered by India's diversity: 14 major languages, more than 200 dialects, many social castes, and 11 major religions.

To improve the effectiveness of its program, in 1976 Indira Gandhi's government instituted a mass sterilization program, primarily for males with two or more children. The program was supposed to be voluntary, based on financial incentives alone. But coercion was allegedly used in a few rural areas to meet sterilization quotas. The resulting backlash played a role in Gandhi's election defeat in 1977. In 1978 a new approach was taken, raising the legal minimum age for marriage from 18 to 21 years for males and from 15 to 18 years for females. After the 1981 census showed that the population growth rate between 1971 and 1981 was no lower than that between 1961 and 1971, the government vowed to increase family planning efforts and funding. Whether such efforts will succeed remains to be seen.

China The People's Republic of China is making impressive efforts to bring its population growth under control. In 1985 China had a population of about 1.042 billion, a birth rate of 19 per thousand,

a death rate of 8 per thousand, and a doubling time of 64 years. At these rates China's population grows by about 11 million persons each year.

The United Nations projects that the population of China may reach 1.2 billion by 2000 and 1.3 billion by 2020. In the 1970s, however, China established the most extensive, and also the strictest, official population control program in the world. Its goal is to achieve ZPG by the year 2000, to stabilize its population at 1.2 billion, followed by a slow decline to a population between 600 million and 1 billion by 2100. Although China has a long way to go, its drop from an estimated birth rate of 32 per thousand in 1970 to 19 per thousand by 1985 is remarkable. This means that China's population is now growing by about 1.1 percent a year, which is comparable to the growth rate in the United States.

China's population control program is built around a number of practices:

1. Strongly encouraging couples to postpone marriage (typical marriage ages are 24 to 28 for women and 26 to 30 for men, with legal minimum marriage ages of 20 for women and 22 for men).

2. Indoctrinating couples in family planning techniques at the time of marriage.

3. Strongly encouraging couples to sign pledges to have no more than one child (by 1982, more than 15 million couples had signed such pledges) and providing those who sign the pledge with salary bonuses or work points for extra food and supplies, larger old-age pensions, better housing, free medical care and school tuition for their child, and preferential treatment for jobs when the child grows up.

4. Requiring those who break the pledge to return all benefits and using intense peer pressure on women pregnant with a third child to have an abortion.

5. Reducing the salary and old-age pension for couples having more than two children and charging them for each "extra" child's rations.

6. Requiring one of the parents in a two-child family to be sterilized (instituted in May 1983).

7. Providing free sterilization and free, easily accessible contraceptives for married couples.

8. Making abortion freely available.

9. Compensating workers for time lost from the job due to sterilization, abortion, and IUD insertions.

10. Using mobile units to bring sterilization and family planning education to rural areas.

11. Training local people to carry on the program.

12. Expecting all leaders to set an example.

Although most countries cannot or do not want to exert the heavy social and political pressures just listed, some elements of China's program could be transferred to other LDCs. Especially useful is the practice of bringing contraceptives and family planning to the people at little or no cost, rather than asking the people to come to special centers. Perhaps the best lesson that other nations can learn from China's experience is that countries should not wait to slow population growth until the only answer is the use of compulsory measures. Forcible approaches can cause a backlash of public resentment and run a high risk of failure. China's population control program has been successful so far. But there are signs of increasing resistance, and the long-term results of this coercive program are not yet known.

Short of thermonuclear war itself, rampant population growth is the gravest issue the world faces over the decades immediately ahead.

Robert S. McNamara

Discussion Topics

1. Should world population growth be controlled? Why or why not?

2. Debate the following resolution: The United States has a serious population problem and should adopt an official policy designed to stabilize its population.

3. Describe the demographic transition hypothesis and give reasons why it may or may not apply to LDCs today.

4. Should federal and state funds be used to provide free or low-cost abortions for the poor? Defend your position.

5. What are some of the ways in which women are deliberately and nondeliberately discriminated against in the United States? On your campus?

6. Should the number of legal immigrants allowed into the United States each year be sharply reduced? Why or why not?

7. Should the number of trained or skilled persons admitted as legal immigrants to the United States from LDCs be sharply decreased or reduced to zero? Why or why not?

8. Should illegal immigration into the United States be sharply decreased? How would you go about achieving this?

PART FOUR

Resources

Our entire society rests upon—and is dependent upon—our water, our land, our forests, and our minerals. How we use these resources influences our health, security, economy, and well-being.

John F. Kennedy

9

Soil Resources

Below that thin layer comprising the delicate organism known as the soil is a planet as lifeless as the moon.

G. Y. Jacks and R. O. Whyte

Strong evidence exists that entire civilizations collapsed because they failed to prevent depletion and loss of the topsoil that supported them. Unless we wish to relearn the harsh lessons of soil abuse, everyone—not just farmers—needs to be concerned with preserving soil and maintaining its fertility. In this chapter we look at the components, formation, and properties of soils. We also consider soil erosion and examine ways to conserve this vital resource.

9-1 Components, Formation, and Physical Properties of Soils

Soil Components   **Soil** is a complex mixture of tiny particles of inorganic minerals and rocks, decaying organic matter, water, air, and living organisms. Carbon, hydrogen, and oxygen, three of the nine chemical elements required by plants in relatively large amounts for healthy growth and reproduction, are obtained from water and from the atmosphere as oxygen and carbon dioxide. The remaining six elements needed in relatively large amounts by plants—nitrogen, phosphorus, potassium, calcium, magnesium, and sulfur—are obtained from *inorganic minerals* in the soil. Iron, manganese, molybdenum, zinc, copper, chlorine, and boron are also essential nutrients for plants but are needed at very low or trace levels.

Ninety percent of all the earth's soils contain a relatively small amount of *organic matter* (1 to 7 percent), consisting mostly of dead leaves, stems, and roots along with insect remains, animal droppings, and worm secretions. Decomposing microorganisms (Section 4-4) break some soil organic compounds as part of various biogeochemical cycles

(Section 5-3) into simpler forms such as nitrate (NO_3^-), phosphate (PO_4^{3-}), and sulfate (SO_4^{2-}) ions that are usable by plants.

Some of the organic chemicals in dead organic matter found in soils are not broken down completely by microorganisms. This partially decomposed organic matter is a dark-colored mixture of organic materials called *humus*. Peat moss, for example, is a humus mixture. Because humus is not soluble in water, it remains in the soil and helps retain water-soluble ions such as potassium (K^+), magnesium (Mg^{2+}), and ammonium (NH_4^+). These ions are produced by bacterial decomposition in the upper soil layers for use by plants.

Another important component of soil is *water*. Rain falling on the soil surface percolates down through the pore spaces of the soil, dissolving minerals and other soluble materials along the way. Some of this water is removed by the plant roots. Through capillary action it is transported upward through the roots, stems, and leaves—carrying nutrients and other materials with it. Some of it may then enter the atmosphere through plant leaves in a process called *transpiration*. A fertile soil holds some water but not too much. Such soils have a high humus content, which helps retain water. With too much rain or irrigation, even fertile soils can become so waterlogged that most useful crops cannot grow.

About 50 percent of a typical soil's volume is made up of pore spaces. These spaces are filled with water in a waterlogged soil and with *air* in an extremely dry soil. Most soils fall between these two extremes. The oxygen gas in the soil air is used by the cells in plant roots to carry out cellular respiration (Section 5-3).

The upper layer of a fertile soil is teeming with *living organisms* such as bacteria, fungi, molds, earthworms, and small insects, and larger burrowing animals such as moles and gophers. A food web of some of the living organisms in soil is shown in Figure 9-1. Most of these organisms act as decomposers, breaking down dead organic matter into plant nutrients. Burrowing organisms help maintain soil porosity by churning, mixing, and

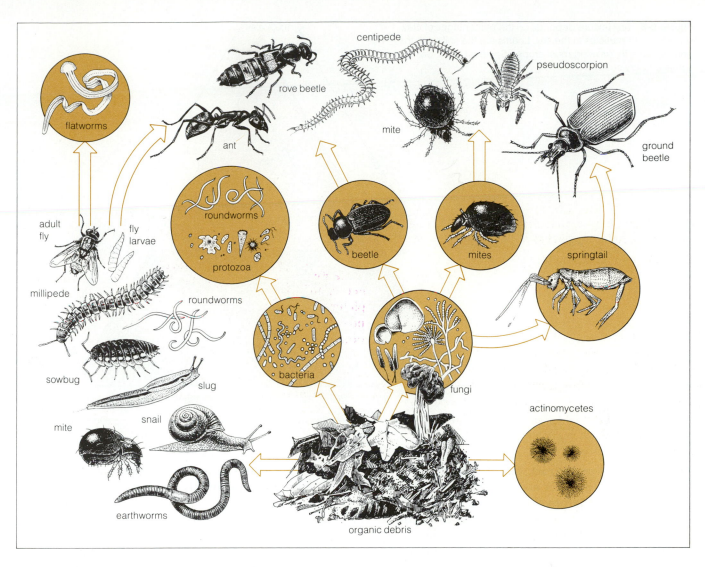

Figure 9-1 Food web of some of the organisms found in soil.

Soil Formation   Soil is formed and changed by a multitude of processes resulting from the interaction of five major factors: parent material, climate, abrasion, organisms, and topography. The main source of the inorganic minerals in soils is the physical and chemical weathering of various types of parent rock. **Physical weathering** involves the breaking down of parent rock into bits and pieces by exposure to temperature changes and the physical action of moving ice and water; growing roots and human activities such as farming and construction also exert this effect. **Chemical weathering** involves chemical attack and dissolution of parent rock by exposure to rainwater, surface

aerating it and by secreting slimes that help hold soil particles together. In addition, their wastes and dead bodies add inorganic and organic material to the soil.

water, oxygen and other gases in the atmosphere, and compounds secreted by organisms.

Temperature, wind, rain, and ice are important climatic factors in the physical and chemical weathering of parent material. In the arid climates of desert biomes, where it is very hot during the day and cold at night, this alternation of heating and cooling can cause rocks to expand and contract and eventually crack and shatter. In cold or temperate climates, an important factor is the disintegration produced by repeated cycles of expansion of water in the pores of parent rock during freezing and contraction when the ice melts.

Unlike physical weathering, chemical weathering changes the chemical composition of the parent material in response to temperature and the presence of water, oxygen, and other materials. In areas where average temperatures are high, these reactions take place relatively rapidly, and in colder regions they take place at a slower rate.

Figure 9-2 Soil texture depends on the percentages of clay, silt, and sand particles in the soil. Loams are the soils with the best textures for growing most crops. (Source: USDA, Soil Conservation Service.)

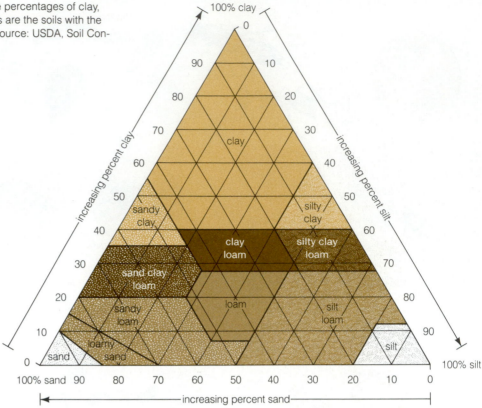

Forces that cause rock particles to move and rub against one another contribute to the process of abrasion and the physical breakdown of parent rock. Moving glaciers cause rock particles to grind together and break into smaller pieces, which are carried along with the moving glacier and deposited when it retreats. Ocean waves are powerful rock grinders. Rivers and flowing streams also move rocks and grind them down. Rocks on the earth's surface are worn down by the sandblasting effect of tiny particles carried by the wind.

Growing roots can exert enough pressure to enlarge cracks in solid rock and eventually split the rock. Plants such as mosses and lichens can penetrate between mineral grains and loosen particles of rock. They can also help trap seeds, dead insects and their excrement, and other windblown debris. Along with bacteria and fungi, they also secrete acid solutions that can slowly dissolve the parent rock.

The slope of the land also has an important effect on the type of soil that can form. When the slope is steep, the action of wind, flowing water, and gravity tends to erode the soil constantly. This explains why soils on steep slopes are often thin and in some cases never accumulate to a depth sufficient to support plant growth. By contrast, soils formed on the valleys below steep mountain slopes receive soil particles, nutrients, water, and organic matter from these slopes and thus are often fertile and highly productive if they are not too wet.

Physical Properties Two important physical properties of soil are *texture* and *structure. Soil texture* refers to the size of the soil's individual mineral particles and the proportion in which particles of different sizes are found in the soil. An international classification system defines particles with a diameter less than 0.002 millimeters as *clay*, those with diameters between 0.002 and 0.02 millimeters as *silt*, and those with diameters between 0.02 and 2 millimeters in diameter as *sand*. The mixtures of clay, sand, and silt lead to the various soil textures shown in Figure 9-2. Variations in soil texture are indicated with terms like *sandy loam* and *loamy silt*.

Soil texture influences the amount of water a soil can hold and the rate at which water percolates through the soil. In sandy soils, so many relatively large pore spaces exist between the sand particles that water and air flow through rapidly. This reduces surface runoff, but such soils drain so well that they retain almost no water. At the other extreme are clay soils, in which the particles are so small and so easily packed together that plant roots cannot penetrate. Clay soils with extra small, closely packed particles are poorly aerated and do

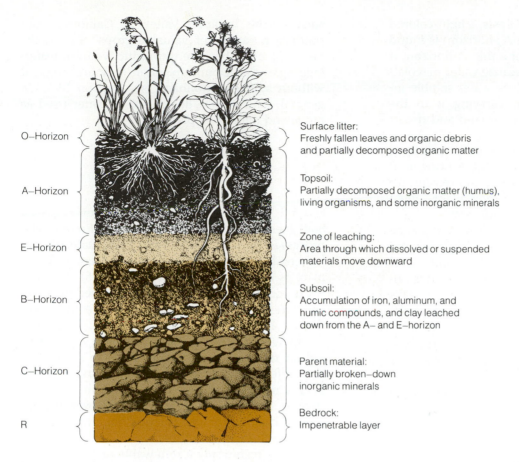

O–Horizon — Surface litter:
Freshly fallen leaves and organic debris
and partially decomposed organic matter

A–Horizon — Topsoil:
Partially decomposed organic matter (humus),
living organisms, and some inorganic minerals

E–Horizon — Zone of leaching:
Area through which dissolved or suspended
materials move downward

B–Horizon — Subsoil:
Accumulation of iron, aluminum, and
humic compounds, and clay leached
down from the A– and E–horizon

C–Horizon — Parent material:
Partially broken–down
inorganic minerals

R — Bedrock:
Impenetrable layer

Figure 9-3 Soil profile showing possible soil horizons. These horizontal layers vary in number, composition, and thickness with different types of soil, as shown in Figure 9-4.

not drain well. Often they are too waterlogged to support many plants.

The best soil texture for growing crops consists of almost equal amounts of sand and silt and somewhat less clay—the so-called *loams*. Such soils have a porosity that allows air circulation and good drainage, yet they retain enough water to support ample plant growth.

The texture of a soil also contributes to *soil structure*—a term describing the way soil particles clump together in larger lumps and clods. The large pores between the clumps allow rapid movement of air and water through the soil. Soil structure depends primarily on the amount of clay and organic material in the soil. Clay soil holds together strongly and when wet can form sizable clumps. This is why the clay used by potters holds together and can be molded.

A good soil for growing plants has a crumbly, spongy quality. When walked on it has a springy feeling. Its particles clump together so that it is about one-half to two-thirds filled with pore space. This abundance of pores provides ample oxygen for plant root cells and retains enough water for roots to absorb without being waterlogged. Since

good agricultural soil is easily compacted, it is important not to walk or drive heavy machinery on it after rains or irrigation.

9-2 Soil Profiles

Most soils consist of a series of distinctive horizontal layers called **soil horizons**, as shown in Figure 9-3. Each horizon has a distinct thickness, color, texture, and composition. A cross-sectional view of the horizons in a soil is called a **soil profile**.

Most mature soils have three or four of the six major horizons: O (surface litter), A (topsoil), E (zone of leaching), B (subsoil), C, and R (bedrock), as shown in Figure 9-3. Below the uppermost *O-horizon* is usually found a porous mixture of humus and mineral particles called the *A-horizon*, commonly referred to as *topsoil*. This layer, ranging in thickness from less than a centimeter on steep slopes to more than a meter in grassland soils, is the most fertile horizon in the entire soil profile. The O- and A-horizons contain most of a soil's living organisms, organic matter, and plant roots.

In many soils formed in forests, a light-colored *E-horizon* (formerly called the A_2-horizon) is found under the O-horizon or under a thin A-horizon. It is called the *zone of leaching* because water percolating through this layer dissolves water-soluble inorganic and organic matter, carrying it to the underlying B-horizon. Most grassland and desert soils do not have E-horizons.

Below the A- or E-horizons is the *B-horizon* or *subsoil*. Although not as fertile as the A-horizon, it provides crucial supplies of water and oxygen for plants with deep roots, and for most plants when the A-horizon dries out or does not exist.

Beneath the B-horizon is the *C-horizon*, a zone of relatively undecomposed mineral particles and rock fragments. In some cases, it is the same kind of material that has slowly broken down (weathered) to form the minerals in a soil's layers. This horizon contains no organic material. The relatively impenetrable layer of *bedrock*, sometimes encountered within a few feet of the soil surface, is referred to by the symbol *R* and is often not shown in soil profiles.

Mature soils in different biomes (Section 4-7) have different soil profiles as shown in Figure 9-4 for five different biomes. Most of the world's crops are grown on grassland soils and the soil exposed when deciduous forests are cleared. The soil under a deciduous forest, however, is less productive than a grassland soil. This is because a deciduous forest soil has a thinner A-horizon than a typical grassland soil and because the higher rainfall in areas where deciduous forest soils are formed leaches many of the nutrients from overlying horizons. Unless these nutrients are replaced by fertilizers, soils of cleared temperate deciduous forests will not grow satisfactory crops after several years of harvest. The soil under a coniferous forest (Figure 9-4) is not very useful for growing crops when the vegetation is cleared.

When tropical rain forests are cleared, the torrential, seasonal rains leach most of the nutrients and remaining minerals from the A-horizon of the exposed soil (Figure 9-4). When exposed to the air and sun, the iron oxide that remains can form red rock called *laterite* or *ironstone*—a rock so hard it is used for paving highways. Once laterite forms, the land can be lost to cultivation permanently. It is estimated that laterite-forming soils are found close to the surface on about 7 percent of the total land area of the tropics—including 2 percent of tropical America, 11 percent of tropical Africa, and 15 percent of sub-Saharan West Africa.

About one-third of the area covered by desert soils is not useful for growing crops because of too little rainfall. With proper soil management and extensive irrigation, soils in certain desert areas, such as the Imperial Valley of California, can produce a variety of valuable crops. Such soils, however, often become unproductive from waterlogging and salt buildup when they are irrigated without sufficient drainage (see Section 10-5). In general, land with a desert soil is better used as rangeland than as cropland.

9-3 Soil Erosion

Natural and Human-Accelerated Soil Erosion Soil does not stay in one place indefinitely. The processes by which earth and rock materials are loosened or dissolved and moved from place to place is called **soil erosion** (Figure 9-5). Blowing wind and flowing water have moved sediments at an extremely slow average rate since the earth formed 4 billion to 5 billion years ago. Before the rise of agriculture, this natural erosion was beneficial for the most part because it led to the formation of fertile deltas and valleys. The average rate of this natural erosion has been significantly increased by poor farming practices, overgrazing, ground clearing for construction, logging, mining, and other human activities.

Soil is technically a renewable resource. However, the average rate of erosion of topsoil per unit of cropland throughout the world greatly exceeds the rate at which it is being formed. According to 1985 estimates by the Worldwatch Institute, the world is now losing an estimated 23.2 trillion kilograms (25.4 billion tons) of soil each year from croplands alone in excess of new soil formation. Each year the amount of topsoil washing and blowing into the world's rivers and oceans would fill a train of freight cars long enough to encircle the entire planet 150 times! At this rate, the world's topsoil is being depleted at 0.7 percent a year—7 percent each decade. This serious situation is worsening as cultivation is being extended into ever more marginal areas of land.

The largest amount of excessive soil erosion is occurring in India, followed in order by China, the Soviet Union, and the United States, which together account for over half the world's food production. Not only does this loss of topsoil reduce soil fertility, it can also reduce irrigation by filling irrigation ditches with sediment, hinder electrical generation by filling dam reservoirs with silt, and obstruct navigable waterways by clogging them with sediment.

The less soil we have, the more food will cost in the future. Thus, what we will pay for food by the end of the century is affected by today's high rates of soil erosion. Furthermore, when one na-

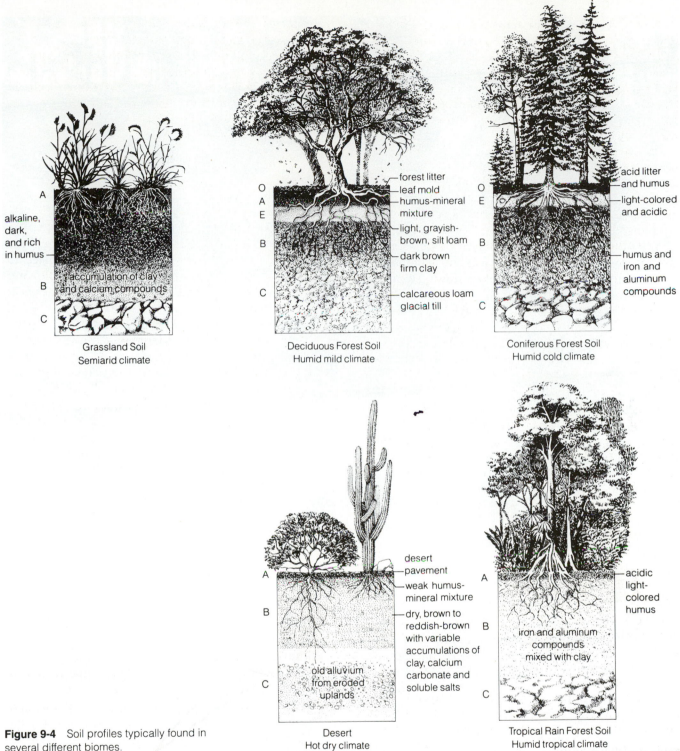

Figure 9-4 Soil profiles typically found in several different biomes.

Grassland Soil
Semiarid climate

alkaline, dark, and rich in humus

accumulation of clay and calcium compounds

Deciduous Forest Soil
Humid mild climate

forest litter
leaf mold
humus-mineral mixture
light, grayish-brown, silt loam
dark brown firm clay
calcareous loam glacial till

Coniferous Forest Soil
Humid cold climate

acid litter and humus
light-colored and acidic
humus and iron and aluminum compounds

Desert
Hot dry climate

desert pavement
weak humus-mineral mixture
dry, brown to reddish-brown with variable accumulations of clay, calcium carbonate and soluble salts
old alluvium from eroded uplands

Tropical Rain Forest Soil
Humid tropical climate

acidic light-colored humus
iron and aluminum compounds mixed with clay

tion loses too much of its topsoil, it must either face famine or import more food. This, in turn, raises the pressure on soils elsewhere in the world. Thus, in coming decades the long-term loss of topsoil through erosion is likely to reduce the rate of economic growth for many nations.

Massive loss of the topsoil is not due to lack of knowledge about how to reduce soil erosion. Instead it results from the political and economic pressures on farmers to produce more in the short run for the world's growing population, with little regard for the long-term consequences. In mountainous regions such as the Himalayas and the Andes, farmers constructed elaborate systems of terraces that allowed them to cultivate steeply sloping land that would otherwise rapidly lose its topsoil. In recent decades, however, growing competition for cropland has pushed farmers further

Figure 9-5 Extensive soil erosion and gully formation due to poor farming practices.

USDA, Soil Conservation Service

up the slopes on steeper hills. In some areas steep slopes are cultivated without building any terraces, causing an irreversible loss of all topsoil within 10 to 40 years. Impoverished farmers may know that plowing a steep slope causes a rapid loss of topsoil, but they have nowhere else to go to grow food for their families.

In the tropics farmers long ago evolved a system of *shifting cultivation* in which a patch of land is cleared, farmed for 2 to 4 years, and abandoned (fallowed) for 10 to 25 years to allow the soil to regain its fertility (Section 2-2). However, in recent decades mounting population pressure in the tropics has caused farmers in many locales to reduce the fallowing period to 2 to 5 years. The result has been a sharp increase in the rate of topsoil erosion.

More than one-third of the earth's land—inhabited by 850 million people—is already classi-

fied as arid or semiarid desert, much of it converted from grassland and forest in a process known as **desertification**. Each year the world's total desert land grows by an area equal to that of South Dakota. Most desertification occurs when shifting sand dunes spill onto cultivatable land found at the edge of large deserts such as the Sahara. However, other causes of desertification include mismanagement of marginal farmland, large-scale deforestation without adequate replanting (Section 12-5), and overgrazing of rangeland (Section 12-6)—all of which can leave the soil subject to erosion.

Surveys by the Soil Conservation Service (SCS) show that the average rate of erosion on cultivated land in the United States is about seven times that at which soil is being reformed. This amounts to an average loss of 2.5 centimeters (1 inch) of soil per acre every 33 years and represents plant nutri-

Figure 9-6 Dust storm approaching Prowers County, Colorado, 1934.

USDA, Soil Conservation Service/Thomas G. Meier

ent losses worth $18 billion a year in 1979 dollars. This average rate of topsoil loss masks much higher losses of five to eight times this rate in heavily farmed areas with the most severe erosion rates. It is estimated that more than one-third of all U.S. cropland is suffering from losses of topsoil that will gradually decrease crop productivity if soil conservation efforts are not greatly increased. Indeed, according to the SCS, *the United States has already lost about one third of the topsoil on the cropland in use today.*

Heavy and increasingly expensive applications of fertilizers hide for a time the rapid depletion of the natural fertility and productivity of U.S. soils; indeed, the resulting sediment is the single largest source of the nation's water pollution (Chapter 22). Most U.S. farmers know about soil conservation techniques but choose to ignore them when they conflict with other techniques that offer a higher yield, and a higher profit over the short term.

The Dust Bowl: A Valuable Ecological Lesson

The Great Plains stretch through 10 states, from Texas through Montana and the Dakotas into Canada. This area is normally dry and very windy and periodically experiences long, severe droughts. Before settlers began planting crops in this region in the 1890s, the extensive root systems of prairie grasses held the rich grassland soil (Figure 9-4) of the Great Plains in place. When the land was planted, these perennial grasses were replaced by annual crops with less extensive root systems. In addition, the land was plowed up after each harvest and left bare part of the year. Thus, the stage was set for severe wind erosion. Accelerated wind erosion and crop failures occurred during the droughts of 1890 and 1910 and returned with great severity between 1926 and 1934.

In 1934, after several years of drought, hot, dry windstorms created dust clouds thick enough in some areas to cause darkness at midday (Figure 9-6); the danger of breathing this dust-laden air was revealed by the dead rabbits and birds left in its wake. During May 1934 the entire eastern half of the United States was blanketed with a massive dust cloud of rich topsoil blown off the Great Plains—2,415 kilometers (1,500 miles) away. Even ships 320 kilometers (200 miles) out in the Atlantic Ocean received deposits of topsoil blown off parts of the Great Plains. This event gave the Great Plains region a new name, the *Dust Bowl*.

An estimated 3.6 million hectares (9 million acres) of farmland was destroyed and an additional 32 million hectares (80 million acres) severely damaged. Thousands of displaced farm families from Oklahoma, Texas, Kansas, and other states migrated westward toward California and to some of the industrial cities of the Midwest and East. Since the nation was still suffering from the Great Depression of the early 1930s, most of these people joined long breadlines upon reaching their destinations.

In May 1934 Hugh Bennett of the U.S. Department of Agriculture addressed a congressional hearing, pleading for new programs to protect the

Figure 9-7 A combination of contour farming and strip cropping can be used to reduce soil erosion.

nation's topsoil. Lawmakers in Washington took action when dust blown from the Great Plains began seeping into the hearing room. In 1935 the United States established the Soil Conservation Service (SCS) as part of the U.S. Department of Agriculture (USDA). With Bennett as its first head, the SCS began promoting good conservation practices in the Great Plains and then in every state, by establishing local soil conservation districts and providing technical assistance to farmers and ranchers. The SCS has since spent more than $22 billion provided by taxpayers to help preserve millions of acres of soil from erosion and destruction.

These efforts, however, did not completely solve the erosion problems of the Great Plains region. From both economic and ecological viewpoints, much of the Great Plains is better suited for grazing than for farming—a lesson its farmers have learned several times since the 1890s and may be forced to learn again. After a moist decade (the 1940s), severe drought returned in the 1950s, and the government had to provide emergency relief funds.

In 1975 the Council of Agricultural Science and Technology (CAST) warned that severe drought could create another dust bowl in the Great Plains, pointing out that despite large expenditures for soil erosion control, topsoil losses then were 2.5 percent worse than in the 1930s.

9-4 Soil Conservation

Soil Erosion Control   The most important step in soil conservation is to hold the soil in place. Methods used to accomplish this goal include **(1)** contour farming, **(2)** terracing, **(3)** minimum-tillage farming, **(4)** strip cropping, **(5)** crop rotation, **(6)** gully reclamation, **(7)** using windbreaks, and **(8)** not planting marginal farming land.

One way to reduce soil erosion up to 50 percent is to use *contour farming*, which involves plowing and planting along the contours of gently sloping land (Figure 9-7) rather than up and down the hills. Each row planted at right angles to the slope of the land acts as a small dam to help hold soil and slow the runoff of water.

Terracing can be used on steeper slopes to reduce soil loss and help retain water. A long, steep slope is converted into a series of broad, level terraces running at right angles to the slope of the land. Water running down the permanently vegetated slope is retained by each terrace, reducing runoff and soil erosion and providing water for crops. In areas of high rainfall, diversion ditches must be built behind each terrace to permit adequate drainage. Terracing is widely used in China, Japan, and the Philippines (especially for ricelands), around the Mediterranean, and in the Andes.

In conventional farming the land is plowed, disked several times, and smoothed to make a planting surface. If plowed in the fall so that crops can be planted early in the spring, the soil is left unprotected during the winter and early spring months. To lower labor costs, save energy, and reduce soil erosion, this intensive tillage approach is being replaced by *minimum-tillage or conservation tillage* in some MDCs. Farmers disturb the soil as little as possible when crops are planted and keep crop residues and litter on the ground instead of turning them under by plowing. Special subsurface tillers are used to break up and loosen the soil without turning over previous crop residues and cover vegetation. However, the technique requires increased use of herbicides to control weeds that compete with crops for soil nutrients. Depending on the soil, this approach can be used for 3 to 7 years before more extensive soil cultivation is needed to prevent crop yields from declining. In some cases special planters are used to inject seeds into the soil without plowing at all (known as *no-tillage farming*). By 1984 minimum tillage practices were used on about 35 percent of all U.S. croplands. The USDA estimates that using such approaches on 80 percent of U.S. cropland would reduce soil erosion by 50 percent or more.

Strip cropping involves planting crops in alternating rows or bands of close-growing plants (such as hay or nitrogen-fixing legumes) and regular crops. The strips of close-growing crops slow water runoff, reduce soil erosion, decrease damage from pests and plant diseases, and help restore soil fertility, especially when legumes are used. This technique can reduce soil losses by up to 75 percent on sloping land when combined with contour farming or terracing (Figure 9-7).

Crop rotation involves annually cycling between areas or strips planted with crops such as corn, tobacco, and cotton, which remove large amounts of nutrients from the soil when harvested, and legumes, which add nitrogen to the soil, or with other crops such as oats, barley, and rye. This method improves soil fertility, reduces erosion by covering land whose fertility is being improved, and reduces pest infestation and plant diseases by annually switching the plants in each area.

On sloping land not covered by vegetation, deep gullies can be created fairly quickly by water runoff, as shown earlier in Figure 9-5. Thus, *gully reclamation* is an important form of soil conservation. Relatively small gullies can be seeded with quick-growing plants such as oats, barley, and wheat to reduce erosion. For severe gullies, small dams can be used to collect silt and gradually fill in the channels. The soil can then be stabilized by planting rapidly growing shrubs, vines, and trees.

Wind erosion from cultivated lands exposed to high winds, such as the Great Plains, can be reduced by using *windbreaks* (Figure 9-8). This technique is especially effective if land not under cultivation is kept covered with vegetation. A large fraction of the windbreaks in the Great Plains have been destroyed to make way for large irrigation systems and farm machinery.

An obvious approach to reducing soil erosion is *not planting crops on marginal lands* that, because of slope, soil structure, presence of high winds and periodic drought, or other factors, are subject to high rates of erosion.

Although all these soil conservation practices are well known to farmers throughout the United States, their use has declined (except minimum-tillage farming). The net result is an increase in soil erosion.

Land-Use Classification To encourage wise land use, the SCS has set up the eight different classes of land, summarized in Table 9-1 and illustrated in Figure 9-9. Soil types and fertility, slope, drainage, erodibility, and other factors are used to classify a particular land area into one of these best-use categories. After land has been classified in this manner, each community has a responsibility to pass laws and use techniques such as zoning to prevent land misuse, as discussed in Section 14-4.

Maintaining Soil Fertility Organic fertilizers and commercial inorganic fertilizers can be applied to soil to restore and maintain plant nutrients lost from the soil by erosion, leaching, and crop harvesting. Three major types of organic fertilizer are animal manure, green manure, and compost. *Animal manure* includes the dung and urine of cattle, horses, poultry, and other farm animals. Application of animal manure improves soil structure, increases organic nitrogen content, and stimulates the growth and reproduction of soil bacteria and

Figure 9-8 Windbreaks are used to reduce erosion on this farm in Trail County, North Dakota.

fungi. It is particularly useful on rotation crops such as corn, cotton, potatoes, cabbage, and tobacco. The use of animal manure in the United States has decreased for several reasons:

1. Most mixed animal- and crop-farming operations have been replaced with separate farms for raising animals (feedlots) and growing crops.

2. Horses and other draft animals that naturally added manure to the soil have largely been replaced by tractors.

3. Animal manure is too expensive to use because of high labor costs and high costs for transporting it from animal feedlots (normally located near urban areas) to rural crop-growing areas.

4. Animal manure has been replaced in most cases by commercial fertilizers that can be more easily stored and applied, and to a lesser extent, by green manure.

Green manure is fresh, green vegetation plowed into the soil to increase the organic matter and humus content available to the next crop. It may consist of weeds that have taken over a field left uncultivated, grasses and clover from a field previously used for pasture, or legumes such as alfalfa or soybeans intentionally grown for use as fertilizer to build up soil nitrogen. The effects of green manure on the soil are similar to those of animal manure. *Compost* is a rich natural fertilizer usually produced by piling up alternating layers of carbohydrate-rich plant wastes (such as cuttings and leaves, animal manure) and topsoil, to provide a home for microorganisms that aid the decomposition of the plant and animal manure layers.

Today the fertility of most soils is partially restored and maintained by the application of *commercial inorganic fertilizers*. The most common plant nutrients added by these products are nitrogen, phosphorus, and potassium. Such fertilizers are designated by numbers. For example, fertilizer

Table 9-1 Land Capability Classification According to the Soil Conservation Service

Land Class	Characteristics	Primary Uses	Secondary Uses	Conservation Measures
Land Suitable for Cultivation				
I	Excellent, flat, well-drained land	Agriculture	Recreation Wildlife Pasture	None
II	Good land, has minor limitations such as slight slope, sandy soil, or poor drainage	Agriculture Pasture	Recreation Wildlife	Strip cropping Contour farming
III	Moderately good land with important limitations of soil, slope, or drainage	Agriculture Pasture Watershed	Recreation Wildlife Urban industry	Contour farming Strip cropping Waterways Terraces
IV	Fair land, severe limitations of soil, slope, or drainage	Pasture Orchards Limited agriculture Urban industry	Pasture Wildlife	Farming on a limited basis Contour farming Strip cropping Waterways Terraces
Land Not Suitable for Cultivation				
V	Use for grazing and forestry, slightly limited by rockiness, shallow soil, wetness, or slope prevents farming	Grazing Forestry Watershed	Recreation Wildlife	No special precautions if properly grazed or logged. Must not be plowed.
VI	Moderate limitations for grazing and forestry	Grazing Forestry Watershed Urban industry	Recreation Wildlife	Grazing or logging may be limited at times.
VII	Severe Limitations for grazing and forestry	Grazing Forestry Watershed Recreation Wildlife Urban industry		Careful management required when used for grazing or logging.
VIII	Unsuitable for grazing and forestry because of steep slope, shallow soil, lack of water, too much water	Recreation Watershed Wildlife Urban industry		Not to be used for grazing or logging. Steep slope and lack of soil presents problems.

6-12-12 contains 6 percent nitrogen, 12 percent phosphorus, and 12 percent potassium. Other plant nutrients may also be present. Ideally, the soil is chemically analyzed to determine the mix of nutrients that should be added.

Inorganic commercial fertilizers are a concentrated source of nutrients and are relatively cheap and easily stored and applied. For these reasons, their use throughout the world increased almost ninefold between 1950 and 1984. The additional food produced by the use of commercial fertilizers now feeds about one out of every three people on earth.

However, commercial fertilizers have their disadvantages. The natural ability of soil to produce nitrogen in forms usable by plants is apparently decreased by the nitrogen in commercial fertilizers. Consequently, the continued yearly application of commercial nitrogenous fertilizers will cause crop yields to decrease unless larger amounts are added each year, at increasing expense and energy use. Commercial fertilizers also reduce the oxygen content of soil by altering soil porosity, so the added fertilizer is not taken up as efficiently. In addition, most commercial fertilizers do not contain many of the micronutrients needed by plants.

Another problem is that the movement of rainwater over and through the surface of land washes some of the plant nutrient compounds in commer-

CLASS **VII** LAND

CLASS **VIII** LAND

CLASS **VII** LAND

CLASS **VI** LAND

CLASS **IV** LAND

CLASS **II** LAND

CLASS **V** LAND

CLASS **I** LAND

CLASS **III** LAND

LAND CAPABILITY CLASSES

	SUITABLE FOR CULTIVATION		NO CULTIVATION—PASTURE, HAY, WOODLAND AND WILDLIFE
I	REQUIRES GOOD SOIL MANAGEMENT PRACTICES ONLY	V	NO RESTRICTIONS IN USE
II	MODERATE CONSERVATION PRACTICES NECESSARY	VI	MODERATE RESTRICTIONS IN USE
III	INTENSIVE CONSERVATION PRACTICES NECESSARY	VII	SEVERE RESTRICTIONS IN USE
IV	PERENNIAL VEGETATION – INFREQUENT CULTIVATION	VIII	BEST SUITED FOR WILDLIFE AND RECREATION

Figure 9-9 Classification of land according to capability; see Table 9-1 for meaning of each class.

cial fertilizers into nearby streams, rivers, and lakes or leaches them into groundwater supplies. An excessive concentration of nitrate, for example, can poison drinking water (especially for infants) and can cause overgrowths of algae in lakes and slow-moving bodies of water (Section 22-4).

Though there are occasional localized successes in efforts to conserve soil, there are no national successes, no models that other countries can emulate. In this respect, soil conservation contrasts sharply with oil conservation, where scores of countries have compiled impressive records in recent years.

Lester R. Brown

Discussion Topics

1. Why should urban dwellers, not just farmers, be concerned with soil conservation?

2. List the following soils in order of increasing porosity to water: loam, clay, sand, and sandy loam.

3. What type of soil texture is best for growing crops? Why?

4. What type of soil structure is best for growing crops? Why?

5. Describe briefly the Dust Bowl phenomenon of the 1930s and explain how it could happen again. How would you prevent a recurrence?

6. Distinguish among contour farming, terracing, and minimum tillage farming, and explain how each can reduce soil erosion.

7. Define strip cropping and crop rotation and tell how each can be used to reduce soil erosion and help restore soil fertility.

8. Go into rural and semirural areas surrounding your campus and attempt to classify the land into the classes shown in Figure 9-9 and Table 9-1. Look for examples of land being used for purposes to which it is not suited.

10

Water Resources

Water is our most abundant substance, covering about 71 percent of the earth's surface. Indeed, life is mostly water, making up 50 to 97 percent of the weight of all plant and animal life and about 70 percent of your body. You might survive a month without food, but only a few days without water. Water, like energy, is also essential for agriculture, manufacturing, and almost every human activity.

With increasing demands for water and with world population increasing by 212,000 each day in 1985, is the world in danger of running out of usable water? What are the present and future water situations in the United States? How can the world's fixed supply of water be managed to get enough of it in the right place, at the right time, and with the right quality? This chapter deals with these questions of water supply; water pollution is discussed in Chapter 22.

10-1 Water's Unique Properties

Much of water's usefulness results from its unique physical properties.

1. *Liquid water has a high boiling point [100°C (212°F)] and solid water or ice has a high melting point [0°C (32°F)].* Otherwise, water at normal temperatures would be a gas rather than a liquid. There would be no oceans, lakes, rivers, plants, or animals.

2. *Liquid water has a very high heat of vaporization* (540 calories per gram). Because water molecules have strong forces of attraction for one another, it takes seven times more energy to vaporize one gram of water than to vaporize one gram of gasoline. Similarly, when a gram of water vapor condenses back to the liquid state, a large quantity of heat is given off. This storage of energy when water is evaporated by solar energy from oceans and other bodies of water, and its release when the water vapor in the atmosphere condenses and falls back to the earth as precipitation, is a major factor in distributing heat from the sun throughout the world. Water's high heat of vaporization is also important in regulating the temperature of your body when water is evaporated from your skin.

3. *Liquid water has an extremely high heat capacity, or ability to store heat.* For example, water absorbs five times as much heat per gram as rock for a given change in temperature. As a result, water heats and cools more slowly than most other substances. This property prevents extremes in temperature changes, helps protect living organisms from the shock of abrupt temperature changes, and makes water effective in removing heat from power plants and other industrial processes.

4. *Liquid water is a superior solvent.* The ability of water to dissolve a greater variety and quantity of substances than any other common solvent enables it to carry nutrients throughout the tissues and organs of plants and animals, to be a good cleanser, and to remove and dilute water-soluble wastes. Because water dissolves so many things, it is also easily polluted.

5. *Liquid water has a very low viscosity, or resistance to flow.* As a result, it flows easily and quickly downhill and is easy to pump.

6. *Liquid water has an extremely high surface tension and an even higher wetting ability.* Together, these properties are responsible for capillarity—the ability of water to be drawn upward from tiny pores in the soil into thin, hollow filaments or capillaries, as in the stems of plants.

7. *Liquid water is the only common substance that expands rather than contracts when it freezes.* Thus, ice floats on water, and bodies of water freeze from the top down instead of from the bottom up. Without this property, most aquatic life as we know it would not exist, and the earth's

surface would be permanently covered with ice. Because water expands on freezing, it also breaks pipes, cracks engine blocks, and cracks streets, soil, and rocks.

10-2 Worldwide Supply, Renewal, Distribution, and Use

World Water Resources The world's fixed supply of water in all forms (vapor, liquid, and ice) is enormous. If it could be distributed equally, there would be enough to provide every man, woman, and child on earth with 292 trillion liters (77 trillion gallons). *However, about 99.997 percent of the world's water supply is not readily available for human use (Table 10-1), and the 0.003 percent that is available is unevenly distributed.*

To understand why only 0.003 percent of all water is available as fresh water for human use, imagine that the total planetary water supply is 38 liters (10 gallons). After we take out the ocean water that is too salty for drinking, for growing crops, and for most industrial purposes, about 1.1 liters (4.5 cups) remains. Of this, about 0.83 liter (3.5 cups) lies too far under the earth's surface or is tied up in glaciers, in ice caps, in the atmosphere, and in the soil as "bound water," unavailable for use by plants or for extraction by users of wells. This leaves only about 0.27 liter (1 cup). When we take out the water that is polluted, relatively inaccessible, and too expensive to get, the supply of usable fresh water is only about 0.001 liter (10 drops).

The Hydrologic Cycle Even the tiny fraction of fresh water that is usable amounts to about an average of 879,000 liters (232,000 gallons) for each person on earth. Furthermore, this supply is continually purified in the natural *hydrologic (water) cycle* (Figure 5-10) as long as we don't pollute it faster than it is replenished. Although all the world's water is recycled eventually, parts of the supply are renewed faster than others, with average recycling rates ranging from 9 days to 37,000 years (Table 10-1). Only the water in the atmosphere, the rivers, and the soil is recycled relatively fast.

The water we use comes from two sources: surface water and groundwater. **Surface water** flows in streams and rivers and is stored in natural lakes, in wetlands, and in reservoirs. Surface water entering rivers and freshwater lakes is called **runoff**. This source of water is renewed fairly rapidly (12 to 20 days) in areas with average precipitation. The flow of surface water on land in a country or region is divided into natural drainage patterns called *watersheds* or *drainage basins*. All precipitation falling anywhere in a given watershed that is not evaporated, stored, or transported out of the area by groundwater flows will run off in the same rivers and streams.

Some water also sinks into the soil, where it may be stored for hundreds to thousands of years in slow-flowing and slowly renewed underground reservoirs. This **groundwater** makes up about 95 percent of the world's supply of fresh water. Figure 10-1 shows the basic features of the groundwater system. Some of the rainwater falling on land slowly percolates through the soil until it reaches a layer, or stratum, of rock that the water cannot penetrate (Figure 10-1). Water builds up in the sand and rock overlying this impervious rock layer and fills all the openings and cracks. The soil and rock become saturated up to a certain level, called the **water table**; above the water table line, the soil is relatively dry. In swamps and areas with high rainfall, the water table may lie at or near the land surface, whereas in dry areas it may be nonexistent, or hundreds to thousands of meters below ground.

Table 10-1 World's Water Resources and Their Average Rates of Renewal

Location	Percentage of World Supply	Average Rate of Renewal
Oceans	97.134	3,100 years (37,000 years for deep ocean water)
Atmosphere	0.001	9 to 12 days
On land		
Ice caps	2.225	16,000 years
Glaciers	0.015	16,000 years
Saline lakes	0.007	10 to 100 years (depending on depth)
Freshwater lakes	0.009	10 to 100 years (depending on depth)
Rivers	0.0001	12 to 20 days
In land		
Soil moisture	0.003	280 days
Groundwater		
To a depth of 1,000 meters (1.6 miles)	0.303	300 years
1,000 to 2,000 meters (1.6 to 3.2 miles)	0.303	4,600 years
Total	100.000	

Figure 10-1 The groundwater system.

Water below the water table flows toward the sea in **aquifers** (Latin for "water bearing"). These permeable underground layers of gravel, sand, or porous rock conduct water slowly from mountain areas with heavy rainfall and snowfall toward the oceans. The rate of flow toward the oceans is determined by the difference in the elevation of the aquifer and sea level and by the permeability of the intervening sediment or rock. Most aquifers represent relatively fixed deposits of water, accumulated over thousands to millions of years, which are normally recharged by precipitation quite slowly. Because of their location deep under the ground, it is difficult—if not impossible—to clean up an aquifer once it has been polluted.

Water Use in the World and in the United States

Between 1940 and 1980 total water use in the world increased almost fourfold and average use per person almost doubled. Today, annual withdrawals of water for human use throughout the world amount to about one-tenth of the renewable supply and about one-fourth of the stable supply typically available during the year. Agriculture accounts for about 70 percent of these withdraw-

als, mostly for irrigation. Between 1950 and 1985 the total area of irrigated cropland worldwide increased almost threefold, with roughly a third of today's harvest coming from the 17 percent of the world's land that is irrigated. The remainder of the water withdrawn is used for industrial and domestic purposes. In India, for example, 92 percent of the water withdrawn each year is used for agriculture, 2 percent for industry, and 6 percent for domestic purposes.

In the United States water use has been increasing rapidly since 1900 and is projected to increase even more between 1985 and 2020 (Figure 10-2). Between 1900 and 1980 U.S. population increased by 320 percent and water use by about 1,250 percent. In 1985 an average of 7,402 liters (1,953 gallons) of water per day per person (Table 10-2) was withdrawn from rivers, lakes, and underground aquifers in the United States. About 34 percent of the water withdrawn each year is used for agriculture, 57 percent by industry (with 47 percent needed to cool electric power plants), and 9 percent for domestic purposes. In 1985 each American used an average of 341 liters (90 gallons) a day for domestic purposes (Table 10-2)—about 3 times the per capita average for the world as a

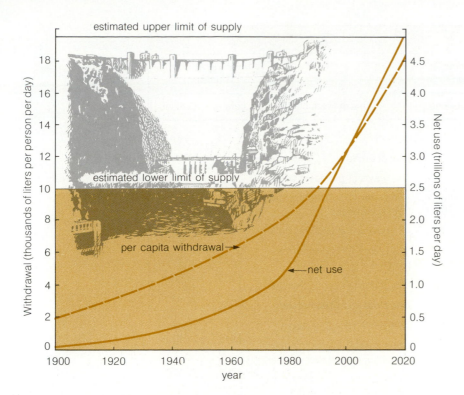

Figure 10-2 The J-shaped curves of present and projected average per capita withdrawal and net water use in the United States. (Source: U.S. Water Resources Council, 1979.)

whole, and 15 to 20 times that of most LDCs. Average daily domestic use per person is 2 to 5 times higher in some cities.

In the United States about 88 percent of the water withdrawn from runoff and groundwater sources each year is returned to rivers and lakes. Much of this comes from electric power plants, which withdraw 47 percent of all water and return all but about 2 percent. Much of the water returned to rivers and lakes, however, is degraded by pollution. In 1979 the U.S. Water Resources Council listed only two states—Kansas and Montana—as having no serious occurrences of surface water pollution and only eight states as having no serious occurrences of groundwater pollution.

10-3 Water Supply Problems

The World Situation Although the average world water supply seems to be sufficient, many areas of the world, including many parts of the United States, are experiencing serious water resource problems. Indeed, a number of experts agree with Gerald D. Seinwill, former acting director of the U.S. Water Resources Council, who considers *the availability of adequate water resources to be the most serious long-range problem confronting the United States and the world*.

At least 80 countries, accounting for nearly 40 percent of the world's population, now experience serious droughts. Africa and Asia face the most serious shortages, although the situation varies considerably with different nations and within nations.

This situation is expected to worsen substantially by the year 2000 because of increasing population and water use and because much of the world's water is located in the wrong place, available at the wrong time of the year, or of poor quality. Increasing conflicts within countries and between countries over shared water supplies are anticipated.

Although water scarcity, droughts, and flooding are serious in some regions, drinking impure water is the major hazard to humans throughout much of the world. The World Health Organization (WHO) estimated that in 1980 only 29 percent of the rural population in LDCs and 75 percent of the urban dwellers in these nations had an adequate supply of potable water (safe to drink). WHO estimates that 25 million people die every year from cholera, dysentery, diarrhea, and other diseases caused by unclean or inadequate water—an average of 68,500 deaths each day (Chapter 20). Such scarcity of drinking water in many LDCs means that every day women and children carry heavy cans or jugs long distances to get untreated,

Table 10-2 Average U.S. Water Use

Left panel

Use or Product	Liters	Gallons
Total Use		
Home Use		
Total per person (per day)	7,402	1,953
Drinking water (per day)	2	0.5
Shaving, water running (per minute)	8	2
	341	90
Shower (per minute)	19	5
Toilet (per flush)	23	6
Cooking (per day)	30	8
Washing dishes, water running (per meal)	38	10
Watering lawn or garden (per minute)	38	10
Automatic dishwasher (per load)	60	16
Bath	135	36
Washing machine (per load)	230	60
Leaky toilet (per day)	90–455	24–120
Leaky faucet (per day)	180–910	48–240
Agricultural Use (Irrigation)		
Total per person (per day)	2,543	671
One egg	150	40
454 grams (1 pound) flour	284	75
Orange	380	100
Glass of milk	380	100

Right panel

Use or Product	Liters	Gallons
Loaf of Bread	570	150
454 grams (1 pound) of corn	645	170
454 grams (1 pound) of sugar from sugarbeets	872	230
454 grams (1 pound) of rice	2,122	560
454 grams (1 pound) of grain-fed beef	3,032	800
454 grams (1 pound) of cotton	7,732	2,040
Industrial and Commercial		
Total per person (per day)	4,518	1,192
Cooling water for electric power plants per person (per day)	3,707	978
Industrial mining and manufacturing per person (per day)	695	183
Refine 3.8 liters (1 gallon) of gasoline from crude oil	38	10
454 grams (1 pound) of steel	133	35
Refine 3.8 liters (1 gallon) of synthetic fuel from coal	1,000	265
One Sunday newspaper	1,060	280
454 grams (1 pound) of synthetic rubber	1,140	300
454 grams (1 pound) of aluminum	3,790	1,000
One automobile	379,000	100,000

Source: U.S. Geological Survey, *Estimated Use of Water in the United States in 1980*, 1984.

often polluted, water from a river or community pump (Figure 10-3).

Water problems often differ between MDCs and LDCs. LDCs may or may not have enough water, but they rarely have the money needed to develop water storage and distribution systems. People must settle where the water is. In MDCs people tend to live where the climate is favorable and then bring in water through sophisticated systems. As Raymond Dasmann put it, "Today people in affluent nations often settle in a desert and demand that water be brought to them, or they settle on a floodplain and demand that water be kept away."

The U.S. Situation The United States has plenty of fresh water, but much of its annual runoff is not in the right place, at the right time, or of high enough quality. In 1979 the U.S. Water Resources Council projected that by 2000 only 3 of the 21 federally designated water regions—New England, the Ohio basin, and the South Atlantic–Gulf area—will have ample water supplies (Figure 10-4).

In the eastern half of the United States the major problem is not a shortage of water, but inability to supply enough water to some urban concentrations of population and industry, coupled with increasing pollution of rivers, lakes, and groundwater by industries and cities (Chapter 22). Some water problems and controversies in the eastern half of the country are:

· Water-rich states in the upper Midwest and Canada are organizing to head off attempts by

United Nations

arid western states to tap water from the Great Lakes, the largest body of fresh water on earth.

- Long Island's 3 million residents must draw all their water from an underground aquifer that is becoming severely contaminated by industrial wastes, leaking septic tanks and landfills, and salty ocean water that is drawn into the aquifer as fresh water is withdrawn.

- New York City officials hope that neither of the two huge tunnels built in 1917 and 1937, which bring water into the city from aqueducts connected to 1,000 streams and 27 artificial lakes throughout the state, will fail before a third $3.5-billion tunnel can be completed, around 2000. The two aging tunnels (which can't be shut down for inspection or repair) and the city's maze of water pipes already leak an estimated 380 million liters (100 million gallons) of water a day.

- Water is being pumped from underground Florida aquifers so rapidly to meet the needs of a mushrooming population that water tables are falling and the ground around Orlando is collapsing upon itself to create huge sinkholes.

The major problem in many of the areas in the western half of the country is a shortage of runoff due to low average precipitation and rapidly declining water tables in many areas, as farmers and cities deplete groundwater aquifers faster than they are recharged. Present water shortages could get much worse if more industries and people move in and if more of this region's precious water

is used to mine coal and oil shale and to tap geothermal energy resources found there (Chapter 17). Some typical water problems and controversies west of the Mississippi are:

- A fight for water has been raging for decades between residents of arid, fast-growing southern California and those in the water-rich northern end of the state. Northern California receives more than two-thirds of the state's rainfall, but about half the state's population lives in the Los Angeles basin, which receives enough rain to support slightly more than 100,000 people.

- Arkansas and Louisiana are fighting efforts by Texas to import water from the Mississippi River to arid, west Texas to replace supplies now being drawn from the underground Ogallala aquifer, which is being depleted rapidly.

- Tribes of Native Americans in Wyoming, Montana, and other states have filed more than 50 lawsuits claiming rights to vast amounts of water in every major western water-user system, based on a Supreme Court ruling that treaties and executive orders dating from the late 1800s entitle Native Americans to enough water to irrigate their reservations.

If present trends continue, most sections of the country are expected to face some form of water shortage by the turn of the century. As a result, you might find Figure 10-4 useful in deciding where to live in the coming decades.

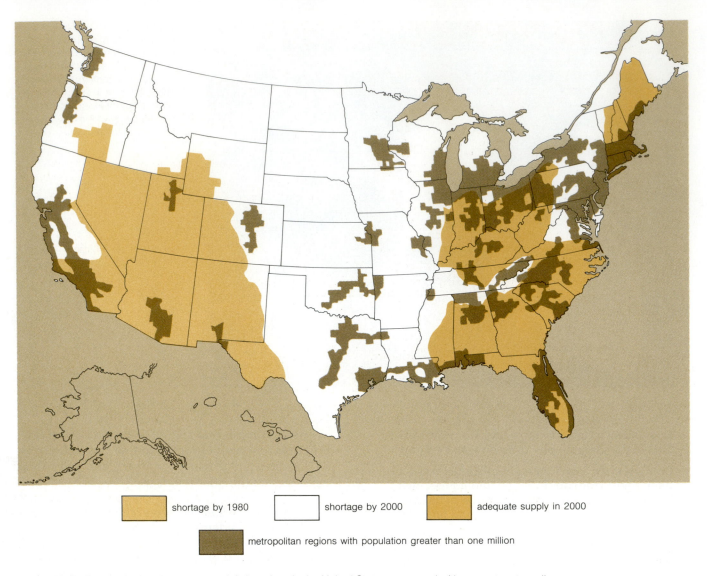

shortage by 1980 shortage by 2000 adequate supply in 2000

metropolitan regions with population greater than one million

Figure 10-4 Present and projected water-deficit regions in the United States compared with present metropolitan regions with populations greater than 1 million. By the year 2000, only 3 of the 21 water regions are expected to have ample water supplies. (Source: U.S. Water Resources Council, 1979.)

Methods for Managing Water Resources Although we can't increase the earth's supply of water, we can manage what we have more effectively (Table 10-3). Some water resource experts favor a "hard path," trying to increase water supplies in an area by building large-scale dams, reservoirs, and canals to transfer water from one water basin to another. Others favor a "soft path," emphasizing voluntary and compulsory water conservation programs coupled with price increases to reflect the true costs of using water for irrigation and for industrial, commercial, and domestic purposes. Others argue that any effective plan for water management must use a combination of all the approaches shown in Table 10-3. In the remainder of this chapter we will look at some advantages and drawbacks of these major methods for managing water resources.

10-4 Building Dams and Water Diversion Projects

Dams and Channelization Ever since people began to irrigate cropland and to live in cities, they have built dams, reservoirs, aqueducts, and pipes to store and carry water. Ideally, the decision to build a dam should involve a careful analysis of potential benefits and drawbacks (Table 10-4). However, many dams in the United States and other nations are built for political reasons, even when analysis shows that cheaper, less harmful, and better alternatives are available. For years environmentalists have opposed completion of a number of big dams and water diversion projects, which they contend would cause more harm than benefits.

Table 10-3 Major Methods for Managing Water Resources

Input: Increase supply in selected areas	Output: Reduce degradation and loss of existing supplies	Throughput: Reduce waste and per capita use
Build dams to create lake reservoirs	Decrease evaporation losses in irrigation	Reduce overall population growth and growth in areas with water problems (arid regions and floodplains)
Divert water from one region to another	Use better drainage for irrigated agriculture to minimize salt buildup in soils	Redesign mining, industrial, and other processes to use less water
Tap and artificially recharge more groundwater	Purify polluted water for reuse (Chapter 22)	Reduce water waste
Desalt seawater and brackish water		Decrease the average per capita use of water
Tow freshwater icebergs form the Antarctic to water-short regions		Increase the price of water to encourage conservation
Control the weather to provide more desirable precipitation patterns		
Control pollution by preventing or limiting the addition of harmful chemicals (Chapter 22)		
Encourage people to move to areas with adequate water		

Table 10-4 Major Advantages and Disadvantages of Using Dams

Advantages	Disadvantages
Reduce the danger of flooding in areas below the dam.	Large acreages of land behind the dam are permanently flooded to form the reservoir. This displaces people and destroys scenic natural areas and wildlife habitats. Abnormally high rains can cause a dam reservoir to overflow its spillways and flood areas below the dam. Because of the mistaken belief that dams protect them from all floods, people build cities and grow crops on floodplains* below the dams. Many experts contend that flood control and reduction of flood losses are better accomplished by reforestation, erosion control (Section 9-3), upstream watershed management, and land-use zoning that prohibits or limits human settlements and farming on floodplains.
Increases local water and food supplies by providing a controllable and reliable flow of water for irrigating areas below the dam.	Often decreases available water supply because water that normally would flow in a river evaporates from the reservoir's surface or seeps into the ground. This evaporation also increases the salinity of the remaining water, decreasing its usefulness for irrigation. Sometimes does not increase food supplies for local residents in LDCs because the irrigation water is used by large landowners to grow cash crops for export.
Creates a large water reservoir that can be used for water recreation.	In the opinion of some, replaces more desirable forms of water recreation (white water canoeing and stream fishing) with less desirable forms (motor and sail boating and lake fishing).
Produces relatively cheap hydroelectric power (Section 19-2) for local and regional residents.	In LDCs the hydroelectric power is sometimes sold to foreign industries rather than being used by local and regional residents. For example, the giant Volta Dam in Ghana is used primarily to supply power to the Kaiser Aluminum Corporation at 5 percent of the average world price for hydroelectric power.
Large dams are safe because none have failed.	Most of the world's large dams are less than 20 years old. The older dams get, however, the higher the risk of failure. Comprehensive risk analyses, like those done for nuclear plants (Section 18-2), are rarely done for large dams. Failure of a large dam could easily kill as many as 200,000 people and cause billions of dollars in damages. Failure of the intermediate-sized Teton Dam killed 14 people and caused $1 billion in damages, and the failure of a small dam above Johnstown, Pennsylvania, killed thousands of people.
	Interrupts the natural flow of a river below the dam, which disrupts fish migration and alters the water's oxygen content (Section 22-3).

*Floodplains—areas vulnerable to periodic flooding—are normally classified as 18-year, 25-year, 50-year, or 100-year floodplains, according to the average interval between major floods. As a result, people mistakenly believe that floods occur only every 18, 25, 50, or 100 years. But this is merely a statistical average; major floods could occur three times in a month or annually for 5 consecutive years.

Another engineering technique used to control floods and soil erosion, drain wetlands, and improve navigation is **channelization** of streams. It consists of straightening, deepening, widening, clearing, or lining existing stream channels. This approach can lead to better flood control in the channelized upstream area, but the greater water flow in these usually increases the amount of flooding, sediment deposits, and bank erosion in downstream areas. The drainage of wetlands as a result of channelization eliminates the habitats of certain wetland plants and animals. To many people the conversion of a winding stream to a straight, open ditch seriously degrades the aesthetic value of a natural area.

Water Diversion Projects Water diversion projects usually involve building huge canal systems to transport water from one river basin to another. A grandiose scheme, called the *Grand Canal concept*, has been proposed recently to solve many of the water supply problems for Canada and the United States. Much of the freshwater runoff now flowing into the Arctic Ocean from the James River Bay in eastern Canada would be used to form a new, dike-enclosed freshwater lake within the bay. From there it would flow by way of rivers into the Great Lakes to reduce the need to flood lands to provide reservoirs. Then, open channels and pumping stations would be used to distribute the water to the major rivers flowing into the Canadian prairies and most of the United States.

The Army Corps of Engineers has also proposed building four separate canals to divert water from the Missouri River and other tributaries of the Mississippi River to the High Plains states now served by the disappearing Ogallala aquifer (Figure 10-6, p. 116). The cost of each canal ranges from $13.4 billion to $40 billion, excluding distribution canals, which could easily double the amount. Nearly one-fourth of the water would be lost through evaporation, spillage, and seepage. Furthermore, the cost of pumping the water through the canals would be at least 10 times what farmers in the area now pay for irrigation water.

In the 1990s the Soviet Union plans to begin work on a massive water diversion project. On paper, the Soviet Union has plenty of fresh water—about 12 percent of the world's total. The trouble is that most of it is in the sparsely populated northern and eastern regions. Therefore the government plans to build 25 large dams to block the flow of at least a dozen north-flowing rivers and pump this water back over mountains to the south. The project is expected to take 50 years and to cost at least $100 billion. Such an expenditure would put

an enormous drain on the Soviet economy and might cost more than its estimated economic benefits.

The project would flood an area larger than western Europe, displacing tens of thousands of people. It could also have potentially serious ecological and climatic effects not only in the Soviet Union, but as far off as North America. Since the rivers scheduled for diversion now flow into the Arctic Ocean, reversing their flow would diminish freshwater flows into the Arctic Ocean and increase its salinity. Some climatologists believe that the increased salinity would lower the freezing point of the Arctic Ocean and cause the ice cap to melt, possibly starting a global warming trend. Other scientists project the opposite effect: that is, global cooling, because a reduced flow of warm fresh water into the Arctic Ocean could cause an increase in the amount of polar ice.

Irrigation Problems: Salinity and Waterlogging
Like past civilizations, we are learning that building dams and diverting surface water from rivers and streams to irrigate land without providing adequate underground drainage eventually destroys the cropland because of excessive soil salinity and waterlogging. As irrigation water flows over and through the ground, it dissolves salts, causing **salinity**. Saline water is again spread over the soil for irrigation, and more salts are left behind when much of the water is lost to the atmosphere by evaporation and transpiration. Unless the salts are somehow flushed or drained from the soil, their buildup stunts crop growth, decreases yields, eventually kills crop plants, forces changes to less profitable crops, promotes excessive water use, and increases capital and operating costs.

A problem that often accompanies soil salinity is **waterlogging**. To keep salts from building up and destroying fragile root systems, farmers often apply more irrigation water than the plant needs to wash or leach salts deeper into the soil profile (Figure 9-4). If drainage isn't provided, irrigation water percolating down and accumulating underground can gradually raise the water table close to the surface and waterlog the roots of plants in toxic saline water. This is a particularly serious problem in areas such as the heavily irrigated San Joaquin Valley in California, where the soils contain a clay layer that is impermeable to water.

According to various estimates, salinity and waterlogging are decreasing crop productivity of 30 to 50 percent of the world's irrigated land, especially in hot, dry climates where evaporation of water from the soil is rapid. Once-fertile areas of southern Iraq and Pakistan now glisten with salt—

Figure 10-5 Because of poor drainage, white salts have replaced crops that once bloomed in heavily irrigated Paradise Valley, Wyoming.

looking like fields of freshly fallen snow. In India, 35 percent of all irrigated land is affected by salinization. In this country, salt buildup is also a potential hazard on half of all irrigated land in 17 western states, already reducing crop production in many areas (Figure 10-5). Based on studies published over the past few decades, food expert Georg Borgstrom estimates that at least 50 percent, and probably close to 65 percent of all presently irrigated land, will be made unproductive from excess soil salinity by the year 2000.

In theory, soil salinity can be controlled if there is enough water that is low in salt content to flush through the soil with adequate drainage. For example, land with excessive salt buildup can be taken out of production for 2 to 5 years and flushed with large quantities of low-salt water. An underground network of perforated drainage pipes can then be installed. However, such flushing and drainage schemes are quite expensive. Furthermore, they only slow the destruction of fertile soil by salinity; they do not stop the process.

Flushing the salts out also increases the salinity of rivers used for irrigation by farmers further downstream. For example, the salinity of the Colorado River increases twentyfold as it passes through irrigated cropland between Grand Lake in north central Colorado and the Imperial Dam in southwest Arizona. The saline water can be drained into evaporation ponds rather than into the nearest river. However, building the ponds, installing an underground network of drain pipes, and pumping the drainage water to such ponds can easily double the fixed costs per acre for a

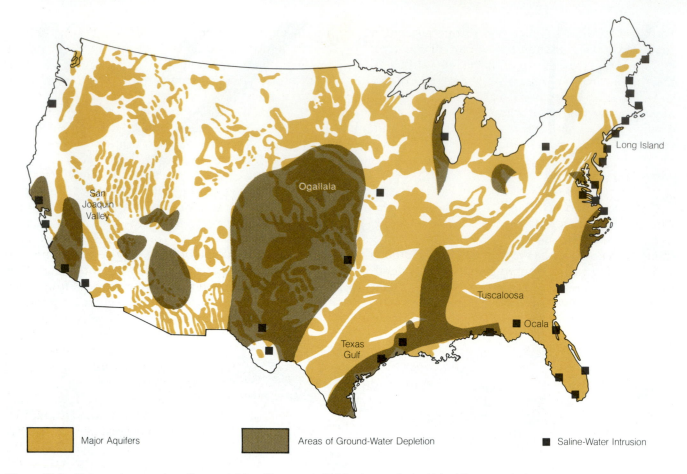

| Major Aquifers | Areas of Ground-Water Depletion | ■ Saline-Water Intrusion |

Figure 10-6 Major underground aquifers containing 95 percent of all fresh water in the United States are being depleted in many areas and contaminated elsewhere through pollution and saltwater intrusion. (Source: U.S. Water Resources Council, 1979.)

farmer. In addition, recent research indicates that these ponds can be a hazard to wildlife, especially waterfowl, from the buildup of toxic levels of selenium and other minerals in the pond water.

10-5 Groundwater Use, Desalination, Towing Icebergs, and Controlling the Weather

Use of Groundwater About half of all U.S. drinking water (96 percent in rural areas and 20 percent in urban areas), 40 percent of the water used for irrigation, and 25 percent of water used for all purposes is withdrawn from underground aquifers (Figure 10-6). One solution to water supply problems is heavier reliance on groundwater. However, there are several problems: **(1)** *aquifer depletion* when groundwater is withdrawn faster than it is recharged by precipitation, **(2)** *subsidence* or sinking of the ground as groundwater is withdrawn, **(3)** *saltwater intrusion* into freshwater aquifers in coastal areas as groundwater is withdrawn faster

than it is being recharged (Figure 10-7), and **(4)** *groundwater contamination* from human activities (Section 22-5).

Groundwater depletion is serious in some areas—especially along the Texas Gulf Coast and in parts of the great Ogallala aquifer (Figure 10-6).

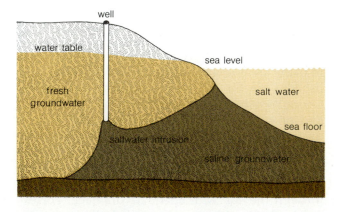

Figure 10-7 Saltwater intrusion along a coastal region. As the water table is lowered, the interface between fresh and saline groundwater moves inland.

Farmers withdraw so much water from this aquifer that the annual overdraft (the amount of water not replenished) is nearly equal to the annual flow of the Colorado River. As a result, the already low water tables in much of the region are falling rapidly, and stream and underground spring flows are diminishing. Experts project that at the present rate of depletion, most of this aquifer could be run dry within 40 years, and much sooner in areas where it is only a few meters deep.

Long before this happens, however, area farmers will be forced to abandon irrigated farming. Already the amount of irrigated land is declining in five of the seven states using this aquifer because of the high cost of energy to operate pumps and the rising cost of pumping water from a depth in some places of 1,825 meters (6,000 feet). Other areas with falling water tables from excessive withdrawal of groundwater include the San Joaquin Valley and the southern part of California, Houston, Texas, and Savannah, Georgia.

One partial solution is to recharge groundwater artificially. Deep groundwater can be pumped up and spread out over the ground to recharge shallow groundwater aquifers. But this can deplete deep groundwater deposits; also, because deep groundwater is often very salty, it can contaminate shallow deposits. Another approach is to recharge aquifers with irrigation water, wastewater, and cooling water from industries and power plants. But much of this water is lost by evaporation, and in many cases it is better and cheaper to reuse cooling water in the industries and power plants themselves.

Another problem is *groundwater contamination* from large animal feedlots, landfills, septic tanks, land disposal of wastewaters, mining, oil and natural gas production, agricultural chemicals, underground injection wells, and other sources (Section 22-4). Once contaminated, groundwater may remain unusable for hundreds to thousands of years. According to the Environmental Protection Agency and a number of water pollution experts, groundwater depletion and contamination is now one of our most serious environmental problems.

Desalination The process of removing dissolved salts from ocean water or inland brackish water that is too salty for irrigation or human use is known as **desalination**. There are two major methods. In *distillation*, energy is used to evaporate fresh water from saltwater to leave the salts behind. In *reverse osmosis*, saltwater and fresh water are separated by a semipermeable membrane, which allows the flow of water but not salts; osmotic pressure forces the pure water to diffuse

through the membrane to dilute the salt solution. In reverse osmosis, a force greater than the osmotic pressure is applied to the salty solution to force fresh water to flow out of the salty water, thus accomplishing desalination.

In 1985 desalting plants produced only about 0.006 percent of the world's daily water and only 0.4 percent of the water used daily in the United States. Most of these plants are small and serve coastal cities in arid, water-short regions where the price of water by any method is high.

Some people see desalination as a major solution to freshwater shortages, but there are economic and ecological problems with this idea. Primarily because of its energy requirements, desalinated water is expensive—costing more than six times as much per volume as water from a tap. Even more energy and money are required to pump desalted water uphill and inland from coastal desalting plants. Most experts project that desalted water will never be cheap enough for widespread use in irrigation—the main use of water throughout the world.

In addition, building and using a vast network of desalination plants would release significant amounts of heat and other air pollutants (depending on the energy source). There would also be the problem of disposing of mountains of salt. If the salt were returned to the ocean, it would increase the salt concentration near the coasts and threaten food resources located in estuarine and continental shelf waters.

Towing Icebergs It may be economically feasible to tow huge Antarctic icebergs to southern California, Australia, Saudi Arabia, Chile, and other dry coastal places. Antarctic icebergs are broad and flat, resembling giant floating table tops or ice islands, and they represent a potential freshwater source equal to five times the world's current domestic use of water, and one-third the world's consumption for all purposes. It would probably take a fleet of tugboats about a year to tow one such berg to California.

This technological feat, however, has a number of problems:

1. No one knows how to lasso, wrap, and tow such a huge object, how to prevent most of it from melting on its long journey through warm waters, or how to get the fresh water from the slowly melting iceberg to shore. Perhaps after spending $100 million, the only thing left would be an empty towline.

2. No one knows whether the project is economically feasible. But because of increased water

demand and dwindling supplies, some economists believe such a project might be attractive
by the year 2000 and should be cheaper and
quicker than most large-scale water diversion
projects.

3. Towing and anchoring such a large, cold mass
   in semitropical areas could cause weather disturbances (possibly creating considerable fog
   and rain) and could have harmful effects on
   marine life.

4. The scheme is a source of political conflict.
   (Who owns the icebergs in the Antarctic?)

5. There would be hazards to other ships if
   chunks of the icebergs under tow were to
   break off and be abandoned in international
   shipping lanes.

Controlling the Weather Several countries, particularly the United States, have been experiment-
ing for years with seeding clouds with chemicals
to produce rain over dry regions and snow over
mountains to increase runoff in such areas. In principle, cloud seeding involves finding a suitable
cloud and flying a plane under or over the cloud or
using ground-mounted burners to inject it with a
powdered chemical. The particles of the chemical
serve as nuclei that cause the small water droplets
in the cloud to coalesce and form droplets or ice
particles large enough to fall to the earth as precipitation. The most effective cloud-seeding chemical
is silver iodide in crystal form, but salt crystals, dry
ice, and clay particles have also been used.

Since 1977 clouds have been successfully
seeded in 23 states covering 7 percent of the U.S.
land area. But we do not know whether we are
increasing total rainfall or merely shifting rain
from one area to another. There are also some
problems:

1. Cloud seeding doesn't work in very dry areas,
   where it is most needed, because there are
   rarely any rain clouds available.

2. Large-scale seeding could change regional or
   even global weather patterns in undesirable
   ways.

3. There could be serious ecological side effects,
   including the unknown effects of silver iodide
   on humans and wildlife, changes in snowfall
   and rainfall, and additional flooding, which
   could destroy or alter wildlife populations, upset food webs, and modify the soil.

4. The likelihood of precipitation can be reduced
   if the seeding is done improperly.

5. There are legal disputes over ownership of
   rights to water in a cloud.

10-6 Water Conservation

Importance of Water Conservation Many water
resource experts argue that water conservation is
the quickest and cheapest way to provide much of
the additional water needed in nations that now
waste much of their input. For example, it is estimated that *at least 30 percent and perhaps 50 percent
of the water used in the United States is unnecessarily
wasted*.

A major reason for this is that artificially low
water prices encourage waste. Another reason is
that riparian rights—a holdover from English common law—allow U.S. landowners to withdraw as
much water as they need from any stream on or
bordering their property. To maintain these rights,
users must continue to withdraw water at rates
they have established over the years: "Use it or lose
it." As a result, cities, big farms, and other heavy
users are reluctant to cut back on wasteful use of
water because this would reduce the amount they
are permitted to withdraw.

Another problem is that the responsibility for
water resource management in a metropolitan area
is divided among many local governments rather
than being handled in terms of an entire water basin. For example, the Chicago metropolitan area
has 349 separate water supply systems and 135
waste treatment plants divided among about 2,000
local units of government over a six-county area.

This is in sharp contrast to the regionalized approach to water management used in England and
Wales. The British Water Act of 1973 replaced more
than 1,600 separate agencies with 10 regional
water authorities based not on political boundaries
but on natural watershed boundaries encompassing one or more river basins throughout the nation. Each water authority owns, finances, and
manages all water supply and waste treatment facilities in its region, including water pollution control, water-based recreation, land drainage and
flood control, inland navigation, and fisheries in
inland waters. Each water authority is managed by
a group of elected local officials and a smaller number of officials appointed by the national government.

Reducing Irrigation Losses Techniques for reducing evaporation and using irrigation water
more efficiently include: **(1)** lining irrigation
ditches to reduce waterlogging, erosion, and seepage losses (although this is expensive and could
decrease the rate of groundwater recharge), **(2)** applying water only when required by actual weather

conditions and in smaller quantities (most farmers use too much), (3) using automated irrigation, in which computers set the water flow, detect leaks, and adjust the amount of water applied based on temperature, wind speed, and soil moisture, (4) using trickle irrigation, in which a network of underground plastic tubes drips irrigation water directly onto the roots of plants rather than flooding entire fields (a method that can double crop yields, reduce growth of weeds, and cut irrigation water use by 50 to 60 percent), (5) capturing and recycling water that would otherwise run off crop fields, (6) shifting to or developing crop varieties that require less water or have a high enough salt tolerance to permit irrigation with saline water, (7) growing more crops in areas with ample water rather than in arid regions that need expensive irrigation, (8) reducing irrigation water runoff by using contour cultivation (Figure 9-7) and terracing, and storing precipitation and irrigation runoff for use in farm ponds or small reservoirs, (9) reducing evaporation of irrigation water by covering the soil with a mulch, (10) allocating more water to those who use conservation measures, and (11) raising water prices to promote conservation (in some areas farmers pay based on the number of acres farmed and thus have no incentive to cut consumption).

Between 1950 and 1980, Israel used many of these techniques to decrease water waste from 83 to 5 percent. According to the Worldwatch Institute, using such methods to raise the worldwide efficiency of irrigation by only 10 percent would save enough water to supply all global residential uses of water.

Wasting Less Water in Homes and Industry

Leaky pipes, water mains, toilets, and faucets alone waste an estimated 20 to 35 percent of the water withdrawn from public supplies before it can be used. There is little incentive to reduce leaks and waste in many cities like New York City because instead of having individual water meters, users are charged flat rates.

Much money is spent to make potable *all* water coming into U.S. homes, buildings, and factories. Yet only about 5 percent of the water in homes and 2 percent of that in industry is used for purposes that require potable water. For example, in a home about 40 percent of the water of drinkable quality is used to flush toilets. Consumers can cut this use considerably by using water-conserving toilets and other water-saving methods and save money in the long run. Specific ways that individuals can save water are given in Appendix 4.

Some U.S. homeowners have set up a dual system with the "gray" water not used in toilets collected in a separate system and used to flush toilets and to water lawns and gardens, reducing average domestic water use by more than 50 percent. However, many health codes forbid such systems even though it has been shown that they are safe with proper installation and controls, and that they can greatly reduce public expenditures for providing water resources and waste treatment. The Pure-Cycle Company in Boulder, Colorado, has developed a complete home recycling system that sits in a small shed outside the residence. It costs about the same as a septic tank installation and is serviced for a monthly fee about equal to that for using most city water and sewer systems.

Much of the water withdrawn in industry is used for cooling and other processes that do not require that it be potable. Much of this water can be recycled several times before it is returned to the natural water cycle. For example, the water used for cooling in electric power plants can be cut by at least 98 percent by recycling it in cooling towers (Section 22-4) rather than removing water from a nearby water source, passing it through the plant once, and returning it to the original source, as is typically done.

Many manufacturing processes (Table 10-2) can use recycled water or be redesigned to use and waste less water. For example, depending on the process used, manufacturing a ton of steel can require as much as 200,000 liters (52,800 gallons) or as little as 5,000 liters (1,320 gallons) of water, and producing a ton of paper from 60,000 to 350,000 liters (15,850 to 92,470 gallons). Manufacturing a ton of aluminum from recycled scrap rather than virgin resources can reduce water needs by 97 percent. However, unless the price of water is increased to reflect its true cost to society, industries have little incentive to recycle water. For example, the cost of water is rarely more than 3 percent of the total manufacturing costs in industries that use large amounts of water, such as food products, pulp and paper, chemicals, metals, and petroleum.

Born in a water-rich environment, we have never really learned how important water is to us. . . . Where it has been cheap and plentiful, we have ignored it; where it has been rare and precious, we have spent it with shameful and unbecoming haste. . . . Everywhere we have poured filth into it.

William Ashworth

Discussion Topics

1. What physical property (or properties) of water:
   a. Account(s) for the fact that you exist?
   b. Allow(s) lakes to freeze from the top down?
   c. Help(s) protect you from the shock of sudden temperature changes?
   d. Help(s) regulate the climate?

2. If groundwater is a renewable resource, how can it be "mined" and depleted like a nonrenewable resource?

3. In your community:
   a. What are the major sources of the water supply?
   b. How is water use divided among agricultural, industrial, and domestic uses? Who are the biggest users of water?
   c. What has happened to water prices during the past 20 years?
   d. What water problems are projected for your community?
   e. How is water being wasted in your community and school?

4. Explain how you and I each can use, directly and indirectly, an average of 7,402 liters (1,953 gallons) of water per day. About how much of this is wasted?

5. Explain why dams may lead to more flood damage than might occur if they had not been built. Should all proposed large dam projects be scrapped? What criteria would you use in determining desirable dam projects?

6. How could we prevent or minimize soil salinity from irrigation and saltwater intrusion in coastal areas?

7. Explain why desalination, although important, will not solve world and U.S. water problems. Using the second law of energy (Section 3-3), explain why desalted water will not be available for widespread use in irrigation.

8. Debate the proposition that all users should pay more for water because its price is too low compared with the increasing costs of providing adequate, usable supplies for the growing U.S. population. What political effects might this have? What effects on the economy? On you? On the poor? On the environment?

Food Resources and World Hunger

Hunger is a curious thing: at first it is with you all the time, working and sleeping and in your dreams, and your belly cries out insistently, and there is a gnawing and a pain as if your very vitals were being devoured, and you must stop it at any cost. . . . Then the pain is no longer sharp, but dull, and this too is with you always.

Kamala Markandaya

Each day there are 212,000 more mouths to feed. Numerous agricultural experts estimate that world food production must double, if not triple, between 1980 and 2015 to feed adequately the 8 billion people projected to be living on this planet by 2015. This means that during a 35-year period that has already started, we must produce as much food as humankind has produced since the dawn of agriculture about 12,000 years ago. How can this be done? Even if enough food is grown, how can it be made available to those who can't afford to buy it? In this chapter we examine these questions to see how food, soil, water, fossil fuel, and fertilizer resources are interwoven with population growth and pollution.

11-1 Food Supply, Population Growth, and World Food Problems

Population Growth and Food Production
Thanks to improved agricultural technologies, practices, and policies, since 1950 food production has risen faster than population growth on all continents except Africa. Despite this success in growing more food, there are some signs of potential trouble. *First*, the rate of increase in the world average per capita food production has been steadily declining each decade: It rose 15 percent between 1950 and 1960, but 7 percent between 1960 and 1970, and only 4 percent between 1970 and 1980. *Second*, plagued by the fastest population growth of any continent (Figure 7-1 and Table 7-2), exten-

sive soil erosion and desertification, a 15-year drought, and underinvestment in agriculture, Africa's average food production per person fell 14 percent between 1970 and 1984. In 1970 most of Africa was self-sufficient in food. By 1984, however, one out of four Africans was fed with grain from abroad and millions were dying from starvation.

Third, although the *percentage* of the population suffering from hunger and malnutrition has declined by several orders of magnitude since 1950, there are more hungry and malnourished people today than in 1950 because of the large increase in population since that time. In 1982 the United Nations estimated that 450 million people—one out of every 10 people on earth—were underfed, mostly in Africa and Asia. The World Bank's estimate was higher: 780 million people—one out of 6.

Finally, 30 to 40 years ago most countries were self-sufficient in food, whereas today most have to import some of their food or are barely self-sufficient. The major food exporting nations now are the United States (which provides about 55 percent of all grain exports), Canada, Australia, New Zealand, France, and Argentina. Prolonged bad weather or economic problems in these countries, especially the United States and Canada, could spell disaster for hundreds of millions of people throughout the world, mostly in many LDCs. This increasing dependence on imports drains the national income of LDCs and reduces opportunities for investments in native food production.

The Geography of Hunger Generally increased worldwide food production hides widespread differences in average per capita food supply between and within nations, and even within a particular family. For instance, although total and per capita food supplies have increased in Latin America, much of this gain has been confined to Argentina and Brazil. For example, the average per capita food supply in Brazil is high, whereas in neighboring Bolivia it is fairly low. In more fertile and urbanized southern Brazil, the average daily per capita food supply is high, but in the semiarid, less fertile northeastern interior, many people are grossly underfed. In the MDCs too there are pockets of hunger. A 1985 report by a task force of physicians estimated that at least 20 million Americans—about 1 out of every 12 people—were hungry mostly because of cuts in food stamps and other forms of government aid.

Food is also poorly distributed within families. Among the poor, young children (ages 1 to 5), pregnant women, and nursing mothers are most likely to be underfed because the largest portion of the family food supply goes to working males.

World Food Problems and Proposed Solutions It is misleading to look at the race between food supplies and population growth as a problem of merely producing more food. If all the grain produced each year were distributed equally among the world's population, everyone would have a more than adequate daily diet. And this estimate does not include other foods, such as beans, fruits, nuts, vegetables, and grass-fed beef.

The world's supply of grain, however, is not distributed equally among the world's people, nor is it likely to be in the future. And even granting equal distribution, people surviving on grain alone would not receive certain essential proteins, vitamins, and minerals. For good health, both *quantity* and *quality* of food are important.

In most LDCs the total amount of food being produced is not the limiting factor causing hunger. Instead, poor people either do not have access to fertile land where they can grow enough food or they do not have the money to buy enough food of the right kind, regardless of how much is available. Thus, *poverty is the chief cause of hunger and malnutrition for individuals throughout the world*.

In addition, people near starvation are often afraid to risk their lives on a strange-looking or strange-tasting food. For example, wheat once sent by the United States to relieve starvation in India was not eaten by people used to rice. Many Americans would probably go hungry if grasshoppers—a delicacy in parts of Africa—were the major food available.

Besides quantity, then, the world's food problem includes a number of complex and interrelated agricultural, economic, social, and ecological problems, as summarized in the box on page 123 along with major solutions proposed by food experts to deal with these problems.

Many of the world's food and food-related problems and their proposed solutions are discussed in other chapters. The rest of this chapter is devoted to understanding the extent of world hunger and malnutrition and major agricultural system types, and evaluating proposals 1 through 6 in the summary box for wasting less food and producing more.

11-2 Human Nutrition, World Hunger, and Malnutrition

Human Nutrition For good health and survival, we require inputs of oxygen and water and five basic groups of nutrients from the food we eat: proteins, carbohydrates, fats, vitamins, and minerals. When food is eaten, digestive processes break down protein into about 22 different amino acid molecules. Carbohydrates are converted into sugars such as glucose, and fats become glycerol and fatty acids, which are more readily used by the body. Our food should also contain a seventh type of material, called *roughage*, consisting mostly of the indigestible cell walls of plants. Although not a nutrient, roughage provides bulk that helps the digestive system function properly and prevents constipation. Some researchers have suggested that the 90 percent decrease in roughage in the U.S. diet since 1900 may have contributed to the increased incidence of heart and circulatory diseases, alimentary tract diseases such as colon cancer (Section 20-3), obesity, and diabetes.

Eight of the 22 amino acids cannot be made in the body and must be obtained from the protein in food. If one or more of these **essential amino acids** are missing from or insufficient in the diet, protein malnutrition can result. **Complete proteins**, or animal proteins such as meat, fish, eggs, milk, and cheese, provide all eight essential amino acids. Plants, however, are **incomplete proteins**, lacking one or more of the essential amino acids; they are also usually low in protein quantity compared to meat and dairy products. However, it is not necessary to consume meat, fish, or dairy products to ensure against amino acid deficiency. The 2 million to 3 million strict vegetarians in the United States get all the essential amino acids they need by eating a proper combination of protein-rich plants such as soybeans, beans, peanuts, and peas and protein-deficient plants such as wheat, potatoes, corn, cassava, and rice. Most vegetarian diets also need to be supplemented with calcium, iron, and vitamins B_{12} and B_2 (riboflavin). However, the diet of most poor people in the LDCs consists primarily of only one or two protein-deficient plants. They cannot afford meat, dairy products, and protein-rich plants.

World Hunger Problems and Possible Solutions

Food and Food-Related Problems

1. *Quantity.* Producing enough food to feed the 4.9 billion humans on earth and the world's livestock, poultry, and pets, which require enough food to feed 16 billion people.

2. *Quality.* Producing food with enough fats, vitamins, critical minerals, and high-quality protein.

3. *Crop protection.* Protecting food before and after harvesting from the pests, diseases, and spoilage that destroy about 45 percent of the food grown each year.

4. *Distribution.* Providing the ships, planes, trains, trucks, storage facilities, roads, and marketing systems needed to store, transport, distribute, and sell food.

5. *Poverty.* Making sure people can afford to grow or buy the quantity and quality of food they need.

6. *Cultural acceptance.* Providing types of food that people with different cultural backgrounds and preferences will buy and eat.

7. *Resource supply.* Having enough fertile soil and fertilizers, pesticides, water, fossil fuels, and nonmineral resources to grow and distribute enough food for the world's human and animal populations.

8. *Economics.* Providing incentives for farmers, especially those in LDCs, to produce and sell more food.

9. *Climate and weather.* Stockpiling enough emergency food supplies to relieve famine when changes in local and global climate and weather patterns reduce food-growing capacity in various parts of the world (Sections 6-3 and 21-5).

10. *Population growth.* Trying to control population growth (Chapter 8) so that world food production and distribution won't have to be doubled every 34 years in the LDCs.

11. *Ecological effects.* Trying to produce, process, and distribute more food without seriously degrading soil, water, air, wildlife, and forests, rangelands, and estuaries.

Proposals for Solving World Food Problems

1. *Simplifying diets and using and wasting less food.* Shortening the food chain by reducing the use of meat and meat products in MDCs; reducing overnutrition and waste of food in industrialized nations.

2. *Using new foods, fabricated foods, and unconventional foods* (see Section 11-5).

3. *Getting more food from the ocean.* Trying to increase the world seafood catch, harvesting ocean algae and krill, and growing and harvesting fish and shellfish in inland ponds (aquaculture) and in fenced-off coastal areas (mariculture).

4. *Adding new farmland.* Cultivating more land by clearing forests, plowing pastures, draining wetlands, and irrigating arid land.

5. *Providing economic incentives.* Making credit available to small farmers, providing subsidies and price supports, and trying to control import and export food prices to encourage farmers to cultivate more land and increase crop yields per acre.

6. *Improving crop yields.* Increasing yields per acre by transferring or adapting industrialized agriculture to LDCs.

7. *Increasing foreign aid.* Having MDCs give to LDCs money and technical assistance designed to help these nations grow more of their own food; finding ways to get this aid to poor rural farmers without having it pass first through the hands of corrupt local officials, who often prevent much of it from reaching the poor.

8. *Instituting land ownership reform.* Encouraging governments in LDCs where land is in the hands of a relatively small number of rich and powerful individuals to distribute more land to the poor.

9. *Controlling population.* Limiting world population growth and size (Chapter 8).

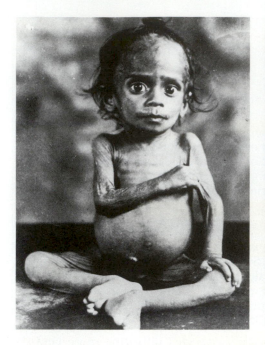

Figure 11-1 Most effects of severe protein–calorie malnutrition or marasmus can be corrected. This 2-year-old Venezuelan girl suffered from marasmus but recovered after 10 months of treatment and proper nutrition.

Plants and Animals That Feed the World Today only 16 plant species feed the world and provide the poor with almost all their food energy and more than three-fourths of their protein. These species are **(1)** five *cereals*: rice, wheat, maize (corn), sorghum, and barley, **(2)** two *sugar plants*: sugarcane and sugar beets, **(3)** three *root crops*: potatoes, sweet potatoes, and cassava, **(4)** three *legumes*: beans, soybeans, and peanuts, and **(5)** three *tree crops*: coconuts, bananas, and nuts. Three of these plants—wheat, rice, and maize (corn)—provide people in LDCs with about two-thirds of their daily energy supply and 55 percent of their protein. So important are these three cereal grains that more than half the world's cropland is devoted to growing them. These 16 basic food crops are supplemented by about 15 major species of vegetables and a like number of fruit crop species, which supply much of the vitamins and many of the minerals necessary to the human diet. Virtually all the domestic meat consumed by humans comes from fish and shellfish and just nine groups of livestock: cattle, sheep, pigs, chickens, turkeys, geese, ducks, goats, and water buffalo.

Undernutrition, Malnutrition, and Overnutrition People living mostly on one or more plants such as wheat, rice, maize, or cassava often suffer from **undernutrition,** or **insufficient caloric intake.** Good nutrition, however, involves more than a daily intake of a certain *quantity* of calories. People whose diets lack one or more essential amino acids, vitamins, or minerals suffer from **malnutrition.** Those who get neither enough calories nor enough protein, vitamins, and minerals are said to be suffering from **protein–calorie malnutrition.**

While an estimated one person in 10 in LDCs dies from undernutrition, malnutrition, or malnutrition-related diseases, about 15 of every 100 people in MDCs suffer from **overnutrition—eating too much of the wrong kinds of food.** In the United States, 10 to 12 percent of children and 35 to 50 percent of middle-aged adults are overweight. These overnourished people exist on diets high in calories, cholesterol-containing saturated fats, salt, sugar, and processed foods, and low in unprocessed fresh vegetables, fruits, and fiber. Partly as a result of their overnutrition, these people are at high risk of diabetes, hypertension, stroke, heart disease, intestinal cancer, tooth decay, and other health problems.

Nutritional Deficiency Diseases Protein–calorie malnutrition has a devastating effect on children under age 5, who need about twice as much protein and energy in relation to body weight as do adults. Most physical effects of severe protein–calorie malnutrition in infants and young children can be remedied if they get treatment and proper nutrition (Figure 11-1). Brain development, however, begins in the uterus and is usually complete by age 2. Some nutrition experts believe that malnutrition during this period can cause mental retardation that cannot be remedied by later corrective nutritional measures. This hypothesis is disputed, but regardless of whether such brain damage is permanent, many infants and young children in poor families throughout the world are

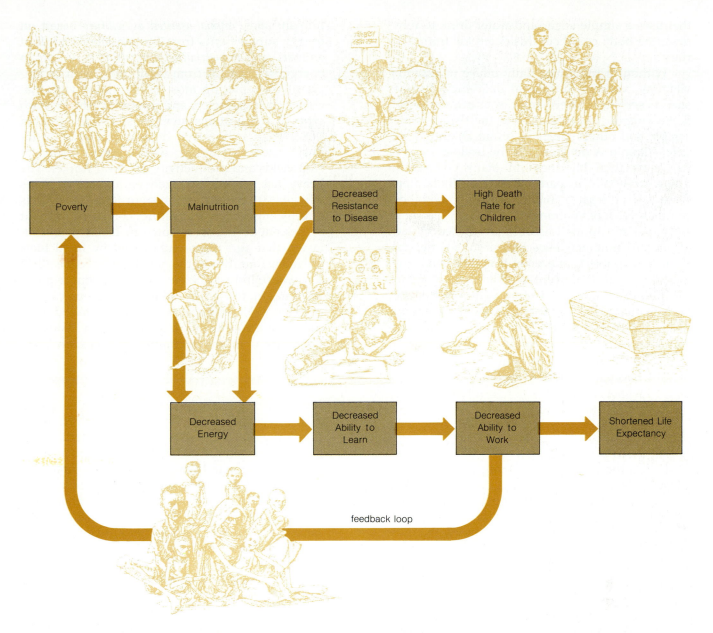

The tragic cycle of interactions among poverty, malnutrition, and disease.

Figure 11-2 The tragic cycle of interactions among poverty, malnutrition, and disease.

not getting the type of diet needed to prevent it. As a result they may be locked into a tragic cycle of malnutrition, infection, and poverty that tends to be passed on to succeeding generations (Figure 11-2).

The two most widespread nutritional deficiency diseases are *marasmus* and *kwashiorkor*. **Marasmus** (from the Greek "to waste away") is a result of a diet that is low in calories and protein. It occurs in poor families where children are not breast-fed or where there is insufficient food after the children are weaned. **Kwashiorkor** (meaning "displaced child" in a West African dialect) occurs with a diet adequate in calories but deficient in protein. It is found primarily in infants and young children soon after they are weaned and placed on

a starchy diet of cassava, maize flour, or other cereals of low protein quantity and quality.

The World Health Organization (WHO) and the UN Food and Agriculture Organization (FAO) estimate that at least *14 million people—half of whom are children under age 5—now die each year from undernutrition, malnutrition, and associated diseases, amounting to an average of 38,400 such deaths each day.* Most of these victims do not starve to death. They die because they become more vulnerable to normally minor infections and diseases such as measles, diarrhea, and flu. For example, the death rate from measles is typically 480 times higher in Ecuador than in the United States. WHO estimates that diarrhea kills at least 5 million children a year. If made available in LDCs, a low-cost treatment that uses a simple sugar-and-water drink to rehydrate the body could cut this death toll from diarrhea.

Without a minimum daily intake of essential vitamins, various nutritional deficiency diseases such as scurvy, beriberi, and rickets can also occur. Some vitamins (such as C and the Bs) are water soluble and others (such as A and D) are fat soluble. Vitamin A deficiency is the leading cause of partial or total blindness in many children in LDCs, especially in parts of Asia. In India alone there are at least 1 million cases of blindness caused by lack of vitamin A.

Other nutritional diseases are caused by lack of certain minerals, especially iron, iodine, calcium, phosphorus, and zinc. Iron-deficiency anemia affects 5 to 15 percent of all adult men, a third of all adult women, and more than half the children in many LDCs. Anemia saps one's energy, makes infection more likely, and increases a woman's chances of dying in childbirth. Every year an estimated 200 million people in the LDCs go deaf or mute because of a lack of iodine.

Importance of Breast-Feeding Breast milk supplies all or most of an infant's nutritional needs during its critical first year of life, and it contains antibodies that help protect infants from diarrhea and other infections. Moreover, breast-feeding is an important form of birth control that protects nursing women from pregnancy for 10 weeks to 26 months after giving birth, depending on the individual mother's health and diet and the intensity of suckling. This effect is particularly important in controlling the population in LDCs, where contraceptives and other means of birth control are not widely available or are too costly for the poor.

In MDCs such as the United States only about 30 percent of the babies born each year are breast-fed, although this percentage is gradually increasing. In LDCs the percentage of breast-fed babies is much higher, ranging from 67 percent in Malaysia to 98 percent in Nepal. However, the practice is decreasing in many LDCs, especially in Latin America.

11-3 World Agricultural Systems

Major Types of Agriculture The three major agricultural systems used in the world today require that the input of solar energy used in photosynthesis be supplemented with other forms of energy. These systems are *simple (subsistence) agriculture* based on an energy supplement from human labor, *animal-assisted agriculture* based on energy supplements from human labor and draft animals, and *industrialized agriculture* based on an energy supplement from fossil fuels, with an emphasis on replacing most human labor with machines such as tractors and combines. The first two types of agricultural system are still widely used in LDCs. MDCs, however, depend primarily on industrialized agriculture, which greatly increases crop yields and productivity per farm worker. The heavy use of industrialized agriculture explains why only 12 percent of the labor force in MDCs is directly engaged in agriculture, compared to about 60 percent of the work force in LDCs. It is estimated that agriculture subsidized by fossil fuels provides four times the yield per hectare (2.47 acres) of unindustrialized agriculture, but requires 100 times more energy and mineral resources.

The major components of industralized agriculture in a country such as the United States are summarized in the accompanying box on page 127.

Industrialized Agriculture in the United States The success of industrialized agriculture when coupled with a favorable climate and fertile soils has been demonstrated by the dramatic increase in food production in the United States (Figure 11-3). Between 1820 and 1985, the percentage of the total U.S. population working on farms declined from about 72 percent to 2.4 percent, but during the same period total U.S. food production approximately doubled, and the output per farmer increased eightfold. In 1985 each U.S. farm worker produced enough food and fiber to meet the needs of 77 people.

With total assets exceeding $1 trillion, agriculture is the biggest industry in the United States—bigger than the automobile, steel, and housing industries combined. Its annual sales represent about one-fifth of the nation's annual gross national product. It is also the nation's largest employer. There are 23 million workers, representing one-fifth of all jobs in private enterprise, in some phase of agriculture from growing food to selling it at the supermarket. U.S. farmers, however, received only about $83.5 billion of the nearly $300 billion Americans spent on food during 1982. The remaining $214 billion went to the food-processing, packaging, advertising, and retailing industries.

Energy Use and Industrialized Agriculture The energy used to grow, process, package, distribute, and prepare food is about 17 percent of total an-

Components of Modern Industrialized Agriculture

1. *Mechanization*. Substituting fossil-fuel-powered machines for manual labor and draft animals. At least two-thirds of the world's cropland is tilled with mechanical power.

2. *Commercial fertilizers*. Using commercially produced fertilizers to increase crop yields and restore soil fertility partially (Section 9-4). One-fourth of the world's annual food output is based on the use of commercial fertilizers.

3. *Irrigation*. Building dams and canals and using fossil-fuel-powered machines to pump water to cropland.

4. *Pesticides*. Using synthetic chemicals to reduce crop losses due to insects, disease, and weed pests (Chapter 23).

5. *Soil conservation*. Establishing and adhering to sound land-use practices to reduce soil erosion (Section 9-4).

6. *Animal feedlots*. Shifting from pastures and open rangeland to feedlots, where hundreds to thousands of domesticated animals are concentrated and fed in a small space to encourage rapid weight gain and more efficient production (Section 11-4).

7. *Genetic selection and hybridization*. Using research to select and develop high-yield and disease-resistant crops, livestock, and poultry (Section 11-8).

8. *Large-scale and specialized production*. Shifting from small, diversified farms to large, specialized farms.

9. *Storage, processing, distribution, and marketing*. Developing storage facilities and extensive transportation, processing, and marketing networks.

10. *Agricultural training and research*. Establishing a system of agricultural schools and research centers and using extension services to expose farmers to new developments.

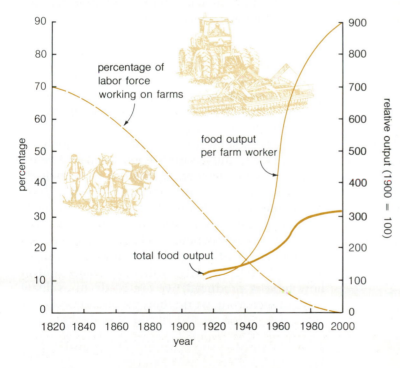

Figure 11-3 Some effects of increased use of industrialized agriculture in the United States.

nual energy use in the United States. Counting the fossil fuel energy inputs used to grow, process, package, transport, refrigerate, and cook all plant and animal food in the United States, *it takes about 9 calories of fossil fuel energy to put 1 calorie of food energy on the table—an energy loss of 8 calories per calorie of food energy*.

To feed the world using the U.S. meat-centered diet with food produced by its energy-intensive agriculture would require twice as much global cropland as is now being cultivated, greatly increase environmental pollution and land degradation, consume 80 percent of the world's annual energy use, and wholly deplete the world's known oil reserves in 13 years. If fossil fuels suddenly became scarce or too expensive, the present agricultural system in industrialized nations would probably collapse, causing a sharp drop in world food production and a corresponding rise in malnutrition and famine.

Organic Farming and Permaculture As fuel prices rise, an increasing number of farmers (about 30,000 in the United States) are using **organic farming**, a method of producing crops and livestock naturally by using only organic fertilizers such as animal manure, green manure, legumes, and compost (Section 9-4) and natural pest control ("good" insects that eat "bad" insects, plants that repel bugs, and crop rotation) instead of synthetic chemical pesticides and herbicides.

A comparison of U.S. organic and nonorganic industrialized farms showed that: **(1)** organic farming provides enough nitrogen for plant growth, but it may not provide adequate phosphorus and potassium; **(2)** pest control without pesticides is moderately successful and may be more successful with increased use of integrated pest management (Section 23-6); **(3)** some organic farms have 5 to 15 percent higher yields per hectare than nonorganic farms, but this varies by crop and region; **(4)** the net financial returns for both types of agriculture are about equal per hectare of farmland, but the indebtedness of organic farmers is usually lower; and **(5)** organic farms use an average of 40 percent less energy per unit of food produced than do conventional farms. A USDA study indicated that a total shift to organic farming in the United States could meet domestic food needs, but there would be some decrease in food available for export. There would also be intense political opposition to such a shift by agricultural chemical industries that sell billions of dollars of fertilizers and pesticides each year.

Experiments with developing self-sustaining agriculture, known as **permaculture**, are being carried out in several places throughout the world including the New Alchemy Institute in Massachusetts (see Appendix 1). Permaculture takes the practices of organic farming one step further by designing an energy-efficient, low-maintenance, high-yielding, interconnected ecosystem in the self-sustaining production of food, fiber, and energy. In such a system some foods are grown outdoors and others in solar-heated greenhouses. Components of the system usually have multiple

functions. For example, large water tanks inside a solar greenhouse **(1)** provide thermal mass to store heat from the sun during the day (for release at night) or during cloudy days (to heat the greenhouse and an attached dwelling), **(2)** are used as ponds to grow fish to provide protein, and **(3)** provide warm, fertile irrigation water for vegetable crops grown inside and outside the greenhouse. Livestock manure is recycled in composting systems, fish ponds, gardens, and orchards.

Home Gardening By 1985, 53 percent of all American households were growing some of their own food, having a retail value of at least $20 billion. Increasingly, manicured lawns are being replaced with small garden plots, and city residents are planting gardens in apartment and condominium grounds, church and school yards, factory and shop lots, public parks, and utility rights-of-way. Window box planters, small greenhouses, and rooftop and patio gardens, where tomatoes and other crops can be grown in small containers, are also widely used. Raised beds are being used to produce high yields of a large variety of vegetables in a small space, a gardening method used for centuries in China. According to USDA estimates, a family of four could be well fed without animal products by using high-yield intensive gardening on only one-sixth of an acre.

Environmental Effects of Producing More Food As agricultural expert Lester R. Brown put it, "The central question is no longer 'Can we produce enough food?' but 'What are the environmental consequences of attempting to do so?'" The entire agricultural system probably has a greater overall environmental impact on the air, land, and water than any other sysytem in modern industrialized societies. The major environmental effects of food production and some proposed solutions are summarized in Table 11-1. More detailed information on these environmental effects and proposed solutions is found throughout much of this text, as indicated in Table 11-1. With this background in how food is produced, we are ready to evaluate ways of increasing world food production.

11-4 Simplifying Affluent Diets and Wasting Less Food and Energy

Human Diets and Food Chain Losses Food chain losses of usable energy (Section 5-2) explain why meat is more expensive than grain and why poor people are forced to live primarily on a diet of grain

Table 11-1 Environmental Effects of Food Production and Proposed Solutions

Effect	Proposed Solutions and Text Discussion
Overfishing	Mariculture and aquaculture (Section 11-6)
Overgrazing	Laws and land-use control (Sections 12-6 and 14-4); less dependence on meat in diets (Section 11-4)
Soil erosion and loss of soil fertility	Minimum-tillage cultivation, crop rotation, and other well-known practices (Section 9-4); land-use control (Section 14-4)
Salt buildup (salination) and waterlogging of irrigated soils	Drainage systems, drip-and-trickle irrigation, and higher water prices (Section 10-6)
Waterborne diseases from irrigation	Education of poor people and research to prevent or cure diseases (Section 20-2)
Loss of forests (deforestation)	Replanting forests (Section 12-5); land-use controls (Section 14-4)
Endangered wildlife from loss of habitat	Laws for wildlife protection (Section 13-3); land-use control (Section 14-4)
Loss of genetic diversity	Genetic storage banks for plants (Section 11-8) and wildlife protection (Section 13-4)
Pollution from pesticides	Increased use of biological control and integrated pest management (Section 23-6)
Water pollution from runoff of fertilizer and animal wastes	Preventing soil erosion and recycling animal wastes to land (Section 9-4)
Climate change from land clearing	Land-use control (Section 14-4)
Air pollution from use of fossil fuels	Reduced use of fossil fuels in agriculture (Section 11-4 and Chapter 21)
Health dangers from food additives	Better enforcement of existing laws (Chapter 20)

rather than meat. This loss of usable food energy primarily as a result of the second law of thermodynamics (Figures 5-3 and 5-6), however, is important only when (1) livestock and poultry animals eat grains or other plants that could be eaten by humans, or (2) these animals are grazing on land or being fed from forage crops (such as hay and alfalfa) grown on land that could be used to grow food crops for humans.

As their incomes rise, people begin to consume more grain indirectly, in the form of meat and meat products, than directly in cereal, flour, and other plant products. Because of this trend toward meat-based diets, about one-third of the world's annual grain production is fed to livestock and poultry to produce meat, eggs, and milk for the affluent. By contrast, only about 3 percent of the available grain supply is fed to livestock in China, India, and most LDCs. Even though there is more hunger and protein malnutrition in Africa than in any other continent, one-third of Africa's protein-rich peanuts are sold as feed for European livestock.

In the United States and Canada, about three-fourths of the grain produced each year is fed to livestock, especially cattle fattened in feedlots a few weeks or months before slaughtering. These animals are fed a rich diet of grain mixed with growth-stimulating chemicals (steroids) and antibiotics. There is growing concern that these antibiotic additives may increase the number of bacteria strains immune to antibiotics in meat-eating humans, thus decreasing our protection against infectious diseases. Animal feedlots also require large inputs of fossil fuel energy—about 20 times that for beef feeding on open rangeland.

Altering Affluent Diets to Reduce Food Waste A number of food experts have called for peoples of affluent nations to alter their diets reduce food waste, thereby making more food available for export to feed the poor. Some analysts have pointed out that theoretically, if cattle in the United States were not fattened in feedlots, enough grain would be available to feed about 400 million people—equal to 80 percent of Africa's population. Even if feedlots were not eliminated, merely decreasing annual meat consumption in the United States by 10 percent theoretically could release enough grain to feed 60 million people. Similarly, commercial fertilizers that are spread each year on U.S. lawns, golf courses, and cemeteries could be used to produce grain for 65 million people.

Pet foods consumed in the United States each year contain enough protein to feed 21 million people. Pets can offset loneliness and help limit population growth by serving as child substitutes. But some experts and animal lovers believe the growth of the pet population in the United States and in many European nations is getting out of control. The United States has the world's highest ratio of pets to people, followed closely by France and Great Britain. Dogs and cats carry 65 diseases transmissible to humans, litter streets with feces and urine, bite humans, and make noise. Furthermore, U.S. taxpayers pay $600 million a year to dispose of 20 million abandoned dogs and cats.

It has also been suggested that education and financial incentives be used to encourage people to throw away less food and to reduce the size of portions served. An estimated 20 percent of all food produced in the United States is wasted; enough food is thrown away each day to feed (theoretically) more than 50 million people a U.S. meat-based diet—twice the number of people below the poverty level in the United States.

However, some analysts point out that grain made available by eating less meat, eliminating feedlots, and so on will not necessarily go to the poor in LDCs or in the United States. About 70 percent of U.S. food exports are used to fatten livestock for the meat-based diets of the world's affluent. Others argue that conserving food in affluent nations could result in less food for the poor because it would lead to food surpluses, price declines, and cuts in production, leading to reductions in exports to other nations.

Furthermore, without feedlots the United States would need to plant more land in forage to keep beef production at the present level. Planting this land would increase energy use up to five times the present levels because of the large inputs of fertilizer and water needed to make these marginal lands usable. In addition, hogs and poultry cannot digest grass and thus cannot be raised on either rangeland or forage. The amount of fossil fuels used to produce food crops can be reduced, but energy savings in one area of food production can lead to increased energy use in another area. For example, little fossil fuel is saved by substituting animal manure for commercial fertilizer unless pastures and feedlots are fairly close to the crop fields. Otherwise, most fossil fuel not used to produce fertilizer is needed to haul the manure long distances.

11-5 Unconventional, Fortified, and Fabricated Foods

Unconventional Foods *Over the next several decades, most of the increase in the world's food supply will result from expanding the supplies of traditional foods—wheat, rice, and maize.* But some suggest the cultivation of other nontraditional plants to supplement or replace such traditional foods in LDCs. Examples include the *winged bean*, containing as much protein as soybeans; *cocoyam*, a native plant of West Africa and Central and South America, as nutritious as the potato; *Ye-ed*, a small bush native to East Africa whose seeds yield a nutritionally balanced diet; and *leucenda*, a protein-rich tropical legume that can grow 3.6 meters (12 feet) high in only

6 months and is also usable for fuelwood and to add nitrogen to the soil.

Some of the insects that compete with us for food crops are also important potential sources of protein. In the Kalahari Desert of Africa cockroaches are a diet staple; crickets and locusts are standard fare in several African countries. Lightly toasted butterflies are a favorite food in Bali. French-fried ants are sold on the streets of Bogotá. Malaysians love deep-fried grasshoppers, and New Guinea residents enjoy eating roasted wood spiders, which taste something like peanut butter. Most of these insects are 58 to 78 percent protein by weight—three to four times as protein rich as beef, fish, or eggs.

Vitamin and Protein Supplements and Fabricated Foods In the United States, bread and flour enriched with vitamins, minerals, and amino acids have helped to eliminate many nutritional deficiency diseases, and adding small amounts of iodine to table salt has virtually wiped out goiter. Similarly, Japan has essentially eradicated beriberi since World War II by enriching its rice with vitamin B_1. Enriching existing foods is relatively inexpensive and does not require people to change their eating habits. A major disadvantage, however, is that these processed foods are not normally available to rural people in LDCs.

In recent years some have suggested that *single-cell protein (SCP)* be used as a protein supplement for humans and especially in animal feed. This high-protein powder can be produced from oil, natural gas, alcohol, sewage, waste paper, and other organic materials by the action of one-celled organisms such as yeast, fungi, and bacteria. A few small-scale pilot factories, mostly in Europe, are producing SCP for use in animal feed, but there are a number of problems. Thus, if SCP can be produced economically on a large scale, most feel it would be best used as a supplement in animal feed, reducing the use of grain.

At present SCP is not useful as a protein supplement for human food because (1) its high ribonucleic acid (RNA) content can cause gout and kidney stone formation; (2) the walls of the cells are indigestible and can cause diarrhea, nausea, and gastric distress; (3) it has a taste many people find unpleasant; and (4) it is fairly costly to produce. Even if these problems can be overcome, some argue that it makes no sense economically or ecologically to use increasingly expensive and dwindling petroleum and natural gas resources to produce edible protein in this form.

Another approach is to supplement the protein in foods with meal or flour made from soybeans, cottonseed, peanuts, coconuts, sunflower seeds, rape seeds, and other oil seeds. Soybean meal has been used as a protein supplement for decades, along with soybean-based soft drinks. Problems arise, however, in making some of these meal supplements palatable and in removing toxic compounds.

LDCs have had low-cost, high-protein meat substitutes, such as Indonesian *tempeh* (produced from soybeans), for hundreds of years. In the United States, oleomargarine and vegetable oils have reduced the use of animal products like butter and lard. In MDCs there has been increasing use of imitation bacon, eggs, chicken, ham, and meat extenders, which contain spun vegetable protein (SVP) fibers, made primarily from soybean concentrate and wheat gluten. If these products were widely accepted, they could dramatically reduce meat consumption in nations such as the United States. Since SVP contains no cholesterol, it could also reduce the incidence of heart disease. To make these products look and taste like meat, a number of dyes, flavors, and other food additives must be used, which some fear might increase risks of cancer and other health hazards (Section 20-4).

11-6   Catching More Fish and Fish Farming

Trends in the World Fish Catch   Fish are the major source of animal protein for more than half the world's people, especially in Asia and Africa. Fish supply about 23 percent of the animal protein worldwide (considerably more than that from beef, twice as much as from eggs, and three times as much as from poultry), and 6 percent of all human protein consumption. Two-thirds of the annual fish catch is eaten by humans. The remaining third, consisting largely of inferior species or waste from the processing of table fish, is processed into fish meal and fed mostly to chickens and hogs.

Between 1950 and 1970, the marine fish catch more than tripled—an increase greater than that occurring in any other human food source during the same period. To achieve large catches, modern fishing fleets use sonar, helicopters, aerial photography, and temperature measurement to locate schools of fish, and lights and electrodes to attract them. Large, floating factory ships follow the fleets to process the catch.

Despite this technological sophistication, the steady rise in the marine fish catch halted abruptly in 1971. Between 1971 and 1976 the annual catch leveled off, and rose only slightly between 1976

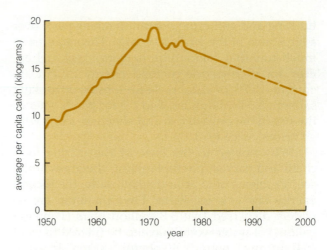

Figure 11-4   Although the total world fish catch has grown between 1950 and 1985 (except between 1971 and 1976), it has not kept up with population growth. As a result, the average per capita world fish catch has declined in most years since 1970 and is projected to decline further between 1985 and 2000 (dashed lines). (Sources: United Nations and Worldwatch Institute.)

and 1985. A major factor in this trend was the sharp decline of the Peruvian anchovy catch, which once made up 20 percent of the global ocean harvest. A combination of overfishing and a shift in the cool, nutrient-rich currents off the coast of Peru were apparently the causes of this decline, which also threw tens of thousands of Peruvians out of work. Meanwhile, world population continued to grow, so between 1970 and 1985 the average fish catch per person declined; it is projected to decline even further back to the 1960 level by the year 2000 (Figure 11-4).

Some scientists believe that the world's annual fish catch could be increased by as much as 43 percent by 2000. But others estimate that it may already be at or near its maximum sustainable yield, and some fear that even the present annual catch may not be sustainable. To avoid overfishing and to leave enough breeding stock for the next year's catch, no more than 40 percent of the available fish in a species should be harvested in a given year. By the early 1980s, overfishing was causing declines in the annual yields of 30 of the leading species of table fish, and 11 of the world's major fishing grounds had been depleted to the point of near-collapse.

Ocean Food Chains   A typical ocean food chain is phytoplankton → zooplankton → mackerel → tuna → humans. Because of the second law of thermodynamics, most favored food fish species are fewer in number than species in lower trophic levels. Thus, catching these fish commercially is a hunting-and-gathering operation taking place over a large area of the world's oceans. This requires so much fuel that caloric inputs from fossil fuel energy for each calorie of food energy caught are enormous: 75 for shrimp, 34 for lobster, 20 for king salmon, 5.3 for flounder, and 1.3 for perch—hence the rise in the price of fish and shellfish, along with oil prices, since 1973.

Why not go to the base of the marine food chain and harvest phytoplankton, the grass of the sea? Problems of taste and smell aside, the second energy law gets us again. Phytoplankton are so widely dispersed that 3.8 million liters (1 million gallons) of water would have to be filtered to yield 454 grams (1 pound) of phytoplankton. In addition, the organisms have to be processed to remove RNA impurities and to break down their cell walls to make them digestible. The extremely large fossil fuel input needed to do all this makes phytoplankton cost more than they are worth as food.

However, we might increase the protein yield of the sea by going one step above plants in the food chain and harvesting krill—tiny, shrimplike crustaceans that are plentiful in Antarctic waters. Krill might be particularly useful as a livestock feed supplement. By 1985 at least eight nations, led by the Soviet Union and Japan, were harvesting krill. Soviet and Japanese scientists believe that each year we could harvest krill equal to the total current annual marine fish catch. But others project that the krill catch may not increase significantly because of present difficulties in making a product that is palatable to consumers, and the high energy and economic costs of harvesting krill, due to their location in distant Antarctic waters and the necessity of straining them from large quantities of seawater. Some scientists also warn that extensive harvesting could endanger aquatic krill eaters, including baleen whales, such as the already threatened blue whale, and several species of fish, seabirds, seals, and penguins.

Fish Farming: Aquaculture and Mariculture If we have trouble catching more fish and shellfish using our present hunting-and-gathering approach, why not raise and harvest fish crops in land-based ponds (*aquaculture* or *fish farming*) or in fenced-in coastal lagoons and estuaries (*mariculture*)? These two approaches already supply about 11 percent of the total world aquatic catch and roughly one-sixth of the seafood consumed directly by humans. Some scientists believe that yields can be increased severalfold by 2000.

Aquaculture is not new. In Asia species such as carp (a species of goldfish), mullet, tilapia, and milkfish have been raised in ponds, canals, and rice paddies for several thousand years. Catfish farms have been common in many parts of the southern United States since the 1960s, and the industry is rapidly expanding. In such land-based aquaculture or fish farming, a complete ecosystem is set up in a pond or small lake. Commercial fertilizers, animal wastes, fish wastes, and even sewage are used to produce phytoplankton. These are eaten by zooplankton and bottom animals, which in turn are eaten by fish.

One of the advantages of fish farming is the high efficiency with which fish convert plant matter to meat. For example, catfish require only 1.5 grams of grain to produce 1 gram of meat, compared to similar ratios of 2.25 for chickens, 3.25 for pigs, and 7.5 for beef cattle fattened in feedlots.

There is no doubt that aquaculture can be a growing source of low-cost, high-quality protein for local consumption in many LDCs—especially those with many lakes, ponds, and marshes. One problem, however, is that the fish can be killed by pesticide runoff from nearby croplands, as has happened in aquaculture ponds in the Philippines, Indonesia, and Malaysia. The rate of growth of aquaculture may be limited, however, by lack of availability of land or by competition with land used to produce other crops and raise livestock.

Japan, the Soviet Union, and the United States have used **mariculture** in estuaries to raise fish and shellfish, especially shrimp, lobster, oysters, and salmon. Estuaries are natural sinks for nutrients flowing from the land to the sea. With controlled fertilization, they can be used to produce large yields of desirable marine species in fenced-off bays, large tanks, or floating cages. Although mariculture could be an important source of protein within 20 to 30 years, many experts doubt that it will live up to its potential. The growing pollution of the sea, particularly along coastal and estuarine zones, threatens both cultivated and wild fish and shellfish (Section 22-6). In addition, coastal wetlands, especially in MDCs, are high in demand for harbors, industry, housing, recreation, and other uses.

11-7 Cultivating More Land

Availability of Arable Land Between 1950 and 1980 the total amount of land used to grow crops worldwide increased by about 28 percent. However, according to the USDA, the rate of growth in the world's cropland dropped from a yearly average of 1 percent in the 1950s to 0.3 percent in the 1970s. The growth rate projected for the 1980s is 0.2 percent; that for the 1990s, 0.15 percent.

Roughly one-third of the world's population now lives in nations where the total area of cropland is shrinking. Furthermore, because of population growth, the average per capita amount of arable land decreased by 21 percent between 1950 and 1980 and is expected to decrease further between 1980 and the end of the century.

However, using the classification in Figure 11-5, some analysts have suggested that we could at least double the world's cropland. Others believe this projection probably will never be realized because of one or more of the following limiting factors: (1) remote location and insect infestation, (2) poor soils, (3) lack of water, (4) conversion of existing cropland to other uses, (5) excessive costs, and (6) lack of economic incentives. Even if more cropland is developed, much of this increase will be used to offset the projected loss of almost one-third of today's cultivated cropland and rangeland through a combination of erosion, overgrazing, waterlogging, salinization, mining, and urbanization. Let's look briefly at each of these limitations.

Remote Location and Insect Infestation as Limiting Factors

About 83 percent of the world's new potential cropland is in the remote and lightly populated Amazon and Orinoco river basins in South America and in Africa's rain forests. Cultivating this land would require massive investments in land clearing and in transportation to ship the harvested crops to places they are needed. In West Africa, potential cropland equal to five times the area now farmed in the United States cannot be used for grazing or farming because it is infested with tsetse flies, which carry sleeping sickness and appear to be essentially ineradicable.

Soil as a Limiting Factor

About 56 percent of the land in the world that theoretically can be converted to cropland (Figure 11-5) is found under the moist jungles and tropical rain forests in Latin America (especially Brazil) and Africa (mostly West Africa). Although blessed with plentiful rainfall and long or continuous growing seasons, the soils of many of these tropical rain forests are not suitable for intensive cultivation because much of the plant nutrient supply is tied up in ground litter and vegetation rather than stored in the soil. Once the forest was cleared, most of the few plant nutrients stored in the soil would be rapidly leached from the soil by heavy rainfall. In addition, when an estimated 5 to 15 percent of these soils are cleared, they bake under the tropical sun into brick-hard surfaces called laterites, which are useless for farming (Section 9-2).

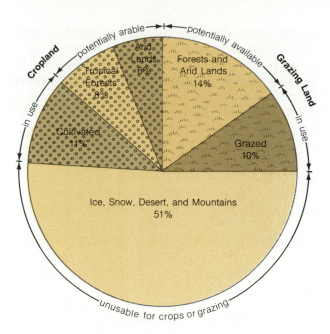

Figure 11-5 Classification of the earth's land. Theoretically, the world's cropland could be doubled in size by clearing tropical forests and irrigating arid lands. But converting this marginal land to cropland would destroy valuable forest resources, cause serious environmental problems, and possibly cost more than it is worth.

Some tropical soils can produce up to three crops of grain per year if massive quantities of fertilizer are applied at the right time, but the costs are high. To make matters worse, the warm temperatures, high moisture, and year-round growing season support large populations of pests and diseases that can wipe out crops. Most pesticides would be washed away by the heavy rains, anyway.

The ecology of tropical ecosystems is less well understood than that of most others, and researchers hope to develop methods for more intensive cultivation in these areas. Some scientists, however, argue that agricultural production in the tropics should be confined primarily to tree products such as those obtained from rubber trees, oil palms, and banana trees adapted to the existing climates and soils.

Irrigation: Water as a Limiting Factor

Almost half the world's food and 30 percent of that in the United States is produced on irrigated acreage (Chapter 10). It is estimated that between 1975 and 2000, the amount of irrigated land will have to at least double for food production to keep up with population growth.

Some doubt that such an increase can be accomplished. More than half the remaining potentially arable land lies in dry areas (Figure 11-5), where water shortages limit crop growth. Large-scale irrigation in such areas would be very expensive, requiring large inputs of fossil fuel energy to pump water long distances. Most of the world's accessible rivers are already dammed to provide irrigation water. Those remaining, such as the Mekong and the Amazon, are not where the water need is greatest. The Amazon's vast width also makes it almost impossible to harness. In addition, the lakes behind dams gradually fill with silt and provide only temporary supplies of irrigation water (Table 10-4). In some areas irrigation water can be pumped up from deep underground aquifers. However, in addition to incurring high energy costs, this can deplete slowly recharged underground aquifers and in the long run decrease crop productivity (Section 10-3).

To complicate matters further, it is estimated that half of the world's irrigated lands are damaged to some extent by salt buildup on the surface (salinization) and waterlogging and large areas have had to be removed from production (Section 10-5). Drainage systems that remove excess water and salt from irrigated fields can be installed. But this is expensive, and the salt removed from these fields can pollute nearby rivers. Some people talk of desalinating ocean water for irrigation, but it takes so much energy to remove the salt and to pump the water inland where it is needed that the method would not be cheap enough for agriculture.

To many observers the key to using irrigation to produce more food does not lie in irrigating more land. Instead, it involves using a variety of water conservation practices to use less water to produce more crops on land already being irrigated (Section 10-6).

Loss of Existing Cropland as a Limiting Factor

Each year the United States loses enough cropland and rangeland to highways, shopping centers, airports, housing developments, factories, reservoirs, and other forms of nonagricultural land use to equal a 0.8-kilometer-wide (0.5-mile-wide) strip stretching from New York to San Francisco. Some marginal land in the United States can be converted to farmland to replace that lost by urbanization and erosion, but such conversion is costly and can increase erosion and water pollution.

Others point out that conversion of farmland to other uses amounts to an annual loss of only about 0.18 percent of the nation's potential cropland and less than 0.24 percent of existing cropland. At this rate, the United States would lose only about 3.6 percent of its potential cropland between 1980 and 2000 and would retain ample acreage to meet domestic and international demands. It is also argued that this loss of cropland is not nearly as serious as that from soil erosion (Section 9-3) and that such conversions may be desirable from an economic standpoint because many nonagricultural uses of cropland have their own value and contribute to the nation's economic growth.

A new threat to the use of cropland to produce food in the United States and in other nations, such as Brazil, is the increasing interest in growing grain or other crops for conversion to ethyl alcohol to make gasohol (gasoline containing 10 to 23 percent ethyl alcohol: see Section 19-5). Fueling all U.S. cars with gasohol would require about 60 percent of the nation's corn crop and could reduce the acreage available for growing food for export. The poor of the world could also lose, because affluent people can afford to pay more for the grain to be used as fuel, thus driving world grain prices up.

Money and Lack of Economic Incentive as Limiting Factors

According to agricultural expert Lester Brown, "The people who are talking about cultivating more land are not considering the cost. If you are willing to pay the cost, you can farm the slope of Mount Everest." Thus, the real questions are: How much will it cost to expand the amount of cropland in many parts of the world, and how will this cost relate to the ability of the poorest people in the world to pay for the food grown there?

Regardless of the amount of arable land available, it will not be used to produce food for other people unless farmers can make some profit on a year-to-year basis. Because of a lack of economic incentive, much of the potentially arable land in both MDCs and LDCs is not cultivated.

Agricultural economics is a tricky business. If a government establishes popular but artificially low food price ceilings and export taxes, farmers have little incentive to produce food to support people living in the cities. Government price supports, cash subsidies, and import restrictions can be used to stimulate crop production by guaranteeing farmers a certain minimum return on their investment from year to year. For example, despite population problems, the average food supply per person in India has been rising for more than two decades. One reason was the green revolution, which enabled Indian farmers to increase crop yields of wheat per area of cultivated land, as discussed in the next section. But the main reason was the removal of government price controls that were keeping wheat prices at artificially low levels and their replacement with price supports that put the price of domestic wheat the same as the import price. With this economic incentive, Indian farmers began producing much more wheat.

On the other hand, if government price supports are too generous, people in the city pay high prices to subsidize the rural farming population, and with good weather farmers may produce more food than can be stored, transported, and sold. Food prices and profits then drop because of the oversupply, and a number of debt-ridden farmers go bankrupt, as was occurring in the United States in the early 1980s.

Some analysts have suggested that the government phase out price supports for crops over several years. The money saved by the government would be used to help farmers pay off their debts, estimated at $212 billion in 1985, provided they agreed to and implemented an approved soil conservation program. Once farmers' debts were whittled down to a manageable level, government payments would stop and farmers would respond only to demands of the actual market. Phasing out crop price supports, however, is difficult to accomplish from a political standpoint.

11-8   Increasing Crop Yields

The Green Revolution   Most experts agree that the quickest and usually the cheapest way to grow more food is to raise the yield per area of existing cropland. This is done by crossbreeding wild and existing strains of crops to develop new varieties that are better adapted to climate and soil conditions and can produce higher yields because they are able to make increased use of fertilizer, water, and pesticides.

Such methods have been used in the United States and most MDCs since the 1950s. The term *green revolution* was coined to describe this approach. The latest green revolution began in 1967. At that time new high-yield dwarf varieties of rice and wheat, specially selected and bred for tropical and subtropical climates through 30 years of painstaking genetic research and trials, were introduced into LDCs such as Mexico, India, Pakistan, the Philippines, and Turkey. The shorter, stiffer stalks of the new varieties allow them to take up to three times as much fertilizer as conventional varieties without toppling over, as long as enough water is available to keep the high levels of fertilizer from killing the crops (Figure 11-6). With sufficient inputs of fertilizer, water, and pesticides at the proper time, wheat and rice yields can be two to three times greater than yields of traditional varieties.

By the 1970s the new wheat varieties had spread to many parts of the world, including China and Bangladesh, and the new rice varieties were adopted in many parts of Southeast and South Asia. Nearly 90 percent of the increase in world grain output in the 1960s and about 70 percent in the 1970s came from increased yields, mostly as a result of the latest green revolution.

Limitations of the Green Revolution   The major factors limiting the increased use of green revolutionary strategies in LDCs are:

1. Without massive doses of fertilizer and water, the new crop varieties produce yields no higher and often less than those from traditional grains.

2. The rise in oil and natural gas prices since 1973 has made water and fertilizer so expensive that many farmers can't afford to use them in large enough quantities to get increased yields.

3. Some of these new varieties are less resistant to insect pests and various diseases than traditional plants.

4. Heavy applications of pesticides used on the new crop strains have caused a sharp increase in malaria in many countries because many strains of mosquitoes have developed genetic resistance to insecticides.

5. High yields are jeopardized unless the crops are planted, irrigated, fertilized, cultivated, and harvested at precise moments in their growth cycle, often requiring technical assistance that is not always available.

6. In some countries storage, transportation, and marketing networks have been inadequate to take advantage of the increased yields.

7. In some nations market prices and other incentives have been insufficient to motivate farmers to produce more crops.

8. Only large landowners and wealthy farmers have the money or the accessibility to credit to buy the fertilizer, irrigation water, pesticides, and equipment that the new seeds require. Thus the effects of the green revolution have not benefited most of the poor peasant farmers who make up 60 to 90 percent of the population of most LDCs.

However, proponents of the green revolution make the following points:

1. Despite its limitations, this development has helped grow more food per capita, so without it world hunger and malnutrition would be much worse.

Figure 11-6 The green revolution. Scientists compare for two Indian farmers a full, older variety of rice (left) and a new, high-yield dwarf variety (right). The dwarf variety can accept much larger amounts of fertilizer.

2. No traditional or new crop variety is perfect, and scientists are now working to create a new green revolution by breeding new high-yield strains that thrive on less fertilizer, make their own nitrogen fertilizer, can do well in salty soils and withstand periods of drought, make more efficient use of solar energy during photosynthesis, and have greater resistance to insects and disease.

3. The green revolution was never intended to solve world food problems and food-related problems, only to buy time for controlling world population growth.

4. The strategy itself should not be blamed for lack of adequate transportation and marketing networks, poor government policies, and the existence of poverty and large landholdings by wealthy elites in poor countries.

Reducing Genetic Diversity Biologist Paul Ehrlich has warned that "aside from nuclear war, there is probably no more serious environmental threat than the continued decay of the genetic variability of crops." When fields of natural varieties are cleared and replaced with monocultures of crossbred varieties, much of the natural genetic diversity essential for developing new hybrids can be reduced or lost forever. For example, a perennial variety of wild corn that, unlike cultivated corn, replants itself each year was recently discovered. This variety is also resistant to a range of viruses and grows well on wet soils. However, the few thousand plants known to exist were found on a hillside in Mexico that was in the process of being plowed up.

Genetic vulnerability was demonstrated in 1970 when most of the U.S. corn crop was of one variety and was wiped out by a blight-causing fungus. Seed companies quickly introduced a resistant seed and recovery was rapid, but the episode made farmers aware that genetic diversity is an important form of insurance against disaster. To preserve genetic variety, the National Academy of Sciences and many other scientists have proposed that naturally growing native plants and strains of food crops throughout the world be collected, and preserved and maintained in genetic storage banks in different areas.

There are two spiritual dangers in not owning a farm. One is the danger of supposing that breakfast comes from the grocery, and the other that heat comes from the furnace.

Aldo Leopold

Discussion Topics

1. Explain how total world food production can keep rising in many countries and regions while average per capita food supply available drops, or rises only slightly.

2. Explain how both these statements can be true: "We averted the threat of famine in the 1970s and early 1980s," and "We now have the largest annual global famine in history."

3. Explain why most people who die from lack of a sufficient quantity or quality of food do not starve to death.

4. Explain how a decline in breast-feeding can lead to an increase in infant deaths from diarrhea and to a rise in the birth rate in LDCs.

5. Debate the following statement: There really is no world food problem because if the grain produced each year were distributed equally among the world's population, everyone would have an adequate diet.

6. Should the United States encourage a partial switch to more home gardens and smaller, intensively cultivated organic farms instead of large, energy-intensive farms increasingly owned by large corporations? Why or why not?

7. Debate the following resolution: The pet population in the United States and other MDCs should be drastically reduced, and spaying or neutering of pets should be mandatory.

8. Summarize the advantages and limitations of each of the following proposals for increasing world food supplies over the next 30 years: **(a)** cultivating more land by clearing tropical jungles and irrigating arid lands, **(b)** catching more fish in the open sea, **(c)** harvesting algae and krill from the ocean, **(d)** harvesting fish and shellfish by using aquaculture and mariculture, and **(e)** increasing the yield per area of cropland.

9. Should price supports and other subsidies paid to U.S. farmers out of tax revenues be eliminated? Why or why not? How would you help ensure that enough farmers make an adequate profit to stay in business?

12

Land Resources: Wilderness, Parks, Forests, and Rangelands

We abuse land because we regard it as a commodity belonging to us. When we see land as a community to which we belong, we may begin to use it with love and respect.

Aldo Leopold

12-1 Land Use in the United States

How Is Land Used? Figure 12-1 shows how the 2.3 billion acres of land in the United States is used. Notice that small and large urban areas consisting of cities and towns with a population of 2,500 or more people account for only about 2 percent of the total. However, the uses of urban and nonurban land are interrelated, because urban areas must be supported by large areas of cropland, rangeland, watersheds, forests, estuaries, and other types of nonurban land. Nonurban land use and resources are discussed in this chapter, and Chapter 14 covers urban land use and land-use planning. Emphasis is on land use in the United States. The principles developed, however, apply to land areas throughout the world.

About one-third of the total land area in the United States is owned jointly by the nation's citizens and is managed for them by the federal government. This land consists mostly of large acreages of forest and grassland but also includes streets and highways, reservoirs, military reservations, and other holdings. About 95 percent of the federally owned land lies in Alaska and in western states, although some is found in nearly every county of every state (Figure 12-2). Management of federally owned land is distributed among the various federal agencies shown in Figure 12-2.

Many of the nation's natural resources are publicly owned but privately used. Individuals use public lands for recreational purposes. Private corporations and individuals cut timber, graze cattle and livestock, and extract oil, natural gas, coal, and

other minerals from federal lands. It is estimated that public onshore lands contain 40 percent of the nation's salable timber, 54 percent of the nation's grazing land, 50 percent of its coal, 80 percent of its recoverable shale oil, 4 percent of its known oil reserves, 6 percent of its known natural gas reserves, 55 percent of all geothermal energy resources, 50 percent of known uranium deposits, and most of its copper, silver, asbestos, lead, molybdenum, beryllium, phosphate, and potash.

Land-Use Ethics In the United States there has been a long history of conflict over how publicly owned lands should be used, as discussed in Section 2-4. Views on how resources on and under publicly owned land should be used fall into four major categories, as summarized in the accompanying box.

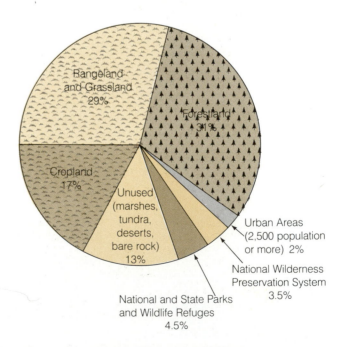

Figure 12-1 Land use in the United States. (Sources: U.S. Bureau of Commerce and the Conservation Foundation.)

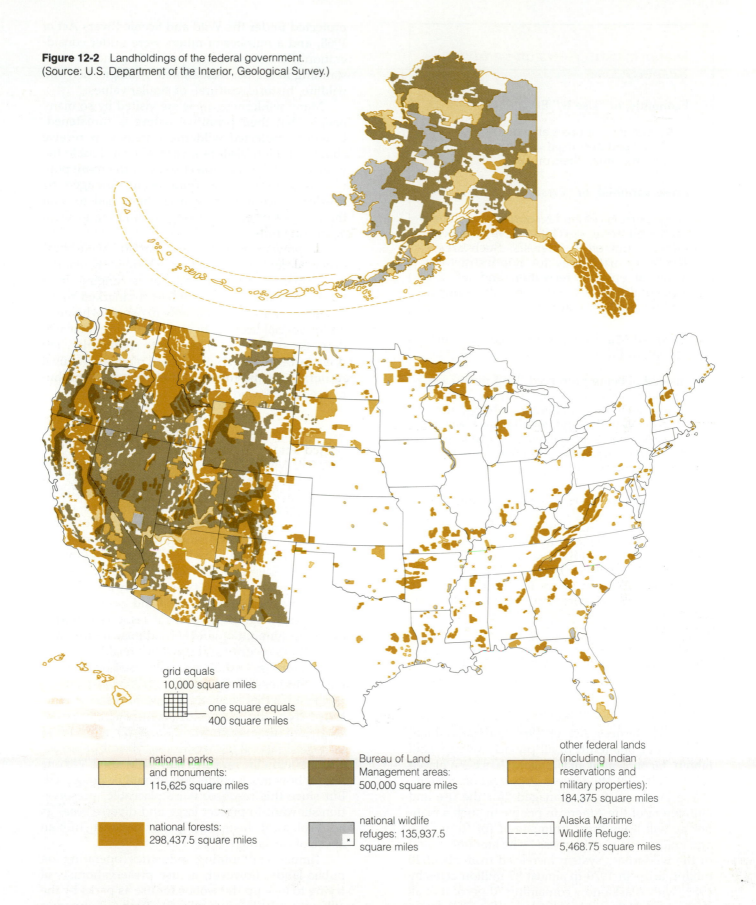

Figure 12-2 Landholdings of the federal government. (Source: U.S. Department of the Interior, Geological Survey.)

grid equals 10,000 square miles

one square equals 400 square miles

national parks and monuments: 115,625 square miles

Bureau of Land Management areas: 500,000 square miles

other federal lands (including Indian reservations and military properties): 184,375 square miles

national forests: 298,437.5 square miles

national wildlife refuges: 135,937.5 square miles

Alaska Maritime Wildlife Refuge: 5,468.75 square miles

12-2 Wilderness

The National Wilderness Preservation System
In the Wilderness Act of 1964 and subsequent laws, about one-tenth of all publicly owned land (about 3.6 percent of all U.S. land) has been placed in the National Wilderness Preservation System. These lands are to be managed "for the use and enjoyment of the American people in such a manner as will leave them unimpaired for future use and enjoyment as wilderness." The amount of land in the wilderness system increased from about 10 million acres in 1970 to almost 83 million acres by 1985, with Alaska now containing 70 percent of all designated U.S. wilderness land. By 1985, more than 60 rivers and river segments had also been

protected under the Wild and Scenic Rivers Act of 1968, and a number of others were under consideration. Rivers chosen for protection must be outstanding in scenic, recreational, geologic, fish and wildlife, historic, cultural, or similar values.

Many wilderness areas are visited by so many people that their primitive nature is threatened. Federally protected wilderness areas now receive about 8 million visitors a year, almost double the number in 1970. To protect some of the more popular wilderness areas from being damaged by overuse, government agencies have had to limit the number of people hiking or camping in them at any one time.

To prevent overuse, historian and wilderness expert Roderick Nash advocates dividing wilderness areas into several categories ranging from those fairly accessible, with clearly marked trails, to remote areas accessible only to people who qualify for special licenses by demonstrating their wilderness skills, to areas that should be left undisturbed as a genetic pool of plant and animal resources for the future of the earth, with no human entry allowed.

Why Preserve Wilderness? Since 1964 the major battle has been over how much land should be added to the wilderness system, or, conversely, whether too much land has already been added to the system. Wilderness enthusiasts have urged Congress to add more areas. They point out that under the guidelines defining wilderness, no more than 4 percent of all the land in the United States excluding Alaska could ever be designated as wilderness. Since this land contains only about 1 percent of the nation's onshore oil and gas and only a small percentage of its mineral resources, setting aside this amount of land should pose no threat to the nation's security and economic health.

Wilderness enthusiasts argue that by preserving wilderness we (1) maintain diverse ecosystems and species that serve as ecological reserves, protected from the ravages of an increasingly populated and urbanized world, and (2) provide an ecological laboratory where the natural processes that maintain life on earth may be studied. No one knows how much wilderness should be preserved. But since this resource is irreplaceable, preservationists want to protect large and diverse areas as an ecological insurance policy against human abuse of the land.

Timber and mining industries operating on public lands, however, accuse preservationists of trying to lock up the woods for use as parks by the affluent and the physically fit. Timber companies also consider it a waste of the nation's natural resources when the National Forest Service and the National Park Service allow large tracts of timber in wilderness areas to go up in smoke because roads and motorized fire-fighting equipment are not allowed in these areas. Wilderness defenders, however, point out that the periodic fires that are caused by lightning clear out dead wood and underbrush, make room for new growth, and keep more mature trees from being destroyed by much hotter fires when flammable underbrush builds to high levels.

12-3 Parks

The Park System in the United States National, state, and local parks are reserved lands that get considerably more use than wilderness areas. Since 1872 the National Park System has grown to include 48 major parks, mostly in the West (Figure 12-2), and 287 national monuments, memorials, battlefields, historic sites, parkways, trails, recreational areas, rivers, seashores, and lakeshores that together make up about 3.5 percent of the nation's total land area. The major national parks provide spectacular scenery on a scale not usually found in state or local parks, preserve wildlife that can't coexist with humans, and serve as buffer zones for wilderness areas. Camping, hiking, fishing, boating, and on-road motorized vehicles are permitted in national parks. However, consumption of resources, as in timbering, hunting, and mining, is not allowed in most park system units.

Stresses on Parks The major problems of national and state parks come from their spectacular success. Between 1950 and 1984, annual visits to the almost 4,000 state parks increased from about 100 million to almost 700 million. During the same period visits to National Park System units increased from about 40 million to almost 335 million, with 18 million overnight stays. Most experts expect this trend to continue, putting increasing stress on already overburdened parks.

Under the onslaught of people during the peak summer season, some parks resemble the cities their visitors are trying to escape. The most popular sites are often overcrowded with cars and trailers and plagued by noise, traffic jams, litter, polluted water, drugs, and crime. Theft of timber, cacti, and petrified wood chips from national parks is a growing problem. In back country areas, hikers and Park Service rangers are sometimes harassed and assaulted by marijuana growers protecting planted areas with armed guards, pa-

trol dogs, electronic warning systems, and explosive booby traps. Park Service rangers, now trained in law enforcement, must spend an increasing amount of their time acting in their capacity as law enforcement officers.

In some parks limited camping spaces can be reserved in advance. Other sites are allocated on a first-come, first-served basis, with some vacationers waiting in a parking lot for two days for a campsite. In some of the heavily used parks the number of hikers and campers in a area at a given time is restricted.

Some people want to turn the most popular and beautiful parks into highly developed recreation and convention centers with luxury hotels, restaurants, and similar facilities. Others argue that visitor centers and such recreational and housing facilities should be located in private or federally owned areas outside the national parks, as has been done for Acadia National Park in Maine. They also believe that the use of motorized vehicles in parks should be discouraged by requiring high entry fees for private vehicles or banning them altogether. In parts of Yosemite, for example, motor vehicles have been banned, and free shuttle buses bring vistors from satellite parking lots to the park interior.

The press of visitors and motor vehicles is not the only stress on national parks. A 1980 survey by the National Park Service revealed that scenic resources were threatened in more than 60 percent of the national parks, while visibility, air and water quality, and wildlife were endangered in about 40 percent of the parks. Some of these threats come from intensive use, but most come from mining, logging, grazing, and land development outside park boundaries.

Government officials and park enthusiasts differ over how increased use of the nation's park system should be handled and how the new threats should be met. Suggestions include (1) adding new parks, (2) shifting scarce funds to better maintenance, conservation, and protection of existing parks rather than purchasing new parkland, (3) transferring ownership of some national parks such as Gateway, Golden Gate, and other recently created urban parks to states and localities, (4) turning more of the management of camping, recreational, and educational activities in the parks to private concessionaires, (5) cutting visitor services to minimal levels at lesser-used park units, and (6) sharply increasing entrance, activity, and concessionaire franchise fees to provide more funds for park maintenance and parkland acquisition. Regardless of the options chosen, the National Park Service faces new challenges in the continuing tension between its two basic purposes of providing for present visitor enjoyment and conserving irreplaceable resources for future generations.

12-4 Importance and Management of Forests

Importance of Forests In addition to their commercial value as producers of wood products, forests have vital ecological functions. They help control climate by influencing the wind, temperature, humidity, and rainfall. They add oxygen to the atmosphere and assist in the global recycling of water, oxygen, carbon, and nitrogen (Section 5-3). Forested watersheds absorb, hold, and gradually release water, thus recharging springs, streams, and groundwater aquifers. By regulating the downstream flow of water, forests help control soil erosion, the amount of sediment washing into rivers and reservoirs, and the severity of flooding. Forests also provide habitats for organisms that make up much of earth's genetic diversity. They also help absorb noise and some air pollutants, and nourish the human spirit by providing solitude and beauty.

Too often economists evaluate forests only in terms of the market value of their products without considering the ecological benefits essential to other economic activities. According to one calculation, a typical tree that lives 50 years provides $196,250 worth of ecological benefits that are only about 0.3 percent of the tree's sale price as wood. For example, over a 50-year period a single tree produces $31,250 worth of oxygen, $62,500 in air pollution reduction, $31,250 in soil fertility and erosion control, $37,500 in water recycling and humidity control, $31,250 in wildlife habitat, and $2,500 worth of protein. While such a calculation is a general estimate, not reflecting actual market values that can be redeemed, it illustrates dramatically how forests are vital to human life and industry everywhere.

Forest Management Just as agriculture is the cultivation of fields, **silviculture** is the cultivation of forests to produce renewable timber resources. Experience has shown that by careful use of the principles of *multiple use* and *sustained yield* (Section 2-4), forest resources can be used for a variety of purposes and harvested in a manner and at a rate that uses and preserves these renewable resources for present and future generations.

The forest management process consists of (1) carrying out an inventory to gather information about the species composition, volume and condition of the trees, soil characteristics, interrelationships with other plant and animal organisms, and the impacts of harvesting on scenic and historical resources, (2) developing a forest management plan by using the inventory data to evaluate the product or use, possible rates of growth, quality of growth needed, other uses to be served by the forested area, and ecological constraints, and (3) deciding the method to be used for harvesting and regenerating the forested area.

The major methods used for harvesting and regenerating mature trees are summarized in the accompanying box.

Protecting Forests In the United States it is estimated that one-fourth of the net annual growth of commercially usable timber is lost to fire, insects, disease, and air pollution. Fires started by people or lightning are the best-known threats to forests.

Tree Harvesting and Regeneration Methods

Selective Cutting

Intermediate-aged or mature trees, either singly or in small groups in a forest stand, are cut at intervals to encourage younger trees to grow and produce an uneven-aged stand with trees of different species, ages, and sizes (Figure 12-3) that over time will regenerate itself. This approach is favored by those wishing to use forests for both timber production and recreation; if the harvest is limited too much, however, there may not be enough commercially desirable timber produced to make the process economically feasible. In addition, the need to

reopen roads and trails periodically for selective harvests can cause erosion of certain soils. A similar method of cutting, not considered to be a reputable forestry practice, is called *high grading*. Here the most valuable commercial tree species are cut without regard for the quality or distribution of remaining trees.

Shelterwood Cutting

Many commercial tree species grow best in sunlight, and for economic reasons are usually grown in an even-aged stand with trees of mostly the same species, size, and age. Such even-aged stands of shade-intolerant species are usually harvested by shelterwood cutting, seed-tree cutting, or clearcutting. Shelterwood cutting involves the removal of all mature trees in an area in a series of cuts over one or more decades. In the first harvest, unwanted tree species and dying, defective, and diseased trees are removed, leaving properly spaced, healthy, well-formed trees as seed stock. In the next stage, 10 or more years later, the stand is cut further so that seedlings can receive adequate sunlight and heat and can become established under the shelter of a partial canopy of remaining trees. Later, a third harvest removes the remaining mature canopy trees, allowing the new stand to develop in the open as an even-aged forest. This method leads to very little erosion and is particularly useful for northern red oak, yellow poplar, basswood, hickories, white ash, and several species of pine. Without careful planning and supervision, however, loggers may take too many trees in the initial cut, especially the most commercially valuable trees.

Seed-Tree Cutting

Nearly all trees on a site are harvested in one cut, with a few of the better commercially valuable trees left uniformly distributed on each acre as a source of seed to regenerate the forest. Seed-tree cutting is sometimes used for harvesting and regenerating southern pine tree species (loblolly, longleaf, shortleaf, and slash). It is not used for most species because of the risk of losing the remaining seed trees as a result of wind and ice, leaving the site without a sufficient seed source to reforest the area.

Clearcutting

All the trees are removed from a given area in a single cutting (Figure 12-4). This is done to establish a new, even-aged stand, usually of a commercially valuable, fast-growing species of shade-intolerant species such as Douglas fir, western white pine, jack pine, loblolly pine, lodgepole pine, black walnut, and black cherry. The clearcut area may consist of a whole stand, a group, a strip, or a series of patches of varying size (Figure 12-5). After clearing, the site may be

reforested naturally from seed released by the harvest and helped by soil disturbance, fire, and rodent control or, increasingly, by the planting of genetically superior seedlings raised in a nursery. Timber companies prefer clearcutting even-aged stands because it permits rapid and efficient regeneration, reduces harvesting costs and road requirements by increasing the volume harvested per acre, and reduces the time of rotation by shortening the time needed to establish a new stand of trees. Clearcut openings and the fringes along uncut areas also improve the forage and habitat for some game species such as deer and elk and some shrubland species of birds. Environmentalists and ecologically oriented foresters recognize that clearcutting can be useful if not overdone. Their concern is usually that the size of the clearcut areas is too often determined by the economics of logging rather than by consideration of forest regeneration. Clearcutting is overused on species that could be cut by less ecologically destructive methods. Excessive use of clearcutting can lead to severe erosion and sediment water pollution if done in large cuts on steeply sloped land. In addition, it creates ugly scars (Figure 12-4) that take years to heal, reduces the recreational value of the forest for years, destroys habitats for many wildlife species, and replaces a genetically diverse stand of trees with a single-age monoculture, increasing the risk of damage from insects and diseases.

Whole-Tree Harvesting

A variation of clearcutting used for harvesting stands for pulpwood or fuelwood involves the use of a machine that pulls a tree from the ground and skids it to a chipping machine. Massive blades reduce the entire tree to small chips about half the size of a matchbox. This approach increases efficiency by using all wood materials in a stand regardless of size. Most ecologists, however, oppose this method because the regular removal will eventually degrade the forest site by taking away most essential nutrients. Research is under way to determine the rates of nutrient depletion at various regions and sites and to ascertain how cutting methods could be modified to reduce the ecological impacts of whole-tree harvesting.

According to the U.S. Forest Service, about 85 percent of all forest fires are started by humans, either accidentally or deliberately.

In the early 1900s Gifford Pinchot, the first head of the National Forest Service, began a policy of trying to prevent and fight all fires in the na-

tional forests. Since the 1950s Smokey the Bear has continued to spread the idea that *all* forest fires are bad. However, research and experience has shown that this is not true and that fire can be an effective forest management tool when properly used.

To understand how fire can be used as a man-

Figure 12-3 Selective cutting of old trees in a climax forest.

Figure 12-5 Patch clearcutting.

Figure 12-4 Clearcutting of redwoods.

agement tool, it is important to distinguish between two types of forest fire. **Crown fires** are extremely hot fires that can destroy all vegetation, kill wildlife, and accelerate erosion. They tend to occur in forests where all fire has been prevented for several decades, allowing accumulations of dead wood and ground litter that burn intensely enough to ignite tree tops. In such forests Smokey the Bear is right in urging us to help prevent fires until forest management can in some way alter conditions and reduce the hazard of crown fires.

Ground fires are low-level fires that typically burn only undergrowth, although they occasionally damage some fire-sensitive trees. They normally do not harm mature trees, and wildlife can generally escape them. Ground fires that occur every 5 years or so, in areas where excessive

ground litter has not accumulated, can help prevent the hotter and more destructive crown fires by burning away this potential fuel. They also help release and recycle valuable nutrients tied up in litter and undergrowth, increase the activity of nitrogen-fixing bacteria, stimulate the germination of seeds of certain conifer species, maintain and help control diseases and insects, and serve some wildlife species by maintaining habitats and providing food from the vegetation that sprouts after fires.

Because of these benefits, ecologists and foresters have increasingly prescribed the use of carefully controlled ground fires as an important tool in the management of some forests (especially those dominated by conifers such as giant sequoia and Douglas fir), and rangelands that would revert to forests without fire treatment. Such *prescribed fires* are carefully controlled, started only when weather and forest conditions are ideal for control and proper intensity of burn, and timed to keep levels of air pollution to an acceptable minimum.

Diseases and *insects* cause much more loss of commercial timber than fires in the United States and throughout the world. Parasitic fungi cause most diseases that damage trees. For example, Dutch elm disease, carried by insects from tree to tree, has killed more than two-thirds of the elm trees in the United States. It was probably brought to the United States accidentally on a shipment of elm logs from Europe. The best methods for controlling tree diseases include (1) banning imported timber that might carry alien parasites, (2) treating diseased trees with antibiotics, (3) developing disease-resistant species, (4) identifying and removing dead, infected, and susceptible trees, and

(5) reducing air pollution—especially sulfur dioxide, ozone, and nitrogen dioxide—which can damage and kill trees and make them susceptible to disease (Section 21-4). The use of fungicides may cause more problems than they cure by retarding tree seed germination and also by killing beneficial fungi and earthworms.

Some highly destructive *insect pests* are the spruce budworm, the gypsy moth, the pine weevil, the larch sawfly, and several species of pine-bark beetles. As discussed in Section 23-6, pest control methods include isolating and removing infested trees, introducing other insects that prey on the pests, using sex attractants to lure insects to traps, releasing sterilized male insects to reduce the population growth of pest species, using integrated pest management, and encouraging natural insect control by preserving forest diversity.

A new threat to many of the world's forests is *air pollution*. The most serious air pollution problem is *acid deposition*, commonly called acid rain. It occurs when sulfur and nitrogen oxide air pollutants released by the burning of fossil fuels in power plants and cars are transformed chemically in the atmosphere to sulfuric and nitric acids and dry acidic particulate matter that fall to the earth in rain, snow, or fog (Section 21-4).

Although the precise mechanism by which acid deposition may be damaging trees is not fully known, evidence is mounting that it is a major threat to the forests in Poland, Czechoslovakia, England, France, East and West Germany, Switzerland, the Scandinavian countries, and eastern North America, as discussed more fully in Section 21-4. The major solution to this problem apparently involves using air pollution control devices on fossil-fuel-burning plants and cars to reduce emissions of sulfur and nitrogen oxides (Sections 21-7 and 21-8).

12-5   Status of World and U.S. Forests

World Forests   Wherever civilization has flourished, forests have been destroyed, reducing the earth's original forested area by at least one-third and perhaps by half. Today about one-fifth of the world's land area is covered with forests, with North America, Latin America, and the Soviet Union containing about three-fourths of the these remaining forests (see map inside front cover). Most of Africa and Asia and parts of Central and South America have little forest. As a result of clearing for farming and grazing and overcutting for lumber and firewood, the world's forests are shrinking by almost 1 percent a year.

Threats to the World's Tropical Forests   Loss of the world's tropical forests is considered to be one of the major environmental problems of the 1980s and 1990s. About one-third of the original expanse of the world's tropical forests has already been cleared or seriously degraded. Losses of original tropical moist forests amount to more than 50 percent in Africa, 42 percent in Asia, 67 percent in Central America, and 37 percent in Latin America.

At the present destruction rate of about 0.6 percent a year—equivalent to an area about the size of Pennsylvania—at least half the world's remaining tropical forests may be gone within the next 50 to 80 years. Ecologists warn that loss of these incredibly diverse biomes could cause the extinction of many thousands of tropical species that may be important in the development of new hybrid food plants needed to support future green revolutions (Section 11-8) and new medicines to fight disease (Section 13-1).

Many of the direct causes of the increased rate of tropical deforestation are a result of tropical LDCs trying to become more economically developed by clearing forests for crop production, livestock grazing, fuelwood, mining, and commercial logging, much of it for international export. Indirect causes of such deforestation are poverty, land ownership patterns that favor the wealthy, unemployment, rapid population growth, and the failure of governments to regulate national and multinational timber companies. Exports of lumber and beef from cattle grazing on cleared forests from LDCs mostly to MDCs provide such LDCs with foreign capital. Without appropriate reforestation and conservation programs, however, the long-term result is exploitation of an important renewable resource that is necessary for the long-term economic development and preservation of environmental quality in the LDCs.

Suggestions by ecologists and foresters for reducing the destruction of tropical moist forest systems include: **(1)** setting aside protected reserves throughout the 49 nations that have most of these forests, with financing provided by annual donations or assessments from MDCs, **(2)** securing commitment by governments to plant many more trees, **(3)** requiring timber companies and consumers in MDCs that benefit most from logging and beef production to bear a greater share of reforestation costs, **(4)** identifying areas in which soils under tropical forests are best suited for various purposes on a renewable basis, and concentrating each type of use on appropriate soils by zoning for the best use, **(5)** using agroforestry techniques by simultaneously planting fast-growing tree crops and food crops on newly cleared forest land, to ensure the establishment of a new forest when the soil is exhausted for agriculture, and (6) greatly expanding education programs for rural peoples to teach them the need for trees and ways to plant them. The Worldwatch Institute estimates that between 1983 and 2000 the rate of tree planting in tropical areas will have to be increased thirteenfold to meet projected needs.

World Fuelwood Crisis More than 2 billion people, including 90 percent of the people in the LDCs, depend on wood as their principal fuel for heating and cooking. This use alone accounts for half the wood harvested in the world, and about 80 percent of the wood cut in the LDCs. One out of four people on earth lives where the collection of wood for fuel outpaces new growth and shortages are expected in the future. The World Bank projects that the rate of fuelwood planting in LDCs (excluding China) must increase fivefold between 1980 and 2000 to avoid enormous ecological and economic costs.

This scarcity of fuelwood accelerates deforestation, especially in areas near villages and cities where commercial markets for fuelwood and charcoal exist. Food shortages are also made worse by fuelwood scarcity when families who cannot obtain wood burn dried animal dung, thus diverting vital fertilizers from the soil.

Some suggestions for dealing with the fuelwood crisis are: (1) promoting community forestry projects in which representatives from each household learn to encourage other villagers to plant, tend, and harvest local woodlots, with the wood distributed among the households and proceeds from any marketable surplus used to support community development projects, as is being done successfully in South Korea, (2) having government foresters act as extension agents to help communities by providing seed or seedlings for fuelwood planting stock and advice on tree care, (3) planting appropriately selected fast-growing fuelwood trees and shrubs in unused patches of land, (4) using agroforestry to grow both crops and fast-growing fuel trees in certain areas, (5) helping to conserve wood energy by developing more efficient wood stoves, solar cookers, and small biogas plants that produce methane gas from organic wastes and leave fertilizer ash as a by-product, and (6) practicing population control to help keep local populations in balance with renewable resource capabilities (Chapter 8).

Forests in the United States During the 250 years since the first colonists arrived, the original forested area in the United States has been reduced by about 45 percent. Since 1920, however, the total forested area has remained about the same. Today about one-third of the land area in the United States is forested (Figure 12-6).

About two-thirds of these forests are classified as commercial forestland, suitable for growing potentially renewable crops of economically valuable tree species. More than 50 percent of this commercial forest area is owned by private individuals and companies, most of which are not forest product corporations. The remaining third consists of noncommercial forests either reserved for use as parks, wildlife refuges, and wilderness, or not capable of producing much commercially valuable timber per acre.

About 18 percent of the nation's commercial forest area, or about 8 percent of the total U.S. land area, is located within the national forests. These national forests, managed under the principles of multiple use and sustained yield, contain 32 percent of our total volume of timber and 51 percent of the softwood sawtimber trees, which provide most of the lumber and plywood used in construction as well as much of the pulp for producing paper.

In 1900 all the lumber used in the United States was cut from private lands. Until 1950 the Forest Service had a custodial, stewardship role in managing the national forests. Since 1950, because of the post–World War II economic boom and the depletion of many private forests, timber companies have increasingly looked to the national forests to supply timber for domestic use and export. Between 1950 and 1983 the percentage of lumber cut from public lands each year and used domestically increased from 15 to 40 percent.

These national forests are also visited at little or no cost by an increasing number of recreational users, hunters, and people seeking escape from the pressure of modern life. Because of these growing and often conflicting demands on national forest resources, the Forest Service has been the subject of heated controversy since the 1960s. Timber company officials complain that they cannot buy and cut enough timber on public lands, especially in the Pacific Northwest's national forests, which contain 35 percent of the nation's total inventory of Douglas fir and other softwood species. Most of the mature timber owned by lumber companies in this region has already been cut. Thus, lumber company officials argue that restricting the cutting in national forests will cause timber shortages, leading to higher prices for houses and wood products. They also argue that increased cutting in national forests—even if it represents a departure from the principle of sustained yield— is needed to renew old-growth forests in the Na-

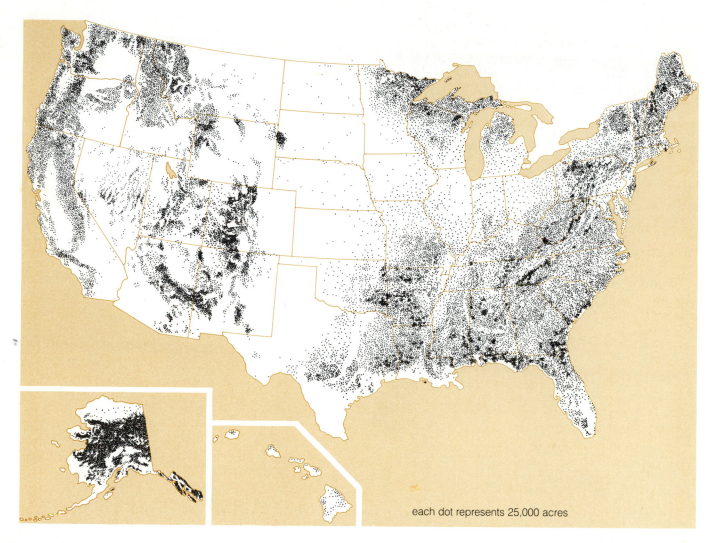

each dot represents 25,000 acres

Figure 12-6 Forest lands in the United States. (Source: Council on Environmental Quality.)

tional Forest System and to give newly planted trees time to mature.

Environmentalists charge that the allocation of funds in the annual budget of the Forest Service has resulted in timber production and mining becoming the favored uses of the national forests. Although the volume of timber the Forest Service sells has not exceeded the overall rate of growth in the national forests, the rate of cutting has exceeded the rate of growth in some areas. Environmentalists argue that the annual cut in the national forests should not exceed sustained yield anywhere and should approach sustained yield only if timber companies are making every effort to reforest their own lands or other private land they cut. They point out that much of the land owned by timber companies is not reforested, even though these companies receive enormous tax benefits intended to encourage better forest management on private lands. In 1984 these tax breaks to the lumber industry cost the U.S. Treasury an amount al-

most equal to the cost of managing the entire National Forest System. To continue to sell timber from the national forests at prices below the cost of growing this publically owned resource, they argue, will only encourage continued exploitation.

12-6 Rangelands

Nature of Rangelands Land on which the vegetation is predominantly grasses, grasslike plants, or shrubs such as sagebrush is called **rangeland**. Most animals, including people, cannot digest the grasses and shrubs, or *forage*, that cover these lands. Ruminant animals like sheep and cattle, however, can digest this vegetation, and they convert it to forms of food that humans can digest. These livestock animals also provide useful nonedible goods such as wool, leather, and tallow.

About one-third of the total land area of the

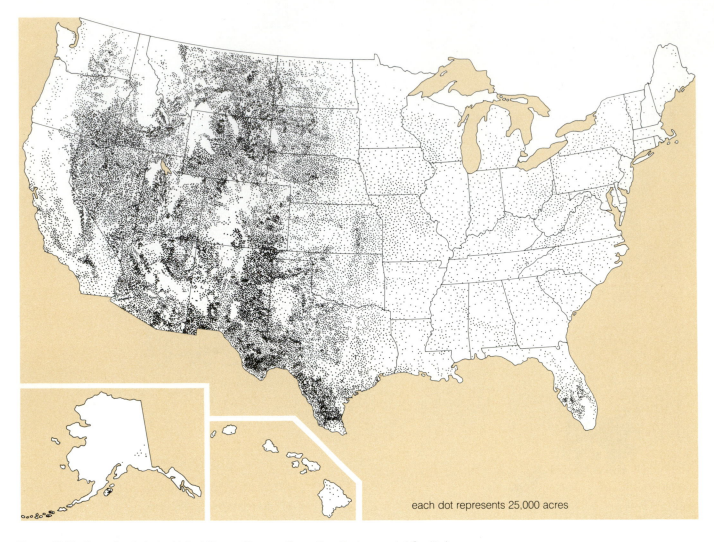

each dot represents 25,000 acres

Figure 12-7 Rangelands in the United States. (Source: Council on Environmental Quality.)

United States consists of rangelands, mostly in the western half of the nation (Figure 12-7). About 34 percent of all U.S. rangeland is owned and managed by the federal government, mostly by the Forest Service and the Bureau of Land Management. About three-fourths of government and privately owned grassland is actively grazed by domestic animals at some time during each year. Mineral and energy resource development—especially oil shale, coal, copper, phosphate, sand, and gravel mining—also takes place on publicly owned rangelands. As the global population increases, it is inevitable that the world's rangelands will be more intensively used and in many cases abused for livestock grazing and other uses.

Characteristics of Rangeland Vegetation Many rangeland weeds and bushes have a single main taproot and can thus be easily uprooted. By contrast, rangeland grass plants have a fibrous taproot system with millions of branches that make such

plants very difficult to uproot. This explains why these grasses help prevent soil erosion (Figure 9-4).

For most plants, when the leaf tip has been eaten, no further leaf growth occurs. By contrast, each leaf of rangeland grass grows from its base, not its tip. Thus, when the upper half of the stem and leaves of rangeland grass has been eaten by livestock or wild herbivores such as deer, antelope, and elk, it can grow back to its original length in a short time. However, the lower half of the plant, known as the *metabolic reserve*, must be left if the plant is to survive and grow new leaves. As long as only the upper half is eaten, rangeland grass is a renewable resource that can be grazed again and again.

Rangeland Carrying Capacity and Overgrazing
Much damage to rangelands has occurred because its carrying capacity is exceeded. When this happens, herbivores start eating the metabolic reserve,

USDA/Soil Conservation Service

Figure 12-8 Fence separates a highly productive rangeland from an overgrazed area.

often clipping the grass to the bare ground. *Overgrazing* (Figure 12-8) can kill the root system of grass. Then either unpalatable weed and shrub invader species take over or all the vegetation disappears, leaving the land barren and vulnerable to erosion. Such overgrazing causes the normal carrying capacity of a rangeland to decrease, sometimes to zero. Restoring overgrazed lands is difficult and expensive. During the 5- to 12-year period required for restoration, the number of animals grazing per acre must be decreased or even reduced to zero.

Climate—especially drought—is also a factor in rangeland damage. Ideally, the number of livestock per acre should be cut back in dry years. A widespread drought, however, can cause higher prices for hay or grain used as supplemental feed and lower market prices for livestock. This encourages ranchers to keep excessive numbers of livestock on a range in the hope that the next year will bring more rainfall and higher meat prices. Sometimes productive rangeland is converted into desert. This process, called *desertification*, is described in Section 9-3.

According to the Council on Environmental Quality, one-third of all U.S. rangeland (excluding Alaska) is in poor condition, and only 15 percent is in good or excellent condition. In 1984 only about 36 percent of the public rangelands were in good or excellent condition. Although much of the overgrazing on public lands took place in the past, the practice continues primarily because public land grazing fees are about five times less than comparable grazing rates on privately owned rangeland. This means that U.S. taxpayers subsidize ranchers with federal grazing permits to the tune of at least $30 million a year—about $2,117 for each rancher using these lands. Earl Sandvig, who used to run grazing programs for the Forest Service, has suggested eliminating the grazing permit system and instead selling grazing rights by competitive bids, as is done for timber-cutting rights in national forests.

Rangeland Management The major tools used in rangeland management are: **(1)** controlling the number and kinds of grazing animals allowed on a given area of rangeland, **(2)** controlling the distribution of grazing animals by herding, the use of fences, rotating stock, and proper placement of supplies of water, salt, and supplemental feed, **(3)** suppressing the growth of undesirable vegetation and encouraging the growth of desirable vegetation by using herbicides, prescribed burning, and in some cases fertilization and reseeding, and **(4)** using poisons, trapping, and hunting to control herbivores such as jackrabbits and prairie dogs that compete with livestock for forage, and livestock predators such as coyotes.

The first two methods are more widely used because experience has shown that the last two approaches are often too expensive, ineffective, or both. Some types of livestock and some breeds of the same type of livestock are better adapted to one type of rangeland than another. For example, sheep normally do better than cattle on rangeland that is steep and hilly and on which shrubs predominate over grasses.

Although fencing and herding are expensive, these measures are effective in moving livestock from one rangeland area to another to prevent overgrazing or to allow overgrazed areas to recover (Figure 12-8). Other highly effective and less expensive methods for distributing livestock over rangeland areas involve the strategic location of waterholes and salt blocks. Livestock need both salt and water, but not together. Placing salt blocks in ungrazed areas away from the water sources causes livestock to spend some time in these areas rather than congregating near streams, wet meadows, and waterholes and overgrazing vegetation there.

Prescribed burning can be used to remove undesirable invader plants and shrubs such as mesquite and to encourage new growth of some desirable grasses. On a long-term basis, however, burning can lead to rangeland degradation by destroying or suppressing the growth of desirable plant species, encouraging invasion by weeds, and promoting soil erosion by destruction of vegetational cover. Herbicides can also be used, normally by aerial spraying, to suppress undesirable plant species. This method, however, is usually too expensive for widespread use over large rangeland areas.

Many ranchers still promote the use of poisons, trapping, and shooting to kill off predatory species that sometimes victimize livestock. For example, since 1940 a highly controversial predator control program has been waged primarily by western ranchers against the coyote. As such predator populations are reduced, however, the populations of rodents and rabbits that they feed on and keep under control grow unchecked and compete with livestock for rangeland vegetation. Thus, in the long run, extermination or drastic reduction of natural predator populations can reduce rangeland productivity and cause economic losses larger than those from livestock killed by predators.

The one process ongoing in the 1980s that will take millions of years to correct is the loss of genetic and species diversity by the destruction of natural habitats. This is the folly our descendants are least likely to forgive us.

Edward O. Wilson

Discussion Topics

1. How does the ecological land-use ethic differ from the idea of balanced multiple use of land and from the preservationist ethic?

2. Should more wilderness areas and wild and scenic rivers be preserved in the United States? Why or why not?

3. Conservationists want people to know and appreciate wilderness, but they don't want wilderness areas intruded on and damaged by large numbers of people. How can this dilemma be resolved?

4. Discuss the pros and cons of each of the following suggestions for the use of national parks:
   a. Higher charges should be made for use.
   b. Reservations should be required.
   c. All private vehicles should be kept out.
   d. Campgrounds, lodges, and other facilities should be moved to nearby areas outside the parks.

5. In what types of forest can prescribed fires be useful, and what are the benefits from such fires? In what types of forest can they be harmful?

6. What would probably be major characteristics of trees that should be (a) clearcut, (b) selectively cut, and (c) shelterwood cut or seed-tree cut?

7. Should trail bikes, dune buggies, snowmobiles, and other off-road vehicles be banned from all national forests, parks, and wilderness areas? Why or why not?

8. Explain how eating a hamburger in a fast-food chain indirectly contributes to the destruction of the world's tropical moist forests. What difference, if any, could the loss of these forests have on your life or on any child you might bring into the world?

9. Should private companies cutting timber from national forests continue to be subsidized by receiving federal payments for reforestation and for building and maintaining roads to areas to be cut? Why or why not?

10. Should fees for grazing on federally owned lands be eliminated and replaced with a competitive bidding system? Why or why not? Why would such a change be politically difficult?

11. Should the poisoning and hunting of livestock predators be allowed on federal rangelands? Why or why not? Try to have both a rancher and a wildlife scientist present to your class their viewpoints on this controversial issue.

13

Wildlife Resources

It is the responsibility of all who are alive today to accept the trusteeship of wildlife and to hand on to posterity, as a source of wonder and interest, knowledge, and enjoyment, the entire wealth of diverse animals and plants. This generation has no right by selfishness, wanton or intentional destruction, or neglect, to rob future generations of this rich heritage. Extermination of other creatures is a disgrace to humankind.

World Wildlife Charter

In the 1850s Alexander Wilson, a prominent ornithologist, watched a single migrating flock of passenger pigeons darken the sky for more than 4 hours. He estimated that this flock of more than 2 billion birds was 384 kilometers (240 miles) long and 1.6 kilometers (1 mile) wide.

By 1914 the passenger pigeon (Figure 13-1) had disappeared forever. How could the species that was once the most numerous bird in North America become extinct in only a few decades? Several reasons for this irreversible loss are: **(1)** the rapid spread of infectious diseases, because these birds nested in dense colonies, **(2)** the inability of the species to recover once flock size has been reduced, because only one egg was laid per nest, and **(3)** the death of millions of birds in severe storms during the annual fall migration to Central and South America.

Two additional reasons for the extinction of the passenger pigeon involve people. When land clearing for farms and cities destroyed habitats and food supplies, the rigid migration patterns of these birds kept them from moving to other breeding areas. The second human factor was hunting. Passenger pigeons made excellent eating and good fertilizer. They were easy to kill because they congregated in gigantic flocks. People used to capture one pigeon alive, sew its eyes shut, and tie it to a perch called a stool. Soon a curious flock alighted beside this "stool pigeon" and were shot or trapped by nets that might contain more than

1,000 birds. Beginning around 1858 massive killing of passenger pigeons became a big business. Shotguns, fire, traps, artillery, and even dynamite were used. Some live birds served as targets in shooting galleries. In 1878 one professional pigeon trapper made $60,000 by killing 3 million birds at their nesting grounds near Petoskey, Michigan. By 1896 the last massive breeding colony had vanished. In 1914 the last passenger pigeon on earth—a hen named Martha, after Martha Washington—died in the Cincinnati Zoo. Her stuffed body is now on view at the National Museum of Natural History in Washington, DC.

Does it really matter that a wild species such as the passenger pigeon became extinct and the existence of wild species such as the whooping crane and the Bengal tiger is threatened primarily because of human activities? In this chapter we examine this question by looking at the importance, protection, and management of wildlife resources.

Figure 13-1 The extinct passenger pigeon. The last known passenger pigeon died in the Cincinnati Zoo in 1914.

13-1 Importance of Wildlife and Wildlife Resources

How Many Species Exist? Worldwide estimates suggest that there may be between 5 million and 10 million different plant and animal species, with insect species far outnumbering all other kinds of life. Some put the estimate as high as 50 million different species—mostly insects. Though scientists are continually adding to the list, so far about 1.5 million different species of plants and animals have been identified worldwide. These millions of different species on earth today are the result of about 3.5 million years of natural evolution.

An estimated 60 to 80 percent of the world's plant and animal species are found only in the tropics. Although tropical moist forests cover less than 7 percent of the earth's surface (see map inside front cover), biologists estimate that they contain about 40 percent of all plant and animal species on earth.

Distinction Between Wildlife and Wildlife Resources All undomesticated species of plants and animals on earth can be classified as **wildlife** or **wild species**, and this broad definition is used in this book. However, "wildlife" is sometimes defined more narrowly to consist only of mammals and birds, excluding species of plants and the vast number of invertebrate animal species—mostly insects.

Wild species that have some useful value to humans are known as **wildlife resources**. These are potentially renewable natural resources if not driven directly or indirectly to extinction or near extinction by human activities. Wildlife resources that provide sport in the form of hunting or fishing are classified as *game species*.

Why Preserve Wild Species? The four basic arguments for not hastening the extinction of the wild species that inhabit the earth are summarized in the accompanying box.

13-2 How Species Become Endangered and Extinct

Extinction of Species It is estimated that 70 to 98 percent of the different species that at one time or another lived on the earth either have become extinct forever or have evolved over many generations into a form sufficiently different to be identified as a new species. Although the disap-

---

Reasons for Preserving Wild Species

1. The Economic Argument

Wildlife resources provide humanity with a wide variety of direct economic benefits as irreplaceable sources of food, medicines, scientific understanding, fuel, fibers, cooking oils, lubricating oils and waxes, dyes, natural insecticides (pyrethrum, rotenone, and caffeine), and building and industrial materials. Pollination by insects is essential for many food and nonfood plant species. Certain natural predator insects, parasites, and disease-causing bacteria and viruses are increasingly being used for the biological control of various weeds and insect pests (Chapter 23). About 40 percent of the prescription and nonprescription drugs now on the world market have active ingredients extracted from plants and animals, and many new drugs will come from presently unclassified species. Other wild species not presently classified as wildlife resources will be needed by agricultural scientists to develop new crop strains that have higher yields and increased resistance to diseases, pests, heat, and drought (Section 11-8). Soon fast-growing trees and bushes not presently classified as wildlife resources may be planted in large *biomass energy plantations* and harvested for use as fuel to supplement or replace the world's dwindling oil supplies (Section 19-5). Less than 1 percent of the earth's plant species have been thoroughly studied to determine their possible usefulness to humans. As tropical moist forests are cleared, many potentially useful plant species and any animal species that depend on them will be wiped out forever—much like throwing away a present before removing the wrapping.

2. The Aesthetic-Recreational Argument

Many wild species are a never-ending source of beauty, joy, and recreational pleasure for humans. It is impossible to place a monetary value on experiences such as observing the leaves change color in autumn, smelling the aroma of wildflowers, or watching an eagle or sea gull soar overhead, or a porpoise glide through the water. A 1980 survey by the Fish and Wildlife Service indicated that

---

nearly 100 million Americans (43 percent of the population) participated in some form of wildlife activity.

3. The Ecological Argument

The most important contributions of the wild species may result from their roles in maintaining the health and integrity of the world's ecosystems. Ecosystem services of wild plant and animal species include provision of food from the soil and the sea, production and maintenance of oxygen in the atmosphere, filtration and detoxification of poisonous substances, moderation of the earth's climate, regulation of freshwater supplies, disposal of wastes, recycling of nutrients essential to agriculture, production and maintenance of fertile soil, control of the majority of potential crop pests and carriers of disease, maintenance of a vast bank of genetic material, and storage of solar energy as chemical energy in food, wood, and fossil fuels. Such ecosystem services are rarely incorporated into cost-benefit decisions regarding development primarily because it is difficult to assign them a dollar figure, and also because most decision makers are unaware how heavily we rely on these services. Because we still know relatively little about the workings of even the simplest ecosystems, we cannot be sure which species play crucial roles today, which ones have genes crucial for our survival and the survival of other species, and how many species can be removed before an ecosystem will collapse or suffer serious damage.

4. The Ethical Argument

Some believe that it is ethically and morally wrong for humans to hasten the extinction of a species. Others go further and believe that each individual wild creature has an inherent right to survive without human interference, just as each human being has the inherent right to survive. In practice, most advocates of the ethical position argue that only a wild *species*—not each individual organism—has an inherent right to survive. This position is based on the belief that the human species is no more important than any other species on earth.

pearance of species is a natural process, there is growing concern that it is being hastened by human activities.

Species heading toward extinction are often classified as either *endangered* or *threatened*. An **endangered species** is one having so few individual survivors that the species could soon become extinct in all or part of its range. Examples are the whooping crane and the California condor (Figure 13-2), and various species of plants—especially cacti and orchids. **Threatened species,** such as the grizzly bear and the bald eagle, are still abundant in their range but likely to become endangered within the foreseeable future because of a decline in numbers.

Conservation specialist Norman Myers warns that between 1975 and 2000 at least 1 million of the world's 5 million to 10 million species will probably have disappeared, primarily because of extensive clearing of the world's remaining tropical moist forests (Section 12-5). Most of these species will be plants and invertebrate animals such as insects, mites, and nematode worms that have yet to be classified and evaluated to determine their use to ecosystems and humans.

Although animal extinctions receive most of the publicity, plant extinctions are more important ecologically. It is estimated that about 10 percent of the world's plant species are already threatened with extinction and 15 to 25 percent of all plant species face extinction by the year 2000.

Major Causes of Extinction Today The major human-related factors that can lead to a species becoming threatened, endangered, or extinct are: **(1)** habitat disturbance and elimination, **(2)** commercial hunting, **(3)** pest and predatory control for protection of livestock and crops, **(4)** collecting specimens as pets, for medical research, and for zoos, **(5)** pollution, and **(6)** accidental or deliberate introduction of a new competing or predatory species into an ecosystem. Some species are driven to extinction or near extinction by a single factor such as habitat loss, while others succumb to a combination of factors. Some species also have biological and behavioral characteristics that make them particularly vulnerable to extinction from human-related activities.

Habitat Disturbance and Loss The greatest threat to plant and animal wildlife is the destruction or alteration of habitat: the area in which species seek food, find shelter, and breed. As the human population grows, it increases its habitat at the expense of the habitats of other creatures. This

Figure 13-2 The whooping crane and the California condor are two endangered species in the United States.

disruption of natural communities can threaten wild species by destroying migration routes, breeding areas, and food sources. Plant species unique to a small locality, and the animals that feed on them, can be eliminated from the earth by a single bulldozer.

Habitat disturbance has been a major factor in the disappearance of some of America's most magnificent bird species, such as the ivory-billed woodpecker, and in the near extinction of the whooping crane and the California condor. The ivory-billed woodpecker formerly inhabited river-bottom hardwood forests in the southeastern United States. As virgin forests have been cut and replaced by plantations of even-aged trees, the insects on standing timber that served as the woodpecker's food source diminished greatly, and no sightings of this bird have been reported since the 1940s.

Commercial Hunting It is important to distinguish between three types of hunting: *commercial hunting, hunting for food,* and *sport hunting primarily for recreation*. At one time commercial hunting in the United States was an important factor in the extermination of the American passenger pigeon and the near extermination of the American bison and the snowy egret.

In the late 1800s the snowy egret, which inhabits coastal regions of the southeastern United States, was hunted almost to extinction because the feathers were used to adorn women's hats. In 1886 the newly formed Audubon Society began a campaign against this slaughter. Texas and Florida passed laws protecting plumed birds, but the laws were mostly ignored. Then in 1900 Congress passed the Lacey Act, which banned interstate traffic in illegally killed wildlife. This cut off the supply, and these beautiful birds began thriving again as parks, preserves, and refuge areas were established for the protection of wildlife species. Some black market trade still exists. However, the economic incentive for selling products from endangered species was sharply reduced in the United States by the Endangered Species Act of 1973, which made it illegal to import or to carry on trade in any product made from an endangered species.

On a worldwide basis, commercial hunting threatens species such as the jaguar, the cheetah, the tiger, and the snow leopard (all hunted for their furs), the alligator (skins), the elephant (ivory tusks), the Asian musk deer (musk oil for perfumes and soaps), and the rhinoceros (horns). For example, rhinoceros horn—actually a mass of compact hair—is worth as much as $5,000 a pound because it is used to make Arabian dagger handles that certify the manhood of young men and is ground into a powder that is used to reduce fever and is thought by some Asians to be a sexual stimulant. Although 60 nations have agreed not to import or export rhino horn, illegal traffic goes on because poacher gangs usually outnumber and sometimes outgun rangers patrolling protected areas.

The sale of a large pair of elephant tusks can provide poor African villagers with the equivalent of 10 years' income. Thus, it is not surprising that some become illegal poachers who kill as many as 10,000 African elephants a year. Even when illegal wildlife smugglers are caught, the penalties are usually too low to hurt overall profits. In 1979 a Hong Kong fur dealer, apprehended for illegally importing 319 Ethiopian cheetah skins valued at $43,900, was fined only $1,540.

At one time *hunting for food* was a major cause of extinction of some species. In the early 1880s, for instance, the eastern elk became extinct in the United States primarily because the species was hunted commercially and privately for food and for its hides. Today such hunting has declined in most areas.

Sport hunting is closely regulated in most nations; it endangers a game species only when protective regulations don't exist or are not enforced. For example, hunters from many parts of the world go to West Central Africa to shoot bull giant sable antelopes, hoping to bring back their magnificent curved horns as a trophy. As a result, only about 500 to 700 of these animals remain in this portion of Africa.

Predator and Pest Control Extinction or near extinction can also occur because of attempts to exterminate pest and predator species that compete with humans and livestock for food. The Carolina parakeet was exterminated in the United States around 1914 primarily because it fed on fruit crops. Its disappearance was hastened because when one member of a flock was shot, the rest of the birds hovered over its body. Large numbers of endangered African elephants have been killed to keep them from trampling and eating food crops. Carnivore predators that sometimes kill livestock are also shot, trapped, or poisoned. Ranchers, hunters, and government employees involved in predator control programs have wiped out the timber wolf, the grizzly bear, and the mountain lion over most of the continental United States. Campaigns to protect rangeland for grazing livestock by poisoning prairie dogs and pocket gophers have just about eliminated their natural predator, the black-footed ferret (Figure 13-3), now one of the rarest mammals in North America.

Pets, Medical Research, and Zoos Worldwide, more than 5.5 million live wild birds are sold each year, most of them ending up in pet-loving nations such as the United States, Great Britain, and Germany. As a direct result of this trade, at least nine

Figure 13-3 The black-footed ferret is one of the most endangered mammalian species in the United States.

bird species are now listed as threatened or endangered. In 1980 more than 128 million tropical fish, 2 million reptiles, and 1 million other individual animals were legally imported into the United States, mostly for sale as pets. Large numbers of these animals die during shipment. Many others die after purchase or are killed or abandoned by their owners.

Some species of exotic plants such as orchids, lilies, palms, and cacti are also endangered because they are collected (often illegally) and sold to decorate houses, offices, and landscapes. Nearly one-third of the cactus species native to the United States are thought to be endangered because they are collected and sold for use as potted plants.

Medical research makes use of large numbers of animals. According to the Office of Technology Assessment, about 71 million animals are used each year in the United States for toxicity testing, biomedical and behavioral research, and drug development. About 60 million of the animals are mice and rats. The remaining 11 million animals that are poisoned, irradiated, suffocated, blinded, driven insane, dismembered, or subjected to other harmful laboratory procedures each year are dogs, cats, monkeys, birds, frogs, guinea pigs, rabbits, and hamsters. Although most test animal species are not endangered, medical research coupled with habitat loss is a serious threat to endangered wild primates, such as the chimpanzee and the orangutan.

Under pressure from animal rights groups, scientists in industry, universities, and government laboratories are trying to find alternative testing methods that either do not subject animals to suffering or—better yet—do not use animals at all. Promising alternatives include the use of cell and tissue cultures, simulated tissues and body fluids, bacteria, and computer-generated mathematical models that enable scientists to estimate the toxicity of new compounds from knowledge of chemical structure and properties.

Public zoos, botanical gardens, and aquariums are under constant pressure to exhibit rare and unusual animals such as the orangutan. For each exotic animal or plant that reaches a zoo or botanical garden alive, many others normally die during capture or shipment. Since 1967 reputable zoos and aquariums have agreed to no longer purchase endangered species, although there are still some abuses. However, in cases of the apparently imminent extinction of a species, efforts may be made to establish a captive breeding stock in the safety of a zoo, as discussed in Section 13-3.

Pollution Chemical pollution is a relatively new but growing threat to wildlife. Industrial wastes, mine acids, and excess heat from electric power plants have wiped out some species of fish, such as the humpbacked chub, in local areas. Oil pollution from tanker and offshore oil rig accidents can affect the survival of some vulnerable marine species such as the black-footed penguin. The Canadian aurora trout now appears to be extinct as a result of pollution of its freshwater habitat from acid deposition (Section 21-4). Chlorinated hydrocarbon pesticides, especially DDT and dieldrin, have been magnified in food chains and have caused reproductive failure and eggshell thinning of important birds of prey, such as the peregrine falcon, the eastern and California brown pelicans, the osprey, and the bald eagle (Section 23-4). The banning of such persistent pesticides in North America and Europe has allowed most of the species to recover. Yet these substances are increasingly being exported for use in LDCs.

Species Introduction A new species that is introduced into an ecosystem can cause a population decrease or even extinction of one or more existing species by preying on them, competing with them for food, destroying their habitat, or upsetting the ecological balance. It can also cause a population explosion of one or more existing species by killing off natural predators. However, not all alien species introductions have had harmful effects. Most major U.S. food crops were deliberately imported from other areas.

Island species are especially vulnerable. For example, the dodo bird, which lived only on the small island of Mauritius in the Indian Ocean, became extinct by 1681 after pigs brought to the island consumed its eggs.

Table 13-1 gives examples of some of the effects of the accidental or deliberate introduction of alien species into the United States. The Department of Agriculture estimates that illegally imported pest species cause at least $10 billion a year in crop damage.

Consider the effects of the accidental introduction of water hyacinths. These plants, which are native to Central and South America, were brought to the United States in the 1880s for an exhibition in New Orleans. A woman took one to plant in her backyard in Florida. Within 10 years the colorful plant was a public menace. Unchecked by natural enemies and thriving on Florida's nutrient-rich waters, the hyacinth, which can double its population in only 2 weeks, rapidly displaced native aquatic plants and blocked boat traffic in many ponds, streams, canals, and rivers. Water hyacinths have also spread to waterways in other southeastern states, aided by canals that crisscross parts of the region.

In 1898 the U.S. Army Corps of Engineers tried unsuccessfully to use a mechanical cropper to remove these plants from navigable waters. Next they tried sodium arsenite. This chemical was somewhat successful, but it was abandoned in the 1930s because the deadly arsenic found its way into the food of spray boat crews. In the mid-1940s a combination of mechanical removal and the herbicide 2,4-D was used, but the water hyacinth continued to spread.

The Florida manatee, or sea cow, feeds on aquatic weeds and in sufficient numbers can control the growth and spread of water hyacinths in inland waterways more effectively than mechanical or chemical methods. But these gentle and playful mammals are threatened with extinction primarily because of slashing by powerboat propellers and entanglement in fishing gear. As a result, only about 800 to 1,000 sea cows now exist in Florida.

In recent years, scientists have brought in other natural predators to help control the hyacinth. For example, a species of weevil that feeds only on hyacinths has been imported from Argentina. Results look promising, but it is too early to evaluate this experiment. The grass carp, or white amur, brought in from the Soviet Union, is also being used to control water hyacinths and other undesirable aquatic weeds. This introduced fish species may solve the water hyacinth problems,

Table 13-1   Damage Caused by Plants and Animals Imported into the United States

Name	Origin	Mode of Transport	Type of Damage
Mammals			
European wild boar	Russia	Intentionally imported (1912), escaped captivity	Destruction of habitat by rooting; crop damage
Nutria (cat-sized rodent)	Argentina	Intentionally imported, escaped captivity (1940)	Alteration of marsh ecology; damage to levees and earth dams; crop destruction
Birds			
European starling	Europe	Intentionally released (1890)	Competition with native songbirds; crop damage, transmission of swine diseases; airport interference
House sparrow	England	Intentionally released by Brooklyn Institute (1853)	Crop damage; displacement of native songbirds
Fish			
Carp	Germany	Intentionally released (1877)	Displacement of native fish; uprooting of water plants with loss of waterfowl populations
Sea lamprey	North Atlantic Ocean	Entered via Welland Canal (1829)	Destruction of lake trout, lake whitefish, and sturgeon in Great Lakes
Walking catfish	Thailand	Imported into Florida	Destruction of bass, bluegill, and other fish
Insects			
Argentine fire ant	Argentina	Probably entered via coffee shipments from Brazil (1918)	Crop damage; destruction of native ant faunas
Camphor scale insect	Japan	Accidentally inported on nursery stock (1920s)	Damage to nearly 200 species of plants in Louisiana, Texas, and Alabama
Japanese beetle	Japan	Accidentally imported on irises or azaleas (1911)	Defoliation of more than 250 species of trees and other plants, including many of commercial importance
Plants			
Water hyacinth	Central America	Intentionally introduced (1884)	Clogging waterways; shading out other aquatic vegetation
Chestnut blight (a fungus)	Asia	Accidentally imported on nursery plants (1900)	Destruction of nearly all eastern American chestnut trees; disturbance of forest ecology
Dutch elm disease, *Cerastomella ulmi* (a fungus, the disease agent)	Europe	Accidentally imported on infected elm timber used for veneers (1930)	Destruction of millions of elms; disturbance of forest ecology

From *Biological Conservation* by David W. Ehrenfeld. Copyright © 1970 by Holt, Rinehart and Winston, Inc. Modified and reprinted by permission.

but it could easily become a major pest itself by eating desirable species of aquatic plants.

Preliminary research by the National Aeronautic and Space Administration (NASA), however, has shown that hyacinths can be: **(1)** introduced in sewage treatment lagoons to absorb toxic chemicals, **(2)** converted by fermentation to a biogas fuel similar to natural gas, **(3)** used as a mineral and protein supplement for cattle feed, and **(4)** used as a fertilizer and soil conditioner.

Characteristics of Extinction-Prone Species
Some species have certain natural characteristics that make them more susceptible to extinction by human activities and natural disasters than other

species, as summarized in Table 13-2. Species with two or more of these characteristics, such as the California condor, are particularly at risk. This large vulture once flourished from Canada to Baja California. By mid-1985 about 9 of these critically endangered birds, including only 1 breeding pair, remained alive in the wild. They were found mainly in a relatively small, mountainous sanctuary north of Los Angeles. Nine additional California condors were being cared for in captivity in the San Diego and Los Angeles zoos.

The decline in the condor population is the result of loss of habitat, food scarcity, shooting for sport or to collect feathers, poisoning, egg collecting, contamination with pesticides, and stress from contacts with humans, coupled with the

Table 13-2 Characteristics of Extinction-Prone Species

Characteristic	Examples
Feed at high trophic levels	Bengal tiger, bald eagle, Andean condor, timber wolf
Large size	Bengal tiger, African lion, elephant, Javan rhinoceros, blue whale, American bison, giant panda
Low reproductive rate	Blue whale, polar bear, California condor, Andean condor, passenger pigeon, giant panda, *Homo sapiens*
Limited or specialized nesting or breeding areas	Kirtland's warbler (nests only in 6- to 15-year-old jack pine trees), whooping crane (depends on marshes for food and nesting), orangutan (now found only on islands of Sumatra and Borneo), green sea turtle (lays eggs on only a few beaches), bald eagle (preferred habitat of forested shorelines), nightingale wren (nests and breeds only on Burro Colorado Island, Panama)
Fixed migratory patterns	Blue whale, Kirtland's warbler, Bachman's warbler
Specialized feeding habits	Ivory-billed woodpecker (beetle larvae in recently dead trees), Everglades kite (apple snail of southern Florida), blue whale (krill in polar upwelling areas), black-footed ferret (prairie dogs and pocket gophers)
Certain behavioral patterns	Passenger pigeon and white-crowned pigeon (nests in large colonies), redheaded woodpecker (flies in front of cars), Carolina parakeet (when one bird is shot, rest of flock hovers over body), Key deer (forages for cigarette butts along highways—it's a "nicotine addict")
Highly intolerant of human presence	California condor, grizzly bear, red wolf, Carolina parakeet
Preys on livestock or humans	Timber wolf, some crocodiles
Valuable for fur, flesh, feathers, or other uses	Cheetah, Bengal tiger, Indian elephant, snow leopard, rhinoceros, green turtle
Found in only one place or region	Woodland caribou, elephant seal, Cooke's kokio, and many unique island species

bird's large size and low reproductive rate. Being big, the condor was an easy target for hunters, ranchers, and farmers who incorrectly blamed the species for the death of lambs, calves, and chickens. Condors, however, are not predators; they feed on animal carcasses.

Condors are monogamous partners for life and usually produce only one offspring every 2 years. Chicks fail to hatch when human activities and noise scare the parents away from the nest during the 42-day incubation period. A newborn chick depends on its parents for up to 2 years and dies if prematurely abandoned. Condors also require a large undisturbed habitat.

13-3 Wildlife Protection

Approaches to Wildlife Protection In general, three main strategies are used to protect endangered and threatened wildlife and to prevent wildlife from becoming endangered: **(1)** *establishing treaties and passing laws* to protect particular species of endangered and threatened wildlife from being killed and to preserve their critical habitats from destruction and degradation, **(2)** *using gene banks, zoos, research centers, botanical gardens, and aquariums* to preserve threatened or endangered species and in some cases to breed individuals of critically endangered species in captivity for eventual return to the wild, and **(3)** *preserving a variety of unique and representative ecosystems* and the variety of species they contain rather than concentrating on individual species.

The Species Approach: Treaties and Laws Organizations such as the International Union for the Conservation of Nature (IUCN), the International Council for Bird Preservation (ICBP), and the World Wildlife Fund (WWF) have identified threatened and endangered species and led efforts to protect them. In 1973, for example, after 10 years of work by the IUCN, representatives from 80 nations drew up the Convention on International Trade in Endangered Species of Wild Flora and Fauna treaty and also developed lists of plants and animals needing protection from international trade. By 1985, 81 nations had agreed to abide by the treaty. Although implementation and enforcement of this agreement vary from nation to nation, it appears to have contributed to the conservation of a number of plants and animals such as sea turtles and crocodiles. In addition, various nations have entered into other international conservation agreements.

The United States has increased efforts to protect native game and nongame endangered species with the passage of several pieces of legislation. The Endangered Species Act of 1973 (including amendments in 1978 and 1982) may be the most far-reaching species protection law enacted by any nation and one of the toughest and most controversial environmental laws ever passed by Congress. This act authorizes the National Marine Fisheries Service of the Department of Commerce to identify and list marine species, and the Department of Interior's Fish and Wildlife Service (FWS) to identify all other plant and animal species that are endangered or threatened in the United States and abroad. Any decision by either agency to add or remove a species from the list must be based solely on biological grounds without economic considerations and must be made within one year of the initial proposal for addition or removal.

The Endangered Species Act also authorizes the departments of Commerce and Interior to design and conduct programs for the recovery of endangered and threatened species, to assist states and other countries to conserve such species, and determine, protect, and—when necessary—purchase *critical habitats* in the United States as required for the normal needs and survival of endangered or threatened species. This act also prohibits interstate and international commercial trade of endangered or threatened *plant or animal species* (with certain exceptions) or products made from such species, and it prohibits the killing, hunting, collecting, or injuring of any protected animal species. It also directs federal agencies not to carry out, fund, or authorize projects that would jeopardize endangered or threatened species or destroy or modify habitats critical to their survival.

The last provision has been highly controversial. In 1975 conservationists filed suit against the Tennessee Valley Authority to stop construction of the $137-million Tellico Dam on the Little Tennessee River because the area to be flooded by the dam reservoir threatened the only known breeding habitat of an endangered fish species, the tiny snail darter. Although the dam was 90 percent completed, construction was halted by the court action for several years.

In 1978 Congress amended the Endangered Species Act to permit a seven-member Endangered Species Review Committee to grant an exemption if it believed that the economic benefits of a project would outweigh the potential harmful ecological effects. At their first meeting, the review committee denied the request to exempt the Tellico Dam project on the grounds that it was economically unsound. In 1979, however, Congress passed special legislation exempting the project from the Endangered Species Act and the dam's reservoir is now full of water. The snail darters that once dwelled there were transplanted to nearby rivers. In 1981 snail darter populations were found in several other remote tributaries of the Little Tennessee River, and in 1983 their status was downgraded by the FWS from endangered to threatened.

Environmentalists complain that the Endangered Species Act is not being carried out as intended by Congress because of budget cuts and administrative rules. For example, in 1983 the federal government spent an amount equivalent to the cost of 12 army bulldozers on endangered species.

The Species Approach: Wildlife Refuges The world's largest system of protected wildlife habitats is in the United States. The first federal wildlife refuge, Pelican Island in Florida, was established through executive order by President Theodore Roosevelt in 1903. Since that time more than 400 refuges have been designated as part of the National Wildlife Refuge System, administered by the FWS. Private organizations like the Nature Conservancy, Ducks Unlimited, and the National Audubon Society also lease or purchase important tracts of wildlife habitats that might be destroyed by commercial development.

Most state and federal wildlife conservation efforts have been focused on the protection and management of waterfowl and big-game species, primarily for the 17 million Americans who hunt. A major reason for this emphasis is that funds used for wildlife conservation come primarily from fees for hunting and fishing licenses and permits and from taxes on hunting and fishing equipment. These revenues from sport hunting and fishing have supported wildlife research, habitat protection and restoration, and reintroduction of wildlife on depleted ranges.

In recent years there has been an increase in state and federal efforts to improve the conservation of nongame animals. Federal wildlife refuges for specific endangered species have helped the endangered Key deer of southern Florida, the trumpeter swan, and the bald eagle to recover.

The Gene Bank–Zoo–Botanical Garden–Aquarium Approach *Gene banks* of most known and many potential varieties of agricultural crops now exist throughout the world, and scientists have urged that many more be established. Despite their importance, gene banks have significant disadvantages and need to be supplemented by preservation of a variety of representative ecosystems throughout the world. Storage is not possible for many species. Accidents, human error, and vandalism can cause irrecoverable losses. Furthermore, stored species do not continue to evolve and thus lack the genetic variability found in a species surviving in the wild.

Zoos, botanical gardens, arboretums, and aquariums are also being used increasingly to preserve

Figure 13-4 The Arabian oryx barely escaped extinction in 1969 after being overhunted in the deserts of the Middle East. Captive breeding programs in zoos in Arizona and California apparently have saved this antelope species from extinction.

and study representatives of critically endangered species that would otherwise become extinct. When it is judged that an animal species will not survive on its own, eggs may be collected and hatched in captivity, or captive breeding programs may be established in zoos or private research centers. Other techniques include artificial insemination of species that don't breed well in captivity, using adults of one related species to serve as foster parents to hatch collected eggs and raise offspring of another species, and removing newly laid eggs from captive or wild individuals for incubation elsewhere. This "egg-pulling" approach is designed to induce parents to renest and lay more eggs.

In some but not all cases, captive breeding and egg incubation programs allow a population to increase sufficiently that the species can be successfully reintroduced into the wild. For example, captive breeding programs at zoos in Phoenix, San Diego, and Los Angeles have been used to rebuild the nearly extinct the Arabian oryx antelope (Figure 13-4), which are now being returned in small numbers to the wild. The number of whooping cranes in the wild has been raised from 15 in 1941 to more than 149 in 1985 by habitat preservation and by captive breeding programs, involving the techniques described above.

Captive breeding can help save some critically endangered species, but it has several disadvantages. It is expensive and requires a minimum captive population of 100 to 150 individuals (half of them born in captivity) to assure long-term survival of a mammalian species in captivity. Most zoos use their limited funds and capacity to display as many different species as possible. Thus, the world's zoos now contain only 20 species with populations of 100 or more individual animals.

Such relatively small captive animal populations also suffer a loss of genetic variability.

Because of limited funds and scarcity of trained personnel, only a few of the world's endangered and threatened species can be saved by laws, wildlife refuges, gene banks, and zoos. At present, efforts to save species are concentrated on "glamour" animal species that find favor with the general public primarily because they are furry and cuddly (koala bears), beautiful (whooping cranes and Bengal tigers), unique (blue whales), or symbolic (bald eagles). Campaigns centered on such well-known animal species can generate sizable public donations and interest. These funds can also be used to help preserve other species with less popular appeal.

An increasing number of wildlife experts, however, have suggested replacing this haphazard approach to saving endangered species with an environmental form of *triage*, a practice developed by Allied doctors during World War I. In triage wounded soldiers were sorted into three groups: those likely to die despite medical care, those likely to recover without medical care, and the remainder, who were treated with the limited medical resources available. The proponents of environmental triage suggest that limited funds for preserving threatened and endangered wildlife be concentrated on species that **(1)** have the best chance for survival, **(2)** have the most ecological value to an ecosystem, and **(3)** are potentially useful for agriculture, medicine, or industry.

The Ecosystem Approach Most wildlife biologists argue that the major threat to most wildlife species today—namely, the destruction of habitat—cannot be halted using a species-to-species approach. Instead, they believe that wildlife conservation efforts should be concentrated on preserving biological diversity in large reserves all over the globe that contain a representative cross section of the world's ecosystems. Once the necessary land had been set aside, adequate and well-protected ecological reserves would help prevent species from becoming endangered by human activities, reduce the need for human intervention to prevent extinction, provide scientists with opportunities for wildlife research, and cost less to run than the management of species one by one.

By 1982, 210 reserves, including 33 in the United States, had been set aside by 55 nations. This is an important beginning, but only about half the world's biogeographical provinces have been included so far, and the quality of protection and management varies widely. It has been proposed that the MDCs set up an international fund to help LDCs protect and manage such reserves. The program would cost about $100 million a year—about what the world spends on military arms every 90 minutes.

13-4 Wildlife Management

Management Approaches Ideally, the goal of a professional wildlife manager is to produce large populations of certain desirable game species that can be harvested each year during hunting season while leaving strong, healthy reproductive populations for the next year. The two major approaches used to manage such species are *population regulation* and *manipulation and protection of habitat*.

Population Regulation by Controlled Hunting Game animals that reproduce rapidly, such as deer, rabbits, squirrels, quail, and ducks, sometimes exceed the carrying capacity of their habitat. For example, a deer population can more than double every 2 years. As the number of deer exceeds the carrying capacity of their range, vegetation is destroyed, the habitat deteriorates, and many animals die of starvation during winter.

Since humans have eliminated most natural predators of deer, carefully regulated hunting can be used to keep the deer population within the carrying capacity of the available habitat. Such sport hunting is usually regulated by game laws. In the United States, for example, populations of game animals such as deer are managed by (1) specifying certain times of the year for the hunting of a particular species, (2) restricting the length of hunting seasons, (3) regulating the number of hunters permitted in an area, and (4) placing limits on the size, number, and sex of animals allowed to be killed.

Habitat Protection and Manipulation When the food habits, predators, and nesting, mating, and cover requirements of a desired game species are known, a wildlife manager can alter and protect the habitat to encourage adequate production of that species for harvesting during hunting season. This is done primarily by controlling the stage of ecological succession (Section 6-1) of vegetation in various areas.

Grizzly bear, wolf, caribou, and bighorn sheep are examples of *wilderness species* that flourish only in relatively undisturbed climax vegetational communities. Their survival depends to a large degree on the establishment of relatively large state and national wilderness areas and wildlife refuges. Wild turkey, marten, Hammond's flycatcher, and gray squirrel are examples *late-successional species* whose habitats require establishment and protection of moderate-sized mature forest refuges.

Midsuccessional species (elk, moose, deer, grouse, snowshoe hare) are found around abandoned croplands and partially open areas created by logging, burning, and grazing and can be protected by the periodic cutting or prescribed burning (Section 12-4) of vegetation to maintain such areas at this stage of succession. *Early-successional species* (rabbit, quail, dove) find food and cover in the weedy pioneer plants that invade an area that has been cleared of vegetation for human activities and then abandoned.

Migratory waterfowl such as ducks, geese, and swans require some special management approaches. Many of these species nest in Canada during the summer and migrate to the United States and Central America along generally fixed routes called *flyways* during the fall hunting season. Thus, international agreements are needed to prevent destruction of the migrants' winter and summer habitats and to prevent overhunting. Waterfowl habitat can be improved by a number of methods, including (1) periodically draining ponds to retard aquatic succession, (2) creating channels and openings in dense marsh vegetation to allow birds to feed and move about, (3) constructing artificial ponds, islands, and nesting sites, and (4) establishing protected waterfowl refuges. In 1934 Congress passed the Migratory Bird Hunting Stamp Act, which authorized the sale of duck stamps to provide funds for the acquisition, maintenance, and development of waterfowl refuges.

13-5 Fisheries Management

Freshwater Fisheries Management The goals of freshwater fish management are to encourage the growth of populations of desirable sport fish species and to reduce or eliminate populations of less desirable species. Techniques for management of freshwater fish species include: **(1)** enacting protective laws to regulate the timing and length of the fishing season for various species, to determine the minimum fish size that can be taken, to establish catch quotas, and to require that commercial fish nets have large enough mesh size to ensure that young fish are not harvested, **(2)** providing and protecting adequate natural and artificial habitats for game fish species by creating hiding places, preventing buildup of debris, organic waste material, and excessive growth of aquatic plants to prevent oxygen depletion, and protecting spawning grounds from pollution, siltation, and predation, **(3)** controlling or eliminating undesirable species and diseases that deplete game fish species, **(4)** restocking depleted areas with fish from hatcheries, **(5)** maintaining migration routes for *anadromous species*, which migrate from fresh water to saltwater and back again, and **(6)** controlling water pollution (Chapter 22).

Marine Fisheries Management The history of the world's commercial fishing and whaling industry is an excellent example of the *tragedy of the commons*—the abuse and overuse of a resource such as ocean fish and mammals that is not owned by anyone and is available for use by everyone. Such depletion of a "common" resource occurs through the ignorance and greed of individuals, industries, or nations as they attempt to secure all of a resource they can in the shortest possible time, with little concern for future supplies or immediate impact.

As a result, many species of commercially valuable fish and whales found in international waters and in the coastal waters have been overfished to the point of *commercial extinction*; that is, they are so rare that it no longer pays to hunt them. For example, the catch of cod, halibut, herring, haddock, and other commercially preferred species in the northwest Atlantic Ocean has been declining for more than a decade. Another threat to commercially important fish species and the other species on which they depend for food and other services is pollution of estuaries and the ocean by toxic chemicals, radioactive wastes, oil, and excess heat (Section 22-6).

Techniques used to manage marine fisheries in the United States include: **(1)** the introduction of food and game fish species (for example, the striped bass along the Pacific and Atlantic coasts), **(2)** the construction of artificial reefs from boulders, building rubble, and automobile tires to attract desirable species, and **(3)** the extension by law of the offshore fishing zone, where foreign fishing vessels are banned without government permission, to 322 kilometers (200 miles) from 19 kilometers (12 miles).

The Whaling Industry The pattern of the whaling industry has been to hunt the most commercially valuable species until it becomes too scarce to be of commercial value and then turn to another species. Of the 11 major species once hunted by the whaling industry, only the sperm, minke, and sei whales remain unendangered, and their days may be numbered.

Probably only a few hundred to a few thousand blue whales remain—perhaps too few for the species to recover. This sharp decline can be attributed to a ruthless, greedy whaling industry as well as to three natural characteristics of the blue whale (Table 13-2). First, they are large and thus easy to spot. Second, they can be caught in large numbers because they tend to congregate in the Antarctic feeding grounds for about 8 months of each year. Third, they multiply very slowly, taking up to 25 years to mature sexually and having one offspring every 2 to 5 years. Once the total population has been reduced below a certain level, mates may no longer be able to find each other, and natural death rates may exceed natural birth rates until extinction occurs. Within the next few decades, blue whales could become extinct, even though, as an endangered species, they are now protected by law.

Today 40 whaling and nonwhaling nations belong to the International Whaling Commission (IWC), established after World War II. The goal of the IWC has been to regulate the annual harvest by setting quotas for hunted species of whales, to ensure a sustainable supply of all commercially important species. Annual quotas for all species have been reduced from 45,000 in 1972 to 6,040 in 1984. Most of these reductions, however, came only after some species had been hunted to commercial extinction.

The United States and a number of other nations in the IWC have repeatedly called for a ban on all commercial whaling. The Greenpeace movement has used bold, commandolike tactics to embarrass whaling nations, stop pirate whalers who ignore the quotas, and gain worldwide publicity and support for protecting whales. In 1982 the IWC voted to end all commercial whaling by the end of 1985, with a gradual phasedown of the catch quota until that date. But Japan, the Soviet Union, and Norway, the three major whaling nations, filed formal protests against the ban. According to IWC rules, such protests free the objecting nation from any obligation to comply with the ruling. In 1984 the U.S. government stated that it would not impose on Japan fishing quotas in U.S. waters if the Japanese would agree to quit all commercial whaling by 1988, 2 years beyond the cutoff date set by the IWC. In mid-1985 a federal judge ruled that this agreement was illegal.

Love the animals, love the plants, love everything. If you love everything, you will perceive the divine mystery in things. Once you perceive it, you will begin to comprehend it better every day. And you will come at last to love the whole world with an all-embracing love.

Feodor Dostoyevsky, The Brothers Karamazov

Discussion Topics

1. Discuss your gut-level reaction to the statement: "Who cares that the passenger pigeon is extinct and the buffalo, the blue whale, the whooping crane, the bald eagle, the grizzly bear, and other species are nearly extinct? They are important only to a bunch of bird watchers, Sierra Clubbers, and other ecofreaks." Be honest about your reaction, and try to organize arguments for your position.

2. Why should an urban dweller be concerned about preservation of wildlife and wildlife habitat?

3. Criticize the idea that since 70 to 98 percent of the species that have existed on earth have become extinct by natural selection, we should not be concerned about the several hundred animal species and thousands of plant species that have become extinct primarily because of human activities.

4. Some argue that all species have an inherent right to exist and should at least be preserved in a natural habitat somewhere on earth. Do you agree or disagree with this position? Why? Would you apply this idea to (a) *Anopheles* mosquitoes, which transmit malaria, (b) tigers that roam the jungle along the Indian–Nepalese border and have killed at least 105 persons between 1978 and 1983, (c) bacteria that cause smallpox or other infectious diseases, and (d) rats that compete with humans for many food sources.

5. Should all coyotes and eagles be exterminated from lands where sheep graze? Why or why not? What are the alternatives?

6. Use Table 13-2 to predict a species that may soon be endangered. What, if anything, is being done for this species? What pressures is it being subjected to? Try to work up a plan for protecting it.

7. Should a ban be placed on the importation of all new plant and animal species into the United States for use as pets, observation in zoos, scientific and medical research, and use as predators or parasites of pests already here? Develop guidelines and criteria that should be met before any new species is introduced.

14

Urban Land Use and Land-Use Planning

Modern cities are centers of employment, education, and culture. But they are also centers of poverty, delinquency, crime, prostitution, alcoholism, and drug abuse. As a rule, cities offer less space, less daylight, less fresh air, less greenery, and more noise.

Georg Borgstrom

14-1 Urban Growth

The World Situation At the beginning of this century less than 3 percent of the world's population lived in **urban areas**—places with a population of more than 20,000 people. By 1985, 41 percent of the world's population lived in urban areas. The United Nations estimates that if present trends continue, by the close of this century at least 50 percent of the people in the world will be urban dwellers, half of them unemployed or underemployed. Accommodating the 3.0 billion

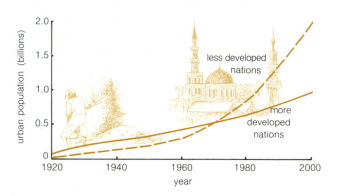

Figure 14-1 Actual and projected urban population growth in more developed and less developed nations between 1920 and 2000. (Source: United Nations.)

people projected to be living in urban areas by the end of this century will be a monumental task.

The rate of urban population growth in LDCs surpassed that in MDCs around 1970 and is expected to increase more rapidly (Figure 14-1), because the LDCs are simultaneously experiencing

Table 14-1 The 10 Largest Urban Areas in the World in 1983 and 2000 (projected)

1983		2000	
Urban Area	Population (millions)	City	Population (millions)
Tokyo–Yokohama	21.1	Mexico City	27.6
New York–northeast New Jersey	18.1	Shanghai	25.9
Mexico City	17.0	Tokyo–Yokohama	23.8
Shanghai	16.5	Peking	22.8
São Paulo	14.0	São Paulo	21.5
Peking	13.5	New York–northeast New Jersey	19.5
Los Angeles–Long Beach	11.6	Bombay	16.3
Rio de Janeiro	10.7	Calcutta	15.9
Greater Buenos Aires	10.5	Jakarta	14.3
London	9.9	Rio de Janeiro	14.2

Source: United Nations.

high population growth *and* rapid migration of people into urban areas. Already at least one-third of the urban dwellers in LDCs live in slums and shantytowns with inadequate drinking water, sanitation, food, health care, housing, schools, and jobs. The population of many of these overburdened slums is doubling every 5 to 7 years.

Unprecedented urban growth in MDCs and LDCs has given rise to a new concept—that of the "*supercity*" or urban area with a population of more than 10 million people. In 1980 there were 10 supercities, four of them located in MDCs (Table 14-1). By 2000, the United Nations projects that there will be 25 supercities. Most will be located in LDCs, and populations exceeding 22 million people are projected for four of them (Table 14-1).

The U.S. Situation According to latest U.S. Census Bureau definitions, Americans live in three major types of areas: **(1) rural areas**, places with a population less than 2,500, **(2) urban areas** consisting of a central city and its suburbs, containing 50,000 or more residents and with an average population density of at least 1,000 per square mile, and **(3)** *other urban places*, small cities and towns not associated with an urban area and having individual populations between 2,500 and 50,000. Sometimes rural areas and other urban places are lumped together and called **nonmetropolitan areas**.

Since 1800 there have been four major population shifts in the United States, as summarized in the accompanying box.

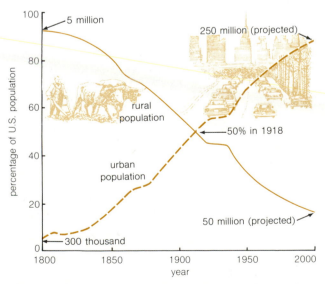

Figure 14-2 Actual and projected percentages of the rural and urban populations in the United States between 1800 and 2000. (Source: U.S. Bureau of the Census.)

Urban Economic and Social Succession Although each city has its own distinct pattern of development, urban experts have developed a generalized model of urban growth and decay that can be used to understand what happens to many urban areas in MDCs such as the United States. A city, like a biological community undergoing secondary succession (Section 6-1), begins as an im-

Major Population Shifts in the United States

1. The *rural-to-urban shift*, with the percentage of the population living in cities and towns of population greater than 2,500 increasing from 5 percent to 74 percent between 1800 and 1985 and projected to reach 80 to 90 percent by 2000 (Figure 14-2). Figure 14-3 identifies the nation's 28 largest urban regions, where more than two out of three Americans now live, and provides details on the seven largest ones.

2. The *central city-to-suburbs shift*, in which large numbers of mostly middle- and upper-class citizens have moved from the central city to the suburbs since 1950. Because of this shift, almost one-third of all Americans now live in the suburbs.

3. The *urban-to-nonmetropolitan shift*, in which people have been moving away from some cities—especially larger ones—in favor of other smaller towns and cities since 1970. This influx of people to rural counties and small towns and cities has increased the stress on many local governments, hindering their ability to provide schools, houses, sewage disposal, and other services, and may eventually cause them to become urban areas, against the wishes of many citizens.

4. The shift in population and economic and political power from *northeastern and midwestern states to southern and western states*, which has been taking place since 1970, and is expected to continue at least until the end of this century. In 1980, for the first time, the population of the South and West exceeded that of the Northeast and Midwest, with particularly high population growth in three of the *Sun Belt* states (California, Texas, and Florida).

c Southern California		f Northern California	
1980 population	12,647,607	1980 population	7,221,281
change in 1970–80	16.2%	change in 1970–80	16.0%
share of U.S. population	5.6%	share of U.S. population	3.2%

Figure 14-3 Urban regions in the United States. (Source: U.S. Bureau of the Census.)

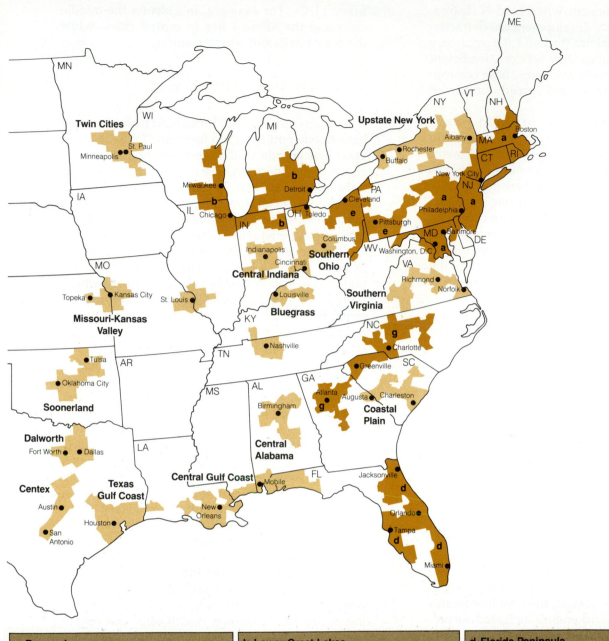

MN						ME	
Twin Cities	WI					NY	VT	NH
Minneapolis ● St. Paul		MI		Upstate New York				Boston
				Rochester ●	Albany ●	MA	a
IA		b	Detroit ●	● Buffalo		CT	RI
	b		e	New York City		NJ	
IL	Milwaukee ●		Cleveland ●	PA	Philadelphia ●	a	a
Chicago ●	IN	OH	Toledo ●	● Pittsburgh	a		
MO		b	Columbus ●		MD ● Baltimore	DE	
		Indianapolis ●	Southern Ohio	WV	Washington, D.C.		
Kansas City	Central Indiana	Cincinnati ●		VA	a		
Topeka ● ●	St. Louis ●		Bluegrass	Richmond ●			
Missouri-Kansas Valley		Louisville ●	Southern Virginia	● Norfolk			
AR	KY			NC	g		
Tulsa ●	TN	Nashville ●		Charlotte ●			
Oklahoma City ●				Greenville ●	SC		
Soonerland	MS	AL	GA	Atlanta ●	Charleston ●		
Dalworth		Birmingham ●	g	Augusta ●	Coastal Plain		
Fort Worth ● ● Dallas	LA	Central Alabama					
Centex		Central Gulf Coast	FL	Jacksonville ●			
Austin ●	New Orleans ●	Mobile ●		d			
Texas Gulf Coast				Orlando ●			
Houston ●				Tampa ● d			
San Antonio ●				d	d		
				Miami ●			

a Boswash	
1980 population	41,786,061
change in 1970–80	0.2%
share of U.S. population	18.4%

b Lower Great Lakes	
1980 population	20,827,980
change in 1970–80	2.8%
share of U.S. population	9.2%

d Florida Peninsula	
1980 population	8,290,959
change in 1970–80	44.7%
share of U.S. population	3.7%

e Cleveburgh	
1980 population	8,141,391
change in 1970–80	-2.7%
share of U.S. population	3.6%

g Piedmont	
1980 population	6,359,479
change in 1970–80	20.4%
share of U.S. population	2.8%

mature pioneer settlement when land is cleared. The early stages of city development are characterized by high productivity (to build the necessary structures and products), few services, inefficient use of matter and energy resources, little community organization, and a rapid spread outward of small structures.

As an urban area continues to spread, its services—especially in the central city—may begin to break down. In the early stages of urban growth, housing, industry, and employment increase rapidly. As the original central city housing deteriorates, many middle- and high-income residents leave and build homes beyond the city core. The resulting loss of business means that many low-income residents left in the central city are unable to find work. Rising unemployment generates the need for welfare, public housing, and other expenditures.

As middle- and upper-income residents leave, property values in the central city decline. These lower property values mean lower tax revenues, because about 87 percent of the operating revenue for local governments in the United States is derived from property taxes. Because there are no new tax revenues to absorb the added financial load, the city government typically raises business and private property taxes and cuts back services and maintenance—causing more businesses and individuals to flee to the suburbs.

Before long, freeways and mass transit systems are built to accommodate commuters—making it easier for more people to live in the suburbs. Freeways and mass transit systems are often built through poor neighborhoods, which usually constitute the cheapest route, because residents do not have enough economic and political power to block the projects. As a result, the poor lose homes and local businesses and endure increased air pollution, noise, and other environmental insults. Each year about 51,000 Americans are displaced from their urban homes to make way for freeways and highways that most of them don't use.

As more freeways to the suburbs are built, transit systems within the central city usually deteriorate. This means that low- and middle-income workers find it more difficult to get to and from work. This accelerating cycle of more unemployment and more poverty breeds despair. In turn, this can lead to an increase in drug use, crime, and other forms of social disruption. These conditions increase the fear and tension for the poor, the elderly, and the handicapped who remain in the central cities.

This typical model of urban growth and decay in many MDCs is altered somewhat in cities in many LDCs. For example, in Calcutta the middle class and the affluent live in central cities, while the poor live in surrounding slums.

14-2 The Urban Environment

Benefits and Stresses Throughout history, people have flooded into cities to find jobs. Those who get good jobs and do well financially often are enthusiastic about living in cities, which provide a variety of goods and services and an exciting diversity of social and cultural activities. In addition, urban social and cultural activities enrich the lives of people who live far from city boundaries.

Not all cities are pleasant places to live. People in many medium-sized to large cities are increasingly subjected to problems related to crowding, such as traffic snarls, housing shortages and high rents, bad service on deteriorating and inefficient public transportation systems, excessive noise, crime, pollution, and long lines everywhere. According to surveys, an increasing number of residents in many central cities in the United States feel isolated, powerless, and entrapped. They are afraid to go out at night, and they surround themselves with locks, alarm systems, and other security precautions. Older people are particularly susceptible to these and other strains of urban life. For many young people, street life and peer group pressure in an urban ghetto tend to reinforce ideas that hard work and traditional employment have few rewards, that crime does indeed pay, and that one who is not part of the predominant society has no obligation to it.

Many low- and middle-income people living in decaying central cities regard attempts to control pollution as largely irrelevant to their most pressing needs: jobs, housing, health care, and education. However, studies in more than 114 major U.S. cities showed that the urban poor have higher rates of illness and death from diseases associated with air pollution than persons living in higher income neighborhoods. The poor in cities also tend to receive poor medical services, and they suffer from infestation by rodents and other pests, little or no public transportation, higher cost for heating because of poorly insulated homes and buildings, and inadequate sanitation and other services. In addition, low- and middle-income workers living in the cities have little choice but to take on jobs in the remaining factories and sweatshops, where they are often exposed to serious noise and health hazards, leading to loss of hearing, black lung, brown lung, and cancer of several forms (Section 20-3).

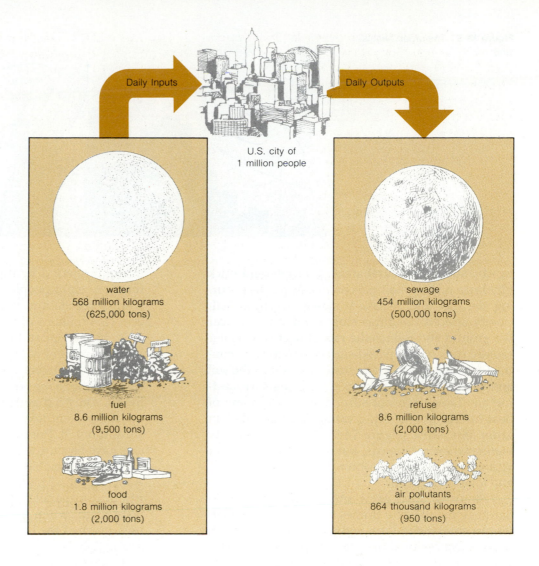

Figure 14-4 Typical daily inputs and outputs for a U.S. city of 1 million people.

Daily Inputs

Daily Outputs

U.S. city of
1 million people

water
568 million kilograms
(625,000 tons)

fuel
8.6 million kilograms
(9,500 tons)

food
1.8 million kilograms
(2,000 tons)

sewage
454 million kilograms
(500,000 tons)

refuse
8.6 million kilograms
(2,000 tons)

air pollutants
864 thousand kilograms
(950 tons)

Many of these problems are intensified in urban areas in less developed countries. Uncontrolled population growth and diminishing job opportunities in overpopulated rural areas in LDCs force people to migrate from the countryside into urban areas. There, too, jobs are seldom available. For this rapidly growing number of migrants, the city becomes a poverty trap, not an oasis of economic opportunity and cultural diversity. The urban poor are forced to live on the streets or to crowd into slums and shantytowns, where they grub for food and fuel. Housing is minimal, sewers are open, drinking water is untreated, and there is little access to schools and hospitals. Ironically, expanded services usually do little to improve conditions, since they attract more of the rural poor to the city.

Despite joblessness and squalor, shantytown residents cling to life with resourcefulness, tenacity, and hope. Most of them are convinced that the city offers, possibly for themselves and certainly for their children, the only chance of a better life.

On balance, most tend to have more opportunities and are often better off than the rural poor they left behind. They also tend to have fewer children, since they have better access to family planning and population control programs (Section 8-3).

Many urban dwellers in LDCs and MDCs are unaware that the cities in which they live are not self-sustaining. They survive only by importing plant and animal food from external ecosystems located throughout the world. Cities also obtain fresh air, water, minerals, and energy resources from outside ecosystems. Instead of being recycled, most urban solid, liquid, and gaseous wastes are discharged to ecosystems that lie mainly outside their boundaries (Figure 14-4).

Urban Heat Islands Another environmental problem in many urban areas is excessive heat buildup as a result of human activities and the nature of urban areas. Anyone who lives or works in a city knows that it is typically warmer there than in

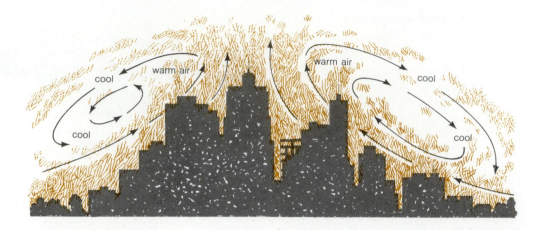

Figure 14-5 The urban heat island.

nearby suburbs or rural areas. Concrete and brick buildings and asphalt pavements absorb heat during the day and release it slowly at night. Tall, closely spaced buildings slow wind velocity near the ground and reduce the rate of heat loss. Water rapidly runs off the paved surfaces in cities, in contrast to rural areas, where water soaks into the soil and then slowly evaporates to cool the surrounding air. Thus, it is not surprising that a dome of heat hovers over a city, creating what is called an *urban heat island* (Figure 14-5). Not only is a city warmer than rural areas, but it typically has lower visibility, more air pollution, less sun, less wind, and lower humidity.

As urban areas grow and merge into vast urban regions, the heat domes from a number of cities can combine to form *regional heat islands*, which could affect regional climate. The prevailing winds that normally would cleanse the center of the dome would already be polluted; hence air pollution levels under the large regional dome could be raised. In addition, summer heat levels in the center could become intolerable, and the use of millions of air conditioners would add even more heat to the atmosphere, increasing the chances of power brownouts and blackouts.

14-3 Urban Transportation

Transportation Options People in urban areas move from one place to another by a mix of three major types of transportation: **(1)** *individual transit* (private automobile, taxi, motorcycle, moped, bicycle, and walking), **(2)** *mass transit* (railroad, subway, trolley, and bus), and **(3)** *paratransit* (carpools, vanpools, and jitneys or van taxis traveling along fixed routes, and dial-a-ride systems).

Motor Vehicles In 1984 there were approximately 470 million motor vehicles in the world, with 700 million projected by the end of the century—one for every eight people. With only 5 percent of the world's population, the United States had 35 percent (164 million) of the world's motor vehicles in 1984. In the United States the car is now used for about 98 percent of all urban transportation and 85 percent of all travel between cities. Today 84 percent of all working Americans go to and from work by car. Only 5.5 percent walk or bicycle to work, and 6.3 percent take mass transit. No wonder British author J. B. Priestly remarked, "In America, the cars have become the people."

The automobile provides many advantages. Above all, it offers those who drive privacy, security, and unparalleled freedom to go where they want to go, when they want to go there. In addition, much of the U.S. economy is built around the automobile. One out of every five dollars spent and one out of every six nonfarm jobs is connected to the automobile or related industries such as oil, steel, rubber, plastics, and highway construction. This industrial complex accounts for 20 percent of the annual gross national product and provides about 18 percent of all federal taxes.

In spite of their advantages, cars and trucks have harmful effects on human lives and on air, water, and land natural resources. By providing almost unlimited mobility, automobiles and highways have been a major factor in the *urban sprawl* that characterizes the highly decentralized cities of the United States today. Because the automobile has allowed a mass exodus from the central city to the surrounding areas, downtown shopping and business areas in most central cities have deteriorated. They have been replaced with a hodgepodge of widely dispersed shopping centers, housing developments, and tangled highway networks.

Urban sprawl has also been a key factor in the decline of mass transit systems in the central cities, where up to 60 percent of the people do not own a car. Without cars and adequate public transportation, almost 100 million poor, young, elderly, and handicapped Americans have very little freedom of travel for work and pleasure.

The world's cars and trucks also kill an average of 170,000 people, maim 500,000, and injure 10 million each year. In the United States about 25 million motor vehicle accidents each year kill about 50,000 people and injure about 5 million, at a cost of about $60 billion annually in lost income, insurance, and administrative and legal expenses. This is equivalent to a death about *every 10 minutes* and a disabling injury *every 16 seconds* of every day. Since the automobile was introduced, almost 2 million Americans have been killed on the highways—about twice the number killed in all U.S. wars.

Large areas of land are also utilized by motor vehicles. Roads and parking spaces take up 65 percent of the total land area in Los Angeles, more than half of Dallas (see photo on page 1), and more than one-third of New York City and the nation's capital. Highways cover a total land area in the United States about equal to the areas of Vermont, New Hampshire, Connecticut, Massachusetts, and Rhode Island combined. Instead of reducing automobile congestion, the construction of thousands of miles of roads has encouraged more automobiles and travel, causing even more congestion. In 1907 the average speed of horse-drawn vehicles through the borough of Manhattan was measured at 18.5 kilometers per hour (11.5 miles per hour). Today, crosstown Manhattan traffic, in cars and trucks with the potential power of 100 to 300 horses, creeps along at an average speed of 8.4 kilometers per hour (5.2 miles per hour).

Bicycles   The leg-powered bicycle won't replace cars in urban areas, but its use could be greatly increased with more bike paths and lanes and secure bike parking to prevent theft. The bicycle, using no fossil fuels and requiring few resources to produce, is very useful for the trips under 8 kilometers (5 miles) that make up about 43 percent of all urban travel. In traffic, cars and bicycles move at about the same average speed.

The ability to use bikes in cities varies with the terrain, weather, and degree of sprawl. Los Angeles is flat, but its sheer distances tend to rule out anything but neighborhood cycling. San Francisco's hills discourage cycling for most people. Davis, California, a city of 38,000 with a favorable climate and flat terrain, has set an example for the rest of the United States. Its 28,000 bicycles account for about one-fourth of all transportation within the city. City employees are provided with bikes to get around, older residents drive large tricycles, and people haul groceries in trailer buggies at-

tached to the rear of their bikes. There are 64 kilometers (40 miles) of bicycle lanes and paths, and some streets are closed to automobiles.

Mass Transit   The number of riders on all forms of mass transit (buses, subways, and trolleys) declined from about 19.5 billion passengers in 1950 to about 8 billion by 1985. Some analysts see the building of new *fixed-rail rapid transit systems* and improving existing systems as a key to urban transportation problems in most large cities. Others argue that such technological solutions are extremely expensive and do not serve many people in today's spread-out cities. They are primarily useful where many people live along a narrow corridor, and even then their high construction and operating costs may outweigh their benefits. Some contend that subway cars and large buses use less energy than cars, but this argument has been challenged. With low average daily loads (typically only 25 percent occupancy), trains, trolleys, and buses use about the same amount of energy per passenger as private automobiles.

Relatively new major fixed-rail rapid transit systems include San Francisco's $1.7 billion, computer-controlled Bay Area Rapid Transit (BART) system and partially completed systems in Washington, DC, Atlanta, and Baltimore. Others are planned or are being built for Miami, Dallas, and Los Angeles. Since its opening in 1972, the BART system has suffered from breakdowns, fires, brake problems, computers that failed in the rain, massive financial losses, and too few riders. The partially completed METRO system of Washington, DC, is better planned than BART and has the advantage of serving a concentrated urban area. But this Rolls Royce of the mass transit systems cost $71.3 million a mile compared to $22.5 million a mile for BART and was built entirely at federal expense. Furthermore, in 1983 it had an operating deficit of $61 million, which is expected to double by 1986.

Other cities such as Buffalo, San Diego, and Portland, Oregon, have built or are building *light-rail trolley systems*—modernized versions of the streetcar systems found in most major U.S. cities in the 1930s and 1940s. Most of these older streetcar systems were purchased and torn up by General Motors and tire, oil, truck, and bus companies in the 1940s to increase sales of buses, trucks, and more cars.

Buses are cheaper and more flexible than trolleys. They can be routed to almost any area in widely dispersed cities. They also require less capital and have lower operating costs than most light- and heavy-rail mass transit systems. But by offering low fares to attract riders, they usually lose money. To make up for losses, bus companies tend to cut service and maintenance and seek federal, state, and local subsidies.

Paratransit Because full-sized buses are cost effective only when full, they are being supplemented by carpools, vanpools, jitneys, and dial-a-ride systems. These paratransit methods attempt to combine the advantages of the door-to-door service of a private automobile or taxi with the economy of a 10-passenger van or minibus. They represent a practical solution to some of the transportation problems of today's dispersed urban areas.

Dial-a-ride systems are in operation in an increasing number of American cities. Passengers call for a van, minibus, or tax-subsidized taxi that comes by to pick them up at the doorstep, usually in about 20 to 50 minutes. Efficiency can be increased by the use of two-way radios and routing by a central computer. These systems are fairly expensive to operate. But compared with most large-scale mass transit systems, they are a bargain, and each vehicle is usually filled with passengers.

In cities such as Mexico City, Caracas, and Cairo, large fleets of *jitneys*—small vans or minibuses that travel relatively fixed routes but stop on demand—carry millions of passengers each day. After laws banning jitney service were repealed in 1979, privately owned jitney service has flourished in San Diego, San Francisco, and Los Angeles and may spread to other cities. Analysts argue that deregulation of taxi fares and public transport fares would greatly increase the number of private individuals and companies operating jitneys in most major U.S. cities.

Discouraging Automobile Use Proposed means of reducing automobile use in cities include: **(1)** refusing to build new highways into and out of the city, **(2)** raising the price of gasoline significantly by adding higher federal and state taxes (as has been done in most European nations, where gasoline costs $2 to almost $4 a gallon), **(3)** setting aside express lanes for buses, streetcars, bicycles, and carpools during peak traffic hours (as in London, Paris, and Washington, DC), **(4)** charging higher road and bridge tolls during peak hours, **(5)** eliminating or reducing road and bridge tolls for cars with three or more passengers, **(6)** taxing parking lots, **(7)** eliminating some downtown parking lots, **(8)** charging automobile commuters high taxes or fees (as in Singapore), and **(9)** prohibiting cars on some streets or in entire areas (as in many European cities). In the United States, most elected officials have been unwilling to risk the wrath of commuters and voters by imposing such measures.

14-4   Urban and Nonurban Land-Use Planning and Control

Methods of Land-Use Planning **Land-use planning** involves deciding the best use for each parcel of land in an area and mapping out suitable locations for houses, industries, businesses, open space, roads, water lines, sewer lines, hospitals, schools, waste treatment plants, and so on. This type of planning involves gathering and using data to determine the best present and future use for each parcel of land. Four major methods are used to make projections and develop a land-use plan: **(1)** extrapolation of existing trends, **(2)** reaction to crisis, **(3)** systems analysis and modeling, and **(4)** ecological planning.

Planning by extrapolating or projecting existing trends into the future is the most widely used approach but one of the least effective. For a 1- or 2-year period this method is normally useful, but for longer terms it can be disastrous because of inability to predict accurately new trends and events. Another widely used and usually ineffective approach is to wait until problems reach the crisis stage before developing a plan of action.

In systems analysis and modeling, data are collected, goals are set, mathematical models are constructed based on different assumptions, and a computer is programmed to project and compare the models. If goals are agreed upon, such models can provide planners with a useful tool for evaluating alternative land-use plans. However, a computer model, like any model, is no better than the data and assumptions on which it is based.

Ecological approaches to land-use planning have also been developed. Although based on classifying land according to its capability, as shown earlier in Figure 9-9, this approach goes much further. The major steps in ecological land-use planning, which has been applied with partial success to many areas, are summarized in the accompanying box.

Problems associated with ecological land-use planning include: **(1)** difficulties in getting reliable scientific, economic, and social data, **(2)** difficulty of weighting the aesthetic and ecological factors, **(3)** lack of effective means for implementing land-use plans, **(4)** political conflicts between those with differing ethical views on how land should be used, and **(5)** undercutting of land-use planning in one area by lack of planning or by planning with opposite goals in surrounding areas. For example, the Greater New York area has 1,476 governmental jurisdictions, each making decisions about land use that affect other jurisdictions.

Unless it can be effectively implemented, a land-use plan merely gathers dust. Major methods for controlling land use include: **(1)** direct purchase of land by a public agency or by private interests, to ensure that it is used for the prescribed purposes, **(2)** zoning land so it can be used only in certain ways, **(3)** giving tax breaks to landowners who agree to use land only for given purposes such as agriculture or open space, **(4)** purchase by public agencies of land development rights that restrict the way the land can be used, **(5)** assigning a limited number of transferrable development rights to a given area of land, **(6)** controlling population growth and land development by limiting building permits, sewer hookups, roads, and other services, and **(7)** using federally required environmental impact statements to stop or delay harmful projects by forcing consideration of adverse impacts and alternatives to all federal land-use projects.

Preserving Urban Open Space Most planners are concerned with preserving open space within and around urban areas. *Urban open space* is any large, medium-sized, or small area of land or water in or near an urban area that can be used for recreational, aesthetic, or ecological functions. One of the most ambitious efforts at preserving *large* open spaces to reduce urban sprawl is the 10- to 16-kilometer (6- to 10-mile) wide greenbelt around London. This example of long-range planning, begun in 1931, has preserved some land-use choices that most cities squandered long ago. But it has failed to halt urban growth; the suburbs have

Steps in Ecological Land-Use Planning

1. *Making an Environmental and Social Inventory* A comprehensive geological, ecological, and social survey of the land is made. This includes an analysis of **(a)** geological variables, such as slopes, soil types and limitations, and aquifer and other hydrologic data, **(b)** ecological variables, such as forest types and quality, ecological value, wildlife habitats, stream quality, estuaries, and historic or unique sites, and **(c)** social and economic variables, such as recreation areas, urban development, social pathology (rates of homicide, suicide, robbery, drug addiction), physical pathology (tuberculosis, diabetes, emphysema, heart disease), pollution, ethnic distribution, illiteracy, overcrowding, housing quality, and industrial plant quality.

2. *Determination of Goals and Their Relative Importance* Experts, public officials, and the general public decide on goals and weight each goal. This is one of the most important and difficult planning steps, in which ethical land-use conflicts (Section 12-1) are resolved.

3. *Production of Individual and Composite Maps* Data for each variable obtained from step 1 are plotted on separate transparency maps. The transparencies are superimposed on one another or combined by computer to give three composite maps—one each for geological variables, ecological variables, and socio-economic variables. Each map shows how the variables interact. Computers and systems analysis models can be used to plot these composite density maps and to weight each value numerically according to decisions made in step 2. Computer modeling can also be used to update maps and to make alternative composites based on different goals and weighting factors.

4. *Development of a Comprehensive Plan* The three composite maps are combined to form a master composite, which shows the suitability of various areas for different types of use. In some cases a computer-generated series of master composites shows the effects of weighting key variables in different ways. Using this technique, it often becomes clear how certain areas of land should or should not be used.

5. *Evaluation of the Comprehensive Plan* The comprehensive plan (or series of alternative comprehensive plans) is evaluated by experts, public officials, and the general public, and a final comprehensive plan is drawn up and approved.

6. *Implementation of the Comprehensive Plan* The plan is set in motion and monitored by the appropriate governmental, legal, environmental, and social agencies.

Figure 14-6 A suburban housing development tract laid out in the usual way.

jumped the belt. There is also continuing pressure to develop some of the space. Another way to preserve large open spaces near an urban area is to build a series of *new towns* beyond the central urban area, as discussed in Section 14-5.

Some cities have had the foresight to preserve open space in moderate-sized to large *municipal parks*. Central Park in New York City and Golden Gate Park in San Francisco are two famous examples. Crisscrossed by six-lane roads and crime infested at night, Central Park is still heavily used. Cities with large municipal parks must continually resist efforts to build freeways through them. Unfortunately, cities that did not plan for such parks early in their development have little or no chance of getting them now.

Since World War II, the typical pattern of a suburban housing development in the United States has been to bulldoze a patch of woods or farmland and build rows of houses, with each house having a standardized lot (Figure 14-6). In recent years a new pattern, known as *cluster development* or *planned unit development* (PUD), has been used with increasing success to preserve moderate-sized

blocks of open space. Houses, townhouses, condominiums, and garden apartments are built on a relatively small portion of land, with the rest of the area left as open space, either in its natural state or for recreation areas (Figure 14-7).

The most overlooked and probably most important open spaces are the small strips and odd-shaped patches of unused land that dot urban areas. The Illinois Prairie Path is a walkway and bridle path running from downtown Chicago to its western suburbs along an abandoned trolley line. Other cities have converted abandoned railroad beds and dry creek beds into bicycle, hiking, and jogging paths. San Antonio, Texas, has revitalized much of its downtown area by developing shops, restaurants, and other businesses along the San Antonio River, which runs through a 21-block area of the city.

Within cities, abandoned lots can be developed as small plazas and vest-pocket parks such as Paley Park in midtown Manhattan—a refreshing refuge for about 3,000 people each day. Abandoned lots cluttered with car hulks, tires, and piles of dirt and rocks can be cleaned up slightly and

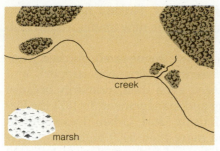
Undeveloped Land

Typical Housing Development

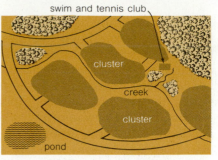
Cluster Housing Development

Figure 14-7 Tract and cluster development forms on the same land area.

used as "adventure playgrounds," which are usually jam-packed with children. These places look dangerous to parents, but studies have shown that fewer accidents occur there than in the more expensive traditional playgrounds, neatly laid out with swings, slides, and steel climbing bars.

14-5 Coping with Urban Problems

Repairing Existing Cities America's older cities have massive maintenance and repair problems—most of them aggravated by neglect. Most of New Orleans's sewers, some of which were purchased secondhand from Philadelphia in 1896, need replacement. When it rains in Chicago, sewage backs up into basements of about one-fourth of the homes. An estimated 46 percent of Boston's water supply and 25 percent of that in Pittsburgh is lost through leaky pipes. One out of every five bridges in the country and two out of every five in urban areas are structurally deficient and considered safe for cars and light trucks only.

The maintenance and repair of existing U.S. bridges, roads, mass transit systems, water supply systems, sewers, and sewage treatment plants could cost a staggering $1.2 trillion or more between 1984 and 2000, according to a 1984 study for the Joint Economic Committee of the House and Senate. These massive repair and maintenance bills are coming due at a time when the federal government, to help hold down record budget deficits, is cutting back funds available for building and maintaining public works.

Revitalizing Existing Cities Billions of dollars have been spent in decaying downtown areas to build new civic centers, museums, office buildings, parking garages, and high-rise luxury hotel and shopping center complexes. Such projects have done much to revive the economy of cities

such as Baltimore and can indirectly benefit the poor by providing additional tax revenues. But they are of little direct benefit to the poor because they mainly provide white-collar jobs for suburbanites and shopping and cultural facilities mostly used by tourists and suburbanites.

In most major cities, older neighborhoods are being revitalized by middle- and high-income residents who buy rundown houses at low prices and renovate them. This trend benefits the city economically by increasing property values and slowing the flight to the suburbs, but it can also displace low-income residents.

A number of cities have set up *urban homesteading programs* to help middle- and low-income individuals improve and live in abandoned housing. Houses and apartment buildings abandoned or acquired by a city because of failure to pay taxes, are resold to individuals or to cooperatives of low-income renters, typically for $1 to $100. The new owners must agree to renovate the buildings and live in them for at least 3 years. In some cases, the city also provides low-interest, long-term loans, which in effect become mortgages on the property. This approach has been particularly successful in Wilmington, Delaware, and in Baltimore.

Many cities have torn down blocks of slum dwellings and replaced them with high-rise public housing projects. Although some of these projects have been successful, many have disrupted families, destroyed protective neighborhood social structures, and eliminated essential neighborhood stores. In addition, the new rents are often too high for the poor. By 1984 nearly a quarter of the country's federally financed public housing authorities were losing money, primarily because of local mismanagement, tenant abuse of property, and such shoddy or nonexistent maintenance that units could no longer be rented.

The outlook for abandoned high-rise, low-income public housing developments, however, is not totally bleak. Some have been converted to schools and commercial buildings. In Jersey City, New Jersey, a rundown high-rise project with a high crime rate was turned around by a tenant committee. The tenants themselves replaced broken windows with unbreakable plastic glass, installed indoor and outdoor unbreakable lights, repaired broken elevators and rubbish chutes, repainted the building, and planted grass and flowers. They also found building managerial and service jobs for some of the tenants.

Some analysts call for increased federal aid to help revitalize the cities. Others argue that such federal aid is often wasted and is controlled by distant bureaucrats who have little knowledge of the needs of local people. Instead, they propose that tax credits be given to private companies who create new jobs and hire the unemployed or disadvantaged in *free enterprise zones* in economically depressed urban neighborhoods. Proponents of this approach point to the example set by Control Data Corporation (CDC), a Minneapolis-based computer firm. Since 1968 CDC has successfully built factories, created jobs, set up career counseling and job training programs, provided day care centers for working mothers, and stimulated renewal in blighted urban areas in Minneapolis, St. Paul, San Antonio, Baltimore, Toledo, and Washington, DC—all without federal aid.

Building New Cities and Towns In 1898 English planner Ebenezer Howard urged that new towns be built to lure Londoners from the city. Since then Great Britain has built 16 new towns and is building 15 more. New towns have also been built in Singapore, Hong Kong, Finland, Sweden, France, the Netherlands, Venezuela, Brazil, and the United States. There are three types: (1) *satellite towns*, located relatively close to an existing large city, (2) *freestanding new towns*, located far from any major city, and (3) *"in-town" new towns*, located in existing urban areas. Typically, such towns are conceived for populations of 20,000 to 100,000 people.

The most widely acclaimed new town is Tapiola, Finland, not far from Helsinki. Designed in 1952, it is being built gradually in seven sections with an ultimate projected population of 80,000. Today many of its more than 30,000 residents work in Helsinki, but the long-range goal is industrial and commercial independence for the town. Tapiola is divided into several villages separated by greenbelts. Each village has its own architect, selected by competition, and consists of several neighborhoods clustered around a shopping and cultural center. Each neighborhood has a social center and contains a mix of high-rise apartments and single-family houses. Finland has drawn up plans for building six more new towns around Helsinki.

Unfortunately new towns rarely succeed without massive government financial support, and some don't succeed even then, primarily because of poor planning and management. In 1971 the Department of Housing and Urban Development (HUD) provided more than $300 million in federally guaranteed loans for developers to build 13 new towns. By 1980 HUD had to take title to nine of these projects, which went bankrupt, and won't fund more new towns. Those built by private interests require the developers to put up large amounts of money to buy the land and install facilities and to pay heavy taxes and interest charges for decades before any profit is made. In the United States two privately developed new towns—Columbia, Maryland, and Reston, Virginia—have been in constant financial difficulty since they were started almost two decades ago, although their situations are gradually improving.

Some planners have proposed guidelines and models for building compact towns and cities that waste less matter and energy resources and are more self-reliant than conventional municipalities. Such self-sufficient cities would be surrounded by farms, greenbelts, and community gardens. Homes and marketplaces would be close together, and most local transportation would be by bus, bicycle, and foot. Buildings would be cooled and heated by sun and wind and wastes would be recycled. Food would be grown locally, and huge factories would be scarce.

Some existing cities have also begun efforts to become more self-sufficient for some of their matter and energy resources. Erie, Pennsylvania, for example, drilled two producing oil wells between 1978 and 1980, and Palo Alto, California, has rezoned almost 5,000 acres within the city as open space to be used primarily for agricultural purposes. Fort Collins, Colorado, runs much of its city transportation system on methane gas generated from its sewage plant and uses nutrient-rich sludge from the sewage plant to fertilize 600 acres of city-owned land to grow corn, which might be converted to alcohol to fuel more city cars.

The city is not an ecological monstrosity. It is rather the place where both the problems and the opportunities of modern technological civilization are most potent and visible.

Peter Self

Discussion Topics

1. Give advantages and disadvantages of emphasizing rural rather than urban development in LDCs. Why do most LDCs emphasize urban development, even though most of their population still lives in rural areas?

2. What life-support resources in your community are the most vulnerable to interruption or destruction? What alternate or backup resources, if any, exist?

3. Explain how each of the following common practices could hasten the decay of a city: **(a)** raising taxes, **(b)** building freeways and mass transit systems to the suburbs, and **(c)** replacing slums with low-cost housing projects. Suggest alternative policies in each case.

4. Massive traffic jams hinder you in getting to and from work each day. Government officials say that the only way to relieve congestion is to build a highway through the middle of a beautiful urban park. As a taxpayer, would you support this construction? Why or why not? What would you suggest to relieve the situation?

5. Debate the pros and cons of **(a)** charging commuters who drive to work alone very high commuting taxes and parking fees, and **(b)** setting aside express lanes on freeways for buses and carpool vehicles.

6. Debate the idea that private landowners have the right to do anything they want with their land.

7. Evaluate land use and land-use planning by your college or university.

8. Make a class survey and draw a map identifying good and poor uses of small, medium-sized, and large open spaces in your area.

15

Nonrenewable Mineral Resources

We seem to believe we can get everything we need from the supermarket and corner drugstore. We don't understand that everything has a source in the land or sea, and that we must respect these sources.

Thor Heyerdahl

The high standards of living enjoyed by most people in industrialized nations are based on the extensive use of copper, lead, mercury, zinc, sand, and stone, and other nonrenewable metallic and nonmetallic minerals, and nonrenewable mineral fuels such as coal, oil, natural gas, and uranium. This chapter discusses nonfuel mineral resources; the next four consider energy resources.

15-1 Abundance, Mining, and Processing of Mineral Resources

Mineral Resource Abundance Figure 15-1 shows the relative abundances of the elements in the earth's crust and in seawater. Notice that only 10 elements make up 99.3 percent of the earth's crust. Most of the remaining elements are found only in trace amounts.

Any natural occurrence of an element, normally in a compound, is called a **mineral deposit**. Although the average abundance of copper in the earth's crust is only 0.007 percent, compounds of copper and many other trace elements are concentrated in certain mineral deposits found at various places in the earth's crust. Any mineral deposit with a high enough concentration of an element to permit it to be mined and sold at a profit is called an **ore**. *High-grade ores* contain relatively large concentrations and *low-grade ores* relatively low concentrations of the desired element. Once ore deposits have been located, the minerals are removed either by subsurface or surface mining.

Subsurface Mining When a metal ore, or a fuel deposit such as coal, lies so deep in the ground that it is too expensive to remove the covering *overburden* of soil and rock, it must be extracted by **sub-**

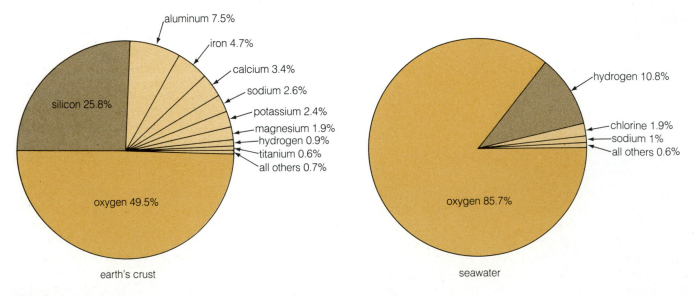

Figure 15-1 Percent by weight of elements in the earth's crust and in seawater.

National Coal Association/Bucyrus Erie Company

Figure 15-2 A giant shovel used for strip mining coal. The cars behind the shovel look like toys.

surface mining. This is usually done by digging a vertical shaft deep into the ground, blasting tunnels and rooms to get to the deposit, and hauling the ore to the surface. Underground mines occasionally collapse, trapping miners. Most cases of mine collapse, however, occur in the more than 90,000 abandoned coal mines in the United States. Such collapses can be prevented by injecting a mud slurry into abandoned tunnels and rooms and letting the water drain away, but this is not often done.

Underground deposits of certain soluble minerals, such as salt, can be removed by *subsurface solution mining*. In this technique, hot water is pumped down an injection well to the deposit, where it dissolves the soluble minerals. The resulting solution is pumped back to the surface, and the salt or other mineral is recovered by evaporation of water. However, this type of mining can contaminate groundwater supplies (Section 22-5).

Surface Mining About 90 percent of the rock and mineral resources used in the United States and more than 63 percent of the nation's coal output (compared to only 9 percent in 1940) are extracted by **surface mining**. In surface mining the overburden is taken away so that underlying mineral deposits can be removed with large power shovels. You can get some idea of such equipment by imagining a shovel 32 stories high, with a boom as long

as a football field, capable of gouging out 152 cubic meters (200 cubic yards) of land every 55 seconds and dropping this 295,000-kilogram (325-ton) load the equivalent of a city block away. This is a description of Big Muskie, a $25-million power shovel used for surface mining coal in the United States (Figure 15-2).

An estimated 52,000 square kilometers (13 million acres) of land in the United States has been disrupted by surface mining of all types. The mining of metals accounted for about 14 percent of this disturbed land, with the remaining 86 percent divided evenly between coal and nonmetals (mostly sand and gravel). So far only about 64 percent of the land area disturbed by coal mining, 26 percent by nonmetals, and 8 percent by metals has been reclaimed. It is projected that by the year 2000 an additional 60,000 square kilometers (15 million acres) will have been disrupted by surface mining.

There are several types of surface mining: **(1)** open pit mining, **(2)** dredging, **(3)** area strip mining, and **(4)** contour strip mining. *Open pit mining* is used primarily for the extraction of stone, sand, gravel, iron, and copper. Sand and gravel are removed from small pits in many parts of the country. Rocks such as limestone, granite, and marble are taken from larger pits called *quarries*. In the Mesabi Range near Lake Superior and in some western states, ores of copper (see the photo on p. 91) and iron are removed from huge open pit mines. Deposits of sand and gravel found in

Figure 15-3 Area strip mining of coal. Bulldozers and power shovels clear away trees, brush, and overburden. Explosive charges loosen the coal deposits, and power shovels or auger drills load the coal onto trucks in the pit area. Strip mining exposes cross sections of the earth's crust (the highwalls), and the overburden from the trench being mined is placed in an adjacent trench that has already been mined.

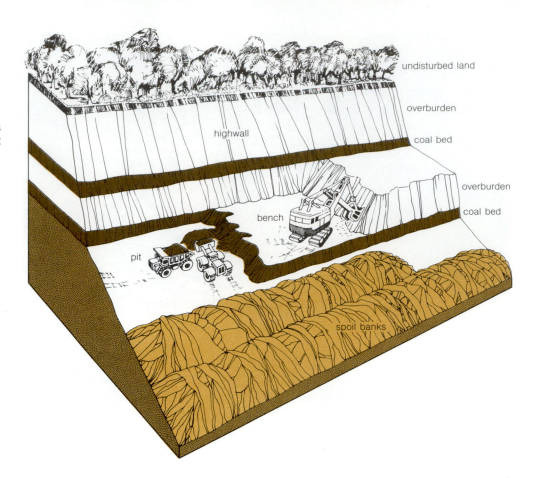

stream- and ocean beds are removed by *dredging*, performed with draglines and chain buckets.

There are two basic types of strip mining: area and contour. In *area strip mining*, which is carried out on flat or rolling terrain, the overburden is stripped away to form a series of parallel trenches (Figure 15-3). Overburden taken from one trench is placed in an adjacent one, leaving a wavy terrain. This technique is used primarily for mining gypsum, for coal in many western and midwestern states (Section 17-3), and phosphate rock—especially in Florida, North Carolina, and Idaho. In *contour strip mining*, a series of shelves, or terraces, is cut into the side of a mountain. The overburden from each new terrace is dumped onto the one below. Contour strip mining is used primarily for extracting coal in the mountainous Appalachian region.

Reclamation of surface-mined land includes revegetation of the disturbed land and control of erosion. With some minerals, the runoff of acids and other chemicals leached out of mine wastes or spoils is also controlled. Temporary sediment basins, terraces, diversion dams, and other techniques can be used to help minimize erosion and acid runoff while mining takes place. After mining activities have ceased, steep slopes are graded and the topsoil is replaced.

The final step in reclamation is the establishment of cultivated crops, perennial vegetation, ponds, or lakes over the entire area. Sometimes annual vegetation is used to reduce soil and wind erosion until perennial vegetation can be established. The productivity of disturbed soil can be improved by adding a layer of nutrient-rich topsoil, or lime, if the soil is acidic, and commercial or natural fertilizers. Reclamation of surface-mined land in the arid and semiarid regions, which contain 90 percent of the nation's low-sulfur coal, is difficult. Problems include fewer plant species adapted to such conditions, lack of water to establish new vegetation, and soils that are often too acidic and contain minerals such as selenium, which are toxic to livestock.

Metallurgical Processes After a metal ore has been mined, physical processes such as crushing and washing are used to remove impurities and obtain the ore in concentrated form. The ore compound must then undergo a chemical reaction to obtain the free metal. For example, aluminum is found in bauxite ore in the form of aluminum oxide (Al_2O_3). Electrical current is passed through molten aluminum oxide to convert it to aluminum metal and oxygen gas ($2Al_2O_3 \rightarrow 4Al + 3O_2$). Iron

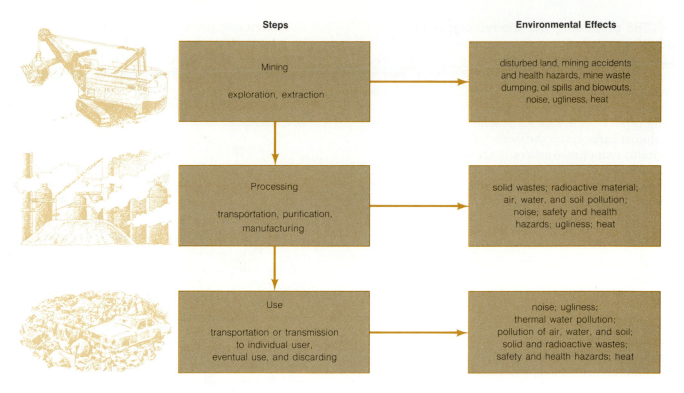

Steps		Environmental Effects
Mining exploration, extraction	→	disturbed land, mining accidents and health hazards, mine waste dumping, oil spills and blowouts, noise, ugliness, heat
Processing transportation, purification, manufacturing	→	solid wastes; radioactive material; air, water, and soil pollution; noise; safety and health hazards; ugliness; heat
Use transportation or transmission to individual user, eventual use, and discarding	→	noise; ugliness; thermal water pollution; pollution of air, water, and soil; solid and radioactive wastes; safety and health hazards; heat

Figure 15-4 Some environmental effects of resource use.

ore in the form of iron oxide (Fe_2O_3) is converted to iron by reaction with carbon monoxide gas ($Fe_2O_3 + 3CO \rightarrow 2Fe + 3CO_2$). Copper ore in the form of copper sulfide (Cu_2S) can be converted to copper metal by heating the ore to high temperatures in air ($Cu_2S + O_2 \rightarrow 2Cu + SO_2$). The sulfur dioxide gas (SO_2) also produced by this process is a dangerous and highly corrosive air pollutant (Chapter 21).

Next the free metal is refined and purified. Then in many cases it is melted and mixed in molten solution with one or more other metals or other elements to form *alloys* with desired properties. Dental amalgam used to fill teeth, for example, is an alloy containing silver mixed with smaller amounts of lead, copper, and mercury; stainless steel is an alloy of iron mixed with smaller amounts of chromium, nickel, and carbon. Often the metal or alloy is finished by heating, polishing, or other methods, to obtain a surface with desired characteristics, such as hardness or corrosion resistance.

Environmental Impact of Resource Use The mining, processing, and use of any energy or nonfuel mineral resource causes some form of land disturbance along with air and water pollution (Figure 15-4). Most land disturbed by mining can be reclaimed to some degree, and some forms of air and water pollution can be controlled (Chapters 21 and 22). But these efforts are expensive and also require

energy, which in being produced and used also pollutes the environment.

For each unit of mineral produced, subsurface mining disturbs less than one-tenth as much land as surface mining and generally produces less waste material than surface mining. But rainwater seeping through surface wastes or spoils and through abandoned mines—especially coal mines rich in sulfur compounds—causes chemical reactions that produce sulfuric acid. This acid can run off into nearby rivers and streams, contaminating water supplies and killing aquatic life. Subsurface mining is also usually more dangerous and more expensive than surface mining.

Environmental problems associated with surface mining include: **(1)** disposal of mine spoils, **(2)** water pollution of nearby rivers and streams from runoff of sediment, acids, and toxic metals from mine spoils, **(3)** pollution of groundwater from leaching of toxic materials from mine spoils, **(4)** air pollution from dust, and **(5)** land disruption. Almost 18,000 kilometers (11,000 miles) of U.S. waterways has been polluted by runoff of sulfuric acid from coal wastes left by both subsurface and surface mining. Surface mining, largely for coal, now consumes about 200,000 acres of farmland a year, especially in the Midwest. Federal and state laws now regulate the strip mining of coal and require that the disturbed land be reclaimed, as discussed in Section 17-3. But there are few regulations for the strip mining of other minerals.

The processing and the refining of ores produce large quantities of waste materials that pollute the air and water. People who work in some metal-processing industries or live near plants with inadequate safeguards have an increased risk of getting some forms of cancer. For example, the lung cancer death rate for arsenic smelter workers is almost three times the expected rate and that for cadmium smelter workers is more than twice the expected rate. Lead smelter workers have higher incidences of lung and stomach cancer (Section 20-3).

15-2 Are We Running Out of Mineral Resources?

Estimating Resource Supplies Estimating how much of a particular mineral resource exists on earth and how long its supply might last is a complex and controversial process. The term **resources** (or *total resources*) refers to the total amount of a particular material that exists on earth. It is difficult, however, to make reliable estimates of the total available amount of a particular resource because the entire world has not been explored for each resource. Even if most deposits of a particular resource were identified, much of the total supply might never be mined because it occurs in such low concentrations that it would cost more to get than it would be worth under almost any conditions.

As a result, most estimates of the available supply of a resource are actually of resource *reserves*—not resources or total supply. The term **reserves** (or *economic resources*) refers to the estimated amount of a particular material in known locations that can be extracted profitably at present prices, using current mining technology. The U.S. Geological Survey also makes estimates of *subeconomic resources*, identified resources that cannot be recovered profitably with present prices and technology but may be converted to reserves when prices rise or mining technology improves. They also make crude estimates of *undiscovered resources* believed to exist, although specific locations, quality data, and amounts are unknown.

Published estimates of the supply of a particular resource usually refer to reserves. The actual supply available in the future is normally higher because reserves increase by discoveries of new supplies, and by improved technology and price increases due to shortages, which permit profitable mining of low-grade deposits initially classified as subeconomic resources. The life of most

nonrenewable mineral resource supplies also can be increased by recycling and reuse, by designing products to last longer, and by using substitutes—all stimulated by rising prices due to shortages. From this discussion you can see why there are so many conflicting estimates of the potential supply of a resource.

Future Mineral Resource Supplies: Optimists Versus Pessimists *In attaining their standard of living, Americans have used more nonfuel minerals and fossil fuels during the past 40 years than all the peoples of the world have used throughout human history.* It is projected that between 1980 and 2000, the world's people will use three to four times as much matter resources as in all of human history.

There is much controversy over whether there will be sufficient supplies of metals and other nonfuel mineral resources to meet these projected needs. One group, called *cornucopians*, holds that we will never run out of needed metallic and nonmetallic minerals. Their position is based on the *economic* idea that reserves of a resource increase indefinitely: Scarcity causes prices to rise, which enables lower grade deposits to be mined and stimulates the search for new deposits and substitutes. This group also believes that new technology can always be developed to mine lower grade deposits and that substitutes can be found for essentially any mineral resource.

The opposing group, called *neo-Malthusians*, believes that affordable supplies of metals and minerals are finite, that there will be shortages of some key materials within the next few decades, and that the environmental effects from using resources at high rates will probably limit resource use in the future even if supplies are adequate. Members of this group emphasize not only discovery of new resources but also recycling, reuse, conservation, reducing average per capita consumption, and slowing population growth. Let's look at the major parts of this controversy in more detail.

Economics and Resource Supply According to standard economic theory, a competitive free market controls supply and demand of all goods and services. If a resource becomes scarce, prices rise; if there is a glut, they fall. Cornucopians contend that rising prices will stimulate new discoveries and the development of more efficient mining technology, making it profitable to mine lower and lower grade ores of key minerals.

However, a number of economists have argued

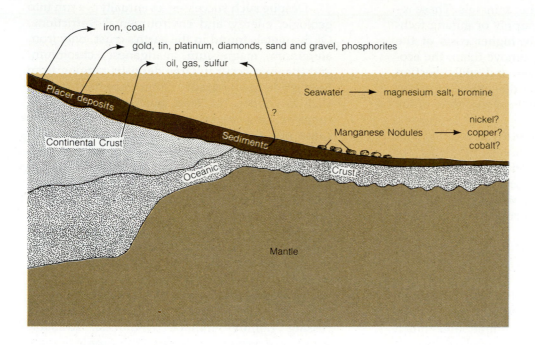

that this idea often does not apply to the supply and demand for nonfuel mineral resources because **(1)** in the United States (and in many other industrial nations) both industry and government have gained enough control over supply, demand, and prices of raw materials and products that a competitive free market doesn't exist, and **(2)** the costs of nonfuel mineral resources account for such a small percentage of the total costs of most goods that resource scarcities are often not reflected in the total costs of such goods.

New Discoveries   There is little doubt that geological exploration—guided by better knowledge about the earth—will extend present known reserves of most minerals. Remote-sensing cameras in orbiting satellites (such as Landsat) now scan the globe for land, forest, mineral, energy, and water resources. According to geologists, some rich new deposits will probably be found in unexplored areas in LDCs; but in the MDCs and in many LDCs most of the easily accessible high-grade deposits have already been discovered. Remaining deposits are harder and more expensive to find and mine and usually are less concentrated.

Exploration for new resources requires a large capital investment and is a risky financial venture. Typically, if geologic theory identifies 10,000 sites where a deposit of a particular resource might be found, only 1,000 sites are worth costly exploration; only 100 warrant even more costly drilling, trenching, or tunneling; and only one is likely to be a producing mine. Even if large new supplies are found, no mineral supply can stand up to continued exponential growth in its use. For example, a 1-billion-year supply of a resource would be exhausted in only 584 years if the level at which it was used increased at 3 percent a year.

Some have pointed out that the oceans contain vast, untapped supplies of key mineral and energy resources. As shown in Figure 15-5, potential resources of the ocean are found in three areas: seawater, sediments and deposits on the shallow continental shelf and slope, and sediments and nodules on the deep ocean floor.

The huge quantity of seawater appears to be an inexhaustible source of minerals, but most of the 90 chemical elements found there occur in such low concentrations that it takes more energy and money to recover them than they are presently worth. For example, to get a mere 0.003 percent of the annual U.S. consumption of zinc from the ocean would require processing a volume of seawater equivalent to the combined annual flows of the Delaware and Hudson rivers. Only magnesium, bromine, and common table salt (sodium chloride) are abundant enough to be extracted profitably from seawater at present prices with current technology.

Offshore deposits and sediments in shallow waters are already important sources of oil, natural gas, sand, gravel, and 10 other minerals. These resources are limited less by supply or mining technology than by increasingly higher costs of the energy needed to find and remove them, the ecological side effects of oil leaks and spills (Section 22-6), and the potentially serious effects of extensive dredging and mining on food resources of the sea.

There is considerable interest in locating and vacuuming up mineral-containing rocks that are believed to be unevenly distributed on the deep ocean floor. These potato-sized nodules contain mostly manganese and iron, with small amounts of copper, nickel, cobalt, molybdenum, and vanadium.

Because of economic and political uncertainties, however, it is unclear whether these resources will be mined in the near future. There is no guarantee that metal prices will be high enough to permit a reasonable profit on such a large and risky investment, especially since ample and much cheaper supplies of the metals involved are expected to be available for many decades. An even greater threat is posed by international legal and political squabbles over ownership of these resources. Since 1968, 160 nations have been working to develop the UN-sponsored Law of the Sea Treaty to govern ocean pollution and exploitation of mineral and living ocean resources in international waters. In 1982, 130 UN member nations approved a final draft of this treaty. But the United States, Turkey, Venezuela, and Israel voted against approval, and 17 other nations abstained and have so far refused to support the treaty. The only major industrial nations voting for the treaty were France and Japan. Because of these legal and economic uncertainties, research and development by the private ocean-mining companies had come to a virtual halt by 1984.

Environmentalists recognize that seabed mining would probably cause less harm than mining on land. They are concerned, however, that vacuuming nodules off the seabed and stirring up deep ocean sediments could destroy seafloor organisms, have unknown effects on poorly understood deep-sea food webs, and pollute surface waters by discharge of sediments from mining ships and rigs.

Improved Mining Technology and Mining Low-Grade Deposits

There is no question that advances in mining technology during the past few decades have allowed the mining of low-grade deposits without significant cost increases. For example, the grade for minable copper ore has been reduced by a factor of 10 since 1900.

Despite such successes, eventually we run into geologic, energy, and environmental restrictions. Of the metals found in the earth's crust, only iron, aluminum, magnesium, manganese, chromium, and titanium occur in large deposits ranging from high grade to low grade. Other important metals, such as copper, tin, lead, zinc, uranium, nickel, tungsten, and mercury, are relatively scarce. Because of the second law of energy (Section 3-3), remaining deposits eventually come to have such low metal concentrations that the costs of digging, transporting, crushing, processing, and hauling away the waste rock have become high enough to discourage mining.

A fundamental assumption of the cornucopian view is an inexhaustible source of cheap energy; most energy experts, however, do not believe such optimism is justified. If energy prices rise in the future as projected, the cost of energy could become a limiting factor in finding, mining, and processing nonfuel mineral resources. Available supplies of water may also limit the supply of some mineral resources because large amounts are needed to extract and process most minerals. Many areas with major mineral deposits are poorly supplied with water.

Substitution

Cornucopians insist that if supplies of minerals run out, human ingenuity and technology will find substitutes. They argue that either plastics, high-strength glass fibers made mostly from silicon (the second most abundant element in the earth's crust), or four of the most abundant metals (aluminum, iron, magnesium, and titanium) can be substituted for most scarce metals. For example, in automobiles, plastics are increasingly substituted for copper, lead, tin, and zinc. Aluminum and titanium are also replacing steel in cars and for some other purposes. Glass fibers are beginning to replace copper wires in telephone cables. The new fibers weigh less, can carry more information and electrical signals, and cost half as much as copper wire. Even if substitutes are not found, cornucopians argue that no material is so vital that its exhaustion would be catastrophic.

Finding substitutes for scarce resources is extremely important, but there are problems: (1) failure to find a substitute could cause serious economic hardships during the adjustment period that would occur when a key material ceased to be available; (2) finding possible substitutes and phasing them into complex manufacturing processes requires costly development programs and long lead times; (3) some materials have unique properties such that adequate substitutes cannot be found (for example, helium has no known substitute), and substitutes for chromium in stainless steel, platinum as an industrial catalyst, gold for electrical contacts, cobalt in magnets and steel alloys, silver for photography, and manganese for making bubble-free steel probably will be inferior; **(4)** some proposed substitutes (for example, cadmium and silver for mercury in batteries) are also scarce; and **(5)** some future technologies may themselves depend on scarce resources that have no known substitutes (for example, nuclear fusion reactors, if ever developed, would put heavy demands on scarce beryllium, niobium, lead, helium, and chromium, and the development of efficient solar photovoltaic cells may require large amounts of scarce gallium).

Recycling Poor people in LDCs have long recognized the need to recycle or reuse almost everything—discarded glass, paper, plastic, rags, tin, and bones. In the MDCs, everyone agrees on the importance of recycling nonrenewable mineral resources. This practice decreases the need for virgin resources, usually saves energy, causes less pollution and land disruption, and cuts waste disposal costs by reducing the volume of solid wastes, thus relieving the pressure on overflowing landfills (Section 24-2).

For example, using scrap iron instead of virgin iron ore to produce steel conserves virgin iron ore and coal. It also requires 65 percent less energy, 40 percent less water, and 97 percent fewer raw materials, and it cuts air pollution by 90 percent and water pollution by 76 percent. Recycling aluminum reduces air pollution associated with its production by 95 percent and requires 92 percent less energy than mining and processing virgin aluminum ore. If returnable bottles replaced the 80 billion throwaway beverage cans produced annually, enough energy would be saved to provide electricity for 13 million people. Recycling half the paper used each year in the United States would save about 150 million trees and conserve enough energy to provide residential electricity for about 10 million people. Despite these advantages, only about one-fourth of the world's steel, aluminum, and paper is recovered for recycling.

In recent years, Americans have heard many slogans ("Waste is a resource out of place," "Urban waste is urban ore," "Trash is cash," "Landfills are urban mines," and "Trash cans are really resource containers") and have been urged to shift from a high-waste society (Figure 3-4) to a low-waste society based on recycling and resource conservation (Figure 3-5). Although surveys indicate that 75 percent of all Americans favor more recycling, only about 10 percent of the waste in the United States is now recycled—compared to 40 to 60 percent in Japan and many European countries.

Recycling in the United States has been hindered because of **(1)** the development of manufacturing processes that use only virgin resources (primarily because of the abundance of cheap raw materials in the past), **(2)** provision of subsidies in the form of tax breaks and depletion allowances to primary mining and energy resource industries to encourage them to find and get resources out of the ground as fast as possible, **(3)** higher railroad and trucking shipping rates for most scrap materials (especially glass and paper) than for virgin materials, **(4)** lack of large and steady markets for recycled materials from mixed wastes, **(5)** the high labor costs involved in recovering materials from mixed wastes, and **(6)** failure to include the cost of disposal in the price of a product. Environmentalist Denis Hayes projects that with proper economic and political incentives about two-thirds of the material resources used in the United States each year could be recycled without significant changes in life-styles.

Reuse and Resource Conservation *Recycling* is the collecting, reprocessing, and refabricating of a resource, whereas *reuse* involves the same product, employed over and over again in its original form. For example, it takes three times more energy to crush and remelt a glass bottle than it does to refill it. Thus, it makes more sense ecologically to replace all nonreturnable glass bottles with returnable bottles that can be used many times before being lost or broken. In practice, however, almost 80 percent of the nation's beer and soft drink containers are nonreturnable, throwaway bottles and cans. The recycling and reuse of aluminum, glass, and paper are discussed further in Section 24-3.

In addition to increased recycling and reuse, environmentalists also call for increased *resource conservation*, especially in MDCs, to reduce the unnecessary waste of matter and energy resources. A glaring example is overpackaging: Product packaging in the United States consumes 65 percent of all paper, 15 percent of all wood, and 3 percent of all energy used. Some grocery store items have containers that cost five times as much as the food that's in them.

Many environmentalists argue that the present one-way flow of matter resources from raw materials to wasted solids found in most industrial societies should be replaced with a sustainable earth or low-waste system that reduces matter and energy consumption. Table 15-1 compares the

Table 15-1 Three Systems for Handling Discarded Materials

Item	For a Throwaway System	For a Resource Recovery and Recycling System	For a Sustainable Earth Resource System
Glass bottles	Dump or bury	Grind and remelt; remanufacture; convert to building materials	Ban all nonreturnable bottles and reuse (not remelt and recycle) bottles
Bimetallic "tin" cans	Dump or bury	Sort, remelt	Limit or ban production; use returnable bottles
Aluminum cans	Dump or bury	Sort, remelt	Limit or ban production; use returnable bottles
Cars	Dump	Sort, remelt	Sort, remelt; tax cars lasting less than 15 years, weighing more than 818 kilograms (1,800 pounds), and getting less than 13 kilometers per liter (30 miles per gallon)
Metal objects	Dump or bury	Sort, remelt	Sort, remelt; tax items lasting less than 10 years
Tires	Dump, burn, or bury	Grind and revulcanize or use in road construction; incinerate to generate heat and electricity	Recap usable tires; tax all tires not usable for at least 64,400 kilometers (40,000 miles)
Paper	Dump, burn, or bury	Incinerate to generate heat	Compost or recycle; tax all throwaway items; eliminate overpackaging
Plastics	Dump, burn, or bury	Incinerate to generate heat or electricity	Limit production; use returnable glass bottles instead of plastic containers; tax throwaway items and packaging
Garden wastes	Dump, burn, or bury	Incinerate to generate heat or electricity	Compost; return to soil as fertilizer; use as animal feed

present throwaway resource system used in the United States, a resource recovery and recycling system, and a sustainable earth or low-waste resource system.

15-3 Key Resources: The World Situation

Depletion Curves and Depletion Rate Estimates

Projection of the *depletion time* of a resource is based on two major sets of assumptions: **(1)** the actual or potential available supply at existing (or future) acceptable prices and with existing (or improved) technology, and **(2)** the annual rate at which the resource is used. Obviously, differing assumptions yield different answers. It is almost certain that no resource will be completely exhausted. Instead, use is normally limited by the prohibitively high cost of mining less accessible and lower grade deposits. For this reason, **depletion time** is defined as the period required to use a certain fraction—usually 80 percent—of the known reserves or estimated total supply of a resource, according to various assumed rates of use.

One estimate, the **static reserve index**, projects the number of years until the known world reserves of a resource will be 80 percent depleted at the present annual rate of consumption. Because of increases in population and resource use, it is

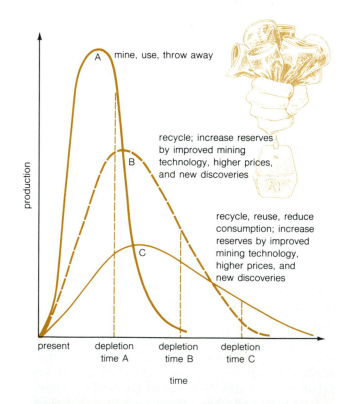

Figure 15-6 Depletion curves for a nonrenewable resource, based on different sets of assumptions. Dashed vertical lines show when 80 percent depletion occurs.

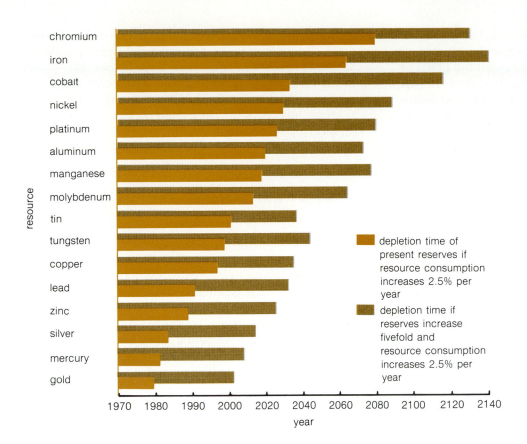

Figure 15-7 Projected times for 80 percent depletion of world reserves of 16 key metal resources based on two sets of assumptions. Note that after the year 2000, the time scale changes.

often more realistic to project the number of years until the known reserves of a resource will be 80 percent depleted if its consumption rate increases annually by a given percentage—typically 2 or 3 percent. Such an estimate is known as the **exponential reserve index**.

A series of projections of different depletion times in the form of *depletion curves* can be obtained by combining the static reserve index or exponential reserve index with additional assumptions about how the resource might be used. A typical set of depletion curves for a hypothetical nonrenewable resource is shown in Figure 15-6. For example, one estimate of depletion time might be based on assuming that the resource is not recycled or reused and that there will be no increase in its estimated reserves, as shown by curve A. A longer depletion time estimate can be obtained by assuming that recycling will *extend* the life of existing reserves and that improved mining technology, price rises, and new discoveries will *expand* existing reserves by some factor, say, 2, as shown in curve B. Curve C projects an even longer depletion time by assuming that reserves will be expanded by an even larger factor by new discoveries and through recycling, reuse, and reduced consumption. Of course, finding a substitute for a resource cancels all these curves and requires a new set of depletion curves for the new resource.

Figure 15-7 shows estimated times for 80 percent depletion of the world reserves for 16 important minerals based on two sets of assumptions. The second more optimistic set of assumptions shows that even if reserves are increased fivefold, the world could run short of tin, tungsten, copper, lead, zinc, silver, mercury, and gold between 2000 and 2040, and could be low on molybdenum, manganese, aluminum, platinum, and nickel between 2060 and 2090.

Figures 15-6 and 15-7 show why there is so much controversy over projected supplies of nonrenewable resources. One can get optimistic or pessimistic projections of the depletion time of a particular resource by making different sets of assumptions. The following guidelines can be used to evaluate different estimates of resource depletion times like those shown in Figure 15-7:

1. Remember that all the curves and estimated depletion times are projections of what *could* happen based on a different set of assumptions, not necessarily what *will* happen.

2. Find out what specific assumptions were used to make each projection.

3. Evaluate the assumptions (and, where possible, the data on which they are based) to see which set seems to be the most reasonable.

Rich Nations Versus Poor Nations: A New International Economic Order Some have charged that the industrialized nations exploit the LDCs by controlling international trade so that the LDCs are forced to sell their resources at a price far below their value. In effect, the MDCs are accused of stripping the world of its fossil fuel and other nonrenewable mineral resources without sufficient regard for the future resource needs of the LDCs. Even though the LDCs are expected to have 79 percent of the world's population in 2000, their projected use of the world's resources is only 23 percent—slightly more than the 20 percent they currently use.

To help correct this imbalance, the LDCs have called for a *new international economic order*—a proposal designed to shift more of the world's wealth from the MDCs to the LDCs. This proposal calls for (1) a substantial increase in aid from industrialized nations to LDCs, with special emphasis on the extension of new credit at favorable terms, (2) removal of trade barriers that restrict LDCs from selling some of their products to MDCs, (3) increasing the prices of raw materials exported from LDCs to MDCs, (4) providing LDCs with a greater say in the running of international lending institutions, such as the World Bank and the International Monetary Fund, and (5) helping the LDCs to pay back some of their massive indebtedness to the MDCs and private banks.

How valid are these charges of resource exploitation of the LDCs by the MDCs? Some analysts argue that when MDCs buy raw materials from LDCs, they provide funds that the LDCs need for their own economic development. Critics of this view counter that if the LDCs continue to sell off their resources at low prices to MDCs to get money for economic development, they may not have enough of these resources left to sustain such development in the future.

Some analysts also argue that the data show that the MDCs are not paying a just price for many resources. For example, average per capita gross national product in the United States has been rising steadily, equaling about $14,090 in 1985. But the average per capita cost of raw materials in the United States is less than $150, a price only slightly higher than at the beginning of the century. Between 1945 and 1977 the United States gave to LDCs about $4 billion in economic aid, $63 billion in loans through international agencies, $3.2 billion in trade concessions, and a great deal of advice and technical help. But Thomas Ehrlich, former president of the International Development Corporation Agency, which coordinates U.S. foreign aid, has pointed out that this aid has been mostly for economic and political rather than for humanitarian reasons and has amounted to only a token in comparison to the profits from sales of products made from raw materials bought cheaply from LDCs. Others argue that the government leaders and rich elites in most LDCs are really exploiting their own people. They contend that much of the aid to LDCs ends up in the pockets of the leaders and the rich instead of reaching the poor.

15-4 Key Resources: The U.S. Situation

Increasing U.S. Dependence on Imports The United States is more self-sufficient in key metals and other nonfuel minerals than any other nation except the Soviet Union. Nevertheless, high consumption rates and cheaper resource supplies in other countries have prompted the United States to import more minerals from resource-rich nations like the Soviet Union, the People's Republic of China, Canada, and several African and South American nations.

Currently the United States stockpiles 93 materials, 80 of them nonfuel minerals, considered to be vital to domestic industry. Strategic stockpiles of these materials are supposed to be large enough to last through a 3-year war, but most supplies are well below this level. Of the 42 most critical nonfuel minerals, the United States was dependent on foreign sources for more than 50 percent of 24 of them by 1984. This included heavy import dependence for chromium, cobalt, manganese, nickel, niobium, platinum, and tantalum, which are particularly important to defense and energy programs, as shown in Table 15-2. Even heavier dependence on imports for these 24 vital nonfuel minerals and others is projected by the year 2000. Japan and many western European nations are even more dependent on imports of vital nonfuel minerals.

Is Import Dependence Good or Bad? Some argue that U.S. dependence on other countries for key resources threatens economic security (if prices increase sharply) and military security (if supplies of vital resources are cut off or severely restricted). Any nation importing a large fraction of its supply of a particular mineral resource would be vulnerable under the following conditions: (1) when most of the world's supply of a resource is held by one country, as is the case with tungsten (China), mercury (Spain), and palladium (Soviet Union), (2) when nations holding most of the world's supply of a resource band together to form an OPEC-style cartel to control supplies and raise

Table 15-2 U.S. Import Dependence for Selected Key Nonfuel Minerals in 1984

Mineral	Percentage Imported	Major Suppliers	Key Uses
Columbium	100	Brazil, Canada, Thailand	High-strength alloys for construction, jet engines, machine tools
Diamonds	100	South Africa, United Kingdom, Soviet Union	Machinery, mineral services, abrasives, stone and ceramic products
Manganese	99	South Africa, Gabon, France	Alloys for impact-resistant steel, dry cell batteries, chemicals
Bauxite-alumina	96	Jamaica, Guinea, Suriname	Aluminum production, building materials, abrasives
Cobalt	95	Zaire, Zambia, Canada	Alloys for tool bits, aircraft engines, high-strength steel
Tantalum	94	Thailand, Malaysia, Brazil	Nuclear reactors, aircraft parts, surgical instruments
Platinum	91	South Africa, United Kingdom, Soviet Union	Oil refining, chemical processing, telecommunications, medical and dental equipment
Graphite	90+	Mexico, China, Brazil	Foundry operations, lubricants, brake linings
Chromium	82	South Africa, Zimbabwe, Soviet Union	Alloys for springs, tools, engines, bearings
Tin	79	Thailand, Malasia, Indonesia	Cans and containers, electrical products, construction, transportation
Rutile	61	Australia, Sierra Leone, South Africa	Paint, plastics, paper, welding-rod coatings
Vanadium	41	South Africa, Canada, Finland	Iron and steel alloys, titanium alloys, sulfuric acid production

Source: U.S. Bureau of Mines, U.S. Office of Technology Assessment.

prices, and (3) when substitute materials are not readily available, as in the cases of chromium, platinum, and manganese. Threats to world peace could also occur in such situations because of conflicts between MDCs competing for scarce supplies.

Mineral experts point out, however, that large percentages of many nonfuel minerals are imported because they are cheaper to extract from the higher grade ores found in other nations than from more plentiful lower grade domestic reserves. This preserves domestic reserves for use later when prices rise. According to the Bureau of Mines and the Geological Survey, the United States has adequate domestic reserves of most key minerals—except chromium, cobalt, platinum, tin, gold, and palladium—for at least the next several decades. However, the Geological Survey estimates that present reserves for most key minerals will not satisfy U.S. needs for more than 100 years without increased recycling, conservation, and a search for substitutes.

Minerals expert Hans Landsberg believes that the formation of cartels for nonfuel mineral resources is unlikely. He points out that stable countries such as Canada, Australia, and possibly Mexico, which are major exporters of strategic minerals, are not likely to form cartels. In addition, less stable nations of Central and South Africa are too poor to do without the income from minerals exports. Landsberg and others also point out that, unlike oil, which is consumed directly, raw materials contribute such a small percentage to the cost of finished products that increases in their prices have relatively small effects. Other analysts argue that mutual import dependence of nations is a stabilizing force for world peace, since conflicts are likely to be resolved by negotiation rather than by military action.

Are Environmentalists Hindering Mineral Exploration? Mining company officials contend that pressures from environmentalists to preserve wilderness, parks, and forest areas (Chapter 12) prevent some areas from being fully explored and mined. Recent studies by the General Accounting Office, however, indicate that only about 25 percent of public lands—mostly military reservations, parks, and public power facilities—have been withdrawn from mineral production.

New leases for mining in wilderness areas have been prohibited since 1983, but these areas had been open for both mineral exploration and development for 80 years previously. Environmentalists also point out that wilderness areas with significant mineral potential were excluded by Congress from protection when the areas were designated and that Congress has the power to open up wilderness areas for mining in cases of national emergency.

Mining interests also contend that environmental and safety regulations have increased the cost of mining and processing of mineral resources and call for a relaxation of these regulations. For example, between 1970 and 1980 eight U.S. zinc smelting plants were shut down because the owners couldn't afford to comply with new strict environmental regulations. As a result, U.S. imports of zinc rose from 25 to 62 percent during this period. Copper mined domestically is more expensive because its price includes the cost of environmental protection, whereas imported copper seldom does. Environmentalists contend, however, that the major factors in increased mining costs are depletion of many high-grade domestic ore deposits, inflation, and higher energy costs and that lack of sufficient pollution control in other mineral-rich nations does not justify relaxing pollution controls in the United States.

Solid wastes are only raw materials we're too stupid to use.

Arthur C. Clarke

Discussion Topics

1. Why should an urban dweller be concerned about the environmental impact from increasing surface mining of land for mineral resources?

2. Debate the following resolution: The United States uses too many of the world's resources relative to its population size, and should cut back on consumption.

3. Summarize the neo-Malthusian and the cornucopian views on the availability of resources. Which, if either, of these schools of thought do you support? Why?

4. Debate each of the following propositions.
   a. The competitive free market will control the supply and demand of mineral resources.
   b. New discoveries will provide all the raw materials we need.
   c. The ocean will provide all the mineral resources we need.
   d. We will not run out of key mineral resources because we can always mine lower grade deposits.
   e. When a mineral resource becomes scarce, we can always find a substitute.
   f. When a nonrenewable resource becomes scarce, all we have to do is recycle it.

5. Use the second law of energy (thermodynamics) to show why the following options are normally not profitable.
   a. Extracting most minerals dissolved in seawater.
   b. Recycling minerals that are widely dispersed.
   c. Mining increasingly low-grade deposits of minerals.
   d. Using inexhaustible solar energy to mine minerals.
   e. Continuing to mine, use, and recycle minerals at increasing rates.

6. Debate the pros and cons of trying to reduce use and waste of minerals by eliminating all tax breaks and depletion allowances for mining industries.

7. Compare the throwaway, recycling, and sustainable earth (or low-waste) approaches to waste disposal and resource recovery and conservation for (a) glass bottles, (b) "tin" cans, (c) aluminum cans, (d) plastics, and (e) leaves, grass, and food wastes (see Table 15-1).

8. Why is it difficult to get accurate estimates of mineral resource supplies? Be sure to distinguish among reserves, subeconomic resources, hypothetical resources, speculative resources, and depletion curves based on static reserve indexes and exponential reserve indexes.

9. Study Figure 15-7 to determine which key metals *might* be in short supply during your expected lifetime. How might such shortages affect your expected life-style?

16

Energy Resources: Types, Use, and Concepts

How quickly we can poison the earth's lovely surface—but how wondrously it responds to the educated caress of conservation.

Donald E. Carr

The amounts and types of *useful* energy available shape not only individual life-styles but also national and world economic systems. This chapter examines types of energy resources and global and U.S. energy use. It also shows how the two energy laws (Chapter 3) can help us evaluate present and future energy alternatives. The next three chapters discuss the advantages and disadvantages of each major energy alternative.

16-1 Types and End Uses of Energy Resources

Primary Energy Resources The *direct* input of essentially inexhaustible *solar energy* alone provides 99 percent of the thermal energy used to heat the earth and all buildings free of charge. Were it not for this *direct* input of radiant energy from the sun, the average temperature outside would be -240°C (-400°F).

Human ingenuity has developed a number of renewable and nonrenewable **primary energy resources** to supplement this direct input of solar energy and to provide the remaining 1 percent of the energy we use on earth. As indicated by Figure 16-1 and Table 16-1, these primary energy resources can be classified as **renewable** (permanent) or **nonrenewable** (temporary). Most people think of solar energy in terms of direct heat from the sun. But broadly defined, *renewable solar energy* includes not only *direct* radiant energy from the sun but also a variety of *indirect* forms of solar energy (Table 16-1).

Conserving energy by improving energy efficiency is also a major source of useful energy. Im-proving energy efficiency means getting more useful energy out of the primary energy we use. It means having an efficient space heating system and a well-insulated, airtight house, not enduring a cold house. For car owners, it means driving a car that typically runs an average of at least 17 kilometers per liter of gasoline (40 miles per gallon), not giving up a car altogether.

End Uses of Energy The *energy flow* or *"spaghetti"* *chart* in Figure 16-1 shows that the world's sources of primary energy can be converted by human ingenuity into electrical, chemical, thermal (heat), or mechanical energy to provide: **(1)** *low-temperature heat* [less than 140°C (284°F)] for heating water and homes and buildings (space heating), **(2)** *high-temperature heat* (up to several thousand degrees) for *industrial processes*, **(3)** *high-temperature heat* used to produce *electrical energy* in a power plant, **(4)** *mechanical energy for propelling vehicles*, obtained when the chemical energy stored in natural gas or liquid fuels is converted by combustion to high-temperature thermal energy, which is then converted to mechanical energy, **(5)** *material items* made from wood, natural fibers, and other forms of *renewable biomass*, and **(6)** *material items* such as plastics, synthetic fibers, pesticides, and many medicines made from *petrochemicals* obtained mostly from natural gas and crude oil.

Global Primary Energy Resources Figure 16-2 shows that by 1984 about 82 percent of the primary energy used throughout the world was provided by the burning of three *nonrenewable* fossil fuels—oil (36 percent), coal (27 percent), and natural gas (17 percent)—and by the nuclear fission of *nonrenewable* uranium atoms to produce electricity (2 percent). The remaining 18 percent of the world's primary energy was provided by burning *renewable biomass energy sources* such as wood, dung, and crop residues (13 percent) and by using *renewable* falling water (hydropower) to produce electricity (5

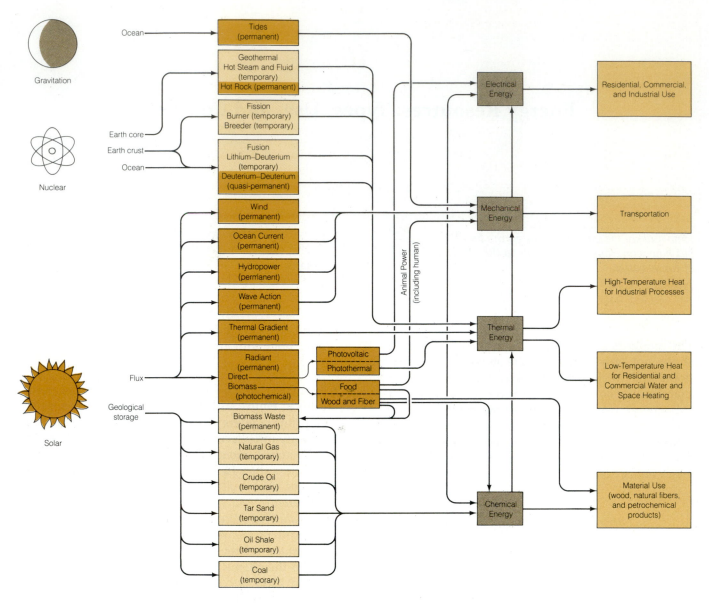

Figure 16-1 Energy flow or "spaghetti" chart showing the earth's major renewable or permanent (dark color) and nonrenewable or temporary (light color) sources of energy and how they can be converted into chemical, thermal, mechanical, and electrical energy. (Source: Adapted from material supplied by the Office of Energy Research and Planning, State of Oregon.)

percent). According to the estimate in Figure 16-2, by the end of the century the world will be less dependent on nonrenewable energy resources (74 percent) and more dependent on renewable energy resources (26 percent).

16-2 Brief History of Energy Use

Energy Use from Primitive to Modern Times
Cultural history has been based on using human ingenuity to increase the average amount of primary energy used per person (Figure 16-3). Aver-

age daily energy use per person increased from about 2,000 kilocalories from food energy for early hunter-gatherers, who had not discovered how to burn wood for heat and cooking, to an average of 12,000 kilocalories per person by early farmers, who burned wood for cooking and heating and used the muscle power of domesticated animals to help them raise grains and vegetables.

By 1850 the direct and indirect average per capita fuel consumption in industrializing nations like Great Britain and the United States had risen to about 60,000 kilocalories per day, with wood providing 91 percent of the energy used in the United States (Figure 16-4). By 1900 coal had replaced

Table 16-1 Nonrenewable and Renewable Primary Energy Resources

Type	Description
Nonrenewable	
Conventional fossil fuels	Underground deposits of crude oil, natural gas, and coal representing the storage of solar energy as chemical energy in geological deposits believed to be formed over millions of years from the decay of dead vegetation and animals under high temperatures and pressures in the earth's crust
Derived fossil fuels	Liquid fuels such as gasoline, diesel fuel, and heating oil obtained by distilling and processing crude oil
Derived synfuels	Synthetic crude oil and synthetic natural gas produced by the liquefaction or gasification of solid coal
Unconventional heavy oils	Underground deposits with an asphaltlike consistency that can be extracted from conventional crude oil deposits by enhanced recovery methods, oil shale rock, and tar sands which can then be upgraded to crude oil and converted to derived fossil fuels
Unconventional natural gas	Deep underground deposits found in tight sands, Devonian shale rock, and coal seams, and dissolved in deep underground deposits of saltwater at very high temperatures and pressures (geopressurized zones)
Conventional nuclear fission (natural fuel)	High-temperature heat released when the tiny centers or nuclei of certain types of uranium atoms found in the earth's crust are split apart by atomic particles called neutrons
Breeder nuclear fission (derived fuel)	Extending the world's uranium supplies by converting a certain nonfissionable type of uranium atom into a type of fissionable plutonium atom not found in nature
Nuclear fusion	High-temperature heat released when temperatures of 100 billion degrees or more are used to fuse together the nuclei of certain types of hydrogen atoms found in water or produced from lithium in the earth's crust to form nuclei of heavier atoms
Trapped geothermal deposits	Low-temperature heat deposited in underground zones of dry steam, hot water, or a mixture of dry steam and hot water primarily from heat released by radioactive substances found in the *mantle* of partially molten rock beneath the earth's crust and the molten rock, called *magma*, located underneath this mantle in the earth's core
Derived hydrogen gas (H_2)	A gaseous fuel produced by using heat or electricity produced by a nonrenewable energy resource to decompose water ($2H_2O + energy \rightarrow 2H_2 + O_2$), with the H_2 gas then burned in air or oxygen to produce energy ($2H_2 + O_2 \rightarrow 2H_2O + energy$)
Renewable	
Tidal energy	Mechanical energy extracted from the abnormally high rise and fall of the tides twice a day in certain parts of the world
Flow-regulated geothermal energy	Medium- to high-temperature heat released by radioactive elements that flows slowly into deep underground deposits of hot dry rocks, partially molten rock (mantle), and molten rock (magma)
Direct solar energy	Radiant energy from the sun that can be used to provide low-temperature heat and hot water for residential and commercial buildings, concentrated to provide high-temperature heat for industrial processes and conversion to electricity, and converted by solar photovoltaic cells directly into electrical energy
Indirect solar energy	Solar energy stored in biomass (trees and other plants), direct and indirect biomass wastes (such as dung, crop wastes, and paper), winds, falling or flowing water (hydro), ocean currents, wave action, and temperature differences (thermal gradients) between the surface and bottom water in tropical oceans, shallow saltwater seas such as the Dead Sea in Israel, and artificially produced saltwater solar ponds
Derived biogas	A form of natural gas produced by the decomposition of biomass wastes
Derived liquid biofuels	Liquids such as ethyl alcohol and methyl alcohol produced by fermentation and anaerobic decomposition of various types of biomass and biomass waste and pyrolysis oil produced by thermal decomposition of biomass and biomass wastes
Derived hydrogen gas (H_2)	A gaseous fuel produced by using direct sunlight to decompose water (see *nonrenewable derived hydrogen gas*)
Improving energy efficiency (energy conservation)	Conserving energy by getting more useful energy out of the primary energy we use

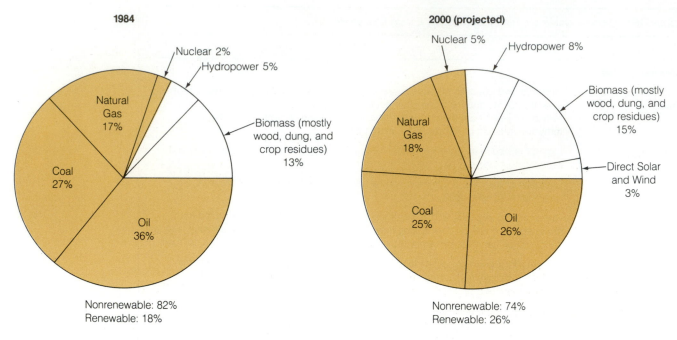

1984

Nuclear 2%

Hydropower 5%

Natural Gas 17%

Biomass (mostly wood, dung, and crop residues) 13%

Coal 27%

Oil 36%

Nonrenewable: 82%
Renewable: 18%

2000 (projected)

Nuclear 5%

Hydropower 8%

Natural Gas 18%

Biomass (mostly wood, dung, and crop residues) 15%

Direct Solar and Wind 3%

Coal 25%

Oil 26%

Nonrenewable: 74%
Renewable: 26%

Figure 16-2 World consumption of primary nonrenewable (shaded) and renewable (unshaded) energy by source in 1984 with projections for 2000. (Sources: U.S. Department of Energy and Worldwatch Institute.)

wood as the major energy source in industrializing European nations and in the United States. By 1950 *crude oil* had become the major energy source and *natural gas* the third largest source of energy in the United States. Use of oil and natural gas increased sharply in the industrialized nations, and in 1984 these resources provided 53 percent of the world's primary energy (Figure 16-2) and 65 per-

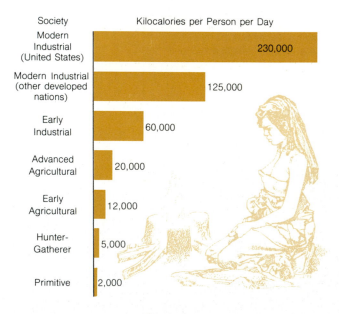

Figure 16-3 Average daily per capita energy use at various stages of human cultural evolution.

Society — Kilocalories per Person per Day

Modern Industrial (United States) — 230,000

Modern Industrial (other developed nations) — 125,000

Early Industrial — 60,000

Advanced Agricultural — 20,000

Early Agricultural — 12,000

Hunter-Gatherer — 5,000

Primitive — 2,000

cent of the primary energy used in the United States (Figure 16-4).

By 1984 each person in most industrial nations used directly and indirectly an average of about 125,000 kilocalories a day—almost 63 times the average per capita energy use at the primitive survival level (Figure 16-3). By 1984 each American directly and indirectly used an average of 230,000 kilocalories of energy per day—twice the level in most industrialized nations and 115 times the survival level. In 1984, 236 million Americans used more energy for air conditioning alone than 1.03 billion Chinese used for all purposes!

Most of the increase in primary energy consumption per person since 1900 has taken place in the MDCs, and the gap in average per capita primary energy use between the MDCs and LDCs has widened (Figure 16-5). In 1984, with about 5 percent of the world's population, the United States accounted for about 25 percent of the world's primary energy consumption. India, with about 15 percent of the world's population, consumed only about 1.5 percent of the world's primary energy.

The 1973 OPEC Oil Embargo and the 1979 Iranian Oil Cutoff Since 1940 the depletion of many low-cost domestic oil deposits has made it cheaper for the United States to import much of its oil. By 1973, about 48 percent of this imported oil came from the 13 nations in the Organization of Petroleum Ex-

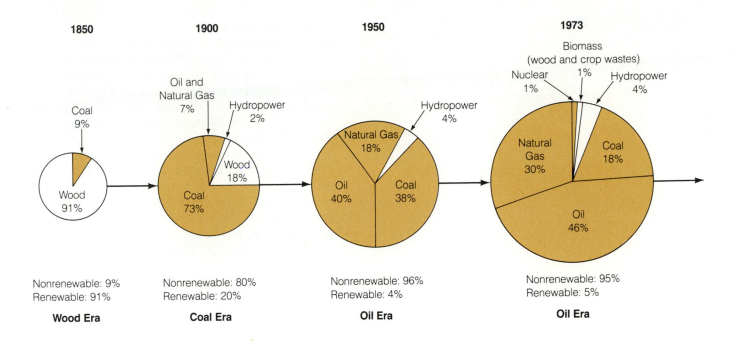

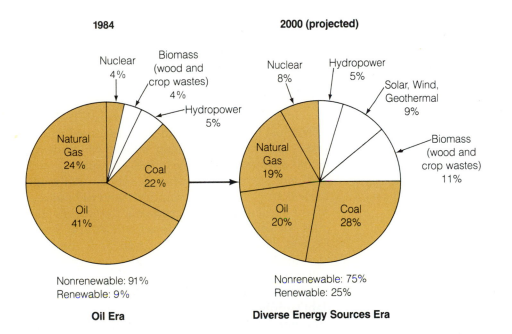

Figure 16-4 Changes in consumption of primary nonrenewable (shaded) and renewable (unshaded) energy resources in the United States between 1850 and 1984 with a projection for 2000. Relative circle size indicates the total amount of energy used. (Sources: U.S. Department of Energy and Center for Renewable Resources; National Audubon Society for year 2000 projection.)

porting Countries (OPEC), which together accounted for 56 percent of the world's oil production and about 84 percent of all oil exports. Other MDCs such as Japan and most western European nations are even more dependent on imported oil than the United States because their domestic supplies are scanty or nonexistent.

On October 18, 1973, the Arab members of OPEC reduced oil exports to Western industrial nations and prohibited all shipments of their oil to the United States because of its support of Israel in the 18-day war waged by Egypt and Syria against Israel. The embargo lasted until March 1974 and

caused a fivefold increase in the average world price of crude oil (from $2.70 to almost $10 a barrel), which helped lead to double-digit inflation, high interest rates, and a global recession. Americans accustomed to cheap and plentiful fuel waited for hours to buy gasoline, and thermostats in homes and offices were turned down. Despite the sharp price increase, however, U.S. oil imports obtained from OPEC nations increased from 48 percent to 67 percent between 1973 and 1978.

Then in 1979 available world oil supplies again decreased as a result of revolution in Iran. The average world price of crude oil rose to nearly $34 a

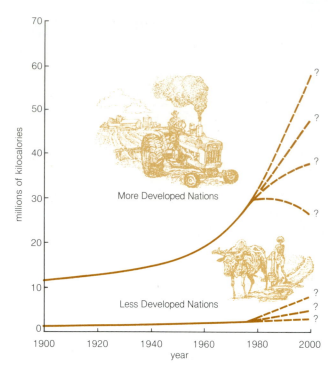

Figure 16-5 Average per capita annual primary energy consumption in the MDCs and LDCs between 1900 and 1984 and projections (dotted lines) to 2000. (Sources: U.S. Department of Energy and International Energy Agency.)

barrel by 1982 before dropping to about $29 a barrel in 1984. Thus, between 1973 and 1984, the price of a barrel of crude oil bought on the world market and imported in the United States increased *974 percent* (from $2.70 to about $29) or *517 percent* adjusted for inflation in constant 1975 dollars (from $3.53 to about $21.77). Thus between 1973 and 1984 there occurred history's largest international transfer of wealth and power from oil-importing to oil-exporting nations. A few oil-exporting MDCs—Norway, the United Kingdom, and the Soviet Union—benefited from higher oil prices, but Mexico and the OPEC nations gained the most. By 1984 the United States was still importing about one-third of the oil it consumed.

By 1984 energy conservation and several years of worldwide economic recession had led to a temporary oversupply, and the average price of crude oil fell from $34 to $27 a barrel between 1982 and 1985. As long as supply exceeds demand, some observers predict a further drop, to $20 to $25 a barrel. Such a drop in oil prices would stimulate economic growth in oil-importing nations such as the United States but would discourage the search for new oil and undermine improvements in energy efficiency and development of other energy alternatives.

Most analysts believe that the so-called oil glut of the 1980s is only temporary because (1) a war or

other crisis such as the collapse of Saudi Arabia's monarchy or a takeover of Mexico's oilfields by Central American guerrillas could cause a sudden cutoff of oil supplies, and (2) the world demand for oil is projected to exceed the rate at which it can be produced between 1990 and 2005. OPEC is projected to increase its share of the world oil market from 27 percent today to about 55 percent in the 1990s. The Congressional Research Service and the International Energy Agency warn that another oil embargo or cutoff of oil supplies because of a Mideast war or revolution in Mexico could cause oil prices to reach $50 to $98 a barrel, reduce the U.S. gross national product by as much as 29 percent, cut jobs by as much as 28 percent, and increase the chances of war.

16-3 How Long Can the Fossil Fuel Era Last?

Supplies of Conventional and Unconventional Crude Oil Unlike supplies of many nonrenewable mineral resources such as copper and aluminum, nonrenewable fossil fuels cannot be recycled or reused. Once burned, the high-quality energy they contain is gone forever.

Until the oil supply disruptions of 1973 and 1979, most nations, rich and poor, assumed that an ample and relatively cheap supply of crude oil would always be available to fuel economic growth and more energy-intensive life-styles. Although these disruptions were temporary and did not represent a true shortage of crude oil, the world's supplies of nonrenewable fossil fuels are finite. Experts disagree, however, over how long the identified and unidentified fossil fuel resources will last.

Cornucopians believe we will have an abundant supply of crude oil, arguing that higher prices for crude oil will stimulate the location and extraction of unidentified crude oil resources, as well as the extraction and upgrading of heavy oils from oil shale and tar sands and from oil too thick to pump out of existing oil wells (Section 17-1).

Other analysts, however, say that people who hold cornucopian beliefs do not understand the arithmetic and consequences of exponential growth in the use of a resource (Section 1-2). According to estimates by the U.S. Department of Energy and the American Petroleum Institute, the world's known *reserves* of crude oil will be 80 percent depleted in 30 years (by 2013) if annual oil consumption remains at the 1984 rate and in 23 years (by 2006) if the annual depletion rate increases by a modest 2 percent because of world economic growth, a drop in oil prices, or both.

The ultimately recoverable supply of crude oil based on undiscovered deposits is estimated to be more than three times today's proven reserves. Even if *all* the estimated supplies of crude oil are discovered and developed—which some experts consider unlikely—and sold at a price many times the 1985 price of $27 a barrel, this ultimately recoverable supply would be 80 percent depleted in 93 years (by 2076) at 1984 usage rates and in 53 years (by 2037) if the rate of oil usage increases by 2 percent a year.

Large supplies of heavy oil that can be upgraded to crude oil are potentially available from oil shale and tar sands and could extend world supplies of crude oil for at least 100 years. However, extracting these heavy oils is expensive and environmentally harmful (Section 17-1).

Assuming that the 1984 rate of crude oil consumption is maintained, **(1)** Saudi Arabia, with the world's largest known crude oil reserves, could supply the world's total crude oil needs for 10 years if it were the world's only source; **(2)** Mexico, which has the world's second largest crude oil reserves, could supply the world's total crude oil needs for about 3 years; **(3)** the estimated crude oil reserves under Alaska's North Slope—the largest deposit ever found in the United States—used by itself, would meet world demand for only 6 months and U.S. demand for 2 to 4 years; and **(4)** if drilling off the East Coast of the United States meets the most optimistic estimates, these potential crude oil reserves could satisfy world crude oil needs for 1 week and U.S. needs for less than 3 months. Thus, anyone who argues that new discoveries will solve world crude oil supply problems is saying that we need to discover the equivalent of a new Saudi Arabian deposit *every 10 years* just to maintain the world's 1984 level of oil use.

Conventional and Unconventional Natural Gas

Supplies of conventional natural gas are projected to last somewhat longer than those of crude oil, but are also limited. The world's identified reserves of natural gas will last about 50 years (to 2033) at 1984 usage rates and 35 years (to 2018) if annual usage increases by 2 percent a year. The world's estimated supply of natural gas that may be found and recovered at much higher prices—the "ultimately recoverable" supply—would last about 200 years (to 2183) at 1984 usage rates and 80 years (to 2063) if annual usage increases by 2 percent a year.

Some energy analysts believe that large additional supplies of natural gas may be obtained from unconventional sources such as tight sands and geopressurized zones (Section 17-2). Others, however, believe that these sources will prove to be disappointing because drilling and extracting

from such deep underground regions may take more energy and money than the resulting gas would be worth.

Coal Based on known reserves, coal is the most abundant conventional fossil fuel in the world and in the United States. Identified world reserves of coal should last for about 276 years (to 2259) at 1984 usage rates and 86 years (to 2069) if annual usage increases by 2 percent a year. The world's estimated ultimately recoverable supply of coal would last about 900 years at the 1984 usage rate and 149 years if annual usage increases by 2 percent a year.

Coal is not only more plentiful, in most countries it is also cheaper to use for producing electricity than oil, natural gas, and nuclear energy, even if coal-burning power plants are equipped with the latest and most effective air pollution control equipment. It is projected to remain cheaper in countries with ample coal supplies well into the next century. For these reasons, some energy analysts suggest that the use of coal be increased to help make the transition from dependence on oil and natural gas to dependence on energy conservation and a mix of renewable energy resources over the next 50 years.

With this historical background in energy use and fossil fuel availability, we are ready to look at some energy concepts that can help us evaluate present and future energy alternatives.

16-4 Energy Concepts: Energy Quality, Energy Efficiency, and Net Useful Energy

Energy Quality and Flow Rates According to the second law of energy (Section 3-3), whenever we use any form of energy, it is automatically degraded to a lower quality or less useful form of energy—usually low-temperature heat that flows into the environment. Thus, *a major factor determining the usefulness of an energy resource is its quality, not its quantity.*

Different forms of energy vary in their quality (Table 16-?) High-quality energy, like that in electricity, oil, gasoline, sunlight, wind, uranium, and high-temperature heat, is concentrated. By contrast, low-quality energy, like low-temperature heat, is dispersed, or dilute. For example, there is more low-temperature heat in the Atlantic Ocean than in all the oil in Saudi Arabia. But this heat stored in the ocean is so widely dispersed that we can't do much with it.

Sunlight does not melt metals or char our clothes because only a relatively small amount of this high-quality energy reaches each square meter

Table 16-2 Energy Quality of Different Forms of Energy

Form of Energy	Energy Quality	
	Relative Value	Average Energy Content (kilocalories per kilogram)
Electricity	Very high	
Very high-temperature heat (greater than 2,500°C)	Very high	
Nuclear fission (uranium)	Very high	139,000,000*
Nuclear fusion (deuterium)	Very high	24,000,000†
Concentrated sunlight	Very high	
Concentrated wind (high-velocity flow)	Very high	
High-temperature heat (1,000°–2,500°C)	High	
Hydrogen gas (as a fuel)	High	30,000
Natural gas (mostly methane)	High	13,000
SNG (synthetic natural gas made from coal)	High	13,000
Gasoline (refined crude oil)	High	10,500
Crude oil	High	10,300
LNG (liquefied natural gas)	High	10,300
Coal (bituminous and anthracite)	High	7,000
Synthetic oil (made from coal)	High	8,900
Sunlight (normal)	High	
Concentrated geothermal	Moderate	
Water (high-velocity flow)	Moderate	
Moderate-temperature heat (100°–1,000°C)	Moderate	
Dung	Moderate	4,000
Wood and crop wastes	Moderate	3,300
Assorted garbage and trash	Moderate	2,900
Oil shale	Moderate	1,100
Tar sands	Moderate	1,100
Peat	Moderate	950
Dispersed geothermal	Low	
Low-temperature heat (air temperature of 100°C or lower)	Low	

*Per kilogram of uranium metal containing 0.72% fissionable uranium-235.
†Per kilogram of hydrogen containing 0.015% deuterium.

of the earth's surface per minute or hour during daylight hours—even though the total amount of solar energy reaching the entire earth is enormous. Wind energy also has a high energy quality, but to perform large amounts of useful work it must flow into a given area at a fairly high rate. Thus, *the overall usefulness of a renewable energy source such as direct sunlight, flowing water, and wind is determined both by its energy quality and by its flow rate (flux)— the amount of high-quality energy reaching a given area of the earth per unit of time.*

Unfortunately, many forms of high quality energy, such as high-temperature heat, electricity, gasoline, hydrogen gas, and concentrated sunlight, do not occur naturally. We must use other forms of high-quality energy like fossil, wood, or nuclear fuels to produce them, to concentrate and store them, or to upgrade their quality.

Matching Energy Quality to End Uses *An important way to reduce energy waste is to supply energy only in the quality needed for the task at hand, using the cheapest possible approach.* From a thermodynamic standpoint, for example, using high-quality electrical energy to heat a home to 20°C (68°F) or to provide hot water at 60°C (140°F) wastes large quantities of high-quality energy. First, at a power plant, high-quality energy from a fossil fuel or nuclear fuel is converted to high-quality heat energy at several thousand degrees, with an automatic loss of some of the heat to the local environment as required by the second energy law. The remaining high-quality heat is used to convert water to steam and to spin turbines to produce high-quality electrical energy, with a further loss of heat to the environment. More degraded energy or heat is lost when electricity is transmitted to a home. There the high-quality electrical energy is converted to relatively low-quality heat energy for space heating or to provide hot water. According to energy expert Amory Lovins, using electricity for such purposes "is like using a chain saw to cut butter."

In 1983 electrical energy provided about 33 percent of all primary energy used in the United States, with coal-burning power plants producing 54 percent of this energy, hydropower 14 percent, natural gas 13 percent, nuclear power 13 percent, and oil only 6 percent. But according to Lovins, the essential uses for electricity in the United States amount to only about 8 percent of annual primary energy needs, mostly for running lights and motors, refining copper, and producing glass, iron, and steel by certain methods. This means that the United States already has more than enough power plants to produce electricity to meet all *essential* needs until 2005—perhaps until 2025. Critics of this idea point out that even though electricity is wasteful of energy when used for certain purposes, people use it in their homes because it is convenient.

From a temperature-matching standpoint, a typical home furnace or hot water heater also wastes large amounts of high-quality energy by burning the oil or natural gas at about 2,000 to 3,000°C (3,632 to 5,432°F) to heat a house to 20°C (68°F) or water to 60°C (140°F). From a thermodynamic standpoint, it is much less wasteful of energy to use active or passive solar collectors to extract heat from the air or from groundwater and use energy-efficient heat pumps to raise the temperature slightly to provide space heating or hot water.

First-Law Energy Efficiency *Another way to cut energy waste and save money—at least in the long run—is to use an energy-conversion device such as a light, home heating system, refrigerator, or automobile engine of maximum energy efficiency.* There are two types of energy efficiency, based on the first and second laws of thermodynamics, respectively.

First-law energy efficiency is the ratio of the useful energy (or work) output to the total energy

(or work) input for an energy-conversion device or process. Normally this energy ratio is multiplied by 100 so that the efficiency can be expressed as a percentage:

$$\text{first-law energy efficiency (\%)} = \frac{\text{useful energy (or work) output}}{\text{total energy (or work) input}} \times 100$$

When energy was cheap, industrialized nations freely used inefficient but relatively inexpensive devices such as incandescent light bulbs. Light bulbs use about 20 percent of America's electricity—about twice the electrical output of the nation's nuclear power plants. Yet, the incandescent light bulb, which is really a heat bulb, used for 95 percent of all home lighting in the United States, has a first-law energy efficiency of only 5 percent. This means that 95 of every 100 kilocalories of electrical energy supplied to the bulb is immediately degraded to low-quality heat, and only 5 kilocalories is converted to light.

Large amounts of energy could be saved by switching to fluorescent light bulbs, with a first-law energy efficiency four times that of the incandescent bulb. New fluorescent light bulbs that look like the screw-in incandescent bulbs are much more expensive than incandescent bulbs of the same wattage, but the new bulbs last 10 years or more compared with about 6 months for incandescent bulbs. This extended life plus its use of 70 percent less electricity than the incandescent bulb means that the fluorescent bulb will be cheaper to buy and run over a 10-year period. Costs should fall with mass production and increased use of these bulbs.

The first-law efficiency of an individual device does not give the whole picture. The overall first-law efficiency of a home heating system, for example, is not determined solely by the first-law efficiency of the furnace. Instead, it is necessary to determine the *net first-law energy efficiency* of the entire energy delivery system. This is done by finding the first-law efficiency for each energy-conversion step in the system, including extracting the fuel, purifying and upgrading it to a useful form, transporting it, and finally using it. Figure 16-6 shows how the net first-law energy efficiencies are determined for heating a home with solar energy and with electricity produced at a nuclear power plant.

Table 16-3 lists the net first-law energy efficiencies for a variety of commonly used space heating systems, and Table 16-4 gives the 1983 average price per 250,000 kilocalories (1 million Btu) for space heating with various fuels in the United

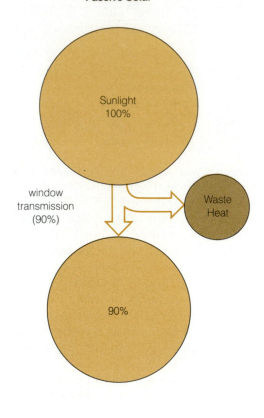

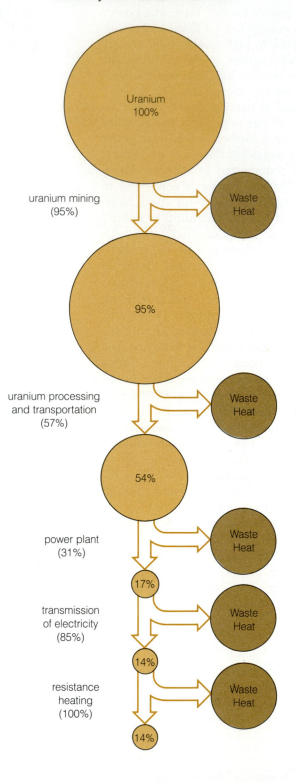

Figure 16-6 Comparison of net first-law energy efficiency for two types of space heating system. Cumulative net efficiency is obtained by multiplying the percentage shown inside the circle for any step by the first-law energy efficiency for that step, as shown in parentheses.

States. Such analysis reveals that superinsulation and a passive solar system (Section 19-1) are the two most efficient and cheapest space heating systems for a new house. For an existing house, the most efficient and cheapest approach is to heavily insulate the house, make it airtight, and use one of the new high-efficiency natural gas furnaces, elec-

tric resistance heating (if you live in one of the few areas where *all* the electricity is produced at a hydroelectric power plant), or an efficient geothermal or solar-assisted heat pump.

Other energy-efficient ways for space heating parts of an existing house are to add simple passive solar window box heaters (which cost about $75 and can be built by most individuals in a few hours) or solariums or solar greenhouses (which can serve as comfortable living rooms or can be used to grow flowers and vegetables throughout the year on the side of the house facing the sun). The least efficient and most expensive way to heat a house is to use the electricity produced by a nuclear power plant for electric resistance heating (Tables 16-3 and 16-4). However, electric quartz radiant heaters (not electric resistance heaters), which are designed to heat people rather than entire rooms, can be used to reduce the overall heating of a house or room.

Table 16-5 gives a similar analysis for heating water for washing and bathing. In terms of the first energy law, the least efficient way to provide hot water in a home is to use electricity produced by a nuclear power plant, and the most efficient method is to use a natural or LP gas-fired *tankless instant hot water heater* that comes on only when the hot water faucet is turned on and heats the water instantly as it flows through a small burner chamber. This is in sharp contrast to conventional natural gas and electric resistance heaters, which keep a large tank of water hot all day and night, whether you need it or not. Tankless instant hot water heaters are widely used in many parts of Europe and are slowly beginning to appear in the United

Table 16-4 Average Costs of Unit of Energy Provided for Space Heating in the United States During 1983

Space Heating System	Average Cost per 250,000 kilocalories (1 million Btu)
Electricity (resistance heating)	$22.86
Kerosene	$11.10
Electric heat pump	$11.00
Active solar new house*	$10.00
Wood (50% efficient stove, $90 a cord)	$ 9.00
Fuel oil	$ 8.80
Propane (LP)	$ 8.65
Natural gas	$ 6.27
Passive solar new house*	$ 5.50
Improving energy efficiency (existing houses)*	$3–$5.50
Superinsulated new house (100% of heat)*	$0–$1.20

Sources: Federal Trade Commission, U.S. Department of Energy, U.S. Office of Technology Assessment, and Worldwatch Institute.
*Average lifetime cost is lower because fuel (solar energy) is free after initial investment has been recovered through fuel savings. Federal and state tax credits also reduce the cost of the initial investment.

Table 16-3 Net First-Law Energy Efficiency for Various Space Heating Systems

Space Heating System	Net First-Law Energy Efficiency (%)
Superinsulated house (100% of heat)	98
Passive solar (100% of heat)	90
Passive solar (50% of heat) plus high-efficiency natural gas furnace (50% of heat)	87
Natural gas with high-efficiency furnace	84
Electric resistance heating (electricity from hydroelectric power plant)	82
Natural gas with typical furnace	70
Passive solar (50% of heat) plus high-efficiency wood stove (50% of heat)	65
Oil furnace	53
Electric heat pump (electricity from coal-fired power plant)	50
High-efficiency wood stove	39
Active solar	35
Electric heat pump (electricity from nuclear plant)	30
Typical wood stove	26
Electric resistance heating (electricity from coal-fired power plant)	25
Electric resistance heating (electricity from nuclear plant)	14

Table 16-5 Net First-Law Energy for Various Hot Water Heating Systems

Hot Water Heating System	Net First-Law Energy Efficiency (%)
Natural gas (tankless instant heater)	75
Passive solar (efficient batch heater for 50% plus instant natural gas heater, 50%)	65
Natural gas (conventional tank heater)	60
Active solar tank (50%) plus instant gas heater (50%)	58
Passive solar efficient batch tank heater (100%)	45
Active solar tank heater (100%)	30
Electric tank heater (electricity from coal-fired plant)	24
Electric tank heater (electricity from nuclear plant)	13

States. A well-insulated conventional natural gas water heater is also fairly efficient.

Table 16-6 lists net first-law energy efficiencies for several automobile engine systems. Note that the net first-law energy efficiency for a car powered with a conventional internal combustion engine is only about 8 percent. In other words, about 92 percent of the energy in crude oil has been wasted by the time it has been converted to gasoline and used to move a car.

An electric engine with batteries recharged by electricity from a hydroelectric power plant has a net first-law efficiency of 25 percent—almost three times that of a gasoline-burning internal combustion engine. But this system cannot be widely used in the United States because most of the major hydroelectric dam sites have already been used (Section 19-2). In addition, present electric-powered cars are expensive (average price: $15,000) and cruise efficiently at no more than 72 kilometers per hour (45 miles per hour). Furthermore, their batteries must be recharged about every 96 kilometers (60 miles) and replaced every 48,300 kilometers (30,000 miles), at a cost of several thousand dollars—although scientists are trying to develop affordable, longer lasting batteries. The second most efficient system is a car with a gas turbine engine. American, Japanese, and several European car makers have prototype gas turbine engines, but they need more development and may turn out to be quite expensive even when mass produced.

Second-Law Energy Efficiency To get a more complete picture of energy loss and waste, we can use **second-law energy efficiency**. This is the ratio of the minimum amount of useful energy needed to perform a task in the most efficient way that is theoretically possible (whether we know how to do it or not) to the actual amount used to perform the task.

$$\text{second-law energy efficiency (\%)} = \frac{\begin{array}{c}\text{minimum amount}\\\text{of useful energy}\\\text{needed to perform a task}\end{array}}{\begin{array}{c}\text{actual amount}\\\text{of useful energy}\\\text{used to perform a task}\end{array}} \times 100$$

This ratio shows how far the performance of an energy device or system falls short of what is theoretically possible according to the second energy law.

Table 16-7 lists estimated second-law energy efficiencies for various energy systems in the United States, and Figure 16-7 shows the overall second-law energy efficiency for all primary energy used in the United States during 1984. In terms of second-law energy efficiency, about 84 percent of all primary energy used each year in the United States is wasted. About 41 percent of this waste is unavoidable and occurs as a result of the second energy law. But at least 43 percent of the primary energy wasted in the United States could

Table 16-6 Net First-Law Energy Efficiency for Automobiles with Various Engine Systems

Engine System	Net First-Law Energy Efficiency (%)
Electric Car with Electricity Produced by	
Hydro	25
Coal	9
Nuclear	5
Combustion Engines	
Gas turbine	12
Diesel	9
Gasoline (conventional internal combustion)	8
Natural gas	8
Steam (Rankine external combustion)	8

Table 16-7 Estimated Second-Law Energy Efficiencies for U.S. Energy Systems

Energy System	Second-Law Energy Efficiency (%)
Space heating	
Heat pump	9
Furnace	5–6
Electric resistance	2.5
Water heating	
Gas	3
Electric	1.5
Air conditioning	4.5
Refrigeration	4
Automobile	8–10
Power plants	33
Steel production	23
Aluminum production	13
Oil refining	9
All systems	10–15*

*Other estimates put this much lower (3–5%).

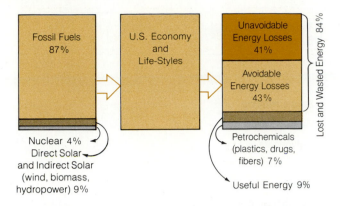

Figure 16-7 Flow and conversion of primary energy to useful energy in the United States based on overall second-law energy efficiency.

be recovered by **(1)** using available, more energy-efficient devices and systems (Tables 16-3, 16-5, and 16-6), **(2)** using well-known conservation techniques, such as insulation, natural ventilation, lower and more comfortable lighting levels, and buildings aligned to take advantage of sunlight and wind (Section 19-7), and **(3)** developing affordable new energy devices such as space heating systems that heat only the occupants of a building rather than the entire inside space, and plastic light pipes and light-emitting diodes that convert 10 to 20 times more of the input of electrical energy to visible light than incandescent or fluorescent bulbs.

Lifetime Versus Initial Costs of Energy Systems

A well-designed, passive solar home or commercial building (Section 19-1) should have lower construction and operating costs than a conventionally built structure of similar size. But the initial costs of active solar heating systems and highly energy-efficient natural gas furnaces, appliances, and automobile engines are sometimes greater than for conventional and less efficient systems. Over their lifetimes, however, most energy-efficient devices save both energy and money. Thus, *it is the lifetime (or life-cycle) cost—not the initial cost—that determines whether an energy-conversion device is a bargain*:

lifetime cost = initial cost + lifetime operating cost

Using Waste Heat

Because of the second energy law, high-quality energy cannot be recycled and is eventually degraded to low-temperature heat that flows into the environment. However, the rate at which waste heat flows into the environment can be slowed. For instance, in cold weather an uninsulated and leaky house loses heat almost as fast as it is produced. By contrast, a well-insulated and airtight house can retain most of its heat for 5 to 10 hours, and a well-designed superinsulated house (Section 19-7) can retain most its heat input for up to 4 days.

In some office buildings, waste heat from lights, computers, and other machines is collected and distributed to reduce heating bills during cold weather, or exhausted in hot weather to reduce cooling bills. Waste heat from industrial plants and electrical power plants can be distributed through insulated pipes and used as a *district heating system* to provide space heating for nearby buildings and homes, greenhouses, or aquaculture ponds.

Waste heat from coal-fired and other industrial boilers can also be used to produce electricity at about half the cost of buying it from a utility company. This electricity can then be used by the plant or sold to the local power company for general use. This combined production of high-temperature heat and electricity, known as **cogeneration**, is widely used in industrial plants throughout Europe. In 1920, 22 percent of the electricity used in the United States was produced by cogeneration at industrial sites. By 1984, however, this percentage was only about 5 percent, primarily because most utility companies charge high backup rates when a cogenerating system is out of operation and pay low prices for cogenerated electricity. If all large industrial boilers in the United States used cogeneration, they could produce electricity equivalent to that from 30 to 200 nuclear or coal-fired power plants (depending on the technology used), at about half the cost. This could reduce the average price of electricity and essentially eliminate the need to build any new large electric power plants in the United States, probably at least through 2020.

Net Useful Energy: It Takes Energy to Get Energy

The two energy laws tell us that the only energy that really counts is *net* useful energy—not the *gross* or *total* energy obtainable from a given quantity of an energy resource. **Net useful energy** can be defined as the total energy produced during the lifetime of an entire energy system minus the energy used, lost, and wasted in making this energy available.

$$\begin{aligned} \text{net useful} \\ \text{energy} \end{aligned} = \begin{aligned} &\text{total energy produced during the} \\ &\text{lifetime of an entire energy system} \end{aligned}$$

$$- \begin{aligned} &\text{energy used, lost, and wasted} \\ &\text{in making this energy available} \end{aligned}$$

For convenience in comparing different energy

systems, **net useful energy** can also be expressed as the ratio of these two factors:

$$\text{net useful energy} = \frac{\text{total energy produced during the lifetime of an entire energy system}}{\text{energy used, lost, and wasted in making this energy available}}$$

The higher this ratio, the greater the net useful energy yield. When this ratio is less than 1, there is a net energy loss over the lifetime of the system.

Net useful energy is like net profit. If you have a business with a total income of $100,000 and annual operating costs of $90,000, your net profit is only $10,000 a year. With operating expenses of $110,000 a year, you would have a net loss of $10,000 a year.

Table 16-8 lists the estimated net useful energy ratios for various energy alternatives that can be used to provide space heating, high-temperature heat for industrial processes, and gaseous or liquid fuels for propelling vehicles. Although refined crude oil burned as fuel oil for space heating or as gasoline in vehicles has a fairly high net useful energy, these yields should decline in the future. Some new high net energy yield deposits of fossil fuels like those in the Middle East may be found in relatively unexplored parts of the world. But in most MDCs it is reasonable to assume that most new large high net energy yield deposits will have lower net useful energy yields. For example, only 19 large oil fields (10 million barrels or more) have been found in the United States during the past 100 years, despite intensive exploration.

To find new oil and natural gas, energy companies must drill deeper into the earth, tap dilute deposits, and extract deposits in remote and hostile areas like the Arctic—far from where the energy is to be used. This means that more money and more of the world's high-quality fossil fuel energy sources are being used to find, deliver, and process new fossil fuel deposits with lower net useful energy yields. The Department of Energy projects that by the year 2000 as much as two-thirds of the primary energy consumption could be used to provide Americans with useful energy. If this happens, there will be less useful energy available from identified reserves and unidentified deposits of nonrenewable energy sources like oil, natural gas, and coal than present estimates indicate (Section 16-3).

Limitations of Thermodynamic Analysis of Energy Alternatives The crude net useful energy estimates in Table 16-8 are not the final word. Some ratios have been disputed because of the complex-

Table 16-8 Net Useful Energy Ratios for Various Energy Systems

System	Net Useful Energy Ratio
Space Heating	
Passive solar	5.8
Natural gas	4.9
Oil	4.5
Active solar	1.9
Coal gasification	1.5
Electric resistance heating (coal-fired plant)	0.40
Electric resistance heating (natural-gas-fired plant)	0.35
Electric resistance heating (nuclear plant)	0.25
High-Temperature Industrial Heat	
Surface-mined coal	28.2
Underground-mined coal	25.8
Natural gas	4.9
Oil	4.7
Coal gasification	1.5
Direct solar (highly concentrated by mirrors, heliostats, or other devices)	0.9
Transportation	
Natural gas	4.9
Gasoline (refined crude oil)	4.1
Biofuel (ethyl alcohol)	1.9
Coal liquefaction	1.4
Oil shale	1.2

Sources: Colorado Energy Research Institute, *Net Energy Analysis: An Energy Balance Study of Fossil Fuel Resources.* 1976; and Howard T. Odum and Elisabeth C. Odum, *Energy Basis for Man and Nature,* 3rd ed., 1981.

ity of such calculations and because all the energy inputs in some systems are not known. Moreover, a consistent set of guidelines has not always been used to determine what energy inputs are to be included or omitted, and there has been disagreement over estimates of some of the energy inputs. As net useful energy analysis improves in coming years, many of these difficulties and conflicting claims should be resolved.

Despite their usefulness in evaluating energy alternatives, first- and second-law energy efficiencies and net useful energy analysis may not always match up with what people and governments consider to be the most efficient social use of an energy resource. For example, homeowners may find electric space heating so convenient and maintenance free (at least at the home site) that they will exert political pressure to maintain artificially low prices for electricity and to have the losses made up by government subsidies, ultimately paid by all taxpayers, including those who do not use electric heating.

Another factor not included in such analyses is the *time* individuals are willing to spend to save money and improve energy efficiency. For example, it takes more time to load, adjust, check periodically, and clean a wood-burning stove than merely to move the dial on a home furnace thermostat. Even more time is required by those who cut and haul their own supply of wood. Similarly, an owner of a typical passive solar energy system must take more time to (1) close insulated shades at night to prevent excessive heat loss and open them early in the morning to capture the sun's heat, (2) close insulated shades, open doors, or open windows throughout the day to prevent excessive heating, and (3) keep windows clean to ensure maximum solar gain. Individuals who get their electricity from panels of solar photovoltaic cells on the roof must also be willing to invest some time to periodically check battery water and levels of electricity stored in batteries and to be sure that the panels of solar cells are clean enough to maximize solar input.

16-5   Evaluating Energy Resources

Questions to Ask   In trying to determine which mix of energy alternatives might provide primary energy for the future, we must think and plan in three time frames that cover the 50-year period normally needed to develop and phase in new energy sources: the *short term* (1986 to 1996), the *intermediate term* (1996 to 2006), and the *long term* (2006 to 2036).

The first step is to decide how much we need of what kinds of primary energy (low-temperature heat, high-temperature heat, electricity, liquid fuels for transportation, etc.). Then we project the mix of energy alternatives—including improving energy efficiency—that can provide the necessary energy services at the lowest lifetime cost and with acceptable environmental impacts. This means that for each energy alternative, we need to know (1) total estimated supply available in each time frame, (2) estimated net useful energy yield, (3) projected costs for development and lifetime use, and (4) potential environmental impacts. Since the 1973 oil embargo, there have been major efforts to gather such information to help nations develop their long-term energy strategies.

Environmental Impact of Energy Alternatives

Using any form of energy or nonrenewable metal or mineral resource has harmful impacts on the environment (Figure 15-4). The faster the rate of energy use or flow, the greater the impact. This is why energy use is directly or indirectly responsible for most land disruption (Chapter 12 and Section 14-3), water pollution (Chapter 22), and air pollution (Chapter 21). For example, nearly 80 percent of all U.S. air pollution is caused by fuel combustion in cars, furnaces, industries, and power plants.

Choosing any energy option or mix of options involves making choices and tradeoffs between several potential environmental impacts. The problem is to choose an option that provides the least environmental damage at an acceptable price, assuming that all other factors are equal.

Hard Versus Soft Paths to a New Energy Era

Because the supply of nonrenewable conventional crude oil is being depleted rapidly, most analysts argue that a transition to dependence on other primary energy resources will be necessary over the next 50 years. Two approaches for providing projected primary energy needs and services in the future are the *hard path* and the *soft path*.

Proponents of the *hard-path energy strategy* emphasize using human ingenuity and tax breaks and other forms of government (taxpayer) subsidies for energy companies to increase supplies of nonrenewable petroleum (conventional and heavy oils), natural gas (conventional and unconventional), coal, and uranium. Emphasis is also placed on providing an increasing amount of electricity for future primary energy needs by building a number of huge, complex, centralized coal-burning and nuclear fission electric power plants between 1986 and 2020. After 2020 there would be a shift from conventional nuclear fission to breeder nuclear fission power plants, which could synthesize some of their nuclear fuel and thus prolong uranium supplies for at least 1,000 years (Section 18-3). After 2050 there would be a gradual shift to almost complete dependence on centralized nuclear fusion electric power plants, if this energy alternative should ever prove to be technologically, economically, and environmentally acceptable (Section 18-4). The political and economic power concentrated in the large energy companies is presently being used to persuade the United States to follow the hard path, to which most annual federal energy research and development expenditures are now devoted.

Proponents of the *soft-path energy strategy*, such as physicist and energy expert Amory Lovins, recognize that most people do not want kilowatt-hours, kilocalories, or barrels of oil. Instead they want the services that energy provides, such as comfortably heated and cooled homes and buildings, reliable transportation, and high-quality manufactured goods—all as cheaply as possible. Lovins and others argue that the quickest, cheapest, and most energy-efficient way to meet projected energy services is through a combination of improving energy efficiency (Section 19-7), decreasing the use of nonrenewable oil, coal, and natural gas, phasing out nuclear power because it is uneconomic, unsafe, and unnecessary (Chapter 18), and increasing the use of a mix of *renewable* direct and indirect solar energy resources.

With the soft path, most *space heating and hot water* for houses and commercial buildings would be provided by a combination of passive and active solar systems, superinsulation, efficient natural gas furnaces, and efficient wood stoves equipped with catalytic converters to reduce air pollution. *Electricity* would be provided by a combination of existing coal-fired plants equipped with the latest air pollution control equipment (Section 21-7), cogeneration at industrial plants, large farms of wind turbines (Section 19-4), solar photovoltaic cells (Section 19-1), and recommissioning of abandoned small hydroelectric power plants (Section 19-2). *High-temperature heat* for industrial processes would be provided by boilers fired by natural gas, forestry wastes, and municipal wastes. Each boiler would be equipped with air pollution control equipment, and waste heat would be used to produce electricity (cogeneration).

Fuel for transportation would be provided by gasoline, biofuels such as alcohols and pyrolysis oils (Table 16-1 and Section 19-5) derived from forestry and farm wastes and from energy crops grown on otherwise unproductive land, and possibly renewable derived hydrogen gas, if efficient and affordable ways of decomposing water with sunlight are developed (Section 19-6). Because they are too costly, energy inefficient, and environmentally harmful compared to other alternatives, little, if any, emphasis would be placed on using *solar hard technologies* such as biomass plantations of fuel crops grown on land that could be used for food crops (Section 19-5), and electricity produced by large solar-powered satellites, solar power plants (Section 19-1), ocean thermal gradient power plants (Section 19-3), and large hydroelectric plants (Section 19-2).

Amory Lovins argues that by 2000, using the soft-path approach, in the United States would (1) cut energy waste in half and provide the same or higher average standard of living, (2) extend the reserves of conventional fossil fuels several dec-ades, because doubling the efficiency of use of a primary energy resource such as oil has the same effect as doubling oil reserves, (3) eliminate or sharply reduce the need for imported oil and natural gas, (4) eliminate or sharply reduce the need to build additional electric power plants of any type, and allow the phaseout of existing nuclear power plants, (5) buy time to phase in a diverse and flexible array of decentralized, renewable primary energy resources depending on local availability of direct sunlight, wind, biomass resources, and falling and flowing water, (6) decrease the overall environmental impacts of primary energy use, (7) be cheaper and quicker to implement than the hard path because direct and indirect solar technologies and ways to improve energy efficiency are already available at affordable prices, require no major technological breakthroughs such as nuclear fusion, and generally can be installed in days, weeks, or months rather than the 7 to 15 years it takes to build a coal-fired or nuclear power plant, and (8) give individuals and communities more control over how they wish to provide desired energy services based on locally available renewable resources.

Lovins and other energy experts also argue that the trend toward large, centralized electric power plants and other energy facilities threatens national security by making the entire U.S. energy system vulnerable to a limited nuclear strike. For example, the detonation of a single, well-placed 1-megaton nuclear bomb at about 500 kilometers (310 miles) above the United States could release an *electromagnetic pulse* (EMP)—a microsecond burst of electromagnetic energy—100 times more powerful than a lightning bolt. Such a high-altitude detonation might not kill anyone directly, but its EMP could overload and burn out every unprotected electronic circuit in the United States, including radios and television sets, computers and computer-controlled equipment, vehicle engines, the electronic control facilities at all power plants and those controlling the flow of fuels through pipelines, and all telephone and electronic communications systems. Electronic equipment can be protected from an electromagnetic pulse. But such protection is costly and is being installed gradually, only for vital military electronic control, communications, and weapons systems.

Lovins and other experts argue that the United States can strengthen its national security and reduce vulnerability to cutoffs of imported oil and to destruction by EMP or other form of nuclear attack of the entire power grid and electronic control and communications equipment by greatly increasing the use of more dispersed, decentralized, small-scale renewable energy resources.

An increasing number of individuals and communities in the United States are shifting to the soft path. But Lovins contends that a more rapid spread of this approach is being hindered by government (taxpayer) subsidies of the hard-path approach, outdated building codes that discourage energy conservation and sometimes require unnecessary backup conventional heating systems, inadequate access to capital for development of solar energy resources, and the false belief that it will be a long time before solar energy can provide a significant fraction of U.S. primary energy.

By 1984 about 18 percent of all primary energy used in the world and 9 percent of that used in the United States came from *renewable* solar energy resources—mostly biomass (wood, dung, and crop residues) and hydropower (Figures 16-2 and 16-4). It is projected that by 2000 direct and indirect renewable solar energy can provide 26 percent of the world's primary energy and 25 percent of U.S. primary energy (Figures 16-2 and 16-4). Several energy analysts go further and project that with a massive development program, direct and indirect solar energy could provide 30 to 40 percent of the world's primary energy by 2000 and 75 percent by 2025.

Next we will use the energy concepts discussed in this chapter to evaluate the various nonrenewable (Chapters 17 and 18) and renewable (Chapter 19) energy alternatives for the world and for the United States.

A country that runs on energy cannot afford to waste it.
Bruce Hannon

Discussion Topics

1. Try to trace your own direct and indirect energy consumption each day to see why it probably averages 230,000 kilocalories per day (Figure 16-3).

2. Explain why most energy analysts urge that improving energy efficiency should form the basis of any individual, corporate, or national energy plan. Does it form a significant portion of your personal energy plan or life-style? Why or why not?

3. Put the following forms of energy in order of increasing energy quality: heat from nuclear fission, normal sunlight, oil shale, air at 500°C (932°F).

4. Explain how using a gas-powered chain saw to cut wood for burning in a wood stove could use more energy than that available from burning the wood. Consider materials used to make the chain saw, its fuel, and periodic repair and transportation.

5. You are about to build a house. What energy supply (oil, gas, coal, or other) would you use for space heating, cooking food, refrigerating food, and heating hot water? Consider long-term economic and environmental impact factors.

17

Nonrenewable Energy Resources: Fossil Fuels and Geothermal Energy

In retrospect, the world may be heavily indebted to OPEC for having raised the price of oil.

Lester R. Brown

17-1 Conventional and Unconventional Oil

Conventional Crude Oil **Crude oil** or **petroleum**, is a gooey, dark greenish-brown, foul-smelling liquid; hydrocarbon compounds make up about 90 to 95 percent of its weight, and about 5 percent by weight is in the form of oxygen, sulfur, and nitrogen compounds. Typically, deposits of crude oil and natural gas are trapped together deep underground, beneath a dome of sedimentary rock such as sandstone and shale, with the natural gas lying above the crude oil. Normally the crude oil is dispersed throughout the pores and cracks of the underground sandstone rock formation, much like a sponge filled with water. If there is enough pressure from water and natural gas under the dome of rock to force some of this crude oil to the surface when the well is drilled, we say that *primary recovery* has occurred. However, such oil wells, called gushers, are relatively rare.

Since oil floats on water, water is injected to force out some of the remaining crude oil when the initial pressure of a gusher has been released or when the well has too little pressure for primary recovery. This is known as *secondary recovery*. Typically the combination of primary and secondary recovery removes only about one-third of the crude oil in a well. This means that two barrels of *heavy oil* with the consistency of asphalt is left in a typical well for each barrel removed by primary and secondary recovery.

As oil prices rise, it may become economical to remove about 10 percent of the heavy oil remaining in a petroleum deposit by *enhanced oil recovery*

processes such as **(1)** forcing steam into the well to soften the heavy oil so it can be pumped to the surface, **(2)** igniting some of the heavy oil to increase the flow rate of the surrounding oil so that it can be pumped to the surface, and **(3)** injecting a chemical such as carbon dioxide or alcohol into the well to dissolve the oil, pumping the mixture to the surface, and extracting the heavy oil. However, these processes are expensive and take energy equivalent to that in one barrel of oil to pump each three barrels to the surface, thus reducing net useful energy yield. Additional energy is needed to increase the flow rate and remove impurities such as sulfur and nitrogen before the heavy oil can be sent via pipeline to an oil refinery. Recoverable heavy oil from known U.S. crude oil reserves could supply U.S. oil needs for only about 7 years at 1984 usage rates.

Figure 17-1 shows the locations of the major petroleum and natural gas fields in the United States. Note that most are in Texas, Louisiana, and Oklahoma and in the outer continental shelf of the Pacific, Atlantic, Gulf, and Alaskan coasts. The federal government leases areas of the outer continental shelf by selling to private companies the rights to explore, develop, and produce crude oil and natural gas. If fuel deposits are found, the buyer of the lease pays the government a royalty based on the amount produced.

Crude oil extracted by primary or secondary recovery from these deposits is *refined* in gigantic distillation columns to separate it into various components, which boil at different temperatures (Figure 17-2). Some of the chemicals extracted from crude oil and natural gas are sent to petrochemical plants for use as raw materials in the manufacture of most industrial chemicals, fertilizers, pesticides, plastics, synthetic fibers, and medicines. About 3 percent of all the fossil fuels used in the world and 7 percent of those used in the United States are used to produce these *petrochemicals*. From Figure 17-2 you can see why most product prices rise when crude oil and natural gas prices rise.

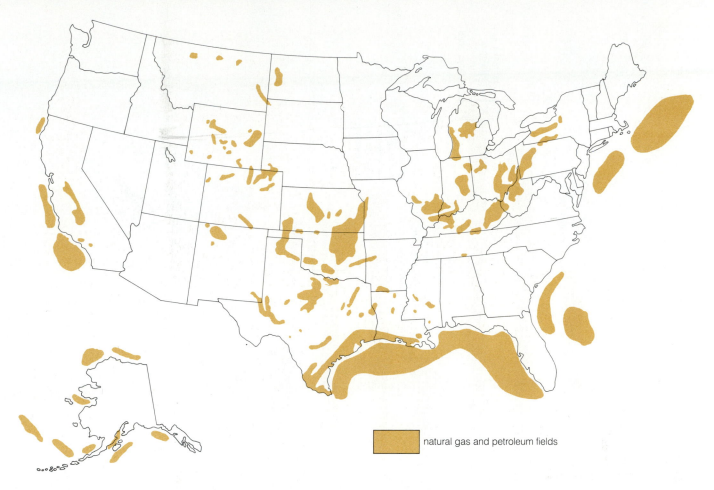

Figure 17-1 Major deposits of natural gas and petroleum in the United States. (Source: Council on Environmental Quality.)

Most of these petrochemicals can also be produced from coal or from the fermentation of plants (biomass), but to convert to a biomass basis the entire chemical industry in an industrialized nation would be an expensive, decades-long process. Presently only Brazil plans to ferment biomass to produce most of its petrochemicals.

Conventional crude oil has been widely used because it **(1)** has been relatively cheap until recently, **(2)** has a high net useful energy yield (Table 16-8), although this is expected to decrease as less accessible and more remote deposits are used, **(3)** can be transported easily within and between nations, and **(4)** is a versatile fuel that can be burned to propel vehicles, provide low-temperature heating of water and buildings, and high-temperature heat for industrial processes and production of electricity.

Its major disadvantages are as follows: **(1)** affordable supplies may be depleted within 40 to 80 years (Section 16-3); **(2)** potential air pollutants such as sulfur oxides, nitrogen oxides, and hydrocarbons are produced when conventional crude oil is burned without adequate pollution control devices (Chapter 21); **(3)** its burning releases carbon dioxide gas, which could alter global climate and gradually raise sea levels (Section 21-5); **(4)** its use can cause water pollution from oil spills (Section 22-6) and contamination of underground water by brine solution reinjected into wells (Section 22-5).

Heavy Oils from Oil Shale and Tar Sands Oil **shale** is a marlstone sedimentary rock consisting mostly of mud and clay mixed with limestone. These deposits also contain varying amounts of a rubbery, solid organic material called **kerogen**, a mixture of heavy hydrocarbon compounds with a high content of sulfur, nitrogen, oxygen, and other impurities. Typically the shale rock is removed by mining, then crushed and heated to about 460°C (900°F) in a container called a *retort* to vaporize the solid kerogen. The vapor from this distillation process is condensed to yield a slow-flowing, dark brown *heavy oil* called **shale oil**. Before it can be piped to an oil refinery, shale oil must be upgraded by increasing its flow rate and hydro-

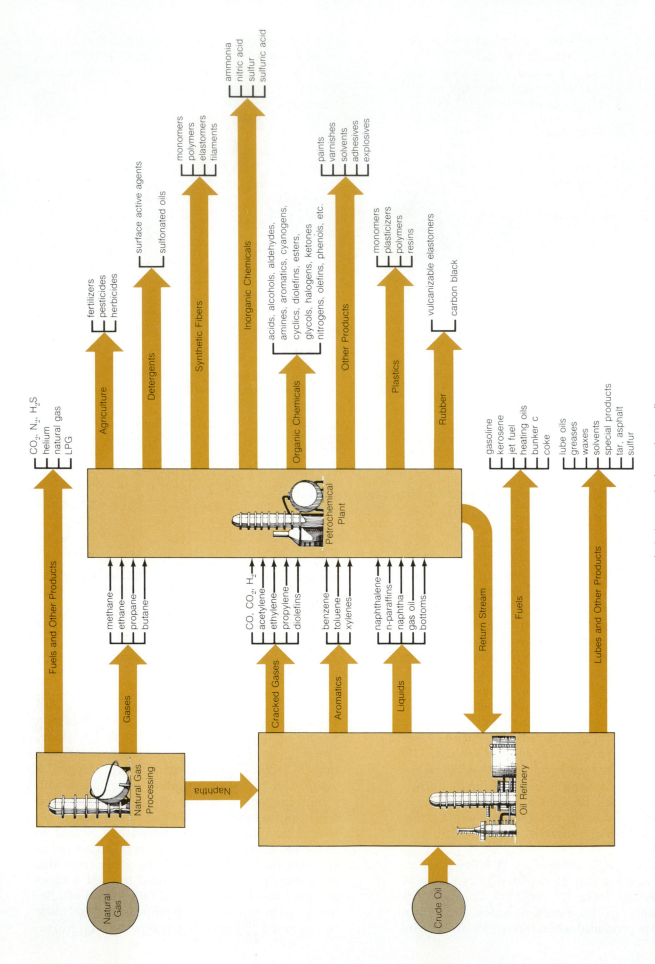

Figure 17-2 Crude oil and natural gas are processed and refined to produce fuels that can be burned, as well as raw materials for manufacturing most of the products in modern society.

gen content and by removing impurities such as sulfur, nitrogen, and metal compounds.

The world's largest known deposits of oil-bearing shale are in the United States, with significant amounts also in Canada, China, and the Soviet Union. It is estimated that the potentially recoverable heavy oil from oil shale in the United States could supply the nation with crude oil for 44 years if consumption remains at 1984 levels, and for 32 years if annual crude oil consumption rises by only 2 percent a year.

However, the estimated cost of making this oil available was about $55 a barrel in early 1985—more than twice the average world price of conventional crude oil. It takes the energy equivalent of about one barrel of conventional crude oil and two to six barrels of water that end up contaminated to produce one barrel of shale oil. This means that without government subsidies, shale oil may always be 1.5 to 2 times more costly than conventional crude oil. This also results in a low net useful energy yield compared to conventional crude oil (Table 16-8). Other problems that may limit shale oil production include: **(1)** the need for large amounts of water for processing, **(2)** the release of potential air pollutants and some possible cancer-causing substances during processing and when burned without adequate pollution control devices (Chapter 21), **(3)** the release of carbon dioxide when processed and burned, **(4)** possible contamination of nearby surface water with dissolved solids (salinity), possible cancer-causing substances, and toxic metal compounds from processed shale rock and possible contamination of groundwater if processed underground (Section 22-5), and **(5)** land disruption from the disposal of large volumes of processed shale rock, which breaks up and expands when heated (somewhat like popcorn).

One way to avoid some of these problems is to remove the shale oil in place (*in situ*) by breaking up the rock with hydraulic pressure or liquid explosives, retorting and distilling the rock while it is still underground, and pumping up the shale oil. By 1985, no oil shale extraction and processing technology had reached the commercial stage. In addition, most pilot-plant projects of major oil companies in the United States have been abandoned because both surface mining and *in situ* methods proved to be too expensive even with large government subsidies.

Tar sands (or oil sands) are swamplike deposits of a mixture of fine clay, sand, water, and variable amounts of a black, high-sulfur, tarlike heavy oil known as **bitumen**. Typically, the tar sand is removed by surface mining and heated with steam at high pressure to make the bitumen fluid enough to float to the top. Like heavy oil from oil shale, the bitumen must be purified and upgraded to synthetic crude oil before being refined. So far it is not technically or economically feasible to remove deeper deposits by underground mining or by extracting the bitumen *in situ*.

The world's largest known deposits of tar sands lie in a cold, desolate area along the Athabasca River in northern Alberta, Canada. Other fairly large deposits are in Venezuela, Colombia, and the Soviet Union. There are smaller deposits in the United States, with 70 percent of these located in Utah. For several years two experimental plants have been supplying about 10 percent of Canada's annual oil demand by extracting and processing heavy oil from Athabasca tar sands at a cost of about $25 to $30 a barrel—equivalent to the average world price for conventional crude oil in early 1985. However, it is estimated that using the same process to produce heavy oil from U.S. deposits of tar sands would cost as much as $45 a barrel. Recoverable deposits of heavy oil from Canadian tar sands could supply that nation's oil needs for about 36 years at 1984 consumption rates. Because domestic reserves of conventional crude oil may be depleted by 1992, Canada hopes to obtain a significant fraction of its crude oil needs from tar sands by 1990.

Large-scale production of synthetic crude oil from tar sands, however, is beset with problems. The net useful energy yield is low. It takes the energy equivalent of one barrel of conventional oil to produce the steam and electricity needed to liquefy and remove three barrels of heavy oil. More energy is needed to remove sulfur impurities and to upgrade the bitumen to synthetic crude oil before it can be sent to an oil refinery. Tar sands are also so abrasive that the teeth of gigantic bucket-wheel excavators half a city block long and as high as a 10-story building are worn out every 4 to 8 hours. In addition, the gooey sands stick to almost everything, clog extraction equipment and vehicles, and slowly dissolve natural rubber in tires, conveyor belts, and machinery parts. Strip mining the tar sands produces even more waste per unit of heavy oil than is produced in mining oil shale. Other problems include the need for large quantities of water for processing, and the release of potential air and water pollutants similar to those produced when oil shale is processed and burned.

17-2 Conventional and Unconventional Natural Gas

Conventional Natural Gas **Natural gas** consists of 50 to 90 percent methane (CH_4) and small amounts of other more complex hydrocarbon compounds such as propane (C_3H_8) and butane (C_4H_{10}). Natural gas lies above underground deposits of crude oil and is found by itself in other underground deposits. When a natural gas deposit is tapped, propane and butane gases are liquefied and removed as *liquefied petroleum gas* (*LPG*) before the remaining gas (mostly methane) is dried, cleaned of hydrogen sulfide and other impurities, and pumped into pressurized pipelines for distribution over land. LPG is stored in pressurized tanks for use in mostly rural areas not served by natural gas pipelines. Very low temperature can be used to convert natural gas to *liquefied natural gas* (*LNG*), which can be transported by sea in specially designed tanker ships.

At the 1984 usage rate, the world's known reserves of natural gas will last until around 2033 and those in the United States until 1993. America's largest known deposits of natural gas lie in Alaska's Prudhoe Bay, thousands of kilometers from natural gas consumers in the lower 48 states. Geologists estimate that up to eight times as much natural gas awaits discovery in the North Slope area. In 1977 the U.S Congress and the Canadian government approved construction of a pipeline to bring this natural gas to San Francisco and Chicago. At a cost that could exceed $43 billion, this 7,725-kilometer (4,800-mile) pipeline will be the most expensive privately financed construction project in history, costing far more than the $9 billion spent on the 1,270-kilometer (794-mile) Alaskan oil pipeline built during the 1970s. If everything goes as planned, the natural gas pipeline could begin delivery to the United States by 1986.

Conventional natural gas is the hottest-burning and cleanest-burning fossil fuel. Its other major advantages include: **(1)** versatility of use, **(2)** ease of transportation in pipelines over land, **(3)** relatively cheap prices (which, however, should rise as price controls are removed), and **(4)** high net useful energy yield. Net useful energy yield is expected to decrease as less accessible and more remote deposits are used, however, and it is reduced by about one-fourth when natural gas is converted to LNG.

Other major disadvantages are **(1)** affordable supplies of conventional natural gas may be depleted within 40 to 100 years (Section 16-3); **(2)** it is difficult, expensive, and dangerous to transport by tanker as volatile, unstable LNG; **(3)** an accidental explosion of the LNG in a tanker could create a massive fireball that would burn everything within its volume, generating radiant energy that would start fires and cause third-degree burns 1.6 to 3.2 kilometers (1 to 2 miles) away; and **(4)** its burning produces carbon dioxide, which could alter global climate and gradually raise sea levels.

Unconventional Natural Gas As conventional natural gas prices rise, some analysts believe that it may become economical to drill *deep* into the earth and extract and process *unconventional natural gas* from **(1)** concrete-hard deep geologic formations of *tight sands*, **(2)** *geopressurized zones* containing deposits of hot water under such high pressure that large quantities of natural gas are dissolved in the water, **(3)** *coal seams*, and **(4)** deposits of *Devonian shale rock*.

Experts agree that there are large deposits of unconventional natural gas but disagree over whether, in practice, the gas can be recovered at affordable prices. If a reasonable amount of this natural gas can be recovered, world supplies of natural gas would be extended for several hundred to a thousand years, allowing natural gas to become the most widely used fuel for space heating, industrial processes, producing electricity, and transportation. However, such a major increase in the annual usage rate of unconventional natural gas could deplete these supplies within a century or two.

Unconventional natural gas has the same advantages and disadvantages as conventional natural gas except that it is more difficult and expensive to recover and process, which means that its net useful energy yield would be only moderate. Furthermore, the deep-drilling technology needed to remove this resource is not fully developed.

17-3 Coal

Mining Conventional Coal Coal has a number of advantages: **(1)** it is the most abundant conventional fossil fuel in the world and in the United States; **(2)** it has a high net useful energy yield for producing high-temperature heat for industrial processes and for the generation of electricity; and **(3)** it is the cheapest way to produce high-temperature heat and electricity in nations with adequate coal supplies. However, three major properties limit its use: It's dangerous to mine, expensive to move, and dirty to burn.

Coal is a solid containing 55 to 90 percent carbon and small amounts of hydrogen, nitrogen, and sulfur compounds. It is formed from fossil remains over millions of years in several stages, with each type of coal having a different carbon content and fuel quality (Figure 17-3). Anthracite releases the largest quantity of heat per unit of weight burned,

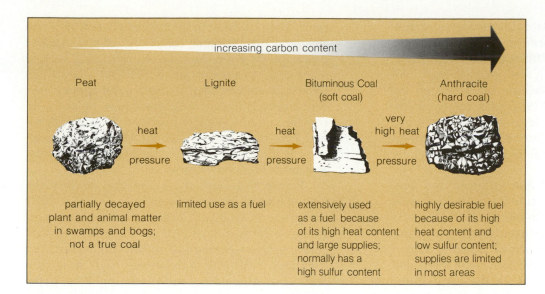

Figure 17-3 Stages in the formation of different types of coal over millions of years.

increasing carbon content

Peat

Lignite

Bituminous Coal (soft coal)

Anthracite (hard coal)

heat

pressure

heat

pressure

very high heat

pressure

partially decayed plant and animal matter in swamps and bogs; not a true coal

limited use as a fuel

extensively used as a fuel because of its high heat content and large supplies; normally has a high sulfur content

highly desirable fuel because of its high heat content and low sulfur content; supplies are limited in most areas

followed by bituminous coal. Anthracite, however, is not as common as the other types of coal and is usually more expensive. Although they have less heat content, lignite and subbituminous coal normally have the lowest sulfur content and thus produce less air pollution from sulfur dioxide per unit of weight burned than anthracite and bituminous coal. Peat, which has a relatively low heat content, is burned for fuel in areas where supplies are plentiful.

Some deposits of coal are found deep underground and must be removed by *underground or deep-shaft mining*. Once an underground mine has been abandoned, surface water entering the mine washes acidic impurities (iron pyrites) from the remaining coal into nearby streams, damaging aquatic plant and animal life. Such acid mine drainage can be controlled by filling sinkholes and rerouting gulleys to prevent surface water from entering abandoned mines or by treating water draining from mines with crushed limestone to neutralize the acidity. When an abandoned mine shaft collapses, a depression in the surface of the earth above the mine, known as *subsidence*, often occurs.

Underground mining is expensive and dangerous because of injuries and deaths from cave-ins and explosions—a single spark can ignite underground air laden with coal dust or methane gas. Between 1900 and 1984 underground mining in the United States killed more than 100,000 miners, permanently disabled at least 1 million miners, and caused at least 250,000 retired miners to spend their last years gasping for breath from black lung disease.

U.S. mine safety laws require ventilating systems, water sprayers, and protective masks to reduce the quantity of coal dust reaching worker's lungs. However, worker safety could be improved significantly by stricter enforcement of these laws and by enactment of tougher new laws requiring a higher level of ventilation, improved control and removal of explosive methane gas and coal dust, and methane detection equipment for all mine shafts.

When the overburden or amount of material on top of a vein of coal is less than 100 meters (328 feet) thick, it can be removed by various types of *surface mining*, described in Section 15-1. This type of mining removes at least 90 percent of the coal in a deposit, costs less per ton of coal removed than underground mining (including the cost of restoring most types of disturbed land), and is less hazardous for miners than underground mining. Without adequate land reclamation, however, surface mining can have a devastating impact on land and also on water supplies and wildlife. By 1983 surface mining accounted for about 62 percent of the coal extracted in the United States and was increasing dramatically in the western half of the country.

To help control the surface mining of domestic coal, the Surface Mining Control and Reclamation Act of 1977 was enacted. According to this act **(1)** surface-mined land must be restored so that it can be used for the same purposes it served before mining; **(2)** surface mining is banned on some prime agricultural lands in the West, and farmers and ranchers can veto mining under their lands, even though they do not own the mineral rights; **(3)** mining companies must minimize the effects of their activities on local watersheds and water quality by using the best available technology, and they must prevent acid from entering local streams and groundwater; **(4)** money from a $4.1 billion fund, financed by a fee on each ton of coal mined, will be used to restore surface-mined land not reclaimed before 1977; and **(5)** responsibility for enforcement of the law is delegated to the states, but the Department of the Interior has enforcement power where the states fail to act and on federally owned lands.

If strictly interpreted and enforced and adequately funded, this law could go a long way in protecting valuable ecosystems. Since the law was passed, however, there has been growing pressure from the coal industry (much of which is owned by the major oil companies), to have it weakened or declared unconstitutional. In the early 1980s, the Reagan administration withdrew federal regulations designed to protect prime farmland from surface mining. The federal inspection and enforcement staff was also cut by 70 percent, thus reducing the effectiveness of the federal government in enforcing the act when states fail to act. To cite only one of numerous examples of laxity in enforcement, in 1983 Utah state officials were carrying out fewer than half the inspections required by law.

Uses of Coal Coal is not as versatile as oil and natural gas and cannot be burned conveniently to propel cars and trucks. About 60 and 70 percent of the coal extracted in the world and in the United States, respectively, is burned in boilers to produce steam used to generate electrical power—providing about 85 and 54 percent of the electricity used in the world and in the United States.

Most of the remaining 40 percent of the coal extracted each year either is burned in boilers to produce steam used in various manufacturing processes or is heated in airtight ovens to produce *coke* (a hard mass of almost pure carbon) for use in converting iron ore to iron and steel. One of the major by-products of the process is *coal tar*, which is used for roofing and road surfacing and in various drugs and dyes.

World and U.S. Coal Supplies World supplies of coal should last from about 100 to 900 years, depending on discovery of unidentified resources and rate of use (Section 16-3). The Soviet Union has an estimated 56 percent of the world's coal reserves, the United States about 19 percent, and China about 8 percent.

Identified coal reserves in the United States are projected to last about 300 years at 1984 usage rates. The rapid rise in oil prices between 1973 and 1980 and the extremely high cost of new nuclear power plants have caused utilities to shift back to using coal. As a result, since 1979 coal usage in the United States has been increasing by about 6 percent a year. If this increase continues, proven U.S. reserves will be depleted around 2032. Unidentified U.S. coal resources could extend these supplies at such high use rates for perhaps 100 years, at a much higher average cost.

Figure 17-4 maps the major coal fields in the United States, located primarily in 17 states. Note that there is relatively little anthracite, the most desirable form of coal. About 45 percent of U.S. coal reserves—containing mostly high-sulfur, bituminous coal with a relatively high heat content—are found east of the Mississippi River in the Appalachian region, particularly in Kentucky, West Virginia, Pennsylvania, Ohio, and Illinois. Most of this coal must be removed by underground mining and has such a high sulfur content (typically 2 to 3.5 percent) that it cannot be burned without using scrubbers or other devices to reduce sulfur dioxide emissions. For these reasons, the percentage of all U.S. coal extracted from fields east of the Mississippi River fell from about 93 percent to 66 percent between 1970 and 1984.

About 55 percent of U.S. coal reserves and three-fourths of the nation's surface-minable coal are found west of the Mississippi River, primarily in North Dakota, South Dakota, Montana, Wyoming, Colorado, New Mexico, Arizona, and Alaska (Figure 17-4). Most of these deposits consist of low-sulfur (typically 0.6 percent) subbituminous and lignite coals with relatively low heating values, which can be surface mined more safely than the underground deposits of bituminous coal found east of the Mississippi. This helps explain why surface mining of western deposits now accounts for about 34 percent of the coal mined in the United States.

However, the heaviest concentration of coal-burning industrial and electric power plants is located east of the Mississippi River—far from these western deposits. When long-distance transportation costs are added, the average cost of coal surface mined in the West and delivered by rail to the East is comparable to that of producing coal by expensive underground mining in the East and delivering it over shorter distances. Slurry pipelines (which transmit a mixture of powdered coal and water) are cheaper than rail for transporting western coal over long distances. However, this method of transportation, along with reclamation of land disturbed by surface mining, is limited by a lack of water in most of the West (Section 10-2). Because about 70 percent of western coal reserves

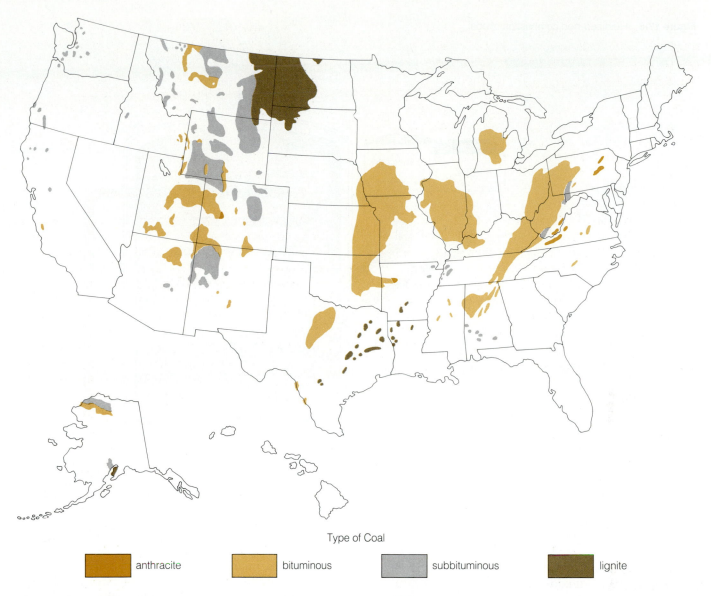

Type of Coal

| | anthracite | | bituminous | | subbituminous | | lignite |

Figure 17-4 Major coal fields in the United States. (Source: Council on Environmental Quality.)

are under federally owned lands, development of these resources will depend primarily on the actions of government agencies and, to a lesser degree, on those of private groups including Native American tribes holding the remaining western coal reserves and water rights.

The future of coal to produce electricity in the United States is uncertain. Primarily because of a declining demand for electricity as a result of energy conservation and economic recession, 75 of the 106 new coal-fired power plants ordered between 1976 and 1984 were canceled.

Air Pollution from Burning Coal Without effective air pollution control devices, emissions of sulfur dioxide, particulate matter, cancer-causing substances, and small amounts of radioactive substances from all U.S. coal-fired electric power plants cause an estimated 10,000 premature deaths, more than 100,000 cases of respiratory disease, and several billion dollars in damage each year to aquatic life, trees, crops, metals, stone, and cement (Sections 21-3 and 21-4). Coal, however, need not be such a dirty fuel to burn. Air pollution emissions from coal-burning power and industrial plants can be sharply reduced by using various pollution control devices (Section 21-7) required by federal law on all U.S. coal-fired plants built since 1978. Because older coal-fired plants are not required to meet these new air pollution control standards, it is estimated that average air pollution control on all coal-burning plants in the United States is only about 50 percent effective, leading to about 5,000 premature deaths a year.

Requiring all U.S. coal-burning plants to have

Figure 17-5 Fluidized-bed combustion of coal.

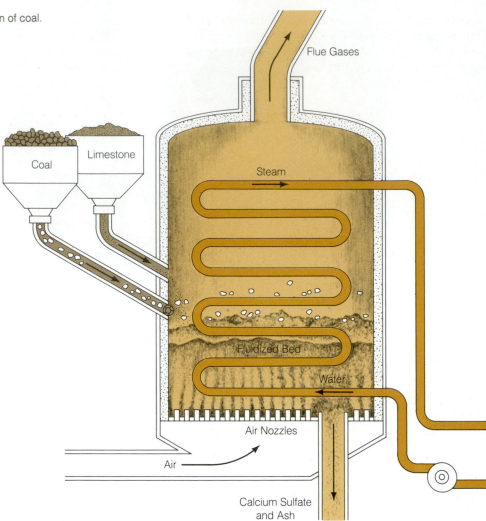

the latest air pollution control devices to remove 95 percent of sulfur dioxide as well as other harmful emissions would reduce the risks from burning coal to about 500 deaths per year for all U.S. coal-burning power plants. But requiring this level of control for *all* harmful air pollutants—not just sulfur dioxide—would probably make it much more expensive to produce electricity in a coal-fired plant than in a nuclear plant with the same output. Another disadvantage of coal is that it produces more carbon dioxide per weight when burned than oil or natural gas. Thus, without expensive pollution controls the increased use of coal could accelerate global climate change (Section 21-5).

Burning Coal More Cleanly and Efficiently   One way to burn coal more efficiently, cleanly, and cheaply is to use **fluidized-bed combustion** (FBC). In FBC a stream of hot air is blown into a boiler from below to suspend a mixture of sand, powdered coal, and crushed limestone. The upward flow of air constantly churns and tumbles this powdered mixture so that it resembles a boiling liquid with the consistency of thin oatmeal—hence the name (Figure 17-5). When the mixture is heated red hot to about 480°C (900°F), the powdered coal is burned very efficiently and the limestone ($CaCO_3$) is converted to calcium oxide (CaO), which reacts with the sulfur dioxide released from the coal to form dry, solid calcium sulfate ($CaSO_4$) or gypsum. This process removes 90 to 98 percent of the sulfur dioxide gas produced during combustion.

It is expected that by the 1990s full-scale FBC boilers will be replacing conventional coal boilers fitted with scrubbers because FBC boilers **(1)** are projected to have lower construction and operating costs, **(2)** can burn a variety of low-grade fuels including rice hulls, heavy oils, wood and wood wastes, urban and industrial trash, sewage sludge, and high-sulfur coal, **(3)** can be retrofitted to conventional boilers, **(4)** are simpler, more reliable, and use less water than scrubbers for sulfur dioxide removal, **(5)** produce fewer nitrogen oxides than conventional coal boilers because the coal is burned at a lower temperature, and **(6)** have a useful by-product, namely solid calcium sulfate, which can be removed and sold as a road subbase, or for cement, soil conditioner, or other uses.

Another approach to burning coal more efficiently and cleanly is *magnetohydrodynamic (MHD) generation*. MHD generation, however, is considered by many to be an engineer's nightmare, and after 25 years of research in the United States, the Soviet Union, and Japan, only a few pilot plants are in operation.

Synfuels from Coal Gasification and Coal Liquefaction

Coal gasification involves converting coal to a gas that can be burned more cleanly as a fuel and, unlike coal, can be transported through a pipeline. In one process coal is converted to coke, which is heated with steam to produce a gaseous mixture of carbon monoxide (CO) and hydrogen (H_2) with a relatively low heating value: H_2O (steam) + C (carbon in coke) $\longrightarrow$ CO + H_2. Because of its low heat content, however, this gas is not worth transporting by pipeline and normally is used only for heating or producing electricity by industries that produce it. A second and more useful process converts coal to synthetic natural gas (SNG), which has a high heating value and can be transported by pipeline.

Coal liquefaction involves converting coal to a liquid hydrocarbon fuel such as methanol or synthetic gasoline. In South Africa a commercial plant has been converting synthetic natural gas made from coal to gasoline and other motor fuels for more than 25 years.

Although converting solid coal to more versatile and cleaner burning synthetic gaseous or liquid *synfuels* is technically possible, it is much more expensive to build a synfuel plant than an equivalent coal-fired power plant fully equipped with air pollution control devices. Other major problems include: **(1)** low net useful energy yield (Table 16-8), **(2)** accelerated depletion of world coal supplies because 30 to 40 percent of the energy content of the coal is lost in the conversion process, **(3)** large water requirements for processing, **(4)** release of larger amounts of carbon dioxide per unit of weight when processed and burned than coal, natural gas, or oil, and **(5)** greater land disruption from surface mining than that from conventional coal because of increased use of coal per unit of energy produced.

Most of these problems (except carbon dioxide emissions, high costs, and low net energy yields) could be avoided or reduced by the underground gasification of coal. The Soviet Union has several underground gasification plants that fuel electric

power plants, and several pilot plants in the United States show promise. However, unless conventional oil and natural gas prices rise significantly, most analysts expect synfuels to play only a minor role in primary energy production until after 2000.

17-4 Geothermal Energy

Nonrenewable Geothermal Energy The decay of radioactive elements deep within the earth generates heat that slowly flows into buried rock formations. Due to intense pressure and lava flow from the molten interior of the earth, some of the earth's geothermal energy escapes through hot springs, geysers, and volcanoes, and some is transferred over thousands to millions of years to normally *nonrenewable* deposits of dry steam, wet steam (a mixture of steam and water droplets), and hot water lying relatively close to the earth's surface.

Geothermal wells can be drilled like oil and natural gas wells to bring this dry steam, wet steam, or hot water to the earth's surface. Although not yet a major component of the world's primary energy budget (Figure 16-2), by 1984 more than 20 countries were tapping such deposits of geothermal energy to produce electricity, to provide low- to moderate-temperature heat for some industrial processes, to heat water in homes and businesses, and to provide space heating. Although nonrenewable, such sources are projected to last for 100 to 200 years in most places. Figure 17-6 shows that most of the accessible and fairly hot geothermal deposits in the United States lie in the western states.

Dry steam deposits are the preferred geothermal resource, but they are also the rarest. Only dry steam wells can be tapped easily and economically at present. This is done by drilling a hole into the reservoir, releasing the superheated steam through a pipe, filtering out solid material, and piping the steam directly to a turbine to generate electricity.

A large natural dry steam well near Larderello, Italy, has been producing electricity since 1904 and is a major source of power for Italy's electric railroads. Two other major dry steam sites are the Matsukawa in Japan and the Geysers steam field, located about 145 kilometers (90 miles) north of San Francisco. The Geysers field has been producing electricity since 1960 more cheaply than fossil fuel and nuclear plants. By 1984 it was supplying more than 2 percent of California's electricity, enough to satisfy the electrical needs of more than a million people, and may supply 25 percent of California's electricity by 1990.

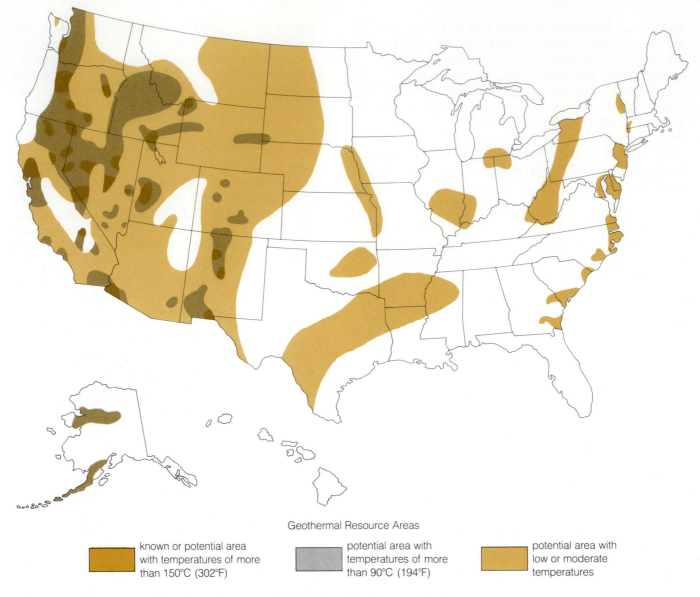

Geothermal Resource Areas

▮	known or potential area with temperatures of more than 150°C (302°F)	▮	potential area with temperatures of more than 90°C (194°F)	▮	potential area with low or moderate temperatures

Figure 17-6   Major deposits of geothermal resources in the United States. (Source: Council on Environmental Quality.)

Underground *wet steam deposits* are more common but are harder and more expensive to convert to electricity. These deposits contain water under such high pressure that its temperature is unusually high (180 to 370°C [356 to 698°F]), unlike water under normal atmospheric pressure, which cannot exceed the boiling point [100°C (212°F)] without being converted to steam. When a geothermal well is drilled to bring this superheated water to the surface, about 10 to 20 percent of the flow flashes into steam because of the decrease in pressure. A centrifugal separator is then used to separate the steam from this mixture of steam and water droplets, and the steam spins a turbine to produce electricity (Figure 17-7). The remaining hot water, which is often high in dissolved salts, and the condensed steam, are usually reinjected

into the earth to prevent buildup of dissolved salts in nearby bodies of water and to reduce subsidence of the ground above the geothermal wells. The largest geothermal electric power plant in the world based on wet steam wells is in Wairakei, New Zealand.

Other wet steam power plants are in operation in Mexico, Japan, and the Soviet Union. Between 1980 and 1984 three demonstration wet steam power plants were built in the United States: two in the Salton Sea area of southern California's Imperial Valley district, and the third at Valles Caldera, New Mexico. Although these plants are producing electricity, drilling problems and corrosion from the salty water have reduced yields, with the energy produced equivalent to that from oil at $40 a barrel.

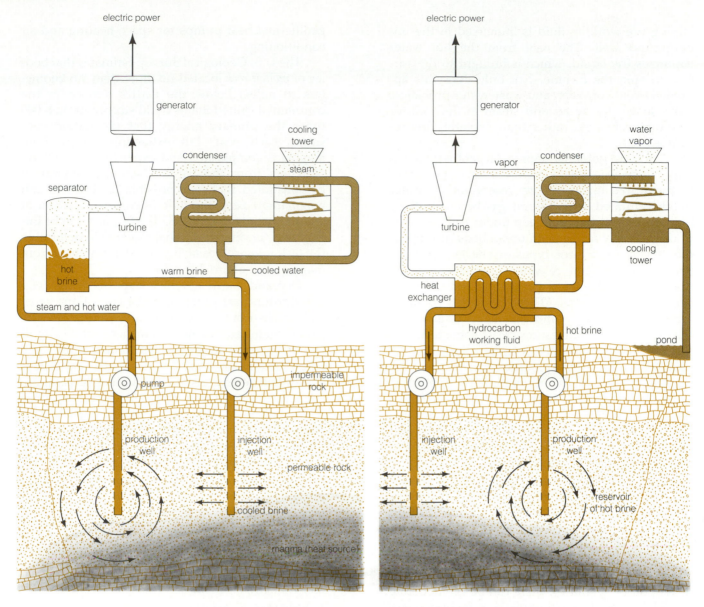

Figure 17-7 Direct-flash (left) and binary-cycle (right) methods for extracting and using geothermal energy to produce electricity.

The third type of nonrenewable geothermal deposit contains *hot water* only, and is even more common than dry steam and wet steam deposits. Almost all the homes, buildings, and food-producing greenhouses in Reykjavik, Iceland, with a population of about 85,000, are heated by hot water drawn from such deep geothermal deposits under the city. In the United States, hot water deposits in more than 180 locations have been used to heat homes in Boise, Idaho, and Klamath Falls, Oregon, and to dry crops and heat farm buildings in South Dakota. If oil prices continue to rise, one expert estimated that just the hot water geothermal resources below southern California's Imperial Valley could produce enough electrical energy to meet the needs of the American Southwest for at least 200 years.

The hot brine water pumped up from such wells can also be used to produce electricity in a *binary-cycle system* (Figure 17-7). The hot brine flows through a heat exchanger containing a working fluid (such as isobutane or a Freon), which boils at a lower temperature than water. This converts the working fluid to a vapor that is used to spin a turbine and then passed through a second heat exchanger, where it is condensed and returned to the first heat exchanger for reuse. The brine is pumped back into the earth to be reheated for future generations.

A demonstration binary-cycle system went into operation in 1984 in the Imperial Valley at Herber, California. The main problem is that the brine corrodes metal parts and clogs pipes. In another method, now being tested, a heat exchanger containing the working fluid is immersed in the underground well. The heat from the hot water vaporizes the liquid, which is brought to the surface to spin the turbine. Not only does this approach avoid corrosion and wastewater problems, but it leaves the water and steam in the well for continual reheating rather than depleting the resource.

A fourth potential source of geothermal energy, *geopressurized zones*, consists of high-temperature, high-pressure reservoirs of water (often saturated with natural gas because of the high pressure), trapped deep under beds of shale or clay, usually far beneath ocean beds. If tapped, they could yield three types of energy: electrical (from high-temperature water), mechanical or hydraulic (from the high pressure), and chemical (from the natural gas). These resources could be tapped by very deep (and expensive) drilling, but the extremely high pressures create some technical problems. At present, this potential geothermal resource is in the early, exploratory phase.

Major advantages of nonrenewable geothermal energy include: **(1)** a 100- to 200-year supply of energy for areas near deposits, **(2)** moderate cost, **(3)** moderate net useful energy yield for large and easily accessible deposits, and **(4)** no production of carbon dioxide.

Two major disadvantages include low overall supply of easily accessible deposits and the impossibility of using this type of energy resource directly to power vehicles. Potentially harmful environmental effects from geothermal energy vary widely from site to site and with the type of geothermal resource being used. Without adequate pollution control, there can be **(1)** moderate to high potential air pollution from hydrogen sulfide, ammonia, radioactive materials, noise, odor, and local climate change, and **(2)** moderate to high water pollution from dissolved solids (salinity) and runoff of various toxic compounds of elements such as boron and mercury. But most experts consider these environmental effects to be less than or at worst about equal to those from fossil fuel and nuclear power plants.

Renewable Geothermal Energy Three potentially *renewable* geothermal energy sources are **(1)** deposits of *molten rock (magma)* at temperatures around 1000°C (1832°F) found in some places not far below the earth's surface, **(2)** *dry hot rock* zones, where magma has penetrated into the earth's crust and is heating subsurface rock to high temperatures, and **(3)** low- to moderate-temperature *warm rock deposits* useful for preheating water and for geothermal heat pumps for space heating and air conditioning.

The U.S. Geological Survey estimates that bodies of *molten rock* located no more than 9.6 kilometers (6 miles) below the earth's surface in the continental United States could supply 800 to 8,000 times the primary energy that the nation consumes each year. But extracting energy from magma is complicated and expensive. In 1981 researchers from the Sandia National Laboratory demonstrated that it is possible to drill through magma and keep the hole open for extraction of geothermal energy. In 1985 the Department of Energy was funding additional research to determine the technological feasibility of extracting heat from magma at an affordable price.

Dry and warm hot rock deposits lying deep underground are potentially the largest and most widely distributed geothermal resource in the United States and in most countries, but they tend to be expensive to locate, tap, and use. By 1985 researchers in the United States and Great Britain had drilled several test wells and had successfully extracted heat, and a demonstration plant was being built in New Mexico. However, before this approach can be developed on a commercial scale, several technical problems must be solved and the economics of the process must be evaluated.

Oil and natural gas will play an important but diminishing role for some time, but how long is less clear. Coal will likely grow in importance, but how much we should burn, considering the serious side effects of its use, is a tough question.

Daniel Deudney and Christopher Flavin

Discussion Topics

1. Explain why you agree or disagree with the following statements.
   a. We can get all the oil we need by extracting and processing the heavy oil left in known oil wells.
   b. We can get all the oil we need by extracting and processing heavy oil from oil shale deposits.
   c. We can get all the oil we need by extracting heavy oil from tar sands.
   d. We can get all the natural gas we need from unconventional sources of natural gas.

2. Why is surface-mined coal gradually replacing coal from deep mines? Do you believe that this is a desirable or undesirable trend? Why? What are the alternatives?

3. Coal-fired power plants in the United States cause at least 5,000 deaths a year, primarily from atmospheric emissions of sulfur oxides and particulate matter. These plants also cause extensive damage to many buildings and to some forests and aquatic systems. Should air pollution emission standards for all coal-burning plants be tightened significantly, even if this raises the price of electricity sharply and makes it cheaper to produce electricity by using conventional nuclear fission? Why or why not?

4. Should all coal-burning power and industrial plants in the United States be required to convert to fluidized-bed combustion? Why or why not? What are the alternatives?

5. Do you favor a U.S. energy strategy based on greatly increased use of coal-burning plants to produce electricity between 1986 and 2020? Why or why not? What are the alternatives?

18

Nonrenewable Energy Resources: Nuclear Energy

We nuclear people have made a Faustian compact [a compact with the devil] with society; we offer an inexhaustible energy source tainted with potential side effects that if not controlled, could spell disaster.

Alvin M. Weinberg

The debate between scientists and engineers over whether nuclear power should be a major energy alternative for producing electricity is intense. Physicist Bernard L. Cohen contends that "Nuclear power is perceived to be *thousands of times* more dangerous than it is. . . . I am personally convinced that citizens of the distant future will look upon it as one of God's greatest gifts to humanity." Phil Bray, head of General Electric's nuclear reactor division in San Jose, California, says, "I could take a reactor, lose every pump, break every valve, blow every electrical unit, melt the core, and eventually bust the containment building, and we still think no one beyond the site boundary would be hurt." Although physicist and strong advocate of nuclear power Alvin M. Weinberg believes that nuclear power can be safe, his remark at the opening of this chapter emphasizes that the widespread use of nuclear power is the greatest single long-term risk ever approached by humankind, one that should be accepted only after intensive public education and debate.

According to Nobel Prize-winning physicist Hannes Alfvén: "Nuclear fission energy is safe only if a number of critical devices work as they should, if a number of people in key positions follow all their instructions, if there is no sabotage, no hijacking of the transport, if no reactor fuel processing plant or repository anywhere in the world is situated in a region of riots or guerrilla activity, and no revolution or war—even a 'conventional one'—takes place in these regions. . . . No acts of God can be permitted." Upon his retirement in 1982, Admiral Hyman G. Rickover, father of the U.S. nuclear submarine program and its director for more than 30 years, told members of Congress:

"The most important thing we could do is have an international meeting where we first outlaw nuclear weapons, then nuclear reactors."

According to energy cost analyst Charles Komanoff, "The fundamental problem facing nuclear power is that it's just too expensive." In 1984 a study group commissioned by the Atomic Industrial Forum, which represents the nuclear industry, concluded: "Nuclear power cannot at this time be considered a viable option on which to base new electric generating capacity in the United States."

Regardless of one's view on nuclear power, its widespread use in the United States and in many other nations appears to be a rapidly fading dream. Originally nuclear power was heralded as a clean, cheap, safe, and already developed source of energy that with 1,800 projected plants could provide as much as 21 percent of the world's primary energy and one-fourth of that in the United States by the year 2000. However, by 1985 after more than 30 years of development, 305 commercial nuclear reactors in 25 countries were providing only 9 percent of the world's electricity—amounting to 2 percent of the world's primary energy (Figure 16-2). In 1985 the largest number of operating commercial reactors was in the United States (88), followed by the Soviet Union (40), France (35), Great Britain (32), and Japan (25). Worldwide, another 210 reactors were under construction or planned, but some of these may never be completed.

Since 1975 the projected future use of nuclear power for producing electricity throughout the world has decreased sharply because of: **(1)** concerns over the safety of nuclear power plants, especially after the much-publicized accident in 1979 at the Three Mile Island nuclear plant in eastern Pennsylvania, **(2)** the need to find safe and politically acceptable methods for storing high-level radioactive wastes for hundreds to thousands of years, **(3)** concern that nuclear fuel will be diverted from civilian reactors by governments and upgraded to make nuclear fission bombs, **(4)** decreased demand for electricity in most industrial-

Table 18-1 Isotopes of Hydrogen and Uranium

Isotope	Natural Abundance (%)	Number of Electrons, e	Number of Protons, p	Number of Neutrons, n	Mass Number, $n + p$
Hydrogen					
Hydrogen-1	99.985	1	1	0	1
Hydrogen-2 (deuterium or D)	0.015	1	1	1	2
Hydrogen-3 (tritium or T)	Negligible	1	1	2	3
Uranium					
Uranium-233	Negligible	92	92	141	233
Uranium-235	0.7	92	92	143	235
Uranium-238	99.3	92	92	146	238

ized nations, **(5)** very high construction and operating costs (a typical new U.S. nuclear reactor costs 2 to 10 times the original estimate, and costs are rising sharply), and **(6)** increasing reluctance of lending institutions to provide large amounts of capital to public utility companies to develop an energy alternative that is not considered to be economic compared to others. Before looking in more detail at these major problems with conventional nuclear fission, as well as those associated with breeder nuclear fission and nuclear fusion, we need to know something about isotopes and radiation.

18-1 Isotopes and Radiation

Atomic Theory and Isotopes Recall from Section 4-4 that an *atom* consists of an extremely small dense center, called the *nucleus*, and one or more negatively charged *electrons (e)* in rapid motion around the nucleus. According to a crude model, the nucleus of all atoms (except one form of hydrogen) contains a mixture of one or more uncharged particles called *neutrons (n)* and positively charged particles called *protons (p)*, each with a relative mass of 1. The number of protons plus the number of neutrons gives the **mass number**. We can use $p + n$ as a measure of the atom's mass because the electrons outside the nucleus are so low in mass that they contribute very little to the overall mass of the atom.

Uncharged atoms of the same element all have the same number of negatively charged electrons outside their nuclei and the same number of positively charged protons inside. For example, the lightest element, hydrogen (H), has one positively charged proton in its nucleus and one negatively charged electron outside. A much heavier element, uranium (U), has 92 protons and 92 electrons.

Atoms of the same element, however, may have different numbers of uncharged neutrons in their nuclei, and thus different mass numbers. These different forms of a particular element with different mass numbers are called **isotopes**. Isotopes of the same element are identified by appending the mass number to the name or symbol of the element: hydrogen-1, H-1; hydrogen-2 (common name, deuterium), H-2; and hydrogen-3 (common name, tritium), H-3. Each of the 92 elements found in nature consists of a mixture of its isotopes, each with a certain percentage abundance. Table 18-1 shows the three naturally occurring isotopes of hydrogen (H) and uranium (U).

Radioactive Isotopes The nuclei of isotopes of a particular element are either *stable* (*nonradioactive*) or *unstable* (*radioactive*). For example, the hydrogen-1 and hydrogen-2 (deuterium) isotopes of hydrogen are nonradioactive (stable) and the hydrogen-3 (tritium) isotope is radioactive (unstable). A **radioisotope** is an isotope whose nuclei spontaneously emit particles, high-energy electromagnetic radiation, or both, at a certain rate, to form a different nonradioactive or radioactive isotope.

Radiation is the propagation of energy through matter and space in the form of fast-moving particles (particulate radiation) or waves (electromagnetic radiation). Ordinary light is a form of low-energy electromagnetic radiation (Figure 4-2). **Ionizing radiation** is high-energy radiation that can dislodge one or more electrons from atoms it hits to form highly reactive charged particles, called *ions*. The most common types of particulate ionizing radiation are high-speed **alpha particles** (positively charged helium nuclei, each with two protons and two neutrons) and **beta particles** (negatively charged electrons) emitted by the nuclei of unstable isotopes and *neutrons* released when an isotope is split apart by nuclear fission. The most common forms of electromagnetic ionizing radiation are high-energy **gamma rays**, from the nucleus of an unstable isotope, and high-energy *X rays* given off when some of the electrons outside the nucleus of an atom release energy after being exposed to a beam of electrons. Nuclei of unstable hydrogen-3 atoms spontaneously emit beta particles, and nuclei of unstable uranium-235 and uranium-238 spontaneously emit both alpha particles and gamma rays.

The rate at which a particular radioactive isotope spontaneously emits one or more forms of radiation is usually expressed in terms of its **half-life**: the length of time it takes for half the nuclei in a sample to decay by emitting one or more types of radiation and, in the process, changing into another nonradioactive or radioactive isotope. Each radioisotope has a unique, characteristic half-life. For example, hydrogen-3 has a half-life of 12.5 years and uranium-238 a half-life of 4.5 billion years. Thus, half a given sample of hydrogen-3 is still radioactive after 12.5 years and one-fourth is still radioactive after two half-lives, or 25 years. Similarly, one-fourth of a given sample of uranium-238 is still radioactive after 9 billion years.

Effects of Radiation on the Human Body Scientists agree that exposure to any type or amount of ionizing radiation has the potential to damage cells in the human body. The two major types of cellular damage are **(1)** *genetic damage*, which alters genes and chromosomes and may show up as a genetic defect in immediate offspring or several generations later, and **(2)** *nongenetic (somatic or body) damage*, which can cause harm during the victim's lifetime. Examples of nongenetic damage include burns, some types of leukemia, miscarriages and cataracts, and bone, thyroid, breast, and lung cancers.

High-energy gamma rays and X rays and high-speed neutrons are so penetrating that they pass through the body easily and inflict damage. They can be stopped only by several inches of lead or about 30 centimeters (1 foot) of concrete. Neither alpha nor beta particles can penetrate the skin. But once radioisotopes emitting these particles have been inhaled or ingested, their emissions can cause considerable damage to vulnerable tissues.

Tissues with cells that divide and reproduce rapidly are normally the most sensitive. Most eas-

Table 18-2 Effects on People of Exposure to High Radiation Doses

Dose (rems)*	Effect
100,000	Death in minutes
10,000	Death in hours
1,000	Death in days
700	Death for 90% within months
200	Death for 10% within months
100	No short-term deaths, but chances of cancer and other life-shortening diseases greatly increased. Permanent sterility in females; 2–3 year sterility in males.

*The rem is a radiation dose unit that measures damaging effect in mammals.

ily damaged are bone marrow (where blood cells are made), the spleen, the digestive tract (whose lining must be constantly renewed), the reproductive organs, and the lymph glands. The fast-growing tissues of a developing embryo are extremely sensitive, and pregnant women should avoid all unnecessary exposure to X rays and radioactivity. Studies have shown that people under 35 have four times the risk of radiation-caused cancer as those over 50. Women are particularly vulnerable between ages 20 and 30.

Two related units used to measure the dose or amount of radiation the body absorbs over time are the *rem* and the the *millirem* (mrem), which is one-thousandth of a rem (1 rem = 1,000 mrem; or 1 mrem = 0.001 rem). A single dosage of 10,000 mrem (10 rems) or less is generally referred to as *low-level radiation*, and a single dosage of 100,000 mrem (100 rems) or more is considered to be *high-level radiation*. Exposure to high levels of ionizing radiation such as that released from the detonation of a nuclear weapon can be fatal (Table 18-2).

Exposure to Ionizing Radiation Regardless of the risks, it is impossible to avoid all exposure to ionizing radiation. Each of us is exposed to a certain amount of *natural or background radiation* (Table 18-3), mostly from **(1)** cosmic rays from outer space, **(2)** naturally radioactive isotopes found in the soil and in bricks, stone, and concrete, and **(3)** radioactivity from natural sources that finds its way into our air, water, and food.

Your exposure to background ionizing radiation varies considerably with factors such as elevation (at high altitudes there is less overlying air to shield you from cosmic rays), soil type, water supply, occupation, and the type of building you

Table 18-3 Estimating Your Average Annual Radiation Dose from Background Radiation and Human Activities

Source of Radiation	Approximate Annual Dose (millirems)
Natural or Background Radiation	
Cosmic rays from space	
At sea level (average)	40
Add 1 mrem for each 30.5 m (100 ft) you live above sea level	_____
Radioactive minerals in rocks and soil: ranges from about 30 to 200 mrem depending on location	55 (U.S. average)
Radioactivity in the human body from air, water, and food: ranges from about 20 to 400 mrem depending on location and water supply	25 (U.S. average)
Radiation from Human Activities	
Medical and dental X rays and tests; to find your total, add 22 mrem for each chest X ray, 500 mrem for each X ray of the lower gastrointestinal tract, 910 mrem for each whole-mouth dental X ray film, 1,500 mrem for each breast mammogram, 8,000 mrem for a barium enema, and 5 million mrem for radiation treatment of a cancer	80 (U.S. average)
Living or working in a stone or brick structure; add 40 mrem for living and an additional 40 mrem for working in such a structure	_____
Smoking a pack of cigarettes a day; add 40 mrem	_____
Nuclear weapons fallout	4 (U.S. average)
Air travel; add 2 mrem a year for each 2,400 km (1,500 mi) flown	_____
TV or computer screens; add 4 mrem per year for each 2 hr of viewing a day	_____
Occupational exposure; varies with 100,000 mrem per year for uranium ore miner, 600 to 800 mrem for nuclear power plant personnel, 300 to 350 mrem for medical X ray technicians, 50 to 125 mrem for dental X ray technicians, and 140 mrem for jet plane crews	0.8 (U.S. average)
Living next door to a normally operating nuclear power plant (boiling water reactor, add 76 mrem; pressurized water reactor, add 4 mrem)	_____
Living within 8 km (5 mi) of a normally operating nuclear power plant; add 0.6 mrem	
Normal operation of nuclear power plants, nuclear fuel processing, and nuclear research facilities	0.10 (U.S. average)
Miscellaneous: luminous watch dials, smoke detectors, industrial wastes, etc.	2 (U.S. average)
	_____
Your annual total	= mrem

Average annual exposure per person in the United States = 230 mrem (with 130 mrem from background radiation and 100 mrem from human activities)

live and work in. Table 18-3 shows that the average American receives about 130 mrem per year from background sources and an additional 100 mrem from human-related activities. You can use Table 18-3 to estimate your own average total exposure to ionizing radiation each year and compare it with average annual exposure per person in the United States of 230 mrem.

How Dangerous Is Ionizing Radiation? Every second of your life you are struck by an average of about 27,000 radioactive particles or pulses of high-energy electromagnetic radiation from a combination of sources. Each of these hits can cause a fatal cancer during your lifetime. We're not all dying of radiation-induced cancer however, because the probability of a single radioactive particle or high-energy electromagnetic wave causing a fatal cancer is very low—about one chance in 50 quadrillion (50,000,000,000,000,000).

According to estimates by the National Academy of Sciences, an average annual exposure of about 200 mrem of ionizing radiation per person in the United States over an average lifetime causes about 1 percent of all fatal cancers and about 5 to 6 percent of all normally encountered genetic defects.

Our largest average exposure to ionizing radiation each year comes from dental and medical X rays and diagnostic tests involving the injection or ingestion of radioactive isotopes (Table 18-3). These important tools save thousands of lives each year and prevent human misery. But some observers contend that many of these X rays and diagnostic tests are taken primarily to protect doctors and hospitals from liability suits. If your doctor or dentist proposes an X ray or a diagnostic test involving radioisotopes, it is suggested that you ask why the procedure is necessary, how it will help find out what is wrong and influence possible treatment, and what alternative tests are available with less risk.

The smallest average exposure to ionizing radiation in the United States comes from nuclear power plants and other nuclear facilities—assuming that they are operating normally. According to the National Academy of Sciences radiation panel and UN Scientific Committee on Effects of Atomic Radiation, exposure to 1 mrem of ionizing radiation (10 times the average from normally operating nuclear power plants and other nuclear facilities) increases the risk that an individual will die from cancer by about one chance in 8 million. Physicist Bernard Cohen has calculated that this risk corresponds to a 1.2-minute reduction in life expectancy and is equivalent to the risk from taking about 3 puffs on a cigarette (each cigarette reduces life expectancy by about 10 minutes) or an overweight person eating 10 extra calories by taking a small bite from a piece of buttered bread. Similarly, the risk of having a genetically defective child because of exposure to 1 mrem of ionizing radiation received before conception is estimated to be about one chance in 40 million. This is about 140 times lower than the estimated genetic risk from drinking 29 milliliters (1 fluid ounce) of alcohol and 2.4 times lower than that from drinking one cup of coffee.

However, radiation experts such as John W. Gofman, E. J. Sternglass, and Helen Caldicott disagree with these conclusions. They cite some animal studies and recently revised estimates of the effects of low-level radiation on survivors of Hiroshima and Nagasaki that indicate that the risk of harm increases with exposure to some types of low-level ionizing radiation. Nuclear power critics also contend that the real danger from nuclear power is not from small, routine emissions of radioactivity but from the extremely small but real possibility of accidents that could result in the emission of large quantities of high-level radiation. With this background in atomic theory and ionizing radiation, let's see how a conventional nuclear fission power plant works.

18-2 Conventional Nuclear Fission

Nuclear Fission Reactors The potential energy locked in the nuclei of atoms of certain elements can be released and converted mostly to high-temperature heat by *nuclear fission*. In **nuclear fission** the nucleus of a heavy isotope such as uranium-235 (Table 18-1) found in uranium ore mined from the earth's crust is split or fissioned apart by a slow- or fast-moving neutron into two lighter nuclei called *fission fragments* (Figure 18-1). Fissions of uranium-235 nuclei can produce any of over 450 different fission fragments or isotopes, most of them *radioactive*.

During fission some of the mass of each uranium-235 nucleus is converted into energy. Each fission also produces two or three neutrons that can be used to fission many additional uranium-235 nuclei if enough are present to provide the *critical mass* needed for efficient capture of the neutrons. These multiple fissions taking place within the critical mass represent a *chain reaction* that releases an enormous amount of energy.

The uranium extracted from uranium ore contains about 99.3 percent *nonfissionable* uranium-238 and only 0.7 percent *fissionable* uranium-235. The concentration of fissionable uranium-235 must be increased from about 0.7 percent to about 3 percent by removing some of the uranium-238 at a fuel enrichment plant before the material can be used to fuel a conventional nuclear fission power plant.

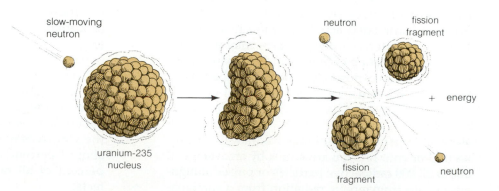

Figure 18-1 Nuclear fission of a uranium-235 nucleus.

slow-moving neutron

uranium-235 nucleus

neutron

fission fragment

+ energy

fission fragment

neutron

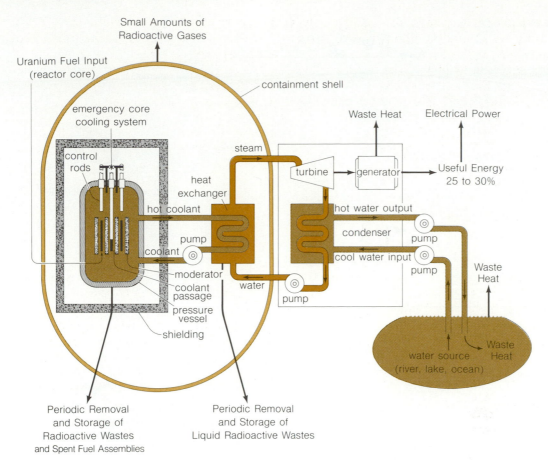

Figure 18-2 A nuclear power plant with a pressurized water reactor (PWR).

In an *atomic or nuclear fission bomb*, a massive amount of energy is released in a fraction of a second in an *uncontrolled* nuclear fission chain reaction. This is normally done by using an explosive charge to suddenly push a mass of fissionable fuel together from all sides to attain the critical mass needed to capture enough neutrons for a massive chain reaction to take place almost instantly. In a *nuclear reactor* used in an electric power plant, the rate at which the nuclear fission chain reaction takes place is *controlled* so that, on the average, only *one* of each two or three neutrons released (Figure 18-1) is used to split another nucleus.

Most nuclear reactors in the United States and throughout the world are *light-water reactors (LWR)*. There are two types of LWRs: the *boiling-water reactor (BWR)* and the *pressurized-water reactor (PWR)*. About 70 percent of the LWRs in the United States and the world are PWRs (Figure 18-2). In both types of LWR the *core* consists of several hundred *fuel assemblies*, each of which contains several hundred long, thin *fuel rods* made of a zirconium alloy and packed with eraser-sized pellets of uranium oxide (UO_2) fuel containing about 3 percent fissionable uranium-235 and 97 percent nonfis-

sionable uranium-238. The uranium-235 fuel in *each* of the fuel rods in a typical reactor can produce energy equal to that from about three railroad cars of coal.

Interspersed between the fuel assemblies are *control rods* made of materials that capture neutrons. These rods are moved in and out of the reactor to regulate the rate of fission and thus the amount of power the reactor produces. To stop the fission process (because of an accident, to make repairs, or to remove spent fuel assemblies), all the control rods must be inserted.

Both types of LWR have water as a *moderator*, circulating between the fuel rods and between fuel assemblies to slow the neutrons emitted by the fission process and to increase their chances of hitting other fuel atoms to continue the chain reaction.* The water also cools the fuel rods to pre-

---

*A non-LWR reactor such as the *Canadian deuterium uranium (CANDU) reactor*, in commercial operation in Canada and in several other countries, uses heavy water (in which the hydrogen atoms are hydrogen-2 instead of hydrogen-1) as a moderator and heat transfer medium. Another non-LWR reactor, known as the *high-temperature, gas-cooled reactor (HTGR)*, uses graphite as a moderator and helium gas to transfer heat out of the core.

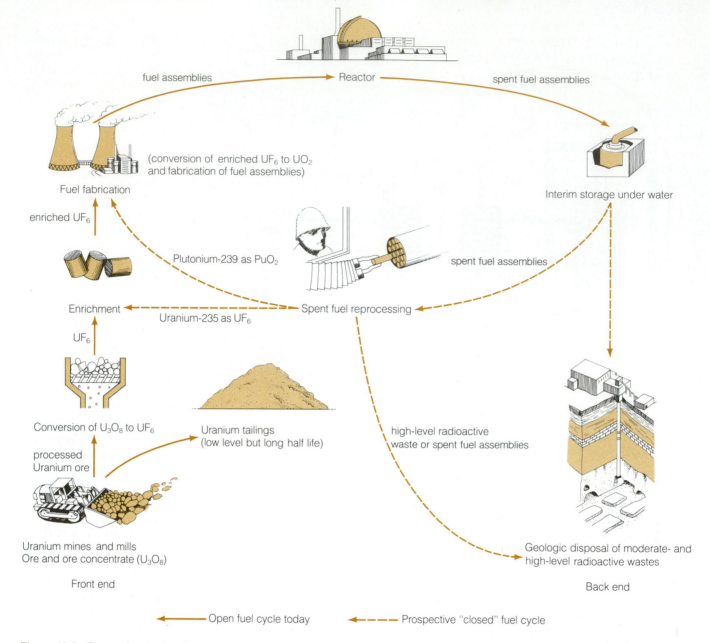

fuel assemblies → Reactor → spent fuel assemblies

(conversion of enriched UF_6 to UO_2 and fabrication of fuel assemblies)

Fuel fabrication

Interim storage under water

enriched UF_6

Plutonium-239 as PuO_2

spent fuel assemblies

Enrichment ← Uranium-235 as UF_6 ← Spent fuel reprocessing

UF_6

Conversion of U_3O_8 to UF_6

Uranium tailings (low level but long half life)

processed Uranium ore

high-level radioactive waste or spent fuel assemblies

Uranium mines and mills
Ore and ore concentrate (U_3O_8)

Geologic disposal of moderate- and high-level radioactive wastes

Front end

Back end

←—— Open fuel cycle today   ←-- -- Prospective "closed" fuel cycle

Figure 18-3   The nuclear fuel cycle.

vent them and other core materials from melting, and it carries the heat produced by the fission chain reaction out of the core and into the heat exchanger.

Each LWR requires an enormous volume of cooling water. The heat removed by this water must either be transferred to the air by using massive cooling towers (Figure 22-2) or returned to bodies of water where it can cause thermal water pollution. Fossil-fuel-burning power plants also require cooling water and can cause thermal water pollution. However, such plants release less heat because they have a higher first-law energy efficiency of 40 percent, compared to the 25 to 30 percent efficiency of an LWR.

Nuclear power plants, each with one or more reactors, are only one part of the *nuclear fuel cycle* (Figure 18-3) necessary for using nuclear energy to produce electricity. *In evaluating the safety and economics of nuclear power, it is necessary to look at this entire cycle—not just the nuclear plant itself.*

Nuclear Reactor Safety   Fission converts some fuel to radioactive fragments, and under intense neutron bombardment the metals in the fuel rods and other metal parts in the core are converted to radioactive isotopes. Because these radioactive fission products also produce lots of heat, they continue to heat the fuel even after the reactor has been shut down. Thus, water must be circulated through the core to prevent a *meltdown* of the fuel rods and the reactor core. An *emergency core-cooling system* serves as a backup to flood the core automatically with water to prevent meltdown of the reactor core.

A light-water reactor cannot blow up like an atomic bomb because neither the fissionable uranium-235 nor the fissionable plutonium-239 produced in the reactor when neutrons bombard nonfissionable uranium-238 is present in sufficient concentration or in the proper geometry to create a runaway chain reaction. However, a reactor core might lose its cooling water through a break in one of the pipes that conduct cooling water and steam to and from the reactor core. If the emergency core-cooling system also failed, *such a loss-of-coolant accident* would cause the reactor core to overheat, eventually melt down, melt through its thick concrete slab, and melt itself into the earth. Depending on the geological characteristics of the underlying strata, the melted core might sink 6 to 30 meters (20 to 100 feet) and gradually dissipate its heat, or it might burn itself deeply into the earth's crust and contaminate groundwater with radioactive materials.

Another possibility is that a powerful gas or steam explosion inside the reactor containment vessel could split the vessel open and release its highly radioactive contents directly to the atmosphere. This cloud of radioactive materials, which would be at the mercy of the winds and weather, could kill and injure many thousands of people and contaminate large areas for hundreds to thousands of years. Because of these very remote but real possibilities for serious accidents, nuclear fission is considered by most analysts to be *potentially* the most hazardous of all energy alternatives.

Advocates of nuclear power, however, argue that a catastrophic nuclear accident is so unlikely that nuclear power is worth the risk compared to the benefits it provides and the even greater hazards from burning coal (Section 17-3). They point out that a meltdown and release of radioactivity to the environment would require the almost simultaneous failure of a series of safety systems, each with an automatic backup system, designed to prevent this from happening.

One of the most extensive reactor safety studies was WASH-1400, or the Rasmussen report, published in 1975. Its authors claimed to have examined and predicted every conceivable sequence of events leading to an accident in a nuclear power plant; they used statistics and computer models to calculate the likelihood of each failure, and thus the probability of a major nuclear power plant ac-

cident. According to this study: **(1)** a meltdown is highly unlikely; **(2)** an "average" meltdown would cause 400 to 5,000 fatalities and an average of $100 million (1975 dollars) in property damages; and **(3)** in the worst possible accident, which is so unlikely that its probability of occurrence is only once during every billion years of reactor operation, there would be an estimated 825 to 13,200 immediate deaths and 7,500 to 180,000 later deaths from cancer (a mean of 50,000 total deaths), 12,375 to 198,000 illnesses, 4,750 to 171,000 delayed genetic effects, contamination with radioactive materials of a 9,000-square-kilometer (3,500-square-mile) area (almost equivalent to the size of Connecticut), and property damage ranging from $2.8 billion to $28 billion (a mean of $15 billion).

However, in January 1979 the Nuclear Regulatory Commission withdrew its endorsement of the Rasmussen report, stating that the NRC as a result of an evaluation of the study by prominent scientists and nuclear safety experts, no longer considered the risk estimates to be reliable. WASH-1400 was also criticized for its failure to assess the risks involved in the entire nuclear fuel cycle (Figure 18-3) rather than only for a nuclear power plant.

The first major sign of trouble occurred in 1975 at the Brown's Ferry nuclear reactor in Alabama when the flame from a candle used by a worker to search for air leaks set off a fire that knocked out five emergency core-cooling systems, seriously reducing the plant's protection against a meltdown. In 1979, what is said to be the worst accident in the history of commercial nuclear power occurred at the Three Mile Island (TMI) nuclear plant in eastern Pennsylvania. The accident was caused by a series of mechanical failures and human errors considered to be essentially impossible by the Rasmussen study. However, the radioactive core of one of the plant's reactors became partially uncovered, and small amounts of radiation were released into the atmosphere. Later examination revealed that the reactor core underwent a partial meltdown and came perilously close to a complete meltdown.

No lives were immediately lost. Furthermore, according to the presidential commission that investigated the accident, there will be few, if any, long-term health effects from the accident. Many nearby residents, however, are not convinced and bear increased anxiety about their health and the long-term health of their children and about the devaluation of their property. In addition, a 1984 study by an independent group of scientists and engineers impaneled by the TMI Public Health Fund concluded that firm projections of the health effects of the accident cannot be made because data published so far on the amount of radiation released during the accident are contradictory and incomplete.

The cleanup of the damaged TMI reactor, which will probably cost at least $1 billion ($300 million more than the cost of building the reactor), threatens the utility with bankruptcy and may not be completed until 1990 or later. Furthermore, confusing and misleading statements about the seriousness of the accident were issued by Metropolitan Edison, the owner of the plant, and by the Nuclear Regulatory Commission. This eroded public confidence in the safety of nuclear power and the ability and desire of authorities to provide accurate information about the potential or actual dangers of nuclear power.

After a lengthy investigation of the accident, the Nuclear Regulatory Commission issued a list of 6,000 new steps that utilities had to take to improve operator training, add safety equipment, and provide and test evacuation plans for all residents within a 16-kilometer (10-mile) radius of any nuclear plant. In response to the TMI accident, the nuclear industry and utilities moved to upgrade reactor safety by (1) establishing a Nuclear Safety Analysis Center to study safety problems, devise solutions, and distribute this information to utility companies, (2) setting up the Institute of Nuclear Power Operations, charged with establishing better management and operator training and standards, (3) installing two telephone hotlines to link each reactor in the country to the NRC's emergency response center, and (4) financing a $40 million ad campaign by the Committee for Energy Awareness to improve the nuclear industry's image and resell nuclear power to the American public.

Many of the advertisements of the Committee for Energy Awareness use the misleading argument that nuclear power is needed in the United States to reduce dependence on imported oil. In fact, since the oil embargo of 1973, the reduction of oil use and oil imports has come mostly from improvements in energy efficiency, increased use of wood as a fuel in homes and businesses, and increased use of coal to produce electricity. By 1984 less than 6 percent of the electricity in the United States was produced by burning oil, and this percentage was dropping sharply as more plants were converted to burn coal.

Disposal and Storage of Radioactive Wastes

Each part of the nuclear fuel cycle produces a mixture of solid, liquid, and gaseous radioactive wastes that must be stored until their radioactivity has dropped to extremely low levels. The largest amount of high-level radioactive wastes is produced at nuclear power plants and at reprocessing plants (Figure 18-3). In addition, large amounts of high-level radioactive wastes are produced by nuclear weapons facilities, and large quantities of low-level wastes are produced by defense facilities, nuclear power plants, hospitals, and at government and university nuclear research laboratories.

Three methods are used to dispose of or store radioactive wastes: dilution and dispersion, delay and decay, and concentration and containment. In *dilution and dispersion*, low-level wastes are released into the air, water, or ground to be diluted to presumably safe levels. *Delay and decay* involves storing medium- to high-level radioactive wastes, such as iodine-131, with relatively short half-lives, as liquids or slurries in double-shell tanks. After 10 to 20 half-lives, they normally decay to relatively harmless levels, at which time they can be diluted and dispersed within the environment.

Concentration and containment is used for high-level radioactive wastes containing radioisotopes with long half-lives. For example, both cesium-137 with a half-life of 27 years and strontium-90 with a half-life of 28 years must be stored safely for about 280 to 560 years. If fuel reprocessing plants are not developed in the United States, spent fuel assemblies containing radioactive plutonium-239 with a half-life of 24,400 years will have to be stored safely for at least 244,000 years (10 times the isotope's half-life). Plutonium-239 can cause lung cancer if only a few particles, each smaller than a speck of dust, are inhaled.

Highly radioactive spent fuel assemblies from U.S. nuclear power plants are being stored temporarily in deep pools of water at nuclear plant sites, pending the development of a method for long-term storage or disposal. High-level liquid wastes from nuclear weapons production (equal in volume to about 200 olympic-sized swimming pools) and from nuclear power plants are also awaiting permanent storage. Presently they are stored in underground tanks in government facilities in Idaho, South Carolina, and Washington State. These tanks must be continuously monitored to prevent or detect corrosion and leaks. More than 1.7 million liters (450,000 gallons) of highly radioactive wastes have already leaked from 20 of the older single-shell tanks built between 1943 and 1965 at the Richland, Washington, storage site. However, a study by the National Academy of Sciences concluded that because of the isolation of the site, the leaks have not caused any significant radiation hazard to public health and would take nearly 1 million years to reach the nearest river. Storage tanks constructed after 1968 have a *double*

Table 18-4 Proposed Methods for Long-Term Storage or Disposal of Nuclear Wastes

Proposal	Possible Problems
Surround waste with concrete or several layers of metal and store in surface warehouses or underground tunnels with careful monitoring until a better solution is found.	Concrete or metal liners might deteriorate; above-ground warehouses may be difficult to guard against sabotage; states may not want the storage sites or the transshipment of wastes; the government might not provide enough funds to investigate more permanent storage methods.
Solidify wastes, encapsulate them in glass or ceramic, place in metal containers, and bury the containers deep underground in earthquake- and flood-proof geological formations, such as dug-out salt or granite deposits.	Occurrence of natural disasters cannot be predicted; heat from radioactive decay might crack glass containers, fracture salt or granite formations so that groundwater could enter the depository, or release from water-containing minerals water that could leach radioactive materials into groundwater supplies; transportation of deadly radioactive wastes to depository sites could be dangerous; wastes might be difficult to retrieve if project fails.
Use rockets or a space shuttle to shoot the wastes into the sun or into space.	Costs would be very high, and a launch accident could disperse deadly radioactive wastes over a wide area. The project may not be technically feasible.
Bury wastes in an underground hole created by a nuclear bomb so that the wastes eventually melt and fuse with surrounding rock into a glassy ball.	Effects unknown and unpredictable; if project fails, wastes cannot be retrieved and could contaminate groundwater supplies.
Bury wastes under Antarctic ice sheets or Greenland ice caps.	Long-term stability of ice sheets is unknown; knowledge about thermal, chemical, and physical properties of large ice sheets is lacking; retrieval could be difficult or impossible if project fails.
Encase wastes in well-designed containers and drop them into the ocean in isolated areas.	No one knows how to design a container that will last long enough; oceans and marine life could become seriously contaminated if containers leak; small currents near the ocean bottom could move the wastes to less isolated sites over hundreds to thousands of years.
Enclose wastes in well-designed containers and drop them into deep ocean bottom sediments that are descending deeper into the earth.	Long-term stability and motion of these sediments are unknown; containers might leak and contaminate the ocean before they are carried downward; containers might migrate back to the ocean or be spewed out somewhere else by volcanic activity; wastes probably could not be retrieved if project fails.
Change harmful isotopes into harmless ones by using high-level neutron bombardment, lasers, or nuclear fusion.	Technological feasibility has not been established; costs would be extremely high; process would create new toxic materials also needing disposal; main effect would be to spread long-lived radioactive isotopes into more dilute radioactive wastes, not to eliminate them.

shell; thus if the inner wall corrodes, the liquid will spill into the space between the two walls, where it can be detected in time to be pumped into another tank. These new tanks have developed no leaks to the environment so far.

Table 18-4 lists the major methods proposed for long-term storage or disposal of high-level radioactive waste. *The safe disposal of high-level radioactive wastes is believed to be technically possible. But after 30 years of research and debate there is still no widely agreed upon scientific and political solution to this problem.* One proposed method is to concentrate the waste, convert it to a dry solid, fuse it with glass or a ceramic material, seal it in a metal canister, and bury it permanently in deep underground salt or rock formations, expected to remain stable for millions of years.

But some geologists and other scientists question this approach, arguing that extensive drilling and tunneling can destabilize such rock structures and that present geologic knowledge is not sufficient to predict the paths of subterranean water flows that could contaminate groundwater drinking supplies with radioactive wastes.

Regardless of the storage method, most citizens strongly oppose the location of a nuclear waste disposal facility anywhere near them. By 1983 at least 22 states had enacted laws banning radioactive waste disposal within their borders, and 7 states had laws prohibiting construction of new nuclear power plants until the government demonstrates a safe method for long-term nuclear waste disposal.

In 1983 the Department of Energy began building the first major geologic repository in the United States, to be used only for storage of high-level radioactive wastes produced by the nuclear weapons program. The $1 billion Waste Isolation Pilot-Plant Project is being built in a bedded salt formation deep under federal land about 40 kilometers (25 miles) east of Carlsbad, New Mexico.

In 1982 Congress passed the Nuclear Waste Policy Act, which **(1)** established a timetable for the

Department of Energy to choose a site and build the nation's first deep underground repository for long-term storage of high-level radioactive wastes from commercial nuclear reactors, **(2)** called for the Department of Energy to design and find a site for a monitored retrievable storage facility for temporary storage of high-level nuclear waste as a backup for the permanent repository, **(3)** established an interim storage program to ease the backlog of spent nuclear fuel at power plants, and **(4)** permitted states and American Indian tribes to turn down repository locations in their jurisdictions unless overridden by simple majorities in both houses of Congress. If intense political opposition by states where proposed repositories are to be located can be overcome and the project does not suffer too many delays, it is conceivable that the United States will have a method for the long-term storage of high-level radioactive wastes between 2000 and 2010.

Decommissioning Nuclear Power Plants After 30 to 40 years of neutron bombardment, a steel reactor vessel becomes unacceptably brittle and the miles of cooling water pipes become too corroded for safe use. Since the high levels of radiation in the reactor vessel prevent repairs, the plant must be shut down in a safe condition. This *decommissioning* process is the final step of the nuclear fuel cycle. By the year 2000, at least 20 presently operating commercial reactors in the United States will probably be candidates for decommissioning.

Scientists have proposed three ways for decommissioning a nuclear power reactor: **(1)** *mothballing*, by removing spent uranium fuel assemblies, draining all slightly radioactive water from the cooling pipes, setting up a 24-hour security guard system to prevent public access for at least 100 years and perhaps 1,000 years, and doing periodic radiological surveys and maintenance, **(2)** *entombment*, or sealing the entire reactor with reinforced concrete after radioactive fuel assemblies, liquid radioactive waste, and surface contamination have been removed to the greatest extent possible, and **(3)** *dismantlement*, the removal of all radioactive materials, which are then stored in a radioactive waste disposal facility.

Dismantlement seems to be the most likely method. The $1 billion price tag for the dismantlement and cleanup of the Three Mile Island reactor suggests that this process will be extremely costly. Adding these costs to the already high price of producing electricity by using the nuclear fuel has led some analysts to believe that nuclear power will never become an economically feasible energy alternative.

Proliferation of Nuclear Weapons Since the late 1950s the United States has been giving away and selling to other countries various forms of nuclear technology. By 1984 at least 14 other countries had entered the international market as sellers. For decades the U.S. government has denied that the information, components, and materials used in the nuclear fuel cycle could be used to make nuclear weapons. In 1981, however, a Los Alamos National Laboratory report admitted: "There is no technical demarcation between the military and civilian reactor and there never was one."

Nuclear weapons can be made from any one of three major fissionable isotopes: highly enriched uranium (HEU) containing 50 to 93 percent uranium-235, relatively pure plutonium-239 (ideally about 94 percent plutonium-239 and 6 percent plutonium-240), and relatively pure uranium-233. Between 4 and 9 kilograms (9 to 20 pounds) of either plutonium-239 or uranium-233, a mass about the size of an orange, and 11 to 25 kilograms (24 to 55 pounds) of uranium-235 are needed to make a small atomic bomb capable of blowing up a large building or a city block and contaminating a much larger area with radioactive materials for centuries. Because the basic principles of building an atomic bomb are well known, there is relatively little doubt that a small group of trained people could make such a "blockbuster" nuclear bomb if they could get enough fissionable bomb-grade material.

Bomb-grade plutonium-239 is very heavily guarded, but it could be stolen from nuclear weapons facilities, especially by employees. Each year about 3 percent of the approximately 126,000 people working with U.S. nuclear weapons are relieved of duty because of drug use, mental instability, or other security risks. By 1978 at least 320 kilograms (700 pounds) of plutonium-239 was missing from commercial and government-operated reactors and storage sites in the United States—enough to make 32 to 70 atomic bombs, each capable of blowing up a city block. No one knows whether this missing plutonium was stolen or whether it represents sloppy measuring and bookkeeping techniques.

Concentrated bomb-grade plutonium fuel could also be hijacked from shipments to breeder nuclear fission plants or stolen from a commercial fuel reprocessing plant or from one of more than 150 research and test reactors operating in about 30 countries. It could be manufactured by using one of the simpler and cheaper technologies for isotope separation presently being developed to concentrate the 3 percent uranium-235 to weapons-grade material.

Some have pointed out that those who steal

plutonium need not bother to make atomic bombs. They could simply use a conventional explosive charge to disperse the stolen plutonium into the atmosphere from atop any tall building. Dispersed in this manner, 2.2 kilograms (1 pound) of plutonium oxide power could theoretically contaminate 7.7 square kilometers (3 square miles) with radioactivity. This radiation, which would remain at dangerous levels for at least 100,000 years, could cause lung cancers among those who inhaled contaminated air or dust in such areas.

Since 1968 more than 120 nations have signed the Treaty on Nonproliferation of Nuclear Weapons. The signatories agreed to reduce their nuclear arsenals (which has not happened) and to assist non-nuclear nations with civilian nuclear power programs if these nations agreed not to use the information and technology to develop nuclear weapons and to place all their civilian nuclear power activities under a system of safeguards administered by the UN-sponsored International Atomic Energy Agency (IAEA). But critics point out that: **(1)** the IAEA has no enforcement power if a diversion of bomb-grade material is detected; **(2)** any member nation can withdraw from the treaty and escape scrutiny with 90 days notice (though none had done so by early 1985); and **(3)** the agency makes on-site inspections only every few months, even though a country could divert plutonium from a civilian nuclear facility and fabricate it into a nuclear weapon within a week.

To make matters worse, by early 1985, 50 nations—including France, China, India, Pakistan, Argentina, Brazil, Israel, and South Africa—had not signed the treaty. By 1985 five nations (the United States, the Soviet Union, Great Britain, France, and China) had built and tested nuclear weapons, four others (India, Israel, South Africa, and Pakistan) were believed to be capable of building nuclear bombs by 1989, and four others (Argentina, Brazil, Iraq, and Libya) are likely to be able to make nuclear weapons by 1994. More than a dozen other nations possess or soon could possess the knowledge and resources to build nuclear weapons, but presently show no interest in doing so.

One suggestion for reducing diversion of plutonium fuel from the nuclear fuel cycle is to contaminate it with other substances that render it dangerous to handle and unfit as weapons material. But so far no acceptable "spiking agent" has emerged that could not be removed by reprocessing or isotope separation.

Some argue that trying to stop nuclear proliferation by reducing the spread of nuclear technology is hopeless because the genie is out of the bottle. Others point out that although nuclear technology is spreading, this does not relieve nuclear nations from the responsibility of keeping its rate of spread as low as possible.

Soaring Costs: The Achilles Heel of Nuclear Power The major factor slowly shutting down the world's nuclear industries is economics. The largest cutback in nuclear power has taken place in the United States. After 34 years of development and a $154 billion investment, including $44 billion in government (taxpayer) subsidies, the 88 nuclear power reactors in operation in the United States in 1985 produced only 14 percent of the nation's electricity and 4 percent of the nation's primary energy—about equal to that provided by wood and crop wastes without significant government subsidies.

Utility companies originally began ordering nuclear power plants in the late 1950s for three major reasons: **(1)** the Atomic Energy Commission and builders of nuclear reactors projected that nuclear power would produce electricity at such a low cost that it would be "too cheap to meter"; **(2)** the nuclear industry projected that the reactors would have an 80 percent *capacity factor*—a measure of the time a reactor is able to produce electricity at its full power potential; and **(3)** the first round of commercial reactors was built with the government paying approximately one-fourth of the cost, using the Price-Anderson Act to protect the nuclear industry and utilities from significant accident liability, and at a fixed cost with no cost overruns allowed. It was an offer utilities could not resist.

Since the construction of this first batch of nuclear power plants, it has become increasingly clear that nuclear power is an extraordinarily expensive way to produce electricity, even when it is heavily subsidized to partially protect it from free market competition. By 1983, producing electricity in a new nuclear plant in the United States had a total cost significantly higher than those of coal, cogeneration, and improved energy efficiency (Table 18-5). By 1990 nuclear power is expected to be even less cost competitive with other methods for producing electricity in the United States.

A 1984 study of 47 nuclear plants by the Department of Energy revealed that 36 cost at least twice as much as initially estimated and 13 cost *four* times as much. A few plants have had astronomical cost overruns. The two-reactor Diablo Canyon power plant in California, originally estimated to cost $450 million, will cost $4.9 billion; and the two-reactor Seabrook plant in New Hampshire, originally budgeted at $1 billion, is now estimated at $9 billion and threatens to bankrupt the utility company.

Table 18-5 Estimated Cost of Electricity Including Construction, Fuel, and Operation from New Power Plants in the United States in 1983, with Projections for 1990

Energy Source	Cost (cents per kilowatt-hour) 1983	Cost (cents per kilowatt-hour) 1990 (projected)
Improved energy efficiency (to reduce quantity of electricity needed)	1–2	3–5
Cogeneration	4–6	4–6
Coal	5–7	8–10
Small hydropower	8–10	10–12
Biomass	8–15	7–10
Nuclear	10–12	14–16
Wind power	15–20	6–10
Solar photovoltaic cells	50–100	10–20

Sources: Charles Komanoff and Worldwatch Institute, January 1984.

Primarily because of declining demand for electricity and rapidly rising costs for building and operating new nuclear power plants in the United States, no new nuclear power reactors have been ordered by utility companies since 1979. In addition, by early 1985 orders for 109 nuclear plants placed between 1972 and 1978 had been canceled. A number of financial analysts and energy experts project that at least half the 39 reactors ordered and still under construction in early 1985 may also be canceled to prevent financial strain or bankruptcy of utility companies. The financial community has become quite skeptical about financing new U.S. nuclear power plants after the Three Mile Island accident showed that they could lose $1 billion or more in equipment in an hour, even without any serious health effects for the public.

In addition, consumer protests are expected to increase because of (1) the sharp rise in electricity rates in areas getting a significant fraction of their electricity from costly new nuclear reactors beginning operation between 1984 and 1990, and (2) attempts by utility companies to have consumers pay for plants under construction and for losses from cancellation of partially completed nuclear reactors. Nuclear plants still under construction in the United States are projected to cost at least $191 billion more than the fuels they will displace over their lifetime. This helps explain why in 1985 Robert Scherer, Chairman of Georgia Power, stated, "No utility executive in the country would consider ordering a nuclear plant today—unless he wanted to be certified insane or committed."

A 1984 report by the Office of Technology Assessment concluded that nuclear power in the United States is not likely to be expanded in this century beyond the reactors already under construction. Nuclear industry officials believe that this bleak outlook for the future of nuclear power in the United States could change if: (1) electricity demand were to rise sharply in the 1980s and 1990s; (2) the costs and time needed to build a nuclear plant were cut significantly by standardizing design and by reducing the paperwork and tests needed to obtain government approval for operating a plant; and (3) public confidence in the safety of nuclear power were restored.

The rate of growth of nuclear power in most other nations has also decreased since the mid-1970s, with nuclear power projected to supply only about 15 to 20 percent of the world's electricity and about 5 percent of the world's primary energy by 2000 (Figure 16-2). Most governments with major nuclear power programs—especially Japan, France, Great Britain, and the Soviet Union—remain strongly committed to using nuclear power to produce a significant amount of their electricity by the end of the century. But public opposition to nuclear power is growing in these nations (except the Soviet Union) and others, costs have been much higher than projected, and plans have been scaled back sharply.

France has the world's most ambitious and cost-efficient plan for using nuclear energy and builds its reactors in less than 6 years (compared to about 12 years in the United States), using standardized government-developed designs. By 1985 it had 35 operating nuclear reactors providing 50 percent of its electricity and 23 more under construction, with the goal of generating 75 percent of its electricity in this way by 1990. However, by 1984 several studies concluded that (1) the government had already built or had under construction more plants than the nation needs; (2) France's nuclear program survives largely through taxpayer subsidies and increasing the national debt rather than through open-market economic competition with coal and other nonrenewable energy alternatives; (3) over the next 20 to 30 years building coal-fired plants would be 35 to 60 percent cheaper than building any new nuclear plants; and (4) about 22 percent of France's foreign debt, which is the third largest in the world, is directly attributable to the government-owned company responsible for building and operating the nuclear plants. Because of this massive debt and a glut of electricity, France reduced its orders for new reactors from six to two per year in 1984 and 1985. The French Planning Ministry concluded that the only reason for not halting all orders for all new reactors until at least

1987 was to preserve jobs at the government-run reactor manufacturing company.

In summary, the major advantages of using conventional nuclear fission to produce electricity are (1) a reactor cannot blow up like an atomic bomb; (2) no carbon dioxide is produced; (3) because of multiple safety systems, a catastrophic accident releasing deadly radioactive material into the environment is extremely unlikely; (4) nuclear reactors do not release air pollutants such as particulate matter and sulfur and nitrogen oxides like coal-fired plants; and (5) water pollution and disruption of land is low to moderate if the entire nuclear fuel cycle operates normally.

The major disadvantages of producing electricity by using conventional nuclear fission are: (1) construction and operating costs are high and rapidly rising, even with massive government subsidies; (2) affordable supplies of uranium may be depleted within 50 to 200 years, depending on rate of development of nuclear power; (3) conventional nuclear fission can be used only to produce electricity and pressurized high-temperature steam; (4) its net useful energy yield is low (Table 16-8); (5) a presumably safe method for storing radioactive wastes for hundreds to thousands of years will not be available until at least 2000, and some scientists doubt that an acceptably safe storage method can ever be developed; (6) its use commits future generations to safely storing radioactive wastes for hundreds to thousands of years even if nuclear fission power is abandoned; (7) its use spreads knowledge and materials that could be used to make nuclear weapons; and (8) although large-scale accidents are extremely unlikely, a combination of mechanical and human errors, sabotage, or shipping accidents could result in release of deadly radioactive materials into the environment.

18-3 Breeder Nuclear Fission

Breeder Reactors At present use rates, the world's supply of uranium should last for at least 100 years. However, some scientists believe that if there is a sharp rise in the use of nuclear fission to produce electricity after the year 2000, *breeder nuclear fission reactors* can be developed to avoid rapid depletion of the world's supply of uranium fuel. Widespread use of breeder reactors could increase the present 50- to 200-year estimated lifetime of the world's affordable uranium supplies for at least 1,000 years and perhaps several thousand years.

A breeder reactor produces within itself new fissionable fuel in the form of plutonium-239 from nonfissionable uranium-238, an isotope that is in plentiful supply. The fuel in a breeder fission reactor consists of a mixture of uranium-238 and an initial charge of fissionable plutonium-239, made by the bombardment of uranium-238 with neutrons inside a conventional fission reactor. Radioactive waste from the conventional reactor is taken to a fuel-reprocessing plant, where the plutonium-239 is separated and purified for use as fuel in a breeder reactor. In the breeder reactor *fast neutrons* are used to fission the nuclei of plutonium-239 and convert the *nonfissionable* uranium-238 into enough *fissionable* plutonium-239 to start up another breeder reactor after 30 to 50 years. Because these devices use fast-moving neutrons for fissioning, they are often called *fast breeder reactors*.

A breeder reactor looks something like the reactor in Figure 18-2, except that its core contains a different fuel mixture and its two heat exchanger loops contain liquid sodium instead of water. Under normal operation a breeder reactor is considered to be much safer than a conventional fission reactor. But in the unlikely event that all its safety systems failed and the reactor lost its sodium coolant, there would be a runaway fission chain reaction, and perhaps a small nuclear explosion with the force of several hundred pounds of TNT. Such an explosion could blast open the containment building, releasing a cloud of highly radioactive gases and particulate matter. A more common problem that could lead to temporary shutdowns, but poses no significant health hazards, involves the leakage of molten sodium, which ignites on exposure to air and reacts violently with water.

Since 1966 several small-scale experimental breeder reactors have been built in the United States, Great Britain, the Soviet Union, and West Germany. Intermediate-scale demonstration breeder reactors with about one-third to one-half the output of a typical commercial nuclear plant have been in operation in Great Britain since 1975 and in France and the Soviet Union since 1977. Since 1977 France has been building Superphénix, a full-sized commercial breeder reactor scheduled to begin operation in 1986. Tentative plans to build full-sized commercial breeders in West Germany, the Soviet Union, and the United Kingdom may be canceled because of the excessive cost of Superphénix (three times the original estimate) and because some studies indicate that breeders will not be competitive economically with conventional fission reactors for at least 50 years.

In 1983, after 13 years of political and scientific debate and a government expenditure of $1.7 billion merely for planning, the proposed Clinch River intermediate demonstration breeder reactor in Tennessee was canceled because (1) it would

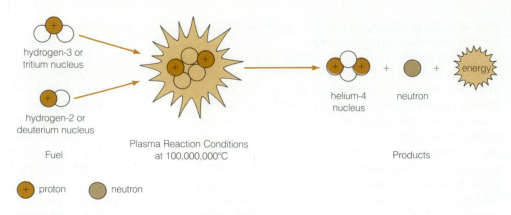

Figure 18-4 The deuterium–tritium (D–T) nuclear fusion reaction.

cost 5 to 12 times the original estimate; **(2)** its design was already outdated by other demonstration breeders; **(3)** numerous studies have shown that the United States will not need breeder reactors until at least 2025, given the slowdown in the building of conventional nuclear fission reactors; **(4)** energy conservation and various renewable energy alternatives can be developed faster and cheaper; and **(5)** there was fear over the worldwide proliferation of nuclear weapons.

18-4 Nuclear Fusion

Controlled Nuclear Fusion In the distant future—probably no sooner than 2050, if ever—scientists in the United States, the Soviet Union, Japan, and a consortium of European nations hope to use *controlled nuclear fusion* to provide an essentially inexhaustible source of energy for producing electricity. In **nuclear fusion**, which takes place in the sun and other stars, the nuclei of light atoms such as hydrogen are forced together at temperatures of 100 million degrees or more until they fuse to form a heavier nucleus. When this fusion process takes place, some of the mass of the light atoms is converted into energy.

Theoretically, at least 100 different nuclear fusion reactions are possible. Presently, however, only the *D–T fusion reaction* (Figure 18-4) is being studied seriously because it has the lowest ignition temperature (about 100 million degrees—roughly five times hotter than the sun's core). In the *D–T fusion reaction*, a hydrogen-2 or deuterium (D) nucleus and a hydrogen-3 or tritium (T) nucleus are fused together to form a larger helium nucleus, a neutron, and energy.

Deuterium is found in about 150 out of every million molecules of water (150 ppm) and can be separated from ordinary hydrogen atoms fairly easily. Thus, the world's oceans provide an almost inexhaustible supply of this isotope. Although there is no significant natural source of tritium, an extremely small quantity can be extracted from seawater. This can be used as an initial charge in a fusion reactor, with the neutrons emitted in D–T fusion reactions used to bombard a surrounding blanket of lithium to breed additional tritium fuel. The scarcity of lithium will eventually limit the use of D–T fusion, but the earth's estimated supply should last for 1,000 to several thousand years, depending on rate of use.

Another possibility is the D–D fusion reaction, in which the nuclei of two deuterium (D) atoms would be fused together to form a helium nucleus. But this reaction requires an ignition temperature about 10 times higher than that for D–T fusion and is not being pursued at this time. If controlled D–D nuclear fusion were developed, the deuterium in the ocean could supply the world with primary energy at many times present consumption rates for 100 billion years. Few scientists, however, expect D–D fusion to become a major source of energy until 2100, if ever.

Achieving Controlled Nuclear Fusion *Uncontrolled nuclear fusion* of deuterium and tritium produced from lithium is presently used in *hydrogen* or *thermonuclear bombs*: The high temperature needed to initiate nuclear fusion is produced by a nuclear fission explosion. The first test explosion of a thermonuclear bomb was carried out by the United States in 1952. However, *controlled nuclear fusion* by the D–T reaction to produce thermal energy that can be converted into electricity is still at the laboratory stage despite almost 35 years of research.

The first step in bringing about a self-sustaining, controlled nuclear fusion reaction is to heat the D–T fuel to about 100 million degrees so that the positively charged fuel nuclei are moving

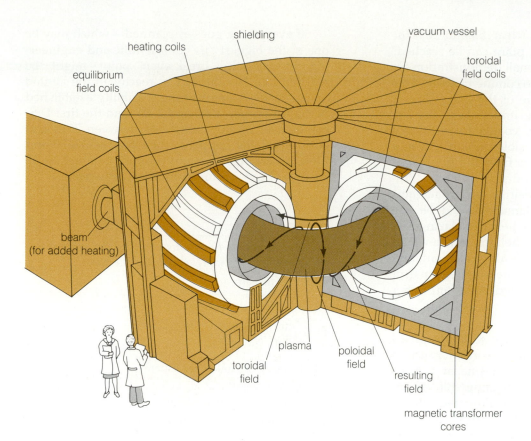

Figure 18-5 The magnetic confinement approach to nuclear fusion in a tokamak reactor.

shielding

heating coils

equilibrium field coils

vacuum vessel

toroidal field coils

beam (for added heating)

plasma

toroidal field

poloidal field

resulting field

magnetic transformer cores

fast enough to overcome their mutual electrical repulsion when they collide. Heating deuterium and hydrogen atoms to such a high temperature creates a gaslike *plasma* in which the deuterium and tritium atoms are so energetic—so hot—that the nuclei lose their electrons. The next step is to find a way to hold and squeeze the plasma together long enough and at a high enough density to ensure that sufficient numbers of positively charged nuclei of the D–T fuel atoms collide and fuse. No physical walls can be used to confine the hot plasma, not only because any known material would be vaporized but also because the walls would contaminate the fuel and instantly cool it below its ignition temperature.

So far the most promising approach is *magnetic confinement*, using powerful electromagnetic fields to confine and force the atomic nuclei in the plasma together within a vacuum. The plasma and vacuum region is surrounded by a wall made of a metal alloy that retains its strength at very high temperatures. A blanket of rapidly circulating liquid lithium would probably be used to remove heat, prevent the metal wall from melting, and breed tritium fuel. The molten lithium coolant would be piped out of the reactor and used to produce steam, which would be used to generate electricity. Outside the blanket, powerful electromagnets, cooled by liquid helium to about −272°C (−458°F) to make them superconducting, would

provide the extremely powerful magnetic fields needed to confine the plasma.

Research so far has concentrated on finding a geometric shape for the confined plasma that minimizes plasma leakage so that effective fusion of the fuel nuclei can take place. One promising magnetic confinement approach is to use electromagnetic fields to squeeze the plasma into the shape of a large toroid or doughnut (Figure 18-5). Such a reactor is known as a *tokamak* (after the Russian words for "toroidal magnetic chamber"), a design pioneered by Soviet physicists, including Andrei Sakharov. In another approach, electromagnetic fields are used to compress the plasma into a long, pipelike shape; at each end, opposing magnetic fields known as tandem mirrors, deflect the plasma and prevent it from leaking out.

By early 1985 none of the several test reactors throughout the world had been able to reach the *break-even point*, where the energy pumped into the reactor equals the energy it produces. Scientists hope to achieve the energy break-even point by 1987 using the tokamak fusion test reactors at Princeton University or the Massachusetts Institute of Technology. If this point is reached in the laboratory, the next, even more difficult step will be to achieve the *burning point*, or true ignition, where the D–T nuclear fusion reaction becomes self-sustaining and releases more energy than is put in.

Building a Commercial Nuclear Fusion Reactor

Assuming that the burning point can be reached, the next step is to build a small demonstration fusion reactor and scale it up to commercial size. This task is considered to be one of the most difficult engineering problems ever undertaken. For example, the electromagnets, cooled to practically the lowest temperature possible on earth, would be located only a few meters from the plasma, at the highest temperature produced on earth. Protecting the extremely sensitive electromagnets from heat and radiation damage would be like trying to preserve an ice cube next to a blazing fire, only much harder. Moreover helium, which would be used to cool the magnets, is a rare element, and supply problems might limit the long-term use of nuclear fusion unless the United States and other nations established a helium conservation and storage program.

Another engineering problem would be posed by the necessity of maintaining the interior section of the reactor containing the plasma at a near-perfect vacuum. More mind-boggling still, the inner walls of the reactor surrounding the lithium blanket must resist constant baths of highly reactive liquid lithium at 1,000°C (1,800°F) and steady bombardment by fast-moving neutrons released when deuterium and tritium fuse. Since neutron bombardment eventually destroys or alters the composition of presently known materials, reactor walls would have to be replaced about every 5 years, at such enormous cost that some scientists doubt whether fusion will ever be economically feasible. Scientists hope to overcome some of these problems by developing special new alloys, but some of the elements used to make these new alloys may be unaffordably scarce.

There are other problems. The neutron bombardment of the walls and other structural materials near the reactor core would convert many of the chemical elements into radioactive materials. As a result, repairs would have to be made by automatic devices, still to be developed, since no human worker could withstand the radiation. There is also concern over the high-level magnetic and electrical fields near the reactor, which might be hazardous to power plant employees.

The estimated cost for a commercial fusion reactor based on presently known approaches is at least two to four times that for a comparable breeder fission reactor and at least four to eight times that for a comparable conventional fission reactor.

If everything goes as planned—which may be one of the biggest "ifs" in scientific and engineering history—the break-even point might be reached in a laboratory test reactor in the United States by 1987, engineering feasibility established perhaps between 1995 and 2005, and the first U.S. commercial reactor completed between 2010 and 2025. If all this happens, then between 2050 and 2150 nuclear fusion might produce as much as 18 percent of U.S. annual primary energy needs.

The United States, the country that led the world into the age of nuclear power, may well lead it out.

Lester R. Brown

Discussion Topics

1. Criticize the following statements:

   a. A conventional nuclear fission plant can blow up like an atomic bomb.

   b. A nuclear fusion plant can blow up like a hydrogen bomb.

2. What method should be used for the long-term storage of high-level nuclear wastes? Defend your choice.

3. Do you favor a U.S. energy strategy based on greatly increased use of conventional nuclear fission reactors to produce electricity between 1985 and 2020? Why or why not?

4. Explain why you agree or disagree with each of the following proposals.

   a. The licensing time of new nuclear power plants in the United States should be halved (from an average of 12 years to 6 years) so that these facilities can be built more economically and can compete more effectively with coal and other renewable energy alternatives.

   b. A crash program for developing the nuclear breeder fission reactor should be developed and funded by the federal government to conserve uranium resources and eventually to keep the United States from being dependent on other nations for uranium supplies.

5. Do you believe that the United States and other industrialized nations should try to reduce the risk of nuclear war by pledging not to sell or give additional nuclear power plants or any related forms of nuclear technology to other nations? Why or why not?

Renewable Energy Resources

Throughout most of human history, people have relied on renewable resources—sun, wind, water, and land. They got by well enough, and so could we.

Warren Johnson

19-1 Direct Solar Energy for Producing Heat and Electricity

Direct Solar Energy for Low-Temperature Heating of Water and Buildings *Passive* or *active solar heating systems* can be used to collect some of the world's *direct* input of renewable solar energy and use it as primary energy to provide from 50 percent to all of the space heating and hot water in homes and commercial buildings. The accompanying box and Figures 19-1 and 19-2 describe the major components of such systems (pages 240–243).

Worldwide by 1985 there were at least 4.7 million homes (1.1 million in the United States) receiving all or part of their heat and hot water by passive or active solar systems. With proper climate-sensitive design, passive or actively heated solar buildings can be built almost anywhere in the United States. By 1985, Israel, with virtually no oil reserves and 320 days of sunshine a year, had installed enough solar collectors on rooftops to meet 40 percent of its hot water needs and expects to use solar energy to heat 60 percent of its hot water by 1990. By 1985, Japan had more than 4 million solar hot water heaters in use, serving 10 percent of its houses, and the goal is to have 7 million installed by 1990.

Because it is thermostatically controlled, an active solar heating system can give more even heating with less involvement by homeowners in opening doors and windows than a passive system. However, active solar heating systems require more materials to build the collectors, pipes, and storage systems than passive systems and thus have a lower net useful energy yield. They are also more expensive than passive systems, both ini-

tially and on a lifetime basis, because they require more maintenance and because eventually collectors, pumps, motors, and fans deteriorate and must be replaced.

In 1984 an active solar heating system for a moderate-sized American home cost $5,000 to $12,000. The cost, however, can be reduced considerably by building a highly energy-efficient house (cuts the solar system size five- to tenfold) and by having the collectors assembled at the building site instead of buying packaged collectors (cuts collector cost by a factor of 2 to 3). Active solar heating systems are economically feasible in much of the United States, however, when lifetime costs, tax credits for installation (up to 70 percent of the cost), and cash rebates from some utilities are considered. To help overcome the initial cost barrier, several firms have begun leasing solar systems or selling the solar output to businesses and other clients.

Well-designed passive solar systems are the simplest, cheapest, most maintenance-free, and least environmentally harmful energy systems for providing from 50 percent to all hot water and space heating needs of a building (Tables 16-3, 16-4, and 16-5). This economic advantage will increase as fossil fuel and electricity prices rise.

Different approaches to passive solar heating include: **(1)** attaching a greenhouse to serve as a solar collector and to grow food and ornamental plants or to serve as a sun room (Figure 19-2), **(2)** reducing temperature variations by storing the solar energy in Trombe walls (Figure 19-2),* **(3)** storing heat in a roof pond exposed to the sun during the day and covered with an insulated panel at night (and vice versa during the summer), and **(4)**

---

*Named after its designer, Felix Trombe, a Trombe wall is a thermal storage wall placed several centimeters inside a large expanse of glass or plastic on the side of a building facing the sun. The wall is either constructed of masonry or filled with water, usually in tall cylindrical columns, and is painted a dark color to absorb heat from the sun. It provides thermal mass, and at night or on cloudy days, it radiates into the house heat collected during sunny periods.

adding simple window box solar collectors, which can be built in a few hours for about $75 each, to all windows facing the sun.

Another option is to build a passively heated and cooled *earth-sheltered* (underground) house (Figure 19-2) or commercial building. By having the only exposed wall facing the sun and using windows or an attached solar greenhouse (or sun room) to capture solar energy, such structures can provide from 50 to 100 percent of the heating and cooling needs of the building. Typically, they cost about 10 to 20 percent more to build initially than a comparable above-ground structure, primarily because of the large amount of concrete needed to carry the heavy load and pressure from the earth. But on a lifetime-cost basis, they should be cheaper than a conventional above-ground house because of reduced heating and cooling requirements, elimination of exterior maintenance and painting,

and reduced fire insurance rates. They also provide more privacy, quietness, and security from break-ins, fires, hurricanes, tornadoes, earthquakes, storms, and nuclear attack than conventional above-ground buildings.

The interior of an earth-sheltered structure can look like that of any ordinary home, and the solar collecting windows or attached greenhouse, sunken atrium, and skylights can provide more daylight than is found in most conventional dwellings. By 1985, at least 4,500 earth-sheltered structures had been built or were under construction in the United States.

During hot weather cooling can be provided in a passive solar structure by using **(1)** heavy insulation, **(2)** deciduous trees and window overhangs or shades on the side facing the sun to block the high summer sun (Figure 19-2), **(3)** evaporative coolers in fairly dry climates, to remove heat when

Major Components of Passive and Active Solar Heating Systems

1. *A solar heat collector*, oriented to face the sun, consisting of double- or triple-paned windows in a *passive system* and a specially designed solar collector normally mounted on the roof in an *active system*. In a typical *active solar heating system*, flat-plate solar collectors, evacuated tubes, or other even more efficient concentrators are mounted on the roof and angled to capture the sun's rays. A typical flat-plate *active solar collector* (Figure 19-1) consists of a coil of copper pipe attached to a blackened metal base and covered with a transparent layer of glass or plastic. Radiant energy from the sun passes through the glass cover, is absorbed by the blackened surface, and is transferred as heat to water, an antifreeze solution, or air pumped through the copper pipe.

2. *A heavily insulated, airtight house* for either system to help temporarily retain as much of the captured heat as possible, often with movable insulated shutters or curtains covering windows to reduce heat loss at night during cold weather or excessive heat gain during the day in hot weather.

3. *A large heat storage system* consisting in a typical passive system of concrete, adobe, brick, or stone walls and floors, water-filled glass or plastic columns or black-painted barrels, or panels or cabinets containing chemicals that

store and release heat through phase changes. Heat storage in a typical active system is provided by insulated tanks of water or crushed stone to store captured heat for slow release at night or during cloudy days unless the structure is superinsulated (Section 19-7).

4. *A heat circulating system* consisting in a passive system of an open design that in cold weather allows heat to flow throughout the house by convection. Active systems use thermostatically controlled pumps and fans to circulate heat by conduction.

5. *An air-to-air heat exchanger* to provide fresh air without significant heat loss or gain, to prevent buildup of excessive moisture and indoor air pollutants in an airtight passive or active solar home.

6. *A batch passive solar water heater* (one or more metal tanks painted black, placed in an insulated box, and covered with glass or plastic, Figure 19-1) or additional active solar collectors to provide hot water.

7. *When necessary, a small backup heating system*, typically consisting of an energy-efficient conventional heating system or wood stove and a conventional well-insulated tank water heater or a tankless instant water heater that operates only as needed.

Active Solar Hot Water Heating System

Figure 19-1 Active and passive solar hot water heaters.

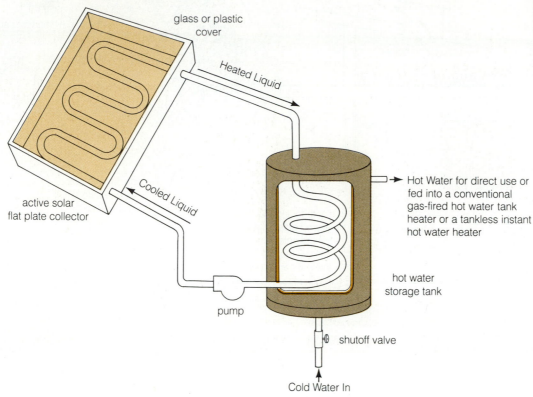

glass or plastic cover

Heated Liquid

Cooled Liquid

active solar flat plate collector

Hot Water for direct use or fed into a conventional gas-fired hot water tank heater or a tankless instant hot water heater

hot water storage tank

pump

shutoff valve

Cold Water In

Passive Solar Hot Water Heating System

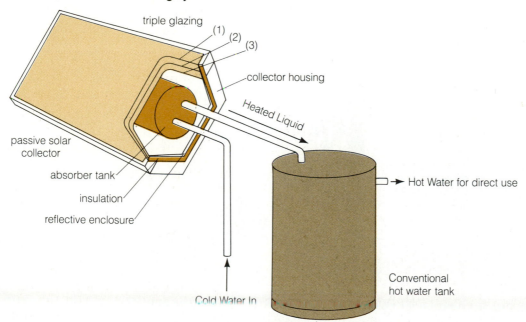

triple glazing

(1) (2) (3)

collector housing

Heated Liquid

passive solar collector

absorber tank

insulation

reflective enclosure

Hot Water for direct use

Conventional hot water tank

Cold Water In

sunlight evaporates water, **(4)** earth tubes, buried 3 to 6 meters (10 to 20 feet) underground where the temperature remains around 13°C (55°F) all year long, to bring in cool and partially dehumidified air (Figure 19-2), and **(5)** a well-designed ventilation system to take advantage of breezes and to keep air moving continuously. According to some experts, the use of climate-sensitive passive solar cooling and natural ventilation should make it possible to construct all but the largest buildings without air conditioning in most parts of the world.

The major advantages of using active or passive systems to collect direct solar energy for low-temperature heating of buildings and water are: **(1)**

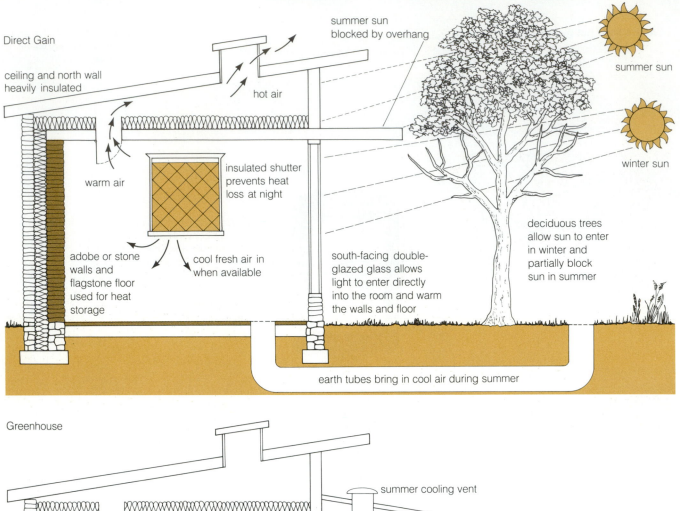

Direct Gain

ceiling and north wall heavily insulated

summer sun blocked by overhang

hot air

warm air

insulated shutter prevents heat loss at night

adobe or stone walls and flagstone floor used for heat storage

cool fresh air in when available

south-facing double-glazed glass allows light to enter directly into the room and warm the walls and floor

summer sun

winter sun

deciduous trees allow sun to enter in winter and partially block sun in summer

earth tubes bring in cool air during summer

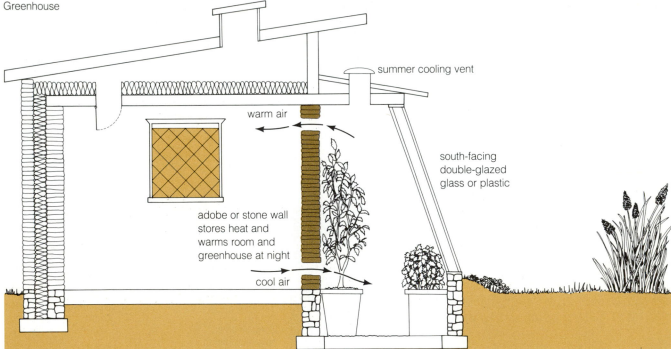

Greenhouse

summer cooling vent

warm air

adobe or stone wall stores heat and warms room and greenhouse at night

cool air

south-facing double-glazed glass or plastic

Figure 19-2 Examples of passive solar design.

the energy supply is free and readily available on sunny days; **(2)** the technology is well developed, fairly simple, and quickly installed; **(3)** passive solar is a cheap way to provide space heating almost anywhere on a lifetime-cost basis, and active solar systems for providing hot water and space heating are cost competitive on a lifetime-cost basis in most areas; **(4)** the net useful energy yield is moderate to high; **(5)** carbon dioxide is not added to the atmosphere; and **(6)** the environmental impacts from air pollution, water pollution, and land disturbance are low, primarily because the systems themselves produce no pollutants during operation and because they can be constructed of far less

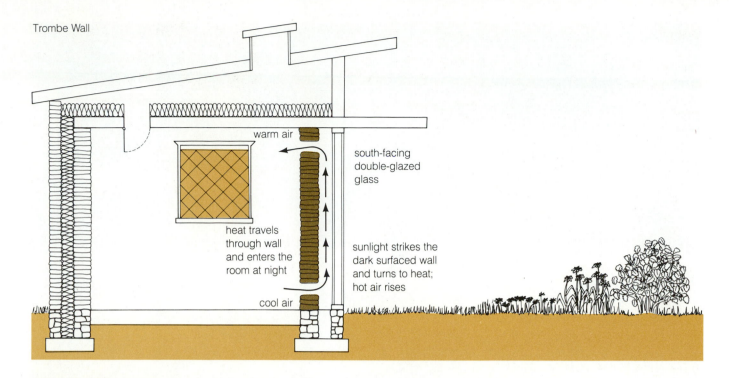

Trombe Wall

warm air

south-facing
double-glazed
glass

heat travels
through wall
and enters the
room at night

sunlight strikes the
dark surfaced wall
and turns to heat;
hot air rises

cool air

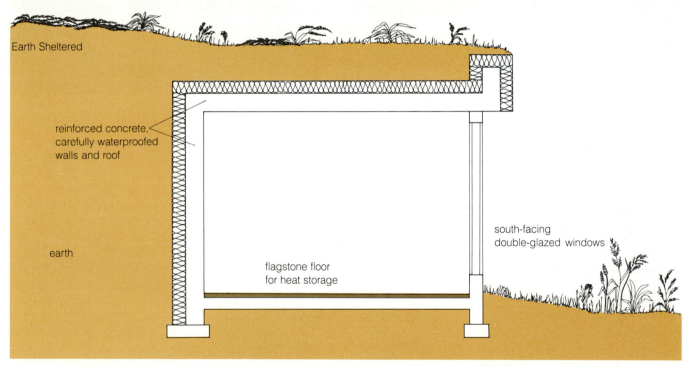

Earth Sheltered

reinforced concrete,
carefully waterproofed
walls and roof

earth

south-facing
double-glazed windows

flagstone floor
for heat storage

materials than are required for large-scale fossil fuel and nuclear power plants.

The major disadvantages of this approach are: **(1)** the energy supply is not available at night and cloudy days, so that except in superinsulated houses, thermal storage systems or small conventional backup heating systems are necessary; **(2)** active systems using liquids for heat transfer can develop leaks and cause water damage to buildings; **(3)** laws are required to guarantee that others cannot build structures that block a user's access to sunlight; **(4)** initial costs are often considerably higher than for conventional systems; and **(5)** passive solar systems require more of the owners' time to regulate heat flow and distribution.

Direct Solar Energy for Producing High-Temperature Heat You can use a magnifying glass to focus sunlight on a small area to produce

Figure 19-3 Solar I power tower used to generate electricity in the Mojave Desert near Barstow, California.

temperatures high enough to boil water and ignite paper or wood chips. A modern system for concentrating direct solar energy to produce high-temperature heat is called a **solar furnace** or a **power tower**.

The world's largest solar furnace, known as the Odeillo Furnace, has been in operation high in the Pyrenees Mountains in southern France since 1970. It contains a gigantic fixed parabolic mirror made up of 9,000 smaller mirrors. Facing it are 63 electronically controlled mirrors or *heliostats* (named after the Greek god of the sun, Helios) that swivel to follow the path of the sun throughout the day. The heliostats beam the sun's rays to specific parts of the fixed parabolic mirror, which focus the rays on a structure in front of the mirror to produce temperatures as high as 2,000°C (5,000°F)—hot enough to melt metals and sand! This solar furnace serves in the manufacture of pure metals and other substances, with the excess heat used to produce steam for spinning a turbine to generate electricity. The electricity not used at the plant is fed into the public utility grid. The cost of such an installation is very high, but unlike coal-fired and nuclear power plants, it produces negligible pollution and

no radioactive materials. Smaller units are being tested elsewhere in France and in Italy, Spain, and Japan.

Several privately and government-financed experimental solar power towers have been built in the United States to produce electricity, and more are under construction. In the Mojave Desert near Barstow, California, Solar I uses 1,818 computer-controlled heliostats to track the sun and focus its rays on a central receiver perched atop a 20-story tower and containing a boiler (Figure 19-3). The next step may involve building a larger (100-megawatt) power tower plant. However, the need to build and maintain the complex of solar collectors, focusing mirrors, plumbing, and other materials reduces the net useful energy yield so much that such systems are projected to have low to moderate net useful energy yields and high costs. It remains to be seen whether they can produce electricity at costs competitive with those of hydro, wind, coal-burning, and conventional nuclear fission power plants.

The environmental impact from air and water pollution for such systems is fairly low to moderate, coming primarily from pollutants released

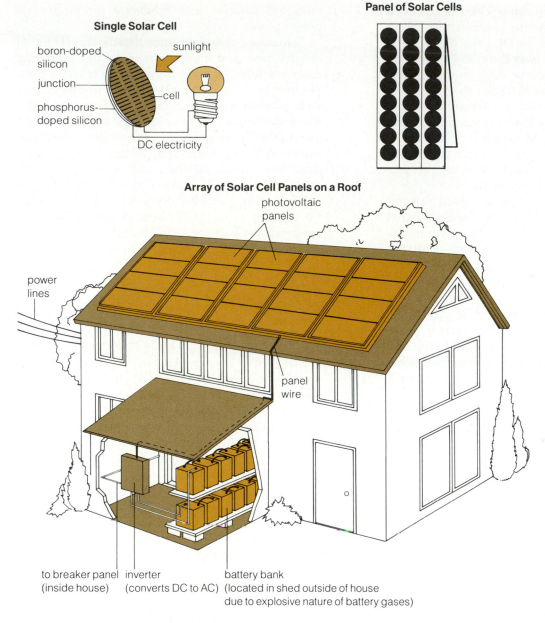

Single Solar Cell

boron-doped silicon

sunlight

junction

cell

phosphorus-doped silicon

DC electricity

Panel of Solar Cells

Array of Solar Cell Panels on a Roof

photovoltaic panels

power lines

panel wire

to breaker panel (inside house)   inverter (converts DC to AC)   battery bank (located in shed outside of house due to explosive nature of battery gases)

Figure 19-4   Use of photovoltaic or solar cells to provide electricity.

during the manufacture of the relatively large amounts of materials used in construction. Land disruption is moderate because of fairly large land area requirements (Figure 19-3).

Using Photovoltaic Cells to Produce Electricity from Direct Solar Energy   The earth's direct input of solar energy can be converted by **photovoltaic cells**, commonly called *solar cells*, directly into *electrical energy* in one simple step. Expensive solar cells are already used to power satellites orbiting the earth and to provide electricity for at least 12,000 homes worldwide (6,000 in the United States), located mostly in isolated areas where the

cost of running electrical lines to individual dwellings is extremely high.

A photovoltaic cell is made of two layers of semiconductor material containing highly purified silicon, separated by junctions. Light energy striking the silicon atoms knocks electrons free, producing a small direct electric current (DC) as the electrons flow across the junctions between the two semiconductor layers (Figure 19-4). Because the voltage and current produced by a single cell are very small, many cells are wired together in a solar panel to provide a generating capacity of 30 to about 100 watts. A number of these panels wired together can be mounted on a roof facing the sun, to produce electricity for a home or building. The resulting current can be used to power (1) electric lights and most common appliances, which can be purchased to run on DC (like those found in recreational vehicles), with the energy stored for use at night and on cloudy days in fairly long-lasting rechargeable batteries (like those in boats and golf carts), (2) converted to alternating current (AC) by an inverter to power conventional lights and appliances, or (3) used to decompose water to produce hydrogen gas, which can be stored in a pressurized tank and burned in fuel cells to provide heat and electricity (Section 19-6).

Several experimental photovoltaic electric power plants are being built that use large banks of solar cells to track the sun. A government-financed, 240-kilowatt, solar-powered plant has been providing electricity for an airport near Phoenix, Arizona, since 1982. A similar system, financed jointly by the Saudi Arabian and U.S. governments, has been providing electricity for three Saudi villages since 1982. At least a dozen similar projects are expected to be built in the next few years, mostly in the western United States and in southern Europe, where ample sunlight is available.

With about 60 percent of the world's population having no access to utility power grids, the worldwide market for solar cells is expected to reach $10 billion by 2000, increasing even more rapidly thereafter. The key to such expansion lies in cost reduction. Some observers believe that solar cells will never become competitive with conventional means of producing electricity if the total costs of the system are included. But the U.S. Department of Energy and solar cell researchers and manufacturers in the United States and Japan project that solar cells should be a competitive electricity source almost everywhere by the mid-1990s.

The major advantages of solar cells are: (1) if solar cells are used according to projections, they could be providing 20 to 30 percent of the world's electricity by 2050, thus eliminating the need to build large-scale power plants of any type and allowing the phaseout of many existing nuclear and coal-fired power plants; (2) they are reliable and quiet, have no moving parts, need little maintenance (occasional washing to prevent dirt from blocking the sun's rays), and should last for 20 to 30 years if encased in glass or plastic; (3) most are made from silicon, the second most abundant element in the earth's crust; (4) if they became cost competitive, they could allow individuals to disconnect from an electric utility company or to use utility company power only as a backup; (5) they produce no air or water pollution during operation, and air pollution from manufacture is low; and (6) land disturbance is very low for roof-mounted systems and moderate for large-scale power plants.

However, there are some drawbacks. (1) Initial construction and equipment costs and operating costs are high, but are projected to become competitive by mid-1990s; (2) the net useful energy yield is moderate to low; (3) depending on design, the widespread use of solar cells may be limited eventually by supplies of expensive or rare elements such as gallium and cadmium used to produce some units; (4) their widespread use could cause economic disruption from bankruptcy of utilities with unneeded large-scale power plants; (5) without effective controls, water pollution is moderate because solar cell manufacture produces fairly large amounts of hazardous chemical wastes.

19-2 Indirect Solar Energy from Falling Water and Ocean Waves

Hydroelectric Power Humans have used falling water as a source of energy for centuries. To produce hydroelectricity (*hydropower*), a dam is built across a stream or river to create a storage reservoir, which provides a year-round supply of water and usually can be used for recreational activities such as boating and fishing. Water from the reservoir is allowed to flow at controlled rates to spin turbines as it falls to the river below the dam. Although the energy contained in falling and flowing water is theoretically a renewable resource (Table 16-1), all hydroelectric power dams have finite lives, typically ranging from 30 to 300 years, because the reservoirs eventually fill with silt.

Falling water can also be used in *pumped-storage systems* to provide supplemental power during times of peak electrical demand. Pumps are used to transfer water from a lake or reservoir to one or more specially built reservoirs at a higher elevation—usually on a mountain. Pumping takes place during low demand periods, when it can be powered by otherwise unneeded electricity from a conventional power plant. When a power company temporarily needs more electricity than can be provided by its conventional hydro, coal-fired, or nuclear plants, water in one or more upper reservoirs is released and passed through turbines to generate electricity as it returns to the lower reservoir. This approach is used to provide 16 percent of the total hydroelectric capacity in the United States. Such a system, however, is costly and has a low net useful energy yield because of the electricity needed to pump water up to the high reservoir. The reservoirs are also normally useless for recreational purposes due to great fluctuations in water levels. Utility companies are finding that peak power demands can be met at much lower cost by loaning their customers at low or zero interest rates money for making improvements that will conserve electricity.

In 1984 hydropower supplied about one-fourth of the world's electricity—twice that from nuclear power—and about 5 percent of the world's total primary energy (Figure 16-2). Among renewable energy resources, only wood presently makes a larger contribution. Some half-dozen nations, including Norway and several countries in Africa, use hydropower to produce essentially all their electricity.

The Worldwatch Institute estimates that projects under construction or in the planning stages will double the output of electricity from hydropower between 1980 and 2000. By 2000 this source could provide an estimated 8 percent of the world's primary energy (Figure 16-2) and perhaps 40 to 50 percent of its electricity, depending on demand. By 1985 LDCs got more than 40 percent of their electricity from hydropower, and their total hydropower capacity is expected to double by 1990. China, with one-tenth of the world's hydropower potential, is likely to become the world's largest producer of hydroelectricity. Work has begun at a site on the Yangtze River that will probably be the world's biggest dam, capable of producing electricity equal to that from 25 large 1,000-megawatt nuclear or coal-fired power plants. This project, however, will force 2 million people to leave their homes. Other nations with plans for significant increases in the use of hydroelectricity include Brazil, the United States, the Soviet Union, Canada, and Austria.

Environmental impact and displacement of people can be reduced sharply by developing small-scale hydroelectric plants on small rivers near villages and other populated areas. This approach is especially useful in supplying electricity at an affordable price for many rural areas in LDCs. China leads the world in the use of small-scale hydropower, with more than 90,000 generators installed since 1952 supplying more than one-third of the electricity in rural areas.

In the United States hydroelectric capacity more than doubled between 1950 and 1983 and in 1984 provided about 13 percent of the electricity and about 5 percent of the total primary energy used. Most large-scale U.S. hydroelectric projects are located in the Southeast and the Northwest in areas with ample rainfall and high elevations. Because most of the sites suitable for large-scale dams in the United States either have been used or are located on rivers protected from development by the Wild and Scenic Rivers Act of 1968, it is pro-

jected that hydroelectric power will supply about 5 percent of the nation's primary energy in the year 2000—the same percentage as in 1984 (Figure 16-4).

In many locales, however, the fraction of electricity supplied by hydropower is being increased by putting some of the nation's 50,000 abandoned small and medium-sized hydroelectric dams and power plants back into service. According to a study by the Army Corps of Engineers, more than 5,100 of these abandoned dams could be put to use, supplying electricity equal to that from 6 to 24 large 1,000-megawatt nuclear or coal-fired power plants. The environmental impact from rehabilitating these existing dams is often small and does not require any new technology. Once rebuilt, such units have a long life, need minimal operating crews, and require little maintenance. Between 1976 and 1984 there were more than 4,000 applications to the government for permits to produce power from small hydroelectric sites in the United States, and more than 2,000 preliminary or final licenses were issued. However, there has been growing opposition by local residents and environmentalists, who contend that some small-scale hydro projects are environmentally harmful.

Another way to increase the use of hydropower in the United States at a relatively low cost and low environmental impact is to upgrade the power-generating capacity of existing dams by adding new or more efficient generators and in some cases slightly increasing reservoir size. Such upgrading could supply electricity equivalent to that from at least 46 large 1,000-megawatt coal-fired or nuclear power plants. However, since the federal government owns most of these dams, the government would have to either fund the projects or allow the dams to be developed by private firms.

The major advantages of hydropower are: (1) there is a large untapped potential in many LDCs, although many sites are in remote locations, far from points of use; (2) the net useful energy yield is moderate to high for conventional systems; (3) the fuel—running water—is free, and operating and maintenance costs are fairly low; (4) no carbon dioxide is added to the atmosphere; (5) the reservoirs for large-scale systems (except those for pumped-storage systems) are usually suitable for boating and fishing; (6) large-scale systems can provide flood control and a regulated flow of irrigation water for areas below the dam; and (7) there are essentially no emissions of air pollutants during operation, and air pollution from construction is comparable to that from a coal-fired or nuclear power plant of similar size.

The major disadvantages of hydropower are: (1) construction costs are high; (2) the net useful energy yield is low for pumped-storage systems; (3) most rivers suitable for large-scale hydropower projects in the United States and Europe have already been developed or are protected from development for environmental reasons; (4) land disruption is high, due to flooding of large areas of farmland, wildlife habitats, mineral deposits, timber areas, and historical and archeological sites above the dam and alteration of ecosystems below the dam; (5) fishing industries below the dam decline; (6) the incidence of waterborne diseases among people in many LDCs farming irrigated land below the dam is increased (Section 20-2); and (7) without proper planning and control, there is the potential for greatly increased soil erosion and sediment water pollution near the reservoir above the dam.

Indirect Renewable Solar Energy from Ocean Waves

Wave energy is derived from wind energy, which in turn is derived from direct solar energy. Capturing this energy has been a dream since at least 1799, when two Frenchmen patented a wave power device. Today research is under way, especially in Japan, Norway, France, and Great Britain, to harness the mechanical energy of wave motion and convert it into electricity or perhaps hydrogen gas fuel (Section 19-6).

Most wave power devices being tested involve using submerged tubes or floating buoys having pneumatic systems that use the mechanical energy from the up-and-down motion created by passing waves to compress air or hydraulic fluids. This extracted mechanical energy can then spin a turbine to produce electricity. So far none of these experiments has led to the production of electricity at an affordable price. But some scientists believe that with sufficient research and development, wave power could supply electricity cleanly and safely and at a competitive price during the 1990s because (1) the source of energy is free; (2) no carbon dioxide is added to the atmosphere; (3) air pollution is very low, with essentially no pollutants emitted during operation; and (4) land disruption is very low, since plants would be floating on the ocean with only electrical transmission lines located on land.

Other scientists, however, doubt that wave power will ever be a significant source of primary energy because of the following disadvantages: (1) it is available only near coastal areas with waves of sufficient height; (2) electrical output varies because of differences in wave height at different times; (3) construction and operating costs are high; (4) the net useful energy yield is low; (5) the output is disrupted and equipment is damaged or destroyed by saltwater corrosion and severe storms; (6) large-scale use could disrupt ocean aquatic life, and chemicals used for cleaning and prevention of buildup of algae and barnacles could contaminate ocean water.

19-3 Indirect Solar Energy from Thermal Gradients in Oceans and Solar Ponds

Ocean Thermal Gradients In tropical areas where the surface water is at least 27°C (80°F), there is a large temperature difference or **thermal gradient** between surface water and deep water where the sun's rays do not penetrate. In such areas, this temperature difference is large enough to permit the extraction of sufficient heat to produce electricity using a large floating *ocean thermal energy conversion (OTEC)* power plant (Figure 19-5). Some 62 nations, mostly LDCs, have national or territorial waters capable of supporting OTEC plants. For the United States, favorable sites include portions of the Gulf of Mexico and offshore areas near southern California and the islands of Puerto Rico, Hawaii, and Guam.

A full-scale OTEC plant would be a gigantic floating platform with massive pipes 30 meters (100 feet) in diameter reaching down as far as 900 meters (3,000 feet) to the ocean bottom. In a typical plant, warm surface water would be pumped through a large heat exchanger and used to vaporize a low-boiling fluid such as ammonia, Freon, or propane (Figure 19-5). As the vapor expanded, it would cause a turbine to turn, generating electricity. The vapor would pass through another heat exchanger, to be cooled by cold water pumped from the ocean's depths and recondensed to the liquid state. Then the liquid would be pumped back to the first heat exchanger for the entire cycle to begin again. The pumps in a moderate-sized 250-megawatt plant would have to be capable of pumping water *each second* greater than the average flow rate of the Mississippi River. The energy they use reduces the net useful energy yield for the system. A large cable might transmit the electricity to shore. Other alternatives include using the electricity produced to desalinate ocean water and extract minerals and chemicals from the sea or to electrolyze water to produce hydrogen gas, which could be piped or transported to shore for use as a fuel (Section 19-6).

Major advantages of this approach are: (1) the source of energy is free and essentially infinite at suitable sites; (2) a costly energy storage and backup system is not required; (3) nutrients brought up when water is pumped from the ocean

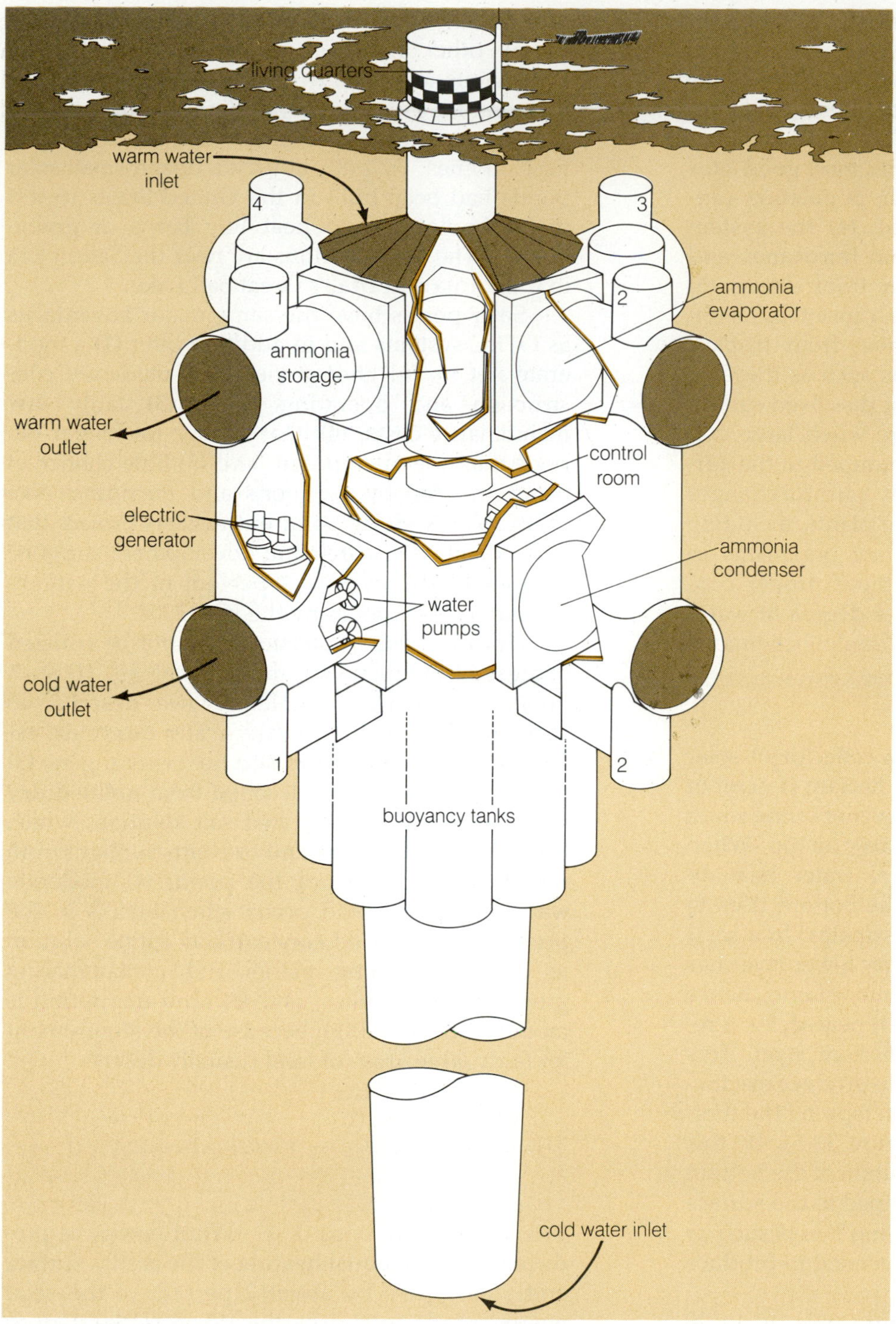

Figure 19-5 Possible design of a large-scale ocean thermal electric plant (OTEC) for generating electricity from the temperature gradient in a tropical ocean.

bottom might be used to nourish schools of fish and shellfish; **(4)** air pollution is low, with essentially no pollutants emitted during normal operation; and **(5)** land disruption is very low, since plants would be floating on the ocean with only electrical transmission lines located on land. Advocates of this approach believe that with enough research and development funding, large-scale OTEC plants could be built within 5 to 10 years, to produce electricity equivalent to that from 10 large

1,000-megawatt coal-fired or nuclear power plants by the year 2000.

Despite large research expenditures by the United States and Japan and the enthusiasm of OTEC experts, many energy analysts believe that large-scale extraction of energy from ocean thermal gradients may never compete economically with other energy alternatives because of the following major problems: **(1)** construction costs are two to three times those of comparable coal-fired plants and operating and maintenance costs are high, primarily due to seawater corrosion of metal parts and fouling of the heat exchangers by algae and barnacles; (2) the net useful energy yield for the entire system is low; (3) OTEC is feasible only in certain tropical ocean sites, with most good sites located far offshore and far from population centers where electricity is needed; (4) the system could be severely damaged from hurricanes and typhoons that periodically sweep tropical seas; (5) alterations in climate could be caused by withdrawal of large amounts of heat from tropical waters and warm water currents such as the Gulf Stream, as well as the release of dissolved carbon dioxide gas into the atmosphere when large volumes of deep ocean water are pumped to the surface; (6) there could be potentially harmful changes in the ecology of ocean areas near plants, including entrapment and killing of marine organisms in cold water intakes; and (7) without proper controls, there would be a moderate degree of water pollution from the release of antifouling chemicals and working fluids.

Solar Ponds A **solar pond** is a collector of solar energy consisting of at least 0.5 hectare (1 acre) of relatively shallow water with a layer of saline water on the bottom and a less saline layer on top. When the sun shines on a body of fresh water, heat absorbed by the deep layers on the bottom rises to the surface, where it quickly dissipates. In a solar pond, however, the higher salinity lower layer has a higher density or mass per unit volume, hence does not rise to the surface when heated. As a result, heat accumulates in the bottom layer. This stored solar energy can be used to produce electricity using the same approach as in tapping the thermal gradient in the ocean (Figure 19-5). In this case, however, the hot brine solution at the bottom of the inland sea or pond is pumped to the surface and used to vaporize a low-boiling liquid such as ammonia or Freon, which is then used to produce electricity.

Although solar ponds capture only 10 to 20 percent of the sunlight striking them, they make up for this with their low cost and ability to store solar energy for continuous use. An experimental solar pond power plant on the Israeli side of the Dead Sea has been operating successfully for several years. By 2000, Israel plans to build a cluster of plants around the Dead Sea to provide most of the electricity it needs for air conditioning and desalinating water.

Artificial solar ponds probably can be built near most desert areas in parts of Africa and the United States by using a dried lake bed or by digging out an area, lining the sides with a black plastic or other material (where the soil is not impervious to water) to absorb heat and raise the temperature of the water, adding lots of salt, and providing a turbine generator and electric transmission lines. By 1985, at least a dozen small solar ponds had been built in the United States to test the technology or to heat hog barns or greenhouses. The best locations are near the Salton Sea in California and Utah's Great Salt Lake.

Solar ponds have the same major advantages as OTEC systems and in addition offer (1) a moderate net useful energy yield, (2) moderate construction and operating costs, (3) fairly low maintenance costs, (4) heat storage for longer periods than the open ocean, and (5) little chance of being battered by typhoons and hurricanes like OTEC plants. Solar pond enthusiasts project that with adequate research and development support, they could supply 3 to 4 percent of the nation's primary energy needs by the year 2000.

However, this projection may not be realized because of the following major problems: (1) such ponds are primarily feasible only near desert areas with ample sunlight; (2) salty water can cause extensive corrosion of pipes and heat exchangers; (3) metals and salts can be leached from unlined dry seabeds by the concentrated salt solution, which turns the gradient dark and prevents sunlight from reaching the bottom of the pond; (4) moderate water pollution could occur when breaks in the plastic liners allowed concentrated saline solution to leak out, killing plants and contaminating groundwater supplies; and (5) land disruption is moderate to fairly high because of the requirement for fairly large areas of land (usually desert).

19-4 Indirect Solar Energy from Wind

Brief History of Wind Use Wind energy is produced by the unequal heating of the earth's surface and atmosphere by about 2 percent of the solar energy reaching the earth. Wind is then given characteristic flow patterns by the earth's rotation. Prevailing winds have been harnessed for centuries to propel ships, grind grain, pump water, and produce electricity (by windmill).

In the early 1900s nearly 6 million relatively small windmills pumped water and generated electricity in rural areas throughout the United States. By the 1940s cheap hydropower, fossil fuels, and rural electrification distribution systems had replaced many windmills, but worldwide there are about 1 million mechanical wind pumps in use, mostly providing water for livestock in Argentina, Australia, and the United States. In the 1950s the promise of cheap nuclear power caused the United States to abandon the development of wind turbines to produce electricity—a trend that is presently being reversed at a rapid pace.

Wind Power Today Today's wind machines range from simple water-pumping devices made of cloth and wood to large wind turbines with metal or fiber glass blades spanning up to 100 meters (328 feet). In the United States such modern wind turbines are being used by individual homeowners to provide most, if not all, of their electricity and by small private companies and utilities in *wind farms*, clusters of up to several hundred wind turbines located mostly in windy mountain passes and connected to existing utility lines. To develop a wind farm in the United States, a private company (1) goes wind prospecting for appropriate sites, (2) arranges to buy or lease the land, (3) negotiates with utilities to get a fair price for the electricity to be produced, (4) lines up investors, who will receive tax credits as well as a share of any profits from the sale of the electricity, and (5) erects the wind machines and produces electricity—usually within a year or two.

Small 3- to 5-kilowatt wind turbines costing from $5,000 to $20,000 completely installed, but with no power storage system, could be the first energy technology that allows a significant number of individuals in favorable wind areas to generate all their own power at an affordable cost. A 1980 study by the Solar Energy Research Institute indicated that 3.8 million homes and hundreds of thousands of farms in the United States are located in areas having sufficient wind speeds to make use of small wind generators.

The fairly high initial cost can be reduced by the 40 to 70 percent tax credits now available from the federal government and in some states. Further savings can be obtained by using reconditioned wind turbines and by selling surplus electricity to the local power company and using the electricity produced at night (when home needs are low) to recharge the batteries in an electric car. The Public Utilities Regulatory Policies Act of 1978 requires electric utilities to buy excess electricity from customers who develop wind, hydro, geothermal, solar, biomass, cogeneration, and other small-scale power-producing systems at a price equal to what the utility saves by not having to produce the electricity. The economic break-even point for wind power will drop sharply as electricity prices rise and projected technological improvements and mass production reduce the initial costs of wind systems.

After a slow start, *wind energy* is turning out to be one of the energy success stories of the 1980s in the United States, despite drastic cuts in federal wind energy research and development funds between 1980 and 1985. Blessed with windy mountain passes and other favorable sites, and aided by federal tax incentives and state tax credits for wind farm investors, California has more than 90 percent of the nation's wind farms. In 1985 at least 70 wind farms in California using almost 10,000 wind turbines produced enough electricity for about 94,000 households. The California Energy Commission projects that wind power will be the state's second least expensive source of electricity by 1990—right behind hydropower—and expects to get 8 percent of its electricity from wind power by the year 2000. By early 1985, wind farms had also been built or were under construction in Vermont, New Hampshire, Montana, Oregon, Hawaii, and the Netherlands, and more than 175 U.S. utility companies had wind energy programs. Other countries planning to make increasing use of wind energy include Denmark, Canada, Argentina, Great Britain, Sweden, West Germany, Australia, and the Soviet Union (with 5,000 wind turbines in operation by the end of 1983).

Some observers believe that developing sophisticated moderate-sized 10- to 50-kilowatt wind turbines to be shared by a small community or group of homeowners makes more economic sense than relying on large-scale wind turbines. Small and intermediate-sized wind turbines are easier to mass produce, and their small rotors are less vulnerable to the stress and metal fatigue that make large wind turbines expensive to build and maintain. In addition, small windmills can produce more power in light winds than large ones, hence can operate a greater percentage of the time. They are also easier to locate close to the ultimate users (thus reducing electricity transmission costs), and they allow greater decentralization of ownership and control, as favored by advocates of the soft energy path (Section 16-5).

Wind Power in the Future Wind power experts project that with a vigorous development program, wind energy could provide 13 to 19 percent of the projected demand for electricity in the United States and 12 percent of the world's electricity by the end of the century. Wind power expert William E. Heronemus suggests that a band of about 300,000 to 1 million giant wind turbines in the high-wind belt from Texas to the Dakotas could provide half the annual electrical needs of the United States. For the heavily populated eastern seaboard, huge wind turbines could float on offshore platforms, with the electricity either transmitted to shore or used to produce hydrogen gas from the electrolysis of seawater, which could be transported to shore in special tankers or perhaps by pipeline for use as fuel for homes, factories, and cars. Heronemus believes that enough hydrogen could be produced to supply all the heating needs of the industrial Northeast.

Advantages and Disadvantages The major advantages of wind power are: **(1)** wind is an almost unlimited, free, and safe source of energy that can be used almost continuously at favorable sites; **(2)** the net useful energy yield for the entire system is moderate; **(3)** wind turbines have low material requirements, and large wind farms can be put into use within 1 to 2 years, much quicker than the 6 to 12 years required for large-scale coal-fired, hydroelectric, and nuclear power plants; **(4)** no carbon dioxide is added to the atmosphere; and **(5)** during use there is essentially no air and water pollution, and the manufacture and use of wind systems produces less air and water pollution than other renewable and nonrenewable energy alternatives.

The major disadvantages of this energy resource are: **(1)** it can be used only in areas with sufficient winds; **(2)** when the wind dies down, it requires backup electricity from a utility company or from a fairly expensive energy storage system (batteries, pumped-water storage, hydrogen production, or a spinning flywheel); **(3)** without proper design, wind turbines emit low-frequency noise pollution and inaudible vibrations and can cause interference with local television reception and microwave communications; **(4)** widespread use of wind farms in some areas might cause interference with the flight patterns of migratory birds; **(5)** damage and injuries from blade loss could limit use in densely populated areas; **(6)** the initial costs are moderate to high and operating costs are still fairly high, although technological improvements and mass production should give wind farms an economic advantage over coal-fired and nuclear power plants in the United States and many other parts of the world by the 1990s; and **(7)** land disturbance is moderate.

19-5 Indirect Solar Energy from Biomass

Biomass Fuel *Biomass fuel* is organic matter that can be burned directly as fuel or converted to a more convenient gaseous or liquid *biofuels* by processes such as fermentation and pyrolysis (heating in the absence of air). In 1984 biomass, mostly from

the direct burning of wood and animal wastes to heat buildings and cook food, supplied about 11 percent of the world's primary energy. By the year 2000 biomass (including biofuels) is expected to provide 15 percent of the world's primary energy (Figure 16-2) and 11 percent of the primary energy in the United States (Figure 16-4).

Energy from Burning Wood and Waste Material
In 1984 wood was the primary source of energy for cooking and heating for about 42 percent of the world's population and for 80 percent of all people in LDCs. However, in 1984 more than 100 million people were unable to get enough fuelwood to meet minimum needs, and another 1 billion people had insufficient firewood. By the year 2000, the United Nations projects that more than 2 billion people will live in areas with inadequate fuelwood supplies.

In MDCs with adequate forest reserves, the use of wood for heating homes and for producing heat and steam in industrial boilers is increasing rapidly because of price rises in heating oil and electricity. By 1983, the use of wood in U.S. industries, especially the paper and wood products industries, accounted for 60 percent of wood energy consumption, with homes and small businesses burning the rest. A few utilities in heavily forested states such as Vermont have begun retrofitting coal-fired power plants to burn wood pellets. These plants are generating electricity for less than $0.07 per kilowatt-hour, which is competitive with most other available energy alternatives. By 1990 at least 25 U.S. utility plants are expected to be burning wood.

By 1985, one out of every five U.S. households was heated entirely or partially with wood, and about 1 million households a year are being added to this category. Compared to electric heating or an oil-burning furnace, a wood stove in a well-forested area can save hundreds of dollars a year in fuel bills (more if a family cuts and hauls their own firewood). In urban areas, where wood must be hauled long distances, many residents pay as much, if not more, for wood as for oil and electricity.

This growth in wood burning, however, can cause some problems, including increased outdoor and indoor air pollution and, without proper controls, excessive deforestation in some areas. Wood burning by industries and utilities causes relatively little air pollution because most large commercial wood boilers come equipped with pollution control systems. Small wood stoves, however, can add significant quantities of pollutants to the atmosphere unless equipped with a catalytic combuster, which burns off most polluting gases. By 1984, Oregon, Colorado, and Montana had passed laws regulating wood stove emissions and at least 20 states were considering such laws. In London and in South Korean cities wood fires have been banned to reduce air pollution. Some observers argue that catalytic combusters should be required on all new and old wood burning stoves to reduce the health hazard.

In agricultural areas, *crop residues* (the inedible, unharvested portions of food crops) and *animal manure* can be collected and burned or converted to biofuels. Hawaii, which plans to produce 90 percent of its electricity from renewable energy by 2005, was burning enough bagasse (the brownish fibrous residue from sugarcane) in 1983 to produce almost 8 percent of its electricity. In most areas, however, plant residues are widely dispersed and require large amounts of energy to collect, dry, and transport—unless collected along with harvested crops. In addition, ecologists argue that it makes more sense to use these valuable nutrients to feed livestock, retard erosion, and fertilize the soil. There is also considerable interest in burning *urban wastes* as a source of energy. But some observers believe that it may be more sound ecologically to compost or recycle these organic wastes.

Another approach is to establish large *energy plantations* or *farms*, where trees, grasses, or other crops would be grown and harvested by automated methods. This biomass would be directly burned, converted to biofuels, or converted to plastics, rubber, and other products of petroleum and natural gas (Figure 17-2). Ideal energy crops would be fast-growing, high-yield perennials that reproduce themselves from cuttings, since seeding requires costly collection and sowing. Grasses would be harvested every few weeks. Trees, planted close together like crops, would be harvested by clearcutting (Section 12-4) every 3 to 5 years on a rotating schedule. However, without careful planning and land-use controls, the use of energy plantations can deplete soil nutrients more rapidly than traditional silviculture (Section 12-4), reduce habitats for many forest plants and animals, require increased use of herbicides and pesticides, and in some cases compete with food crops for prime farmland.

Another suggestion is to create *petroleum plantations* of some of the 2,000 varieties of plants of the genus *Euphorbia*, which store energy in hydrocarbon compounds (like those found in oil) rather than as carbohydrates. After harvesting, the oillike material would be extracted (much as grapes are crushed to make wine) and refined to produce gasoline, and the unused woody plant residues would be converted to alcohol fuel. Such plants could be grown on semiarid, currently unproductive land. The recent discovery of the properties of another tree, the copaiba (*Copaifera langdorfii*), might eliminate the refining step, since every 6 months the copaiba plant produces oil that can be used directly to power a diesel engine. The net useful energy yield, however, could be low. In addition, the economic feasibility of this approach has not been determined, and because of insufficient water, large-scale plantations may be out of the question in many areas where such crops grow best.

Another category of potential biomass resources is the large-scale growth and harvesting of *aquatic plants* such as algae, water hyacinths, and kelp seaweed. Collecting, drying, and processing these plants, however, might require so much energy (and money) that net useful energy yields would be too low and costs too high. In addition, large-scale harvesting could disrupt ocean and freshwater ecosystems.

Advantages and Disadvantages The major advantages of obtaining energy from burning wood and biomass waste material are: (1) there are essentially unlimited supplies of trees and plants in areas with adequate forests and replanting programs and moderate to large supplies of urban wastes in heavily populated areas without large-scale recycling programs; (2) when collected and burned directly and efficiently near its source, biomass has a moderate to high net useful energy yield, although such yields are low for some aquatic plants (which must be dried), urban wastes, and some crop wastes that must be collected from large areas; (3) the costs are moderate, but may be high for aquatic plants and urban waste; (4) as long as trees and plants, which use carbon dioxide in photosynthesis, are not harvested faster than they grow back, burning biomass will not add to the net amount of carbon dioxide in the atmosphere; and (5) because biomass and biofuels are low in sulfur, they cause less air pollution from sulfur dioxide than the burning of coal and oil.

Major disadvantages include: (1) without adequate reforestation, large-scale use of wood for fuel causes deforestation, soil erosion, siltation, flooding, and loss of wildlife habitat; (2) because wood has a much lower energy content per unit of weight than coal, oil, or natural gas, large quantities must be burned to get the amount of energy provided by a much smaller quantity of fossil fuel; (3) wood contains a lot of water that must be removed by drying, often by burning fossil fuels; (4) wood-burning stoves increase the chances of home fires; **(5)** whole-tree removal and removal of forest, crop, and animal wastes for burning or conversion to biofuels could deplete the soil of valuable plant nutrients and cause increased soil erosion; **(6)** if urban wastes are burned, paper and other materials cannot be recycled (Section 24-3); **(7)** widespread use of energy plantations could lead to competition with food crops for prime farmland, possibly raising food prices; **(8)** air pollution is moderate because without adequate control, wood burning and the burning of urban wastes can create public health problems from emissions of soot, small particles, and cancer-causing substances; **(9)** water pollution is moderate to high, because unless done with great care, intensive tree and energy-crop farming can pollute water with sediments from soil erosion and runoff of fertilizers and pesticides; and **(10)** land disturbance is moderate to high without adequate reforestation, and land requirements for energy plantations are large.

Biofuels Plants, organic wastes, and other forms of solid biomass can be converted by bacteria and various chemical processes into gaseous and liquid biofuels such as **biogas** (a gaseous mixture of methane and carbon dioxide), liquid methanol (methyl alcohol or wood alcohol), and liquid ethanol (ethyl alcohol or grain alcohol), and other liquid fuels.

In China millions of biogas digesters convert organic plant and animal wastes into fuel for heating and cooking. After the biogas has been removed, the remaining solid residue can be used as fertilizer on food crops or, if contaminated, on nonedible crops such as trees. When they work, biogas digesters are highly efficient. However, they are somewhat slow and unpredictable, and vulnerable to low temperatures, acidity imbalances, and contamination by heavy metals, synthetic detergents, and other industrial effluents. Because of these problems, few MDCs use anaerobic digestion on a large scale and their use in LDCs such as India and China is declining.

Anaerobic digestion also occurs spontaneously in the estimated 20,000 landfill sites around the United States. Los Angeles has tapped into this source to heat some 3,500 homes. In 1976 Calorific Recovery by Anaerobic Processes (CRAP) began providing Chicagoans with methane made from cattle manure collected from animal feedlots. Converting to methane all the manure that U.S. livestock produce each year could provide nearly 5 percent of the nation's total natural gas consumption at 1983 levels. But collecting and transporting the manure for long distances would require a large energy input. Recycling this manure to the land to replace artificial fertilizer, which requires

large amounts of natural gas to produce, would probably save more natural gas.

Anticipating the depletion of crude oil supplies, the world must find a liquid fuel substitute for gasoline and diesel fuel. Some analysts see the biofuels, methanol and ethanol, as the answer to this problem, since both alcohols can be burned directly as fuel without requiring other additives to boost octane ratings. Although air pollution from burning methanol and ethanol is lower than that from gasoline, their combustion produces acetaldehyde, which smells bad, harms vegetation, and at high concentrations can irritate the skin and eyes and damage the lungs. These harmful emissions could be eliminated by better design or by equipping cars with pollution control devices. As with other forms of biomass, the use of these biofuels does not lead to a net increase in carbon dioxide in the atmosphere if trees and crops from which the fuels are made are replanted as fast as they are harvested.

Wood, wood wastes, other woody crop residues, sewage sludge, garbage, and coal (Section 17-3) can be gasified and converted to *methanol*. So far methanol has had little use as a fuel for conventional vehicles because it blends poorly with gasoline and corrodes rubber, plastic, and some metal parts in conventional internal combustion engines. However, automotive engineers point out that mass-produced engines redesigned to burn methanol should cost no more than gasoline engines, and road tests in West Germany, the United States, and Brazil have shown that methanol-powered engines perform at least as well as their gasoline-powered counterparts. Unless more efficient production methods can be developed, the use of pure methanol as a fuel will not be economically feasible until gasoline prices reach about $0.79 to $1.06 per liter ($3 to $4 a gallon)—already a reality in many European nations, where gasoline taxes are high to encourage conservation. A mixture of diesel fuel with 15 to 20 percent by volume methanol, called *diesohol*, is being tested and could lower the emissions of nitrogen oxide pollutants that are a drawback of regular diesel fuel.

A mixture of ethanol, water, and marketable carbon dioxide gas can be produced relatively simply from the fermentation of sugars found in sugarcane, sugar beets, sorghum, cassava, and corn. The resulting mixture is then distilled to produce relatively pure *ethanol*. Pure ethanol can be burned in today's cars with little engine modification. Gasoline can also be mixed with 10 to 23 percent by volume ethanol to make *gasohol*, a form of unleaded gasoline that burns in conventional gasoline engines.

By 1984, ethanol accounted for 43 percent of automotive fuel consumption in Brazil, with more than 1.2 million of Brazil's 10 million cars running on pure ethanol and the remaining 8.8 million cars using a gasohol mixture containing 23 percent ethanol. Brazil produces its ethanol by the fermentation of sugarcane and cassava and plans to use ethyl alcohol to produce some industrial chemicals now produced as petrochemicals (Figure 17-2). This ambitious program helped Brazil cut its oil imports in half between 1978 and 1984. The nation plans to triple its production of ethanol fuel by 1993.

By 1984, gasohol containing 10 percent by volume ethanol accounted for 5 percent of gasoline sales in the United States. Most users don't even know they are using gasohol when they fill up with this type of high-octane unleaded fuel. Presently gasohol is competitive with unleaded gasoline because it is exempt from the federal gasoline tax and from varying amounts of state gasoline taxes in at least 30 states. New energy-efficient distilleries are reducing the costs of producing ethanol, and soon this fuel may be able to compete with other forms of unleaded gasoline without tax breaks.

Some observers have argued that if large areas of U.S. cropland are used to produce corn or other grains for conversion to ethanol, the United States may not have an exportable surplus of grain to help feed the world's growing population and to help counter American trade deficits from importing oil. This problem would be greatly reduced if grain-producing nations fed more of their cattle and other grass-eating livestock on rangeland or perhaps on the grain residues produced in alcohol distilleries (if these residues are enriched by yeast proteins) instead of grain. Another problem is that the distillation process used to produce ethanol produces large volumes of a toxic waste material known as "swill," which if allowed to flow into waterways would kill algae, fish, and plants.

The net useful energy yield for producing ethanol for use as a fuel is moderate to low and can even be zero or negative when the alcohol is produced in an older oil- or natural gas-fueled distillery. However there is a net useful energy yield of 1.5 to 1 for recently completed distilleries using modern technology and powered by coal, wood, lignite, or solar energy.

19-6 Tidal Power and Hydrogen Fuel

Tidal Power   In a few places in the world, *gravitational attraction* between the moon and the rotating earth, and to a lesser extent between the sun and the rotating earth, causes a rise and fall of tides large enough to constitute a source of *renewable (tidal) energy*. Twice a day high tides cause large volumes of water to flow inland, and twice a day low tides cause these large volumes of water to flow back into the sea.

However, only about two dozen places in the world have enough change in water height between tides to produce electricity at an affordable cost. The largest tidal fluctuation in the world—16 meters (52 feet)—is found along the Bay of Fundy in Canada. If the opening to such a body of water is narrow enough to be obstructed by a dam with gates that can be opened and closed, the energy in the tidal flow can be used to spin turbines to produce electricity. Since there are two high tides and two low tides a day, this process can produce electricity only during four periods each day.

Since 1968 a small 250-megawatt commercial tidal power station has been in operation in France on the Rance River with tides up to 13.5 meters (44 feet), and another commercial station is in operation in the Soviet Union. Although operating costs are fairly low, the French project cost about 2.5 times more than a conventional hydroelectric power plant station with the same output built further up the Rance River.

Two possible locations for tidal power stations in the United States are the Cook Inlet in Alaska and Passamaquoddy Bay in Maine. However, government studies concluded that tidal power projects at these two locations would not be economical at 1983 electricity prices. By early 1984, Canada had completed a small-scale tidal demonstration project on the Bay of Fundy. If this project is successful, a number of large-scale commercial tidal power stations may eventually be built.

The major advantages of using tidal energy as a source of primary energy are: **(1)** the source of energy (tides) is free; **(2)** operating costs are fairly low; **(3)** carbon dioxide is not added to the atmosphere; **(4)** the net useful energy yield is moderate; **(5)** air pollution is very low, with no emissions produced during operation; and **(6)** land disturbance is low because the plant is located on water.

The major disadvantages of this energy source are: **(1)** there are few suitable sites, and many of these are located far from point of use, thus requiring costly electric transmission lines; **(2)** construction costs are high; **(3)** the system is subject to seawater corrosion and storm damage; **(4)** power is produced only periodically, and the time of day that power is generated varies daily with changes in times for high and low tides; and **(5)** water pollution is moderate, with possible damage to ecology of bays and estuaries by alteration of water flows and increases in the average tidal height in the bays.

Hydrogen Gas as a Fuel   Some scientists have suggested the use of *hydrogen gas (H_2)* to fuel cars

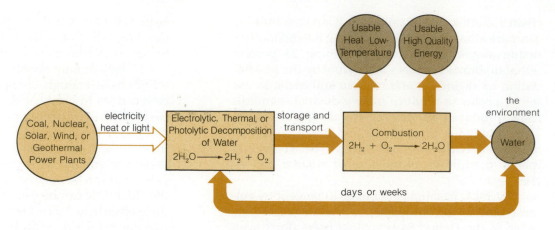

Figure 19-6 The hydrogen energy cycle has a number of advantages over the present fossil fuel energy system. But since hydrogen gas doesn't occur in nature, it must be produced by using electricity, heat, or perhaps solar energy to decompose water—thus reducing the net useful energy yield and increasing the cost of hydrogen fuel.

and heat homes and buildings when oil and natural gas run out. Although hydrogen gas does not occur in significant quantities in nature, it can be produced by chemical processes from nonrenewable coal or natural gas or by using heat, electricity, or perhaps sunlight to decompose water (Figure 19-6). Seawater could be used to provide an almost infinite supply of hydrogen gas, in sharp contrast to fossil fuels.

Hydrogen gas can be burned cleanly by reaction with oxygen gas to produce water. This reaction could take place in a power plant, a specially designed automobile engine, or in a *fuel cell* that converts the chemical energy produced by the reaction into direct-current electricity. Burning hydrogen in pure oxygen yields no emissions of air pollutants. However, burning in atmospheric oxygen results in small amounts of nitrogen oxides formed as by-products, as occurs when fossil fuels are burned. Although hydrogen gas is highly explosive, most analysts believe we could learn how to handle it safely, as we have learned to manage gasoline and natural gas.

The major catch to the widespread use of hydrogen gas is that it is found in nature in trace amounts and thus must be produced by using energy from another source. This raises the cost and results in a low to moderate net useful energy yield, depending on source of energy used to decompose water—nuclear fission, direct solar, or wind, for example. Burning hydrogen does not add carbon dioxide to the atmosphere if direct or indirect solar energy or nuclear power is used for water decomposition. However, carbon dioxide would be added to the atmosphere if electricity from coal-burning or other fossil-fuel-burning power plants were used to decompose water. The use of nuclear power to decompose water would increase the chances of adding radioactive materials to the environment and produces materials that could be used to make nuclear weapons.

Scientists are trying to develop *photoelectrochemical cells* in which light or solar energy can split water molecules into hydrogen and oxygen gases with reasonable efficiency. Even if affordable materials are used and reasonable efficiencies are obtained, it may be difficult and expensive to develop large-scale commercial cells. According to the most optimistic projections, affordable commercial cells for using solar energy to produce hydrogen will not be available until after 2000.

19-7 Energy Conservation: Improving Energy Efficiency

Advantages and Disadvantages Improving energy efficiency has more important advantages than any other energy alternative. Specifically, this approach **(1)** is an essentially unlimited source for providing energy by reducing unnecessary waste of energy obtained from any source, **(2)** has a high net useful energy yield, **(3)** saves money, with initial costs often paid back by energy savings within one to several years, **(4)** reduces the environmental impacts of all other energy resources by reducing the use and waste of energy, **(5)** extends supplies of petroleum, natural gas, coal, and other nonrenewable energy resources, **(6)** reduces dependence on imported energy resources, **(7)** greatly reduces or eliminates the need for any new coal-fired and nuclear power plants, **(8)** reduces international competition for oil, hence diminishes the chances of conventional and nuclear wars, **(9)** adds no carbon dioxide to the atmosphere, **(10)** buys time and frees capital for developing and phasing in new energy alternatives, and **(11)** is a labor-intensive alternative that provides jobs and promotes economic growth.

Major disadvantages of improving energy efficiency are as follows: **(1)** it requires individuals to consider more complex but realistic lifetime costs for energy systems and devices rather than just initial costs; **(2)** high initial costs for some equipment usually prevent low- and moderate-income individuals from taking advantage of this option in the absence of tax breaks or other forms of government assistance; **(3)** replacing most existing energy-inefficient houses, buildings, and industries with new energy-efficient units would take 30 to 50 years, and improving the average efficiency of the automobile fleet and building mass transit systems would take about 10 years; and **(4)** some conservation measures require people to alter their life-styles, which may be judged inconvenient or impractical by individuals.

Methods for Improving Energy Efficiency Because of the important advantages of improving energy efficiency, it is not surprising that between 1979 and 1984 energy conservation in the United States saved more than 100 times more energy than was obtained from finding new supplies of oil, natural gas, coal, and uranium and from phasing in new nuclear power plants. Despite these important improvements in energy efficiency, there is a long way to go: Most cars on American roads still get relatively poor gas mileage; most existing houses and buildings are still underinsulated and leaky; and most new buildings do not take advantage of available energy-efficient construction techniques.

Major methods for improving energy efficiency are given in Table 19-1, and Appendix 5 lists ways individuals can save energy. A number of studies have shown that improvements in energy efficiency such as those shown in Table 19-1 could cut the 1985 average per capita primary energy use in the U.S. by 50 percent and perhaps *ultimately* by 90 percent without decreasing the quality of life. American buildings, for example, typically consume about 50 to 90 percent more energy than they would if oriented, designed, insulated, and lighted to use energy more efficiently.

A monument to energy waste is the 110-story World Trade Center in Manhattan, which uses as much electricity as a city of 100,000 persons. Not a single window in its walls of glass can be opened to take advantage of natural warming and cooling; its heating and cooling systems work around the clock, even when no one is in the building. This is in contrast to Atlanta's 24-story Georgia Power Company building, which uses 60 percent less energy than conventional office buildings. Energy-saving features include: **(1)** innovative design (each floor extends over the one below to allow heating by the low winter sun while blocking out the higher summer sun to reduce air conditioning costs), **(2)** a computer programmed to turn off all lights at 6 P.M. unless instructed otherwise, **(3)** efficient lights to focus on desks rather than illuminating entire rooms, and **(4)** an adjoining three-story building where employees can work at unusual hours, thereby saving energy that would be necessary to service the larger building.

Building a *superinsulated house* is the cheapest and most effective way to improve the efficiency of home energy use, especially in cold climates. Such a house has the features of a passive solar home (Section 19-1) plus massive amounts of insulation. Even in Saskatchewan, Canada, where winter temperatures may average -40°C (-40°F), a well-designed superinsulated house can get *all* its space heating and hot water without a conventional backup system, typically from a combination of passive solar gain (about 59 percent), waste heat from appliances (about 33 percent), and the body heat of the occupants (about 8 percent). Such houses retain most of this heat input for at least 100 hours, and inside temperatures probably would never fall below 10°C (50°F) even in extremely cold weather. By the end of 1983, there were more than 5,000 superinsulated houses in Canada (3,000 in the province of Saskatchewan) and at least 5,000 in the United States. The number of such houses is growing rapidly as consumers, architects, and builders become familiar with their advantages and construction techniques.

To encourage greater energy conservation in buildings, several energy analysts have suggested a law that makes it impossible to sell a house that has not been weatherproofed up to certain minimum standards; also required would be an energy audit by a certified inspector to be sure that required standards are met. A program similar to this went into effect in Portland, Oregon, in 1984. It has also been suggested that federal and state governments maintain and extend existing tax credits for homeowners and businesses that use energy-conserving devices, materials, and construction (including passively heated or cooled above- or below-ground buildings and superinsulated buildings).

About one-third of the electricity generated in the United States and other industrial nations is used to power household appliances. If all U.S. appliances were replaced by the most efficient models available, the demand for electricity would fall by more than the amount of electricity gener-

Table 19-1 Major Ways to Improve Energy Efficiency

Method	Examples
Tightening existing buildings	Insulating; caulking; weatherstripping; adding storm windows and doors; having "house doctors" make energy audits
Retrofitting existing buildings	Passive solar heating by adding south-facing windows, solar greenhouses or sun rooms, solar window box heaters, and batch hot water heaters; active solar heating by adding rooftop collectors for space heating and providing hot water
New buildings	Building superinsulated, passive solar, earth-sheltered, or active solar homes and buildings
Efficient appliances	Purchasing new high-efficiency appliances such as furnaces, refrigerators, stoves, washers, freezers, dryers, air conditioners, heat pumps, and lighting
Efficient transportation	Shifting to vehicles with high fuel efficiency; improving driving habits to reduce gasoline use; using mass transit; walking or riding bicycles where possible; shipping more freight by rail, water, and pipeline rather than by truck or airplane
Computerization and substituting low-energy communication systems for high-energy transportation	Using computerized monitors and controls to improve automobile fuel efficiency and home and factory heating, cooling, and lighting systems; using industrial robots and computerized communications systems to reduce business commuting, meetings, and other forms of travel and to save money and time through video conferencing, electronic mail, and allowing many employees to work at home (the electronic cottage concept)
Retooling of factories	Switching to more energy-efficient manufacturing processes, materials, and electric motors
Industrial heat recovery and cogeneration	Recovering waste heat for use in heating water and space heating; using high-temperature waste heat to generate electricity (cogeneration)
Matter recycling	Recycling steel, aluminum, paper, and other forms of matter to reduce use of more energy-intensive virgin matter resources (Section 15-1)
Urban planning	Building efficient mass transit and paratransit systems and bike paths (Section 14-3); building more compact and self-sufficient towns and cities (Section 14-5); developing municipally or neighborhood-owned district heating systems based on locally available renewable energy resources; altering building codes, using zoning ordinances, and providing tax breaks to require or encourage more insulation and minimum energy-efficiency standards for new houses and buildings, energy audits of all buildings, active and passive solar heating of buildings and hot water, and high-density housing developments; providing tax breaks and free or low-cost loans for improving the energy efficiency of housing occupied by low-income individuals

ated by the 88 nuclear power plants operating in the United States in 1985.

Industry accounts for close to half the worldwide use of electricity. Improvements in energy efficiency in industry include phasing in a new process for producing aluminum that cuts electricity use by 25 percent and one for recycling aluminum that saves 90 percent of the power needed to produce it. Almost two-thirds of the electricity used in industry runs electric motors. New, improved designs reduce the electricity use of motors by 30 to 50 percent. Phasing in these new motors could probably reduce overall electricity use in most nations by at least 10 percent.

Japan, Sweden, and most industrialized western European nations with standards of living equal to or greater than that in the United States use only one-third to two-thirds as much energy per person. This is due to a combination of factors including greater emphasis on energy conservation and fewer passenger miles traveled per person (made possible because of more compact cities), as shown in Table 19-2. However, national energy use patterns vary, and approaches that save energy in one nation can't always be used in other countries.

19-8 Developing an Energy Strategy for the United States

Overall Evaluation of U.S. Energy Alternatives Table 19-3 evaluates energy alternatives for the United States. Energy experts will continue to argue over these and other futuristic projections, and new data may affect some of the information

Table 19-2 Energy Use in the United States and Sweden

United States	Sweden
Average per capita energy use is 230,000 kcal/day.	Average per capita energy use is 150,000 kcal/day.
Average gross national product per person in 1985 was $14,090.	Average gross national product in 1985 was $12,400.
Energy use for transportation is high because the nation is large, cities are dispersed, mass transit is not widely used, and the average car gets relatively low gas mileage.	Energy use for transportation is one-fourth that in the United States because the country is smaller, cities are compact, mass transit is efficient and widely used. Many Swedes walk and use bicycles for short trips, and the average car gets better gas mileage.
Energy waste is encouraged by low taxes on gasoline and low tariffs on oil imports.	Energy conservation is encouraged by high taxes on gasoline and high tariffs ($10 per barrel) on oil imports.
Mass transit within cities is inefficient and is not available to many suburbs (Section 14-3). Intercity rail service has been cut sharply and is not available to most areas.	Mass transit within cities is highly efficient and the intercity system provides rapid and direct service to most areas.
Large amounts of energy are used to produce a given amount of steel, copper, oil, cement, paper, and most chemicals because most aging American industries still use older, less energy-efficient technologies.	Less energy is used to produce the same quantity of steel, copper, oil, cement, paper, and most chemicals because most Swedish industries use newer, more energy-efficient technology.
Building codes in most areas do not require adequate insulation.	Building codes require adequate insulation.
Low priority is given to energy conservation in granting housing loans.	High priority is given to energy conservation in granting housing loans.
Cities are not required to develop a government-approved energy plan.	Cities are required to develop a government-approved energy plan.
Municipally owned district heating systems do not exist.	Municipally owned district heating systems provide heat for 30 percent of the population.
Kitchen stoves are used for cooking only.	Often a coil of pipe passing through the oven carries stove-heated water to the kitchen sink and to a bathroom.
Water is kept hot 24 hours a day in 150- to 225-liter (40- to 60-gallon) tanks.	Hot water is often supplied only when needed by use of tankless demand hot water heaters.
Large frost-free refrigerators use about 1,700 kW-hr of electricity per year.	Smaller nonfrost-free refrigerators use about 550 kW-hr of electricity per year.
The government has no long-range energy plan, government expenditures for energy conservation and development of renewable energy sources have been sharply reduced, and nuclear power receives most of the federal energy funds.	By 2005 the government plans to use one-third to one-fourth as much energy as in 1982 through energy conservation, and to eliminate dependence on imported oil, phase out nuclear power, and increase dependence on renewable energy (from 20 percent in 1982) to 70 percent.

shown in this table. But it does provide a useful framework for making decisions based on presently available information.

Four major conclusions can be drawn from the detailed analysis of each major energy alternative given in this chapter and the three preceding ones, as summarized in Table 19-3:

1. The best short-, intermediate-, and long-term alternative for the United States and other nations is to reduce unnecessary energy waste by improving the efficiency of energy use.

2. Total systems for future energy alternatives in the world and in the United States will probably have low to moderate net useful energy yields and moderate to high development costs. Since there may not be enough capital available to develop all alternative energy systems, projects must be carefully chosen so that capital will not be depleted on systems that yield too little net useful energy or prove to be unacceptable economically or environmentally.

3. Instead of depending primarily on one nonrenewable energy resource like oil, the world and

Table 19-3 Evaluation of Energy Alternatives for the United States (shading indicates favorable conditions)

Energy Resource	Estimated Availability*			Estimated Net Useful Energy of Entire System†	Projected Cost of Entire System	Actual or Potential Environmental Impact of Entire System‡
	Short Term (1986–1996)	Intermediate Term (1996–2006)	Long Term (2006–2036)			
Nonrenewable Resources						
Fossil fuels						
Petroleum	High (with imports)	Moderate (with imports)	Low	High but decreasing§	High for new domestic supplies	Moderate
Natural gas	High (with imports)	Moderate (with imports)	Low	High but decreasing§	High for new domestic supplies	Low
Coal	High‖	High‖	High‖	High but decreasing§	Moderate but increasing	Very high‖
Oil shale	Low	Low to moderate	Low to moderate	Low to moderate	Very high	High
Tar sands	Low	Fair? (imports only)	Poor to fair (imports only)	Low	Very high	Moderate to high
Biomass (urban wastes for incineration)	Low	Low	Low	Low to moderate	High	Moderate to high
Synthetic natural gas (SNG) from coal	Low	Low to moderate	Low to moderate	Low to moderate	High	High (increases use of coal)
Synthetic oil and alcohols from coal and organic wastes	Low	Low	Low	Low to moderate	High	High (increases use of coal)
Nuclear energy						
Conventional fission (uranium)	Low to moderate	Low to moderate	Low to moderate	Low to moderate	Very high	Very high
Breeder fission (uranium and thorium)	None	None to low (if developed)	Moderate	Unknown, but probably moderate	Very high	Very high
Fusion (deuterium and tritium)	None	None	None to low (if developed)	Unknown	Very high	Unknown (probably moderate)
Geothermal energy (trapped pockets)	Poor	Poor	Poor	Low to moderate	Moderate to high	Moderate to high
Renewable Resources						
Conservation (improving energy efficiency)	High	High	High	Very high	Low	Decreases impact of other sources
Water power (hydroelectricity)						
New large-scale dams and plants	Low	Low	Very low	Moderate to high	Moderate to very high	Low to moderate
Reopening abandoned small-scale plants	Moderate	Moderate	Low	Moderate to high	Moderate	Low

Table 19-3 Evaluation of Energy Alternatives for the United States *(continued)*

Energy Resource	Estimated Availability*			Estimated Net Useful Energy of Entire System†	Projected Cost of Entire System	Actual or Potential Environmental Impact of Entire System‡
	Short Term (1986–1996)	Intermediate Term (1996–2006)	Long Term (2006–2036)			
Tidal energy	None	Very low	Very low	Unknown (moderate)	High	Low to moderate
Ocean thermal gradients	None	Low	Low to moderate (if developed)	Unknown (probably low to moderate)	Probably high	Unknown (probably moderate)
Solar energy						
Low-temperature heating (for homes and water)	Moderate	Moderate to high	High	Moderate to high	Moderate to high	Low
High-temperature heating	Low	Moderate	Moderate to high	Moderate	Very high initially (but probably declining fairly rapidly)	Low to moderate
Photovoltaic production of electricity	Low to moderate	Moderate	High	Moderate	High initially but declining fairly rapidly	Low
Wind energy						
Home and neighborhood turbines	Low	Moderate	Moderate to high	Moderate	Moderate to high	Low
Large-scale power plants	None	Very low	Probably low	Low	High	Low to moderate?
Geothermal energy (low heat flow)	Very low	Very low	Low to moderate	Low	High	Moderate to high
Biomass (burning of wood, crop, food, and animal wastes)	Moderate	Moderate	Moderate to high	Moderate	Moderate	Moderate to high
Biofuels (alcohols and natural gas from plants and organic wastes)	Low to moderate?	Moderate	Moderate to high	Low to moderate	Moderate to high	Moderate to high
Hydrogen gas (from coal or water)	None	Low	Moderate#	Unknown (probably low to moderate)#	Unknown (probably high)#	Variable#

*Overall availability based on supply and technological, net useful energy, economic, and environmental impact feasibility.
†Rough estimates only. Better and less conflicting estimates will be available when standard guidelines for net energy analysis are adopted by all investigators.
‡See summaries for each alternative in Chapters 17, 18, and 19 for details.
§As accessible high-grade deposits are depleted, more and more energy and money must be used to find, develop, upgrade, and deliver remote and low-grade deposits.
‖Coal's very high environmental impact can be reduced to an acceptable level, however, by methods that are technically and economically feasible today.
#Depends on whether an essentially infinite source of heat or electrical energy (such as fusion, breeder fission, wind, or the sun) is available to produce hydrogen gas from coal or water. Net useful energy, costs, and environmental impact will depend on source of heat or electricity. A breakthrough in using sunlight to break down water directly (photolysis) could occur at any time, yielding dramatic advantages.

the United States will probably shift to much greater dependence on improving energy efficiency and a mix of renewable energy sources over the 50 years it typically takes to develop and phase in new energy resources (Figures 16-2 and 16-4). To provide flexibility and to reduce chances of shortages, each major energy alternative in the future may provide 10 to 20 percent of the world's or a nation's primary energy.

4. As improvements in energy efficiency are made and dependence on renewable primary energy resources is increased, primary energy production will become more localized and variable, depending on local climatic conditions and availability of renewable energy resources.

Attempts to Develop a U.S. Energy Plan After the 1973 oil embargo, Congress was prodded to pass a number of laws (see Appendix 3) to deal with the country's energy problems. Most energy experts agree, however, that these laws do not represent a comprehensive short-, intermediate-, or long-term energy strategy for the United States.

In 1981 the National Audubon Society proposed an intermediate national energy strategy that represents a mix of the hard and soft energy paths without sacrificing the nation's economic growth or requiring major changes in life-style. With this plan: **(1)** improvements in energy efficiency would be used to keep the total amount of energy used by all Americans in 2000 about equal to the amount used in 1982; **(2)** dependence on imported oil would be reduced from 15 percent of all energy used and 35 percent of all oil used in 1982 to only 3.8 and 19 percent, respectively, in 2000; **(3)** the share of primary energy obtained from renewable energy sources would be increased from 9 percent in 1982 to 25 percent in 2000 (Figure 16-4); and **(4)** nuclear and coal-fired power plants under construction in 1980 would be completed, but no new nuclear power plants would be built and a few new coal-fired power plants would be built, if needed.

This strategy would require investments of $675 billion (roughly $150 per capita per year) in energy conservation and $570 billion (about $130 per capita per year) on development of renewable energy resources between 1980 and 2000. According to the Audubon Society, this $1.2 trillion investment over 20 years is much smaller than that necessary to produce the equivalent amount of energy from new oil and gas supplies, synthetic fuels, and new coal-fired and nuclear power plants.

While politicians, energy company executives, and environmentalists argue over which path to follow, an increasing number of citizens have taken energy matters into their own hands and chosen the soft path. They are insulating, caulking, and making other improvements to conserve energy and save money, building new passively heated and cooled solar homes, adding passive solar heating to existing homes, and growing more of their own food.

Similarly, local governments in a growing number of cities, including Portland, Oregon; Seattle, Washington; San Diego, Los Angeles, and Davis, California; St. Paul, Minnesota; and Wichita, Kansas, are developing their own successful programs to improve energy efficiency. They are showing that local initiative is a faster and cheaper way to improve energy efficiency and to develop direct and indirect solar energy resources than waiting for action at the federal and state levels.

In the long run, humanity has no choice but to rely on renewable energy. No matter how abundant they seem today, eventually coal and uranium will run out. The choice before us is practical: We simply cannot afford to make more than one energy transition within the next generation.

Daniel Deudney and Christopher Flavin

Discussion Topics

1. Give your reasons for agreeing or disagreeing with the following propositions, which have been suggested by various analysts.

   a. The United States should cut average per capita energy use by at least 50 percent between 1986 and 2000.

   b. A mandatory energy conservation program should form the basis of any energy policy for the United States.

   c. To solve world and U.S. energy supply problems, all we need to do is recycle energy.

   d. Federal subsidies for all hard- and soft-energy alternatives should be eliminated so that all choices can be evaluated in an open, competitive marketplace.

   e. All government tax breaks and other subsidies for conventional fuels (oil, natural gas, coal), synthetic natural gas and oil, and nuclear power should be removed and limited subsidies granted for the development of energy conservation, and solar, wind, and biomass energy alternatives.

   f. Development of solar and wind energy should be left up to private enterprise, without help from the federal government.

   g. A network of large-scale solar electric power plants should be built to solve present and future U.S. energy problems.

2. List 20 ways in which you unnecessarily waste energy each day, and try to order them according to the amount of energy wasted (see Table 19-1 and Appendix 5). Draw up a plan showing how you could eliminate or reduce each type of waste. Which ones are the most difficult to reduce? Why?

3. Make an energy-use study of your campus or school, and use the findings to develop an energy conservation program.

4. What are some of the major political roadblocks to the transition to a world depending primarily on renewable direct and indirect solar energy for most of its energy needs?

PART FIVE

Pollution

Humans of flesh and bone will not be much impressed by the fact that a few of their contemporaries can explore the moon, program their dreams, or use robots as slaves, if the planet Earth has become unfit for everyday life. They will not long continue to be interested in space acrobatics if they have to watch them with their feet deep in garbage and their eyes half-blinded by smog.

René Dubos

The Environment and Human Health: Disease, Food Additives, and Noise

Though their health needs differ drastically, the rich and the poor do have one thing in common: both die unnecessarily. The rich die of heart disease and cancer, the poor of diarrhea, pneumonia, and measles. Scientific medicine could vastly reduce the mortality caused by these illnesses. Yet, half the developing world lacks medical care of any kind.

William U. Chandler

When there is an upset of the complex, delicate balance that normally exists between our bodies and the environment, we can become afflicted with a *disease*. The upset may result from factors in the physical environment (air, water, food, and sun), the biological environment (bacteria, viruses, plants, and animals, including humans), the social environment (work, leisure, and cultural habits and patterns, such as smoking, diet, or excessive drinking), or any combination of these three sources.

In this chapter we examine some major types of disease in LDCs and MDCs and the environmental factors that can cause, aggravate, or hasten the spread of such diseases. The potential threats to human health from some of the thousands of food additives in processed foods and from exposure to excessive noise are also discussed in this chapter. Threats to the health of humans and other organisms and other harmful effects from air pollution, water pollution, pesticides, and solid waste and hazardous wastes are discussed, respectively, in Chapters 21, 22, 23, and 24.

20-1 Types of Disease

Infectious and Noninfectious Diseases Human diseases can be broadly classified as *infectious* and *noninfectious*. An **infectious disease** occurs when

we are *host* to disease-causing living organisms called *agents*, such as bacteria, viruses, and parasitic worms. Infectious diseases can be classified according to the method of transmission. **Vector-transmitted infectious diseases** (Table 20-1) are carried from one person to another by a living organism (usually an insect), the *vector*.

Non-vector-transmitted infectious diseases are transmitted from person to person without an intermediate carrier. Examples include the common cold, tuberculosis, cholera, measles, mononucleosis, syphilis, and gonorrhea. Transmission of such diseases usually takes place by one or a combination of methods: **(1)** close physical contact with infected persons (syphilis, gonorrhea, mononucleosis, leprosy), **(2)** contact with water, food, soil, or other materials contaminated by fecal material or saliva from infected persons (cholera, typhoid fever, infectious hepatitis), or **(3)** inhalation of air containing tiny droplets of contaminated fluid expelled when infected persons cough, sneeze, or talk (common cold, influenza, tuberculosis).

An infectious disease that is carried by many hosts without leading rapidly to many deaths is said to be *endemic* to the population. For example, mononucleosis is *endemic* in the U.S. population, and many high school and college students have it unknowingly. An *epidemic* occurs when a sudden, severe outbreak of an infectious disease such as influenza affects many people in a population and leads to many deaths. An infectious disease like influenza becomes *pandemic* when it spreads worldwide to infect and kill a large number of people. Because of the extensive movement of people and food throughout the world, only strict sanitation and public health measures can protect against pandemics from non-vector-transmitted infectious diseases. Vector-transmitted infections can become pandemic only when vector organisms like fleas and bats are transferred throughout the world and can survive under a variety of climates and conditions.

Table 20-1 Major Vector-Transmitted Infectious Diseases

Disease	Infectious Organism	Vector	Estimated Number of People Infected (millions)
Malaria	*Plasmodium* (parasite)	*Anopheles* (mosquito)	500*
Schistosomiasis	*Schistosoma* (trematode worm)	Certain species of freshwater snails	200
Filariasis (elephantiasis and onchocerciasis, or river blindness)	Several species of parasitic worms	Certain species of mosquitoes and blood-sucking flies (elephantiasis); female black flies (onchocerciasis)	270
Trypanosomiasis (African sleeping sickness and Chagas' disease)	*Trypanosoma* (parasites)	Tsetse fly (African sleeping sickness); kissing bugs (Chagas' disease)	100

*250 million new cases each year.

Noninfectious diseases are not relayed by disease-causing organisms, and except for genetic diseases are not transmitted from one person to another. Examples include cardiovascular (heart and blood vessel) disorders, cancers, diabetes, chronic respiratory diseases (bronchitis and emphysema, Section 21-3), allergies, nerve and other degenerative diseases (cerebral palsy and multiple sclerosis), and genetic diseases (hemophilia and sickle cell anemia). Many of these diseases have several, often unknown causes and tend to develop slowly and sometimes progressively over the years. Typically they are caused by: **(1)** exposure to certain chemicals (some cancers and emphysema), ultraviolet energy from the sun (some forms of skin cancer), and pollen and other materials found in air, water, and food (asthma and hay fever), **(2)** inherited genetic traits (hemophilia and sickle cell anemia), **(3)** a combination of environmental and genetic factors (emphysema), and **(4)** changes in body chemistry triggered by unknown causes (cerebral palsy and diabetes).

Acute and Chronic Diseases Diseases can also be classified according to their effect and duration. An **acute disease** is an infectious disease such as measles or typhoid fever from which the victim either recovers or dies in a relatively short time. A **chronic disease** lasts for a long time (often for life) and may flare up periodically (malaria), become progressively worse (cancer and cardiovascular disorders), or disappear with age (childhood asthma). Chronic diseases may be infectious (malaria, schistosomiasis, leprosy, tuberculosis) or noninfectious (cardiovascular disorders, cancer, diabetes, emphysema, hay fever). Table 20-2 sum-marizes major characteristics of acute infectious diseases, chronic infectious diseases, and chronic noninfectious diseases.

The Social Ecology of Disease Although there is considerable room for improvement, impressive achievements have been made in human health in both MDCs and LDCs over the past 25 years. Since 1960 the average life expectancy has increased from 41 years to 50 years in LDCs and from 70 years to more than 75 years in MDCs.

Table 20-2 shows that the prevalence and mortality of the three major categories of diseases differ among the MDCs and LDCs. The four leading causes of death in LDCs are: **(1)** respiratory infections such as pneumonia (18 percent of all deaths), **(2)** heart disease and strokes (18 percent), **(3)** diarrheal infections and parasites (17 percent), and **(4)** cancer (8 percent).

The populations of LDCs tend to have a shorter average life span than those in MDCs, largely because of the complex interactions among poverty, malnutrition, and infectious diseases (Figure 11-2). Poor people in these nations are more likely to come into contact with infectious organisms because of contaminated water and food, crowding, and poor sanitation. According to the World Health Organization, about 80 percent of all infectious disease is caused by unsafe drinking water and inadequate sanitation.

The tropical or equatorial location of most LDCs also increases the chances of infection, because hot, wet climates and the absence of winter enable disease vectors to thrive year round. In addition, poor people—especially infants—tend to be more susceptible to diseases because they are

Table 20-2 Comparison of Acute Infectious, Chronic Infectious, and Chronic Noninfectious Diseases

Characteristic	Acute Infectious: Measles, typhoid fever, whooping cough, smallpox	Chronic Infectious: Malaria, schistosomiasis, tuberculosis	Chronic Noninfectious: Cardiovascular disorders, cancer, diabetes, emphysema
Cause	Living organism	Living organism	Usually several, often unknown environmental and/or genetic factors
Transmission	Usually nonvector	Vector and nonvector	Not transmitted directly but some may be transmitted genetically
Time for development (latent period)	Short (hours or days)	Long (usually years)	Long (usually years)
Duration	Usually brief (days)	Long (often for life)	Long (often for life)
Effects	Usually temporary or reversible	Usually irreversible	Usually irreversible
Age group	Children and adults	Adults, middle to old age	Adults, middle to old age
Prevalence	High in less developed nations, low in more developed nations	High in less developed nations, low in more developed nations	High, especially in more developed nations where longer life spans allow diseases to develop
Mortality	High in less developed nations, low in more developed nations	High in less developed nations, low in more developed nations	High in more developed nations
Prevention	Sanitation, clean drinking water, vaccination	Sanitation, clean drinking water, vaccination, vector control	Control of environmental factors such as smoking, diet, and exposure to polluted air, water, and food

more likely to be weakened by malnutrition. Thus, infectious diseases from which the rich usually recover—whooping cough, typhoid fever, pneumonia, diphtheria, measles, dysentery, and diarrhea—tend to kill the poor, most of whom do not live long enough to contract chronic noninfectious conditions such as heart disease, cancer, or emphysema.

One indicator of the prevalence of infectious diseases in a country is its *infant mortality rate*. In 1985 the average infant mortality rate in the MDCs was 18 deaths of infants less than a year old per 1,000 live births, compared to an average of 90 deaths per 1,000 live births in LDCs. Finland, Denmark, Canada, Hong Kong, Norway, Iceland, France, Japan, Sweden, and several other nations have very low infant mortality rates—from 7 to 10 infants out of every 1,000 born die in their first year of life. In the United States, which in 1985 ranked eighteenth in the world in infant mortality, about 10.5 infants out of every 1,000 live births died in their first year. By contrast, in some African and Asian nations the infant death toll during the first year is 100 to 200 out of every 1,000 live births.

Fortunately, major improvements in the health of people in the LDCs can be made with preventive and primary health care measures at a relatively low cost. The most important of these are providing (1) better nutrition and birth assistance for pregnant women in LDCs, where half of all births are delivered without any assistance from a trained midwife or doctor, (2) family planning, greatly improved child care (including the promotion of breast-feeding), to reduce the infant mortality rate, and (3) clean drinking water and sanitation facilities to the third of the world's population that lacks them. Extending such primary health care to all the world's peoples would cost an additional $10 billion a year, one-twenty-fifth as much as the world spends each year on cigarettes.

In most MDCs, safe water supplies, public sanitation, adequate nutrition, and immunization have nearly stamped out many infectious diseases. In 1900 the infectious diseases pneumonia, influenza, tuberculosis, and diarrhea were the leading causes of death in the MDCs. By contrast, the four leading causes of deaths in MDCs today are heart disease and strokes (48 percent), cancer (21 percent), respiratory infections (8 percent), and accidents, especially automobile accidents (7 percent).

These deaths are largely a result of environment and life-style rather than invasion of the body by an infectious agent. Except for auto accidents, these deaths result from chronic diseases that take a long time to develop, have multiple causes, and are largely attributable to the area in which we live and work (urban or rural), our work environment, our diet, whether we smoke, and the amount of alcohol we consume. At least 50 percent of all deaths and injuries from automobile accidents in the United States are related to alcohol. Analysts argue that these alcohol-related deaths

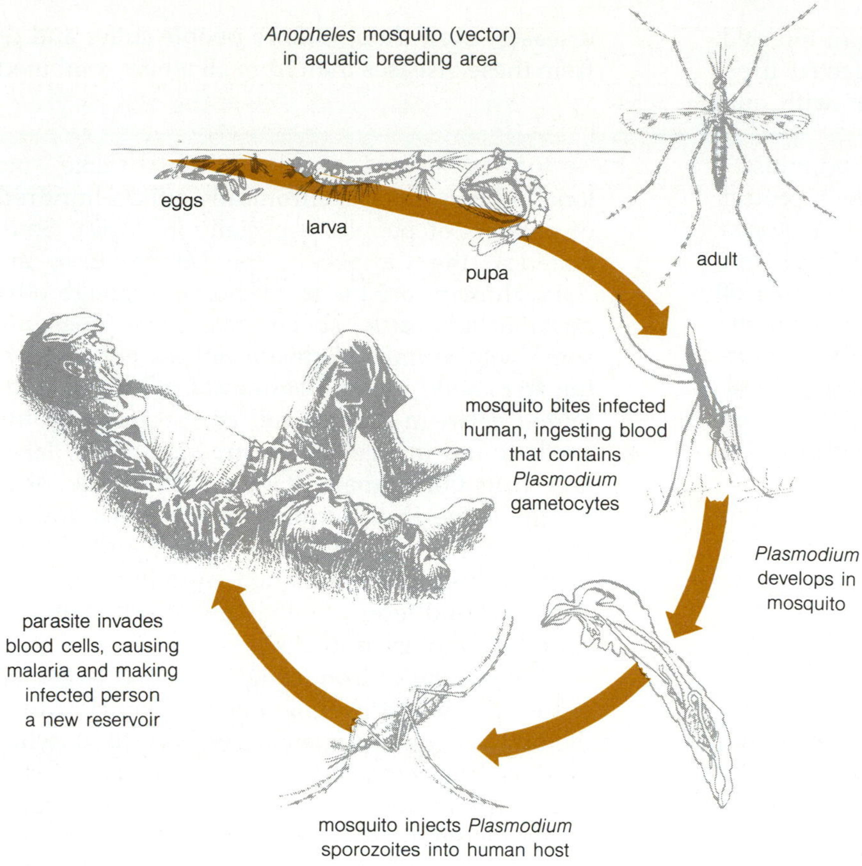

Anopheles mosquito (vector)
in aquatic breeding area

Figure 20-1 The life cycle of malaria.

eggs

larva

pupa

adult

mosquito bites infected
human, ingesting blood
that contains
Plasmodium
gametocytes

Plasmodium
develops in
mosquito

parasite invades
blood cells, causing
malaria and making
infected person
a new reservoir

mosquito injects *Plasmodium*
sporozoites into human host

could be reduced sharply by requiring all passengers to use seat belts (as is done in essentially all MDCs except the United States), raising the drinking age to 21, increasing the enforcement and penalties for driving under the influence of alcohol, and discouraging happy hours, chug-a-lug contests, and other practices that foster excessive alcohol consumption over a short period of time.

Even though heart disease, stroke, cancer, and pulmonary disease are among the leading causes of death, the American public is getting healthier both in terms of lower death rates and in terms of a lower incidence of disability due to illness. Average life expectancy for Americans born in 1984 was 78.2 for women and 70.9 for men, compared to averages of 68.2 in 1950 and 47.3 in 1900.

This rise in average life expectancy since 1950 is the result of the decline in death rates from infectious disease primarily because of a combination of **(1)** improved nutrition and better hygiene and sanitation, **(2)** modern medical care—especially the use of antibiotics to treat infections, and **(3)** more recently, an increase in the number of Americans who have modified their life-style to prevent illness by not smoking, by using alcohol in moderation or not at all, by exercising more, and by getting better nutrition. Despite these hopeful trends, there is a long way to go. About 40 percent of all cancer deaths and 25 percent of all heart dis-

ease deaths in the United States occur in persons under age 65. Some analysts believe that perhaps half these deaths could be postponed by preventive medicine involving changes in life-style and early diagnosis.

20-2 Infectious Diseases: Malaria, Schistosomiasis, and Cholera

Vector-Transmitted Malaria   People in the United States and in most MDCs tend to view malaria as a disease of the past. But *in the tropical and subtropical regions of the world, malaria is still the single most serious health problem—killing from 2 million to 4 million people a year and incapacitating many millions more.* Today more than half the world's people live in malaria-infested regions.

Malaria's symptoms come and go; they include fever and chills, anemia, an enlarged spleen, severe abdominal pain and headaches, extreme weakness, and greater susceptibility to other diseases. Caused by one of four species of protozoa (one-celled organisms) of the genus *Plasmodium*, the disease is transmitted from person to person by the bite of several species of *Anopheles* mosquito, which act as vectors (Figure 20-1). Malaria can also be transmitted when a person receives the blood of an infected donor and when an infected drug user shares a needle with an infected user. For this reason heroin is usually "cut" with quinine, an antimalarial drug.

One way to control malaria is to administer antimalarial drugs like chloroquine, which protect people against infection from bites of *Anopheles* mosquitoes. Antimalarial drugs are helpful, but they cannot be used effectively to rid an area of malaria. The cost is too high because people in infected areas would have to take the drugs continuously throughout their lives. In addition, new strains of carrier mosquitoes eventually develop that have genetic resistance to any widely used antimalarial drug.

Another approach is vector control—trying to get rid of the mosquito carriers by draining swamplands and marshes and by spraying breeding areas with DDT and other pesticides. During the 1950s and 1960s the World Health Organization made great strides in reducing malaria in many areas, eliminating it in 37 countries by widespread spraying of DDT and the use of antimalarial drugs. In India malaria cases were cut from 100 million in 1952 to only 40,000 in 1966, and in Pakistan cases were reduced from 7 million in 1961 to 9,500 in 1967.

Since 1970, however, malaria has made a dramatic comeback in many parts of the world. By 1978 the number of cases had risen to 50 million in India and to 10 million in Pakistan. According to the WHO, there are now at least 250 million new cases of the disease each year. Epidemiologists at the U.S. Centers for Disease Control estimate that the true figure may be close to 800 million new cases a year. In Africa alone the disease kills 1 million children under the age of 5 each year. This tragic resurgence has occurred because of (1) increased genetic resistance of mosquito carriers to DDT and other insecticides and to antimalarial drugs, (2) rising costs of pesticides and antimalarial drugs (between 1974 and 1975 the price of DDT tripled, primarily because of rising oil prices), (3) the spread of irrigation ditches, which provide new mosquito breeding grounds, (4) the physical impossibility of reaching and spraying all mosquito-infested areas, and (5) reduction of budgets for malaria control due to the belief that the disease had been controlled.

Research is being carried out to develop biological controls for *Anopheles* mosquitoes and to develop antimalaria vaccines, but such approaches are in the early stages of development and lack adequate funding. The WHO estimates that only 3 percent of the money spent each year on biomedical research is spent on malaria and other tropical diseases, even though more people suffer and die from these diseases than from all others combined.

Vector-Transmitted Schistosomiasis Like malaria, schistosomiasis chronically afflicts hundreds of millions of people, especially in Africa, South America, the Caribbean, the Middle East, and Asia. Humans are the major hosts, although other hosts include cattle, sheep, goats, cats, dogs, and some wild animals. Schistosomiasis is caused by the trematode worm *Schistosoma*, which is transmitted between human and animal hosts by tiny snails found in freshwater streams, rivers, lakes, and irrigation canals (Figure 20-2). The adult worms lodge in the human host's veins and deposit eggs in surrounding organs and tissues, causing chronic inflammation, swelling, and pain. The urine and feces of newly infected humans can start the entire cycle again.

Victims suffer from cough, fever, enlargement of the spleen and liver, a general wasting away of the body, filling of the abdomen with fluid (which produces the characteristic pot belly), and constant pain; they are more susceptible to other diseases, and are often too weak to work. Although the disease itself is rarely fatal, persons who are severely malnourished or severely infected may die.

In rural areas of Africa and Asia it is difficult for villagers to avoid contact with infested water because they collect it for drinking, cooking, and washing clothes and bathe and swim in it. They are also infected while washing cattle, fishing, planting rice, working in irrigation ditches, and engaging in *wadu*, the ritual washing that devout Muslims perform five times a day before praying. The building of dams for hydroelectric power and irrigation, such as the Aswan High Dam in Egypt, tends to intensify the spread of the disease because the irrigation ditches are ideal breeding places for the snails that transmit the disease. As workers urinate and wade in the ditches, schistosomiasis spreads widely.

Schistosomiasis can be reduced or even eradicated in an area by: (1) preventing human excreta from reaching the snails through improved sanitation, (2) preventing people from swimming or washing in contaminated water, (3) protecting people who farm or fish in contaminated waters by the use of boots and protective clothing, (4) developing drugs to kill the worms in the body, (5) killing the snails with chemical poisons (molluscicides), and (6) using a combination of engineering approaches including draining marshlands where snails breed, increasing the water velocity in irrigation canals to prevent infestation by snails,

Figure 20-2 The life cycle of schistosomiasis.

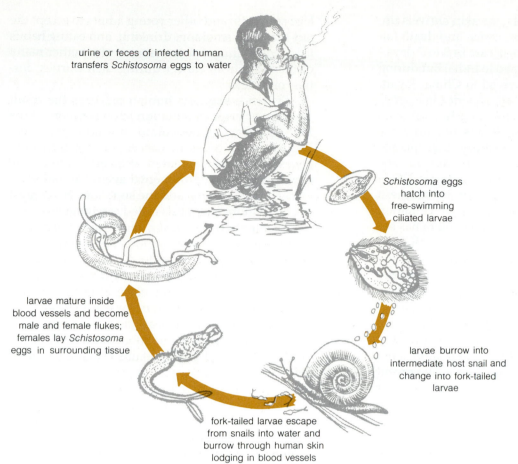

urine or feces of infected human transfers *Schistosoma* eggs to water

Schistosoma eggs hatch into free-swimming ciliated larvae

larvae mature inside blood vessels and become male and female flukes; females lay *Schistosoma* eggs in surrounding tissue

larvae burrow into intermediate host snail and change into fork-tailed larvae

fork-tailed larvae escape from snails into water and burrow through human skin lodging in blood vessels

draining irrigation projects to prevent stagnant pools and seepages, removing aquatic vegetation from irrigation ditches that serve as snail habitats, and keeping irrigation ditches relatively far from houses.

In practice, however, schistosomiasis is a very difficult disease to control. The first three approaches are expensive, difficult to put into effect, or both. In the fourth approach, several drugs have been developed that can kill most of the worms in the human body. But so far none works on all people or on all species of the parasite. Such drugs either are expensive or have serious and occasionally lethal side effects, must be administered continually because they do not prevent reinfestation, and fail after a few years because of drug resistance in the parasite or relaxation of use when the incidence of the disease has fallen off.

So far, the fifth approach has had only limited success because the chemicals are too expensive for the almost continuous use that is required. A few snails that survive by burrowing into the mud, or the influx of snails from outside an area, can quickly rebuild the population to high levels. In addition, the chemicals can kill fish, poison the water, and have other bad ecological and health side effects. However, there is some hope. Preliminary results indicate that the dried, ground berries of the endod, used by villagers in Ethiopia as a detergent for washing clothes, may be effective in killing the snails without hurting other animals and plants. Projects in Israel, Japan, China, and the Philippines have shown that the sixth approach can also help prevent an increase in the incidence of the disease. However, as long as so little money is spent worldwide on schistosomiasis research, reducing the incidence of this disease in the LDCs is going to be very difficult.

Non-Vector-Transmitted Cholera One of the most frightening infectious diseases goes by the name of *cholera*—a term used to describe a collection of infections that result in severe diarrhea and dehydration. The bacteria causing cholera are transmitted from person to person through water and food supplies contaminated with sewage. Within a half-day the infected individual suffers from severe diarrhea and vomiting, which leads to rapid dehydration. Unless the victim is treated with antibiotics to combat the infection and with fluids and salt to counter the effects of dehydration, the blood pressure falls, the skin shrivels up, and severe muscular cramps, coma, and death follow. This sequence of events occurs in 2 to 7 days.

Cholera was once confined to India; but during the nineteenth century it spread to China, Japan, East Africa, Europe, and North America in a series of six pandemic plagues, each lasting 10 to 20 years each and killing thousands. It was not until the middle of the nineteenth century that people learned that contaminated water spread the dreaded disease. After the last pandemic (1865-1875), cholera was pushed back into its southern Asia homeland as a result of improved sanitation and vaccines. Since 1961, however, cholera has begun to march across continents for a seventh time; today it kills people all over Asia, Africa, and the Middle East. There is fear that an infected traveler may bring this scourge to Latin America, where poor sanitation would allow it to spread rapidly.

Cholera is a major threat to the urban poor who live in crowded, unsanitary conditions and have no access to vaccines. Even vaccination is not a cure-all, since cholera vaccine gives a 50 percent chance of protection, and that protection lasts for about 6 months. The only effective way to combat cholera—along with a host of other waterborne infectious diseases—is to improve water supplies and sanitation throughout the world, especially in the tropics. Tropical disease experts estimate that chlorinating and filtering water supplies and using fairly simple sanitation measures like pit latrines and simple privies could reduce the incidence of cholera by 60 to 90 percent.

20-3 Chronic Noninfectious Diseases: Cancer

Nature and Effects of Cancer   **Cancer** is the name for a group of more than 100 different diseases—one for essentially each of the major cell types in the human body. It is characterized by the uncontrolled or malignant growth of cells in body tissues, leading to the formation of malignant tumors that tend to grow rapidly and spread to other parts of the body in a process called *metastasis*.

Cancer is usually a *latent disease*, requiring a time lag of 15 to 40 years between the initial cause and the appearance of symptoms. The long latency period, along with the number of different types of cancer, makes it extremely difficult to identify the specific cause or causes of a particular cancer. This time lag also prevents many people from taking simple precautions that would greatly decrease their chances of getting the disease. For instance, it is difficult for healthy high school students, col-

lege students, and other young adults to accept the fact that their smoking, drinking, and eating habits *today* will be major determinants of whether many of them die from cancer during their thirties, forties, or fifties.

Evidence suggests human cancer is the result of a genetic error or mutation in one or more of the 100,000 normal genes found in each human cell. Because most types of cancer are not inherited, genes are usually mutated after birth usually by exposure to an environmental agent like radiation (X rays, radioactivity, and the sun's ultraviolet rays) or one or more chemicals called **carcinogens**.

A mutation, however, does not necessarily lead to cancer. Most mutations are likely to be in genes that have nothing to do with cancer. The malignant growth of cells can occur only if the mutation occurs in the right place in certain genes. Even then, cancer may not occur because most evidence indicates that at least one more mutation, and in most cases many more mutations, must take place in a cell that contains the first mutation before the cancer process can begin. The first mutation, produced by a step called *initiation*, somehow sensitizes the cell to the actions of a second chemical, called a *promoter*. Many chemicals in the human environment can act as promoters, with most affecting only certain types of tissue. For example, certain kinds of dietary fat appear to be indirect promoters acting on cells of the breast and colon, and cigarette tars contain both carcinogens that can initiate lung cells and promoters that can act on the initiated lung cells. Exposure to the promoter somehow allows the initiated, mutated cell to proliferate more than do neighboring noninitiated cells. This increases the population of initiated cells relative to other cells in the same tissue so that the number of them available for a second mutation gradually increases, as well.

This unlikely, complex, and still poorly understood sequence of events helps explain why we are all not dying of some form of cancer and why it takes years to develop a malignant tumor. It also suggests why cancers are most common in tissues that reproduce rapidly such as the skin and the lining of the uterus and gastrointestinal tract.

Incidence and Geography of Cancer   Although cancer is mostly a disease of middle and old age because of its long latency period, it is second only to accidents as the leading cause of death in children between the ages of 5 and 14 in the United States. Males and females in the United States have different incidences and death rates from different types of cancer (Figure 20-3). In the United States the incidence of cancer in men begins to increase

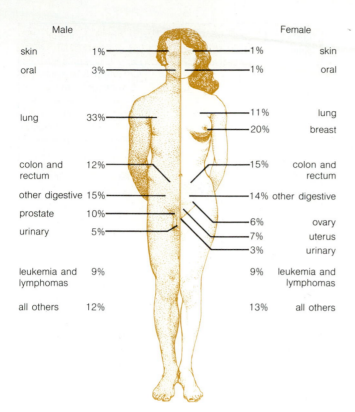

Male				Female
skin	1%		1%	skin
oral	3%		1%	oral
lung	33%		11%	lung
			20%	breast
colon and rectum	12%		15%	colon and rectum
other digestive	15%		14%	other digestive
prostate	10%			
urinary	5%		6%	ovary
			7%	uterus
			3%	urinary
leukemia and lymphomas	9%		9%	leukemia and lymphomas
all others	12%		13%	all others

Figure 20-3 Where fatal cancer strikes; percentages of all U.S. cancer deaths in 1977.

sharply at 35 and peaks at 65. For women it rises rapidly at age 20 and peaks at age 60.

Lung cancer is the leading cause of cancer deaths in the United States. In 1984 lung cancers caused by smoking killed about 147,000 Americans. Because lung cancer is difficult to detect in its early stages, only about one out of eight victims are saved. People with a nagging cough or hoarseness should see a doctor before it's too late. The best way to avoid lung cancer is prevention—primarily by not smoking.

The leading cause of cancer in women is breast cancer, although within a few years lung cancer is projected to be the number one killer of women. In 1984 about 38,000 American women died as a result of breast cancer. Unlike lung cancer, breast cancer can be detected in its early stages. A victim has an 87 percent chance of surviving 5 or more years if the breast cancer is detected before it has spread to other parts of the body. Otherwise, the chance of recovery drops to only 47 percent. Monthly self-examination of the breasts for lumps or thickening of the tissue can usually lead to detection of this type of cancer at an early stage. Treatment now rarely involves radical surgery or breast removal.

Many of the 58,000 deaths of Americans from bowel or colonrectal cancer in 1984 could have been prevented if victims had carried out a simple, at-home detection test every year after age 50, had a doctor make a proctoscopic examination of the rectum every 3 to 5 years after age 50, and consulted a physician when they experienced any unexplained change in bowel or bladder habits. Similarly, many of the 10,000 deaths of American women in 1984 from uterine cancer could have been prevented by regular pelvic examinations by a physician and by use of the Pap test for detection of this form of cancer.

Most cases of skin cancer, except a form known as malignant melanoma, can be cured if detected early enough by regular examination of the skin and by reporting to a physician immediately any skin sore that does not heal quickly or any mole or wart that begins to increase in size or change in color. Even if cured, however, skin cancers often leave disfiguring scars. Outdoor workers are particularly susceptible to cancer of the skin on the face, hands, and arms. White Americans who spend long hours in the sun or under sun lamps (which are even more hazardous than direct exposure to the sun) greatly increase their chances of developing skin cancer and also tend to have wrinkled, dry skin by age 40. Sunbathers can reduce this risk greatly and still get a tan (although more slowly) by using lotions containing sunscreen agents, which block out most of the harmful ultraviolet rays of the sun and sun lamps.

Cancer death rates vary geographically between countries and within countries. The rates for breast cancer and bowel cancer are four to five times higher in the United States than in Japan, whereas stomach cancer rates in Japan are almost seven times higher than in the United States.

Diagnosis and Treatment Many people automatically think of death when they hear the word *cancer*. Oncologists, physicians who specialize in cancer, consider that any patient who survives for 5 years after treatment and shows no trace of the disease is cured. On this basis, *about 40 to 50 percent of all Americans who get cancer can now be cured because of a combination of early detection and improved use of surgery, radiation, and drugs (chemotherapy) as treatment (Table 20-3), compared to only a 25 percent cure rate 30 years ago.** Survival rates for many types of cancer now range from 67 to 87 percent (Table 20-3).

Despite this important progress, Table 20-3 shows that there has been little success in increasing the survival rates for victims of cancers of the pancreas, esophagus, lung, stomach, and brain. Only in one form of lung cancer, known as "oat cell" carcinoma, have drugs shown substantial promise.

Today treatment consists of surgery to remove cancerous growths, and radiation and drugs to kill cancer cells. Radiation and today's anticancer drugs are somewhat like shotgun blasts—they kill not only the cancerous cells but also hair follicles, gut cells, and other normal fast-growing cells. This is why extreme fatigue, diarrhea, weight and hair loss, severe nausea and vomiting, reduction in the body's resistance to infection, and destruction of blood platelets are common side effects of the use of radiation and chemotherapy to treat cancer.

Experts are exploring wholly new kinds of treatment that, if successful, should lead to greatly increased survival rates and fewer harmful effects. One approach is *immunotherapy*, using chemicals and monoclonal antibodies made in the lab by genetic engineering techniques to stimulate the body's own disease defenses to attack cancer cells. Scientists hope to develop a series of *monoclonal antibodies* for each type of cancer. Such antibodies would be used to carry drugs or radioactive isotopes directly to tumors without harm to healthy tissues. This rifle-shot approach to treatment could eliminate most side effects associated with present chemo- and radiotherapy approaches. In 1982 one

---

*To obtain information from the National Cancer Institute about cancer and to find out where to go for help, call the toll-free number, 1-800-4-CANCER.

Table 20-3 Chances of Recovering from Cancer in the United States

Type of Cancer	Chance of Survival for 5 Years or More (%)	
	Diagnosed 1960–1963	Diagnosed 1973–1980
Among Adults		
Lining of uterus	73	87
Testis	63	82
Skin (melanoma)	60	79
Breast	63	73
Bladder	53	72
Hodgkin's disease	40	70
Uterine cervix	58	67
Prostate	50	67
Colon	43	50
Kidney	37	50
Rectum	38	48
Ovary	32	37
Leukemia	14	32
Brain	18	22
Stomach	11	15
Lung	8	12
Esophagus	4	4
Pancreas	1	3
Among Children		
Hodgkin's disease	52	84
Wilm's tumor	57	75
Acute lymphocytic leukemia	4	58
Brain and central nervous system	35	53
Neuroblastoma	25	47
Bone	20	46
Non-Hodgkin's lymphoma	18	46
Acute granulocytic leukemia	3	22

of the first patients to be treated with monoclonal antibodies was cured of a rare type of lymphatic cancer that had spread to his liver, blood, bone marrow, and spleen. Preliminary tests using monoclonal antibodies are also being conducted in patients with colon cancer, breast cancer, and leukemia.

Cancer Risk Factors Hereditary factors are involved in an estimated 10 to 30 percent of all cancers. Examples of genetically transferrable cancer risks include leukemia in children with Down's syndrome, breast cancer, ovarian cancer, malignant melanoma, retinoblastoma (a rare form of eye cancer), and lung cancer (although heredity is not nearly as important as smoking).

Environmental factors—including such aspects of life-style as diet and smoking—are believed to contribute to or directly cause the remaining 70 to 90 percent of all cancers. For example, studies have shown that U.S. Mormons, who do not smoke or drink alcohol or beverages containing caffeine, have a rate of cancer incidence one-half that of the average U.S. white population. Thus, each of us can greatly reduce the risks of developing cancer by trying to find work in a less hazardous environment, not smoking, drinking in moderation (maximum consumption two beers or drinks a day) or not at all, and adhering to a healthful diet.

Cancer and Smoking   There is widespread agreement and overwhelming statistical evidence from more than 30,000 studies that the leading cause of cancer deaths (mostly from lung cancer) is *cigarette smoking*—causing an estimated one-third of all cancer deaths in the United States and three-fourths of all lung cancer deaths in American men. Smoking kills one in four of all smokers, primarily from heart disease (240,000 deaths a year), cancer (147,000 deaths a year), and respiratory diseases (61,000 deaths a year). The overall health cost of smoking in the United States, including direct health effects and lost work productivity, exceeds $40 billion a year. This amounts to a cost paid for by both smokers and nonsmokers of $2 per pack of cigarettes.

The effects of smoking on cancer rates can also be increased by synergistic interaction with other factors. For example, more lung cancer is found among cigarette smokers living in a polluted industrial urban environment than among those living in relatively unpolluted rural areas. Alcohol consumption by itself causes an estimated 3 percent of all cancer deaths in the United States, especially liver cancer. Alcohol and smoking can also interact synergistically. In the United States, for example, moderate smokers who drink heavily are 25 times more likely to develop cancer of the esophagus than smokers who don't drink. Heavy drinkers are two to three times more likely to develop cancer of the mouth and larynx, with smoking multiplying the risk.

A British government study showed that adolescents who smoke more than one cigarette have an 85 percent chance of becoming smokers. Other studies have shown that a child is about twice as likely to become a smoker if either parent is one.

There is some good news, however. Studies show that about 10 to 15 years after former smokers quit, they have about the same risk of dying from lung cancer as those who never smoked.

Suggestions that have been made for reducing the cancer risks from smoking include: **(1)** trying to prevent young people from getting hooked on cigarettes, **(2)** banning all cigarette advertising, or restricting it to simple black-and-white print ads, so that slick presentations cannot be used to create the impression that smokers are young, attractive, sophisticated, healthy, and sexy, **(3)** passing and enforcing strict laws to protect nonsmokers in public places and encouraging offices and businesses to allow smoking only in designated areas, **(4)** eliminating all financial subsidies to the tobacco industry, and offsetting the resulting loss of income and jobs by providing aid and subsidies that would allow farmers to grow more healthful crops, **(5)** specifying for all cigarettes maximum allowable levels of tar, nicotine, and other health-threatening chemicals, **(6)** taxing cigarettes in the United States at $2 a pack to discourage smoking and to have smokers—not nonsmokers—pay for the health and productivity losses now borne by society as a whole, and **(7)** increasing research to develop tobacco substitutes and less harmful cigarettes.

Sweden has led the way by putting into effect a program whose goal is to make children born after 1974 the country's first nonsmoking generation. Intensive education on the hazards of smoking will be introduced at all school levels beginning in kindergarten, no one under 16 will be allowed to buy cigarettes, the sale of cigarettes in vending machines will be banned, cigarette prices will be raised gradually, cigarette advertising will be phased out, and all cigarette packs must carry tar and nicotine levels along with 16 different warnings on the specific hazards of smoking.

Cancer and Diet   A second major cause of cancer is *improper diet*, causing an estimated 35 percent (with a range of 10 to 70 percent) of all cancer deaths. However, evidence linking specific dietary habits to specific types of cancer is difficult to obtain and is controversial. The major factors—especially in cancers of the breast, bowel (colon and rectum), liver, kidney, stomach, and prostate—may be fats, fibers, nitrosamines, and nitrites. The incidence of cancers of the colon, rectum, and female breast is about five times higher in Americans than in Japanese, who have low-fat diets. Third-generation offspring of Japanese immigrants, however, have about the same incidence of these types of cancer as other native Americans. A high-fat, high-protein diet may also be a factor in cancers of the breast, prostate, testis, ovary, pancreas, and kidney.

Others have proposed that high incidences of bowel cancer may result from the low fiber content in Western diets, a by-product of the use of highly refined foods. Studies have given conflicting results, however, and the low-fiber hypothesis is still controversial. More recent work indicates that the incidence of bowel cancer is lower in test animals and apparently in individuals whose diet includes daily intakes of cruciferous vegetables such as cauliflower. There is also some evidence that the incidence of stomach and bowel cancer may be reduced by adequate amounts of beta-carotene, vitamin C (and perhaps vitamin E), and trace minerals such as selenium and zinc.

High levels of nitrate and nitrite food preservatives (Section 20-5) found in smoked and cured meats and in some brands of beer may increase the risk of stomach cancer because the body converts them to carcinogenic nitrosoamines. This has been implicated in the very high incidence of stomach cancer in Japan, where the diet consists of large amounts of dried, salted, pickled, and smoked fish.

A 1982 study by the National Academy of Sciences concluded that the risk of certain types of cancer like those of the lung, stomach, colon, and esophagus could be significantly reduced by a daily diet that (1) consists of no more than 30 percent saturated and unsaturated fats (compared to the 40 percent fat diet of the average American) by cutting down on fatty meat, whole-milk dairy products, butter, cooking oils, and fats, (2) includes fruits, especially oranges and grapefruit, which are rich in vitamin C, (3) includes carotene-rich, dark green and deep yellow vegetables and cruciferous vegetables, (4) includes whole-grain fiber from raw bran (the cheapest source), bran in cereals, and fibers in vegetables and fruits, (5) includes a dietary intake of selenium not exceeding 200 micrograms per day, and (6) includes very few foods that contain large amounts of nitrates and nitrites as preservatives (sausages, smoked fish and ham, bacon, bologna, salami, corned beef, hot dogs, and so on).

Cancer and the Workplace The third major cause of cancer, *occupational exposure* to carcinogens and radiation, causes about 5 percent of cancer deaths according to some health scientists and up to 20 percent according to others. This represents from 22,000 to 88,000 deaths each year. Most of these deaths could be prevented by stricter rules and stricter enforcement of existing rules governing exposure of workers to radiation and dangerous chemicals.

The major job risk is exposure to *asbestos*, which has been related to cancers of the lung, bowel, stomach, and the linings of the lungs and stomach (mesothelioma).* It also greatly increases the risk of noncancer lung diseases like emphysema and asbestosis. Asbestos is a major threat to asbestos miners and plant workers, auto mechanics (because of brake linings), steamfitters, carpenters, tile setters, and insulation and construction workers. An estimated 100,000 U.S. workers have died prematurely as a result of exposure to asbestos, and 350,000 additional deaths are projected over the next 30 to 35 years.

Asbestos fibers carried home from factories on the clothes and bodies of workers have also caused lung cancer in about one-third of their family members. School children, office workers, and other members of the general public can also be exposed to asbestos, primarily as a result of its use as insulation from 1900 to 1980 in many buildings and schools.

Other major job risks include exposure to (1) *vinyl chloride* (3.5 million workers), which increases the risks of a rare liver cancer (hemangiosarcoma), brain cancer, and lung cancer, respectively, 200, 4, and 2 times, (2) *benzene* (2.9 million workers), which increases the risk of leukemia by a factor of 2 to 7, (3) *arsenic* (1.5 million workers), which increases the risk of lung cancer by a factor of 3 to 8, and (4) *coal-tar pitch* and *coke oven emissions* (60,000 workers), which increases the risk of cancers of the lung, larynx, skin, and scrotum by a factor of 2 to 6.

Reducing job-related exposure requires strict control of all mining and manufacturing processes involving known or suspected carcinogens. Major methods for controlling exposure to hazardous industrial chemicals include: (1) eliminating the chemical, (2) segregating the hazard in a room or building, (3) covering bins, conveyors, and vats containing toxic substances, (4) installing ventilation and exhaust systems that draw work-place pollutants away from workers and replace the exhausted air with fresh air, and (5) rotating workers in and out of hazardous jobs to keep exposure below unsafe levels.

If enforced, the Occupational Safety and Health Act of 1970 and the Toxic Substances Control Act of 1975 (Chapter 24) could be important factors in establishing such controls. However, po-

---

*The serious cancer risk from exposure to asbestos is greatly increased by smoking. An asbestos worker who smokes has 11 times the risk of dying from lung cancer as a co-worker who doesn't smoke, and 55 times the risk of a person who neither works with asbestos nor smokes. For this reason, the Johns-Manville Corporation decided in 1977 not to hire smokers to work in their asbestos mines and plants.

litical pressure by industry officials has hindered effective enforcement of these laws. Maximum permissible standards set by the Occupational Safety and Health Administration (OSHA) for many major pollutants in the work place range from 2 to 100 times higher than those set by the EPA for the general population. Such double standards for protection exist in most other industrialized countries, but occupational standards for most work-place pollutants are much more stringent in the Soviet Union and in many East European countries than in the United States.

Furthermore, according to occupational health expert Samuel Epstein, workers in general are denied knowledge of the types and concentrations of the chemicals they are exposed to and are often not informed whether chemicals have been adequately tested and whether they are toxic or carcinogenic under conditions in the plant.

Opponents of stricter government control of worker exposure to potentially dangerous chemicals sometimes state that almost anything can cause cancer, so why pick on one particular chemical. Health scientists argue, however, that this is a misleading argument because most chemicals tested don't cause cancer. By 1977 about 700 of the 1,500 chemicals that had been thoroughly tested were found to cause cancer in test animals, and only 26 (18 of them found in the work place) had been directly linked to cancer in humans.

Cancer and Pollution A fourth cause of cancer is *air and water pollution*, estimated to contribute to between 1 and 5 percent of cancer deaths in the United States. This contribution may be more significant, however, for certain individuals, including: **(1)** those living in airtight, energy-efficient housing without air-to-air heat exchangers (because of abnormally high levels of indoor air pollution—Section 21-1), **(2)** nonsmokers who work or live in an environment that exposes them to cigarette smoke, and **(3)** residents of cities where chlorinated water can interact with industrial organic compounds and bromine to form halomethanes, which have been linked to abnormally high incidences of rectal, bladder, and colon cancer (Section 22-2).

20-4 Food Additives

Use and Types of Food Additives In LDCs many rural and urban dwellers consume harvested crops directly. In MDCs and in a growing number of cities in LDCs, harvested crops are used to produce *processed foods* for sale in grocery stores and restau-

rants. A large and increasing number of natural and synthetic chemicals called **food additives** are deliberately added to processed foods to retard spoilage, to enhance flavor, color, texture, and to provide missing amino acids and vitamins. Although some food additives are useful in extending shelf life and preventing food poisoning, most are added to improve appearance and sales. For example, the following letter lists only a few of the 93 different chemicals that may be added to the "enriched" bread you buy in a grocery store.

To the Editor of the [Albany, New York] *Times-Union*:

Give us this day our daily calcium proprionate (spoilage retarder), sodium diacetate (mold inhibitor), monoglyceride (emulsifier), potassium bromate (maturing agent) calcium phosphate monobasic (dough conditioner), chloramine T (flour bleach), aluminum potassium sulfate acid (baking powder ingredient), sodium benzoate (preservative), butylated hydroxyanisole (antioxidant), monoisopropyl citrate (sequestrant); plus synthetic vitamins A and D.

*Forgive us, O Lord, for calling this stuff BREAD.**

J. H. Read
Averill Park

All food, of course, is just a mixture of chemicals, but today at least 2,800 different chemicals are deliberately added to processed foods in the United States. Each year the average American consumes about 55 kilograms (120 pounds) of sugar, 7 kilograms (15 pounds) of salt, and about 4.5 kilograms (10 pounds) of other food additives. Table 20-4 summarizes the major classes of food additives. The most widely used groups of additives—coloring agents, natural and synthetic flavoring agents, and sweeteners—have the sole purpose of making food look and taste better.

Most food additives are probably harmless or at least pose so little risk relative to their benefits, like preventing food spoilage and food poisoning, that we accept their use. However, a handful of once widely used additives such as red dyes no. 2 and 4 have been banned in the United States because of their potential harm to humans. Most chemicals added to U.S. foods have not been adequately tested for links to cancer, genetic mutations, and birth defects.

The extremes of the controversy over food additives range from "Essentially all food additives are bad" and "We should eat only natural foods" to "There's nothing to worry about because there is no absolute proof that chemical X has ever harmed a human being." As usual, the truth probably lies somewhere between. Some additives are necessary and safe, but others are unnecessary,

---

*Used by permission of the *Times-Union*, Albany, New York.

Table 20-4 Commonly Used Food Additives and Food Processes

Class	Function	Examples	Foods Typically Treated
Preservatives	To retard spoilage caused by bacterial action and molds (fungi)	Processes: drying, smoking, curing, canning (heating and sealing), freezing, pasteurization, refrigeration Chemicals: salt, sugar, sodium nitrate, sodium nitrite, calcium and sodium propionate, sorbic acid, potassium sorbate, benzoic acid, sodium benzoate, citric acid, sulfur dioxide	Bread, cheese, cake, jelly, chocolate syrup, fruit, vegetables, meat
Antioxidants (oxygen interceptors, or freshness stabilizers)	To retard spoilage of fats (excludes oxygen or slows down the chemical breakdown of fats)	Processes: sealing cans, wrapping, refrigeration Chemicals: lecithin, butylated hydroxyanisole (BHA), butylated hydroxytoluene (BHT), propyl gallate	Cooking oil, shortening, cereal, potato chips, crackers, salted nuts, soup, toaster tarts, artificial whipped topping, artificial orange juice, many other foods
Nutritional supplements	To increase nutritive value of natural food or to replace nutrients lost in food processing*	Vitamins, essential amino acids	Bread and flour (vitamins and amino acids), milk (vitamin D), rice (vitamin B,), corn meal, cereal
Flavoring agents	To add or enhance flavor	Over 1,700 substances, including saccharin, monosodium glutamate (MSG), essential oils (such as cinnamon, banana, vanilla)	Ice cream, artificial fruit juice, toppings, soft drinks, candy, pickles, salad dressing, spicy meats, low-calorie foods and drinks, most processed heat-and-serve foods
Coloring agents	To add aesthetic or sales appeal, to hide colors that are unappealing or that show lack of freshness	Natural color dyes, synthetic coal tar dyes	Soft drinks, butter, cheese, ice cream, cereal, candy, cake mix, sausage, pudding, many other foods
Acidulants	To provide a tart taste or to mask undesirable aftertastes	Phosphoric acid, citric acid, fumaric acid	Cola and fruit soft drinks, desserts, fruit juice, cheese, salad dressing, gravy, soup
Alkalis	To reduce natural acidity	Sodium carbonate, sodium bicarbonate	Canned peas, wine, olives, coconut cream pie, chocolate eclairs
Emulsifiers	To disperse droplets of one liquid (such as oil) in another liquid (such as water)	Lecithin, propylene glycol, mono- and diglycerides, polysorbates	Ice cream, candy, margarine, icing, nondairy creamer, dessert topping, mayonnaise, salad dressing, shortening
Stabilizers and thickeners	To provide smooth texture and consistency; to prevent separation of components; to provide body	Vegetable gum (gum arabic), sodium carboxymethyl cellulose, seaweed extracts (agar and algin), dextrin, gelatin	Cheese spread, ice cream, sherbet, pie filling, salad dressing, icing, dietetic canned fruit, cake and dessert mixes, syrup, pressurized whipped cream, instant breakfasts, beer, soft drinks, diet drinks
Sequestrants (chelating agents, or metal scavengers)	To tie up traces of metal ions that catalyze oxidation and other spoilage reactions in food; to prevent clouding in soft drinks; to add color, flavor, and texture	EDTA (ethylenediamine-tetraacetic acid), citric acid, sodium phosphate, chlorophyll	Soup, desserts, artificial fruit drinks, salad dressing, canned corn and shrimp, soft drinks, beer, cheese, frozen foods

*Adding small amounts of vitamins to breakfast cereals and other "fortified" and "enriched" foods in America is basically a gimmick used to raise the price. The manufacturer may put vitamins worth about 0.5¢ into 340 grams (12 ounces) of cereal and then add 45 percent to the retail price. Vitamin pills are normally far cheaper sources of vitamins than fortified foods. The best way to get vitamins, however, is through a balanced diet.

unsafe, or of doubtful safety. The key questions are: **(1)** What food additives are necessary? **(2)** What food additives are safe? **(3)** How well are consumers protected from unnecessary, unsafe additives?

Natural Versus Synthetic Foods *The presence of synthetic chemical additives does not necessarily mean that a food is harmful, and the fact that a food is completely natural is no guarantee that it is safe.* A number of natural, or totally unprocessed, foods contain potentially harmful and toxic substances.

Polar bear or halibut liver can cause vitamin A poisoning. Lima beans, sweet potatoes, cassava (yams), sugarcane, cherries, plums, and apricots contain glucosides, which our intestines convert to small amounts of deadly hydrogen cyanide. Eating cabbage, cauliflower, turnips, rutabagas, mustard greens, collard greens, or brussels sprouts can cause goiter in susceptible individuals. Certain amines that can raise blood pressure dramatically are found in bananas, pineapple, various acid cheeses (such as Camembert), and some beers and wines. Safrole (a flavoring agent once used in root beer) and a component of tarragon oil cause liver tumors in rats. Three chemicals that could cause cancer are formed when parsnips, celery, figs, and parsley are exposed to light. Aflatoxins produced by fungi that are sometimes found on corn and peanuts are extremely toxic to humans and are not legal in U.S. food at levels above 20 ppb.

Clams, oysters, cockles, and mussels can concentrate natural and artificial toxins in their flesh. In addition, natural foods can be contaminated with food-poisoning bacteria, such as *Salmonella* and the deadly *Clostridium botulinum*, through improper processing, food storage, or personal hygiene. The botulism toxin from *Clostridium botulinum* is one of the most deadly chemicals known. As little as one-ten-millionth of a gram (0.0000001 gram) can kill an adult, and it is estimated that 227 grams (half a pound) would be enough to kill every human being on earth. Because of modern food-processing methods, however, there are only about 10 to 20 cases of botulism annually in the United States.

Not all the news is bad. Natural foods such as citrus fruits and carrots apparently contain natural anticarcinogens. Examples include vitamins C and E, selenium, and carotene, which may inhibit or block the action of gene-altering substances that can cause some types of cancer. Many synthetic food additives, such as vitamins, citric acid, and sorbitol, are identical to chemicals found in natural foods. *Whether natural or synthetic, a chemical is a chemical is a chemical.* As long as the chemical is pure, it makes no difference whether you get vitamin C from eating oranges or from taking synthetic vitamin C tablets. And if you are poisoned, it makes no difference whether the substance is a natural chemical or a manufactured chemical.

Because there are potentially harmful chemicals in both natural and synthetic foods, the question boils down to whether enough of a chemical is present to cause harmful effects, and whether the effects are cumulative. The answers are not simple because individuals vary widely in their susceptibility to chemicals. Some chemicals are harmful at any level, while others are harmful only above certain thresholds. In addition, a chemical that has been thoroughly tested and found to be harmless by itself may interact synergistically with another substance to produce a hazard.

Testing a single food additive or drug may take up to 8 years and may cost $200,000 to $1 million. Thus, since it is essentially impossible to test the many thousands of natural and synthetic chemicals for possible synergistic interactions, we must determine whether the *benefits* of using a particular chemical in our food outweigh the *risks*. This involves scientific research, but it also involves economic, political, and ethical judgments that go far beyond science.

20-5 Consumer Protection from Food Additives

FDA and the GRAS List In the United States the safety of foods and drugs has been monitored by the Food and Drug Administration (FDA) since its establishment by the Pure Food and Drug Act of 1906, which was amended and strengthened by the 1938 Food, Drug, and Cosmetic Act. Yet not until 1958 did federal laws require that the safety of any new food additive be established by the manufacturer and approved by the FDA *before* the additive was put into common use. Today the manufacturer of a new additive must carry out extensive toxicity testing costing upward of a million dollars per item, and the results must be submitted to the FDA. The FDA itself does no testing but merely evaluates data submitted by manufacturers.

However, these federal laws did not apply to the hundreds of additives that were in use before 1958. Instead of making expensive, time-consuming tests on additives, the FDA drew up a list of the food additives in use in 1958 and asked several hundred experts for their professional opinions on the safety of these substances. A few substances were deleted, and in 1959 a list of 415 substances was published as the "generally recognized as safe" or *GRAS* (pronounced "grass") *list*.

Since 1959 further testing has led the FDA to ban several substances that were on the original GRAS list, including cyclamate sweeteners (1969), brominated vegetable oil (1970), and a number of food colorings, such as red dye no. 2 (1976). The ban on the most widely used artificial food color, red dye no. 2, came 5 five years after Soviet scientists reported that it caused cancer in laboratory mice. Between 1969 and 1980 the FDA reviewed all 415 items on the GRAS list and made the following conclusions: 371 of the additives were considered to be safe as currently used, 19 additives (including caffeine, BHA, and BHT) needed further study, 7 (including salt and 4 modified starches) can be used but at restricted levels. In addition, 18 items were recommended for removal from the GRAS list. By 1982 the FDA had also reviewed all other food additives approved for use since the publication of the original GRAS list.

As a regulatory agency, the FDA is caught in the crossfire between consumer groups and the food industry. It is criticized by consumer groups as being overly friendly to industry and for hiring many of its executives from the food industry—a practice the FDA contends is the only way it can recruit the most experienced food scientists. At the same time, the food industry complains that the FDA sometimes gives in too easily to demands from consumer groups. Both sides have criticized the agency for bureaucratic inefficiency.

Some Controversial Food Additives   Three types of allegedly harmful additives are *food colorings made from coal-tar dyes, nitrates and nitrites, sulfur dioxide*, and *sulfites*. About 3.1 million kilograms (6.8 million pounds) of food dyes is consumed in the United States each year. Although natural pigments and dyes exist, most artificial food dyes are derived from coal tar. In 1900 about 100 artificial dyes were in use in the United States, but evidence began accumulating that many of them cause cancer in test animals. About half the food additives that have been banned by the FDA have been coal-tar dye food colorings, and by early 1985 only six coal-tar dyes were still approved for use in food. Nevertheless, public exposure to coal-tar food dyes can be extensive. The FDA estimates that 10 percent of all U.S. children have eaten more than 454 grams (1 pound) of these dyes by age 12. Critics charge that the six remaining dyes should be banned because studies have linked them to cancer in laboratory test animals. For the past 20 years, however, the FDA has listed these dyes in an in-

terim "provisional listing" category, under which they can be used pending a final decision.

Another controversial group of additives consists of nitrates and nitrites (mostly as sodium nitrate and sodium nitrite), added to most processed meats to retard spoilage, prevent botulism, enhance flavor, and improve color. Studies have shown that they may form nitrosoamines, which can cause cancer of the stomach and esophagus in animals. In 1978 the FDA and the Department of Agriculture created a stir when they proposed the gradual phasing out of these preservatives. Food industry officials and scientists, however, point out that this might not help because humans are exposed to even larger concentrations of nitrosoamines from other sources. For example, a pack of 20 filter cigarettes provides a nitrosoamine intake of about 17 micrograms and a single cigar 11 micrograms, compared to an average dietary intake of about 1.1 micrograms per day.

Nitrates and nitrites continued to be used in 1985 because of opposition to the ban from the food-processing industry and because no substitutes had been found that are as effective as nitrates in preventing spoilage of some foods. While the search for such substitutes goes on, some scientists urge Americans to limit their consumption of cured and smoked meats. Even if nitrates and nitrites turn out to be harmless, they are high in sodium and fat, two components that contribute to cardiovascular disease.

Many restaurants sprinkle sulfites such as sodium bisulfite on fruit and vegetable salads and potatoes to prevent darkening. Sulfur dioxide and sulfites are also used as preservatives in the processing of wine, beer, grape juice, dehydrated potatoes, imported shrimp, dried apricots and prunes, some baked goods and snacks, and some drugs. As many as 1 million of the 10 million Americans suffering from asthma can have an acute allergic reaction to sulfites. This allergic reaction can cause weakness, difficulty in swallowing, severe wheezing, labored breathing, coughing, extreme shortness of breath, blue discoloration of the skin, and loss of consciousness. Some nonasthmatics can also have allergic reactions to sulfites. Between 1982 and 1984, four deaths were linked to the consumption of food containing sulfites.

In 1983 the FDA was planning to declare most sulfites safe. But after protests by consumer and physician groups the FDA recommended that state officials who have the responsibility for regulating food wholesalers, groceries, and restaurants insist that these businesses either cease using the additives or alert consumers to their use of sulfites by signs on grocery bins or notes on menus. The National Restaurant Association has urged its members not to use sulfites, or at least to notify customers of their use. But many still use sulfites, without warning customers. By early 1985 the FDA had refused to ban or severely restrict the use of sulfites, and only a few states had required restaurants to notify customers that they use sulfites.

The Delaney Clause   One powerful weapon the FDA has is the *Delaney clause.** This 1958 amendment to the food and drug laws prohibits the deliberate use of any food additive that has been shown to cause cancer in animals or humans. The FDA must evaluate the evidence linking an additive to cancer; if the FDA finds a risk, however slight, it must ban the chemical. The language of the amendment allows for no extenuating circumstances or consideration of benefits versus risks. Between 1958 and 1985, the FDA used this amendment to ban only nine chemicals.

Critics say the Delaney clause is too rigid and is not needed, since the FDA already has the power to ban any chemical it deems unsafe. In general, the food industry would like to see it removed, while some scientists and politicians favor modification to allow consideration of benefits versus risks. Others point out that the clause allows chemicals to be banned even if they cause cancer in test animals at doses 10 to 10,000 times greater than the amount a person might be expected to consume. These critics also argue that results of cancer tests in animals don't necessarily apply to humans.

Supporters of the Delaney amendment say that since humans can't serve as guinea pigs, animal tests are the next best thing. Such tests don't prove that a chemical will cause cancer in humans, but they can strongly suggest that a risk is present. Moreover, all but two substances known to cause cancer in humans also cause cancer in laboratory animals. Supporters also argue that the high doses of chemicals administered in animal tests are necessary to compensate for the relatively short life spans of test animals and for their relatively fast rates of metabolism and excretion. Tests using low doses not only would be inaccurate but would also require thousands of test animals to demonstrate that an effect was not due to chance. Such tests would be prohibitively expensive. Supporters also

favor the rigidity of the law. They argue that a carcinogen should be automatically banned because threshold levels for cancer-causing agents have not yet been established and carcinogens may be nonthreshold agents.

Indeed, instead of revoking the Delaney clause, some scientists feel it should be strengthened and expanded. Some even argue that the clause gives the FDA too much discretion, including the right to reject the validity of well-conducted animal experiments that do show carcinogenicity. These critics cite the FDA's infrequent use of the clause as evidence that the law is too weak. It is also argued that the absence of discretion in deciding whether to invoke the law protects FDA officials from undue pressure from the food industry and politicians. If the FDA had to weigh benefits versus risks, political influence and lobbying by the food industry could delay the banning of a dangerous chemical while it underwent years of study. The long delays and failure to ban other potentially harmful chemicals not covered by the Delaney clause illustrate this problem.

What Can the Consumer Do?   It is almost impossible for a consumer in an affluent nation to avoid all food additives. Indeed, as we have seen, many additives perform important functions, and there is no guarantee that natural foods will always be better and safer. However, to minimize risk, we can do the following:

1. Try to eat a balanced daily diet, consuming less sugar, salt, and animal fats, and more vegetables, fresh fruits, and whole grains. Foods rich in vitamin C (tomatoes, peppers, and citrus fruits) seem to lower the risk of cancer of the esophagus or the stomach. Vegetables like carrots, spinach, and broccoli are rich in carotene and may reduce the chances of getting cancer of the lung, breast, bladder, and skin.

2. Become informed about additives and natural foods that have come under suspicion, and try to avoid them until their safety has been established. The National Academy of Sciences suggests that we cut back on salt-cured and smoked foods such as ham, bacon, bologna, and frankfurters to reduce the risks of getting cancer of the esophagus and stomach. The Center for Science in the Public Interest and other consumer health groups suggest that consumers read food labels carefully and avoid any foods containing brominated vegetable oil (BVO), butylated hydroxyanisole (BHA), butylated hydroxytoluene (BHT), propyl gallate, quinine, saccharin, sodium nitrate, sodium ni-

---

*Named after Representative James J. Delaney of New York, who fought long and hard to have this amendment passed despite great political pressure and heavy lobbying by the food industry.

trite, sulfur dioxide, sodium bisulfite, and food colorings such as citrus red 2, orange B, red 3, yellow 5, and yellow 6.

20-6 Noise Pollution

Sonic Assault According to the Environmental Protection Agency, nearly half of all Americans, mostly urban dwellers, are regularly exposed in their neighborhoods and jobs to levels of noise that interfere with communication and sleeping. The EPA also estimates that about 25 million Americans are exposed to noises of sufficient duration and intensity to cause some permanent loss of hearing, and this number is rapidly rising. Industrial workers head the list, with 19 million hearing-damaged people out of a work force of 75 million. Workers who run a high risk of temporary or permanent hearing loss include boilermakers, weavers, riveters, bulldozer and jackhammer operators, taxicab drivers, bus and truck drivers, mechanics, machine shop supervisors, bar and nightclub employees, and performers who use sound systems to amplify their music. Millions of young people and adults who listen to music at loud levels using home stereos, "jam boxes" held close to the ear, and earphones are also incurring hearing damage.

Measuring and Ranking Noise To determine harmful levels of noise, sound pressure measurements in **decibels (db)** can be made using a small sound pressure level meter (called a *decibel meter*). A mathematical equation can be used to convert sound pressure measurements to loudness levels.

Sound pressure and loudness, however, are only part of the problem. Sounds also have pitch (frequency), with high-pitched sounds annoying us more than low-pitched ones. Normally, sound pressure is weighted for high-pitched sounds and reported in dbA units, as shown in Table 20-5. Sound pressure becomes painful at around 120 dbA and can kill at 180 dbA. Because the db and dbA sound pressure scales are logarithmic, a tenfold increase in sound pressure occurs with each 10-decibel rise. Thus, a rise in sound pressure on the ear from 30 dbA (quiet rural area) to 60 dbA (normal restaurant conversation) represents a thousandfold increase in sound pressure.

Effects of Noise Excessive noise is a form of stress that can cause both physical and psychological damage. As noise control advocate Robert Alex Baron reminds us, "Air pollution kills us slowly but silently; noise makes each day a torment."

Table 20-5 Effects of Common Sound Pressure Levels

Example	Sound Pressure (dbA)	Effect with Prolonged Exposure
Jet takeoff (25 meters*)	150	Eardrum rupture
Aircraft carrier deck	140	
Armored personnel carrier, jet takeoff (100 meters), earphones at loud level	130	
Thunderclap, textile loom, live rock music, jet takeoff at 61 meters, siren (close range), chain saw	120	Human pain threshold
Steel mill, riveting, automobile horn at 1 meter, jam box stereo held close to ear	110	
Jet takeoff at 305 meters, subway, outboard motor, power lawn mower, motorcycle at 8 meters, farm tractor, printing plant, jackhammer, garbage truck	100	Serious hearing damage (8 hours)
Busy urban street, diesel truck, food blender, cotton spinning machine	90	Hearing damage (8 hours), speech interference
Garbage disposal, clothes washer, average factory, freight train at 15 meters, dishwasher	80	Possible hearing damage
Freeway traffic at 15 meters, vacuum cleaner, noisy office or party	70	Annoying
Conversation in restaurant, average office, background music	60	Intrusive
Quiet suburb (daytime), conversation in living room	50	Quiet
Library, soft background music	40	
Quiet rural area (nighttime)	30	
Whisper, rustling leaves	20	Very quiet
Breathing	10	
	0	Threshold of hearing

*To convert meters to feet, multiply by 3.3.

Continued exposure to high sound levels permanently destroys some of the microscopic hairlike cells (cochlear cells) in the fluid-filled inner ear, which wave back and forth to convert sound energy to nerve impulses. Permanent hearing loss begins with exposure for 8 hours or more to 85 dbA levels of sound pressure, a condition often encountered with the use of stereo headphones to listen to music at levels above a normal conversational level. Such hearing loss also occurs from exposure to 92 dbA for more than 6 hours, 95 dbA for more than 4 hours, 100 dbA for more than 2 hours, 105 dbA for more than 1 hour, 110 dbA for more than 30 minutes, 115 dbA for more than 15 minutes, and to more than 115 dbA for any length of time. Sound experts suggest that sound levels are high enough to cause permanent damage to your ears if you need to raise your voice to be heard above racket, if a noise causes your ears to ring or feel full, or if voices begin to sound as if they are coming from a barrel.

In addition to causing psychic shock, sudden noise automatically constricts blood vessels, dilates pupils, tenses muscles, increases heartbeat and blood pressure, and causes wincing, holding of breath, and stomach spasms. Constriction of the blood vessels can become permanent, increasing blood pressure and contributing to heart disease. Migraine headaches, gastric ulcers, and changes in brain chemistry can also occur.

What Can Be Done? Noise control can be accomplished in three major ways: **(1)** reducing noise at its source, **(2)** substituting less noisy machines and operations, and **(3)** reducing the amount of noise entering the listener's ear. Workers can shield themselves from excessive noise by wearing hearing protectors, ranging from simple plugs to bulky headsets to custom-made plastic inserts with valves that close automatically in response to noise. Noisy factory operations can be enclosed or partially enclosed. Houses and buildings can be insulated to reduce sound transfer and also to reduce energy waste. Quieter trucks, motorcycles, vacuum cleaners, and other noisy machines are available.

The Soviet Union and many western European and Scandinavian nations are far ahead of the United States in reducing noise and in establishing and enforcing noise control regulations. Europeans have developed quieter jackhammers, pile drivers, and air compressors that do not cost much more than their noisy counterparts. Most European nations also require that small sheds and tents be used to muffle construction noise, and some countries reduce the clanging associated with garbage collection by using rubberized collection trucks. Subway systems in Montreal and Mexico City have rubberized wheels to reduce noise. In France, cars are required to have separate highway and city horns.

The government standard for overexposure to noise in any U.S. work place has been set at 90 dbA for 8 hours a day—still significantly above the standard of 85 dbA considered to be the maximum safe level. Industry officials have opposed lowering the standard to 85 dbA because implementation would cost an estimated $20 billion. By 1984 the Environmental Protection Agency had issued noise level regulations for air conditioners, buses, motorcycles, power mowers, some trucks and trains, and some construction equipment, as required by the Noise Control Act of 1978. So far, however, enforcement of these regulations has been almost nonexistent because the law merely fines violators and the EPA's $14-million program to curb noise pollution was eliminated by budget cuts in 1981.

The EPA has proposed a national plan for noise control. But noise control experts point out that unless Congress strengthens existing noise control laws and penalties and increases the amount of money appropriated for noise research and control, this plan will continue to collect dust in bureaucratic files.

Health and a good state of body are above all gold, and a strong body above infinite wealth.

Ecclesiastes 40:15

Discussion Topics

1. Why are infectious diseases more common in LDCs? Why do so many infants and young children in LDCs die from measles, diarrhea, and other common childhood diseases?

2. Discuss the life cycle, mode of transmission, effects, possible control, and side effects of control of malaria and schistosomiasis.

3. Should DDT and other pesticides be banned from use in malaria control? Why or why not? (See Chapter 23.)

4. How can cholera be brought under better control? Why are there often outbreaks of cholera in areas struck by earthquakes, floods, or other natural disasters?

5. Give some possible reasons for the death rate from breast cancer in the United States being almost six times higher than in Japan, while the death rate for stomach cancer in Japan is almost seven times higher than in the United States.

6. Analyze your life-style and diet to determine your relative risks of developing some form of cancer before you reach age 55. Which type of cancer are you most likely to get? How could you significantly reduce your chances of getting this cancer?

7. Give your reasons for agreeing or disagreeing with each of the seven proposals listed in the text designed to reduce the health hazards of smoking.

8. Do you agree or disagree with the following suggestions made by some environmentalists and health scientists: **(a)** all new and presently used additives should be reviewed and tested not only for toxicity and carcinogenic effects but also for their ability to induce birth defects and long-term genetic effects; **(b)** all testing of additives should be performed by a third party, independent of the food industry; **(c)** all additives, including specific flavors, colors, and sodium content should be listed on the labels or containers of all foods and drugs; **(d)** all additives should be banned unless extensive testing establishes that they are safe and that they enhance the nutritive content of foods or prevent foods from spoilage or contamination by harmful bacteria and molds. Defend your position.

9. **a.** Using Table 20-4 and the label of a food from the grocery store, try to classify the additives listed as necessary and safe, unnecessary but safe, or unnecessary and potentially harmful. Compare evaluations of different common foods by other class members.

   b. Compare brands. Are there some that don't contain additives that are controversial, either because of safety or their usefulness?

   c. Evaluate the additives found in baby foods and recommend whether they should be allowed or banned.

10. Explain the fallacies in the following statements:

    a. All synthetic food additives should be banned, and we should all return to safe, nutritious natural foods.

    b. All foods are chemicals, so we shouldn't worry about artificial additives.

    c. Since some natural foods contain harmful chemicals, we should not get so concerned about synthetic food additives.

    d. Food additives are essential and without them we would suffer from malnutrition, food poisoning, and spoiled food.

11. Do you believe that the Delaney clause should be revoked, left as is, altered to allow an evaluation of risks and benefits, or strengthened and broadened? Give reasons for your position.

12. As a class or group project, try to borrow one or more sound pressure decibel meters from the physics or engineering department or from a local stereo or electronic repair shop. Make a community survey of sound pressure levels at several times of day and at several locations; plot the results on a map. Include measurements of earphones and a room with a stereo, and take readings at an indoor concert or nightclub at various distances from the speakers of the sound system. Correlate your findings with those in Table 20-5.

Air Pollution

Tomorrow morning when you get up take a nice deep breath. It will make you feel rotten.

Citizens for Clean Air, Inc. (New York)

Air pollution alerts in heavily populated and industrialized Tokyo have become a way of life. In Mexico City air pollution on an average day is five to six times greater than maximum safety levels set for U.S. citizens by the Environmental Protection Agency. The air pollution capital of the world may be Cubato, Brazil, a petrochemical center where essentially no birds or insects remain, most trees are blackened stumps, 40 of every 1,000 babies are stillborn, air pollution monitoring machines break down from contamination, and the mayor refuses to live in the city. Since the passage of the Clean Air Act in 1970, the air in most parts of the United States has become measurably cleaner, but there is still a long way to go. In this chapter we examine the types, sources, and effects of air pollution and the methods used to help control air pollution.

21-1 Types and Sources of Air Pollution

Composition of the Atmosphere and Air Pollution The atmosphere, or gaseous envelope surrounding the earth, is arbitrarily divided into several zones (Figure 21-1). About 95 percent of the mass of the air is found in the innermost layer of the atmosphere known as the **troposphere**, which extends only 8 to 12 kilometers (5 to 7 miles) above the earth's surface. Clean air in the troposphere consists of a mixture of 78 percent by volume nitrogen gas (N_2), 21 percent by volume oxygen (O_2), and 1 percent by volume other gases.

Air pollution is normally defined as air that contains one or more chemicals or possesses a physical condition like heat in high enough concentrations to harm humans, other animals, vegetation, or materials. As with all types of pollution,

we run into the problem of conflicting views of what constitutes *harm* and what levels of risk we are willing to accept.

Most potential air pollutants are added to the troposphere in the form of compounds produced by the combustion of fossil fuels such as coal, natural gas, fuel oil, and gasoline. Whether such compounds build up to harmful levels in a given area depends on climate, topography, population density, the number and type of industrial activities, and the organisms or materials being affected. Because they contain large concentrations of cars and factories, cities normally have higher air pollution levels than rural areas. However, prevailing winds can spread air pollutants produced in urban and industrial areas to the countryside and to other downwind urban areas.

Air Pollution in the Past Air pollution from human activities, of course, is not new. Our ancestors had it in their smoke-filled caves and later in their cities. In 1273 Edward I of England declared the first known air quality laws, forbidding the use of high-sulfur coal. In 1911, 1,150 Londoners died from the effects of coal smoke. The author of a report on this disaster coined the word *smog* for the mixture of smoke and fog that often hung over London. An even more deadly London air pollution incident killed 4,000 people in 1952. Additional air pollution disasters in 1956, 1957, and 1962 killed a total of about 2,500 people. These deadly episodes triggered a massive air pollution control effort that has given London much cleaner air today.

In America the Industrial Revolution brought air pollution as coal-burning industries and homes filled the air with soot and fumes. In the 1940s the air in industrial centers like Pittsburgh and St. Louis became so thick with smoke that automobile drivers sometimes had to use their headlights at midday. The rapid rise of the automobile, especially since 1940, brought new forms of pollution such as photochemical smog and lead from the burning of leaded gasoline (Section 24-5).

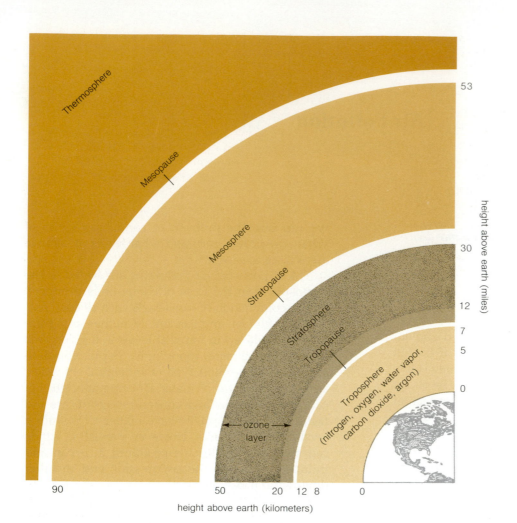

Figure 21-1 Structure of the earth's atmosphere (not drawn to scale). About 95 percent of the air is in the troposphere.

The first known U.S. air pollution disaster occurred in 1948, when fumes and dust from steel mills and zinc smelters became trapped in a stagnated air mass over Donora, Pennsylvania. Twenty people died and more than 6,000 became ill. In 1963 an air pollution disaster in New York City killed about 300 people and injured thousands. The periodic occurrence of serious air pollution episodes during the 1960s in New York, Los Angeles, and other large cities led to efforts to reduce air pollution levels nationwide. As a result, the United States now has one of the strongest air pollution control programs in the world.

Major Types of Air Pollutants The major types of air pollutants are **(1)** carbon oxides: carbon monoxide (CO) and carbon dioxide (CO_2), **(2)** sulfur oxides: sulfur dioxide (SO_2) and sulfur trioxide (SO_3), **(3)** nitrogen oxides: nitrous oxide (N_2O), nitric oxide (NO), and nitrogen dioxide (NO_2), **(4)** hydrocarbons (organic compounds consisting of various combinations of hydrogen and carbon), **(5)** photochemical oxidants such as ozone (O_3), and organic aldehydes and peroxyacyl nitrates (PANs)

formed in the atmosphere by the reaction of oxygen, nitrogen oxides, and hydrocarbons under the influence of sunlight, **(6)** particulate matter such as dust, soot, asbestos, metals and metal compounds (Section 24-5), and suspended droplets of oil, sulfuric acid (H_2SO_4), and nitric acid (HNO_3), **(7)** other gaseous inorganic compounds such as hydrogen fluoride (HF), ammonia (NH_3), and hydrogen sulfide (H_2S), and **(8)** various radioactive substances (Chapter 18).

These chemicals can be classified as either *primary* or *secondary* air pollutants. A **primary air pollutant** is a harmful concentration of a chemical added *directly* to the air. Examples include sulfur dioxide, nitric oxide, nitrogen dioxide, carbon monoxide, and soot, dust, and some other forms of particulate matter.

A **secondary air pollutant** is a harmful concentration of a chemical formed in the atmosphere through a reaction among primary pollutants or between a primary pollutant and one or more normal components of the atmosphere. For example, sulfur dioxide can react with oxygen gas in the air to form the secondary pollutant sulfur trioxide, which can react with water vapor in air to form droplets of sulfuric acid, another secondary air pollutant, as discussed in more detail in Section 21-2.

Indoor Air Pollution To escape polluted air you might go home, close the doors and windows, and breathe in what you believe to be clean air. But a number of scientists have found that the air inside some homes is often more polluted and dangerous than outdoor air on a smoggy day. This is especially true in energy-efficient, relatively airtight houses that do not use air-to-air heat exchangers to bring in sufficient fresh air and in the nation's almost 5 million mobile homes. Mobile homes have a smaller volume and lower air-exchange rates than conventional homes, use a larger proportion of plywood and other materials containing volatile organic compounds, and are more likely to use propane as a fuel.

Indoor air today is much cleaner than that found decades ago when houses were heated with leaky coal-burning furnaces, but there is still cause for concern. Some of the major indoor air pollutants include: **(1)** *radioactive radon gas* released from earth and rock beneath homes and stone, brick, sand, and concrete block used for slab or inside construction, **(2)** *formaldehyde* released from materials such as foam insulation, particle board, plywood furniture, and drapes and carpets made from synthetic fibers, **(3)** *asbestos* particles released from some older types of wall, ceiling, pipe, and boiler insulation, vinyl floor material, and some patching compounds, and **(4)** *carbon monoxide, nitrogen dioxide, and particulate matter* released primarily by the combustion of fuels in unvented kerosene heaters and gas stoves and wood stoves.

Measurements in one relatively airtight, extremely energy-efficient home without an air-to-air heat exchanger revealed high levels of formaldehyde throughout the house and indoor radioactivity levels more than 100 times the natural outdoor background level. Measurements in a number of energy-efficient homes without sufficient air-exchange rates revealed radioactivity from radon ranging from 10 to 2,700 times normal background levels. Premature lung cancer deaths from indoor exposure to radon and its decay products are estimated at 5,000 to 20,000 each year in the United States.

Sources of Air Pollutants Natural decay processes, winds, and volcanic eruptions add to the atmosphere most of the nitrous oxide, carbon monoxide, carbon dioxide, methane, terpenes, radioactive radon and tritium, and particulate matter, and much of the hydrogen sulfide, carbon monoxide, and dust. However, these natural inputs are usually widely dispersed throughout the world. When they do reach harmful levels, as in the case of volcanic eruptions, they are usually taken care of by natural chemical cycles (Section 5-3) and weather patterns (Section 4-5), which also help control levels of potentially dangerous chemicals added to the atmosphere by human activities.

The primary pollutants produced by human activities, which account for more than 90 percent of the air pollution problems in the United States, are carbon monoxide, sulfur oxides, volatile organic compounds, nitrogen oxides, and particulate matter. Emissions of sulfur dioxide, nitric oxide, and nitrogen dioxide from human activities dwarf natural emissions by a factor of 50 to 100. U.S. power plants and industrial plants alone produce 91 percent of the total weight of emissions of sulfur oxides, 82 percent of the particulate matter, 59 percent of the nitrogen oxides, 25 percent of the hydrocarbons, and 17 percent of the carbon monoxide released into the atmosphere each year.

In terms of the *total amount* of primary pollutants emitted each year in the United States, *carbon monoxide* is number one, and transportation is its major source (Table 21-1). But it is more meaningful to rank pollutants on the basis of their potential or actual harm to humans and other forms of life, as shown in Table 21-1. On this basis, sulfur oxides and particulate matter rank as the top two pollutants, and carbon monoxide drops to last place. In terms of air pollution sources, stationary fuel combustion (primarily at coal-burning power plants) is the most serious, with industry (especially pulp and paper mills, iron and steel mills, smelters, petroleum refineries, and chemical plants) and transportation in second and third places, respectively (Table 21-1).

21-2   Industrial and Photochemical Smog

Types of Smog Table 21-2 shows how various types of air pollutant can be classified as either *industrial smog* or *photochemical smog*. Although the distinction between these two types of smog is convenient, most urban areas suffer from both types of air pollution, often with one type predominating. Sulfur dioxide and particulate matter are the main ingredients of *industrial smog*. It tends to predominate in cities like London, Chicago, Baltimore, Philadelphia, and Pittsburgh, which usually have long, cold, and wet winters and depend heavily on coal and oil for heating, manufacturing, and producing electric power.

Photochemical smog contains a mixture of primary pollutants such as nitrogen oxides and carbon monoxide and secondary pollutants such as ozone, peroxacyl nitrates (PANs), and aldehydes such as formaldehyde (CH$_2$O). Cities in which photochemical smog tends to predominate usually have sunny, warm, dry climates, and their main source of air pollution is the internal combustion engine. Examples include Los Angeles, Denver, Salt Lake City (see p. 263), Sydney, Mexico City, and Buenos Aires. Let's examine in more detail how these two different types of smog are formed in the troposphere.

The Sulfur Cycle and the Formation of Industrial Smog

Sulfur is transformed to different compounds and circulated through the ecosphere in the *sulfur cycle* (Figure 21-2). It enters the atmosphere from natural sources as (1) hydrogen sulfide, a colorless, highly poisonous gas with a rotten-egg smell, from active volcanoes and the decay of organic matter in swamps, bogs, and tidal flats, (2) sulfur dioxide, a colorless suffocating gas, from active volcanoes, and (3) particles of sulfate salts like ammonium sulfate from sea spray. Except for occasional volcanic eruptions, these natural additions of sulfur compounds to the atmosphere are widely dispersed and usually are diluted so that they do not exceed the normal levels in clean, dry air.

About one-third of all the sulfur compounds and 99 percent of the sulfur dioxide reaching the atmosphere from all sources comes from human activities. About two-thirds of the human-related input of sulfur dioxide into the atmosphere results when sulfur-containing coal and oil are burned to produce electric power. The remaining third comes from industrial processes such as petroleum refining and the heating of sulfur-containing ore compounds of metals such as copper, lead, and zinc in air to convert them to the free metal.

When oil or coal is burned, about 97 percent of the sulfur impurities reacts with oxygen gas in the atmosphere to produce sulfur dioxide gas: S + O$_2$ → SO$_2$. The remaining 3 percent of these impurities reacts with oxygen to form sulfur trioxide: 2S + 3O$_2$ → 2SO$_3$. If these sulfur oxides are not removed by scrubbers or other devices (Section 21-7), they spew out of chimneys and smokestacks and enter the troposphere. Within a few days, 5

Table 21-1 Relative Importance of Primary Pollutants and Their Sources in the United States

	Annual Emissions		Estimated Relative Health Effect	
	Percentage of Total	Rank	Percentage of Total	Rank
Pollutant				
Sulfur oxides	15	2	34	1
Particulate matter	5	5	28	2
Nitrogen oxides	13	4	19	3
Volatile organic compounds	14	3	18	4
Carbon monoxide	53	1	1	5
Total	100		100	
Source				
Stationary fuel combustion	21	2	44	1
Industry	16	3	27	2
Transportation	55	1	24	3
Solid waste disposal	2	5	3	4
Miscellaneous	6	4	2	5
Total	100		100	

Table 21-2 Basic Types of Smog

Characteristic	Industrial Smog	Photochemical Smog
Typical city*	London, Chicago	Mexico City, Los Angeles
Climate	Cool, humid air	Warm, dry air and sunny climate
Chief pollutants	Sulfur oxides, particulates	Ozone, PANs, aldehydes, nitrogen oxides, carbon monoxide
Main sources	Industrial and household burning of oil and coal	Motor vehicle gasoline combustion
Time of worst episodes	Winter months (especially in the early morning)	Summer months (especially around noontime)

*All these cities have both types of smog, but the type listed makes the major contribution.

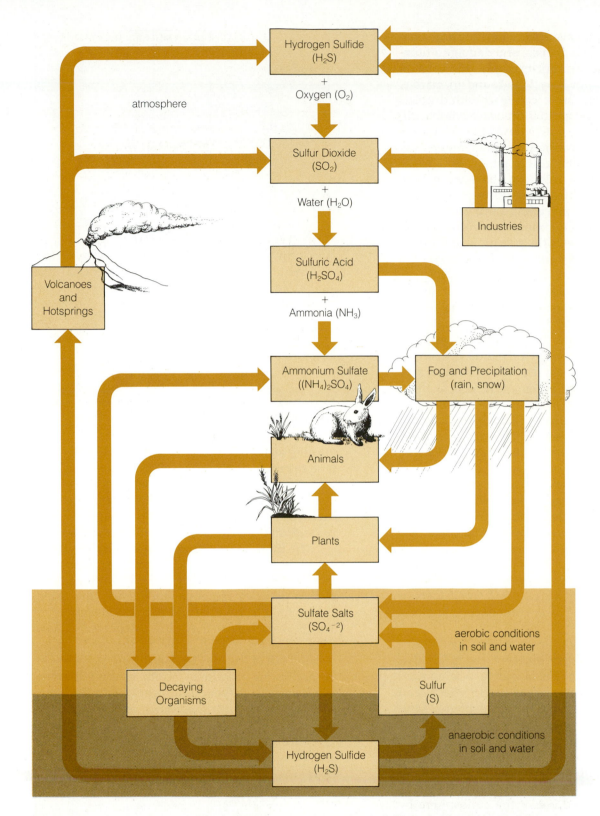

Figure 21-2 The sulfur cycle, showing both natural and human-related additions of sulfur to the atmosphere.

percent or more of the sulfur dioxide in the atmosphere reacts with oxygen gas in the air to form sulfur trioxide. This gas reacts almost at once with water vapor in the air or on the ground to form an aerosol mist or fog of tiny droplets of sulfuric acid suspended in air: $SO_3 + H_2O \rightarrow H_2SO_4$. This atmospheric mist of sulfuric acid droplets falls to the earth as acidic rain or snow as a component of *acid deposition*, commonly known as acid rain, as discussed in more detail in Section 21-4.

This highly corrosive acid can irritate and damage the lungs, corrode metals, and damage a number of other materials. For example, the calcium carbonate ($CaCO_3$) or limestone found in marble, mortar, and building stone can react with droplets of sulfuric acid in the air to form water-soluble calcium sulfate ($CaSO_4$): $CaCO_3 + H_2SO_4 \rightarrow CaSO_4 + H_2O + CO_2$. Rain washes the $CaSO_4$ away, leaving a building or statue visibly pitted. Because of this reaction, famous ruins on the Acropolis in Athens have deteriorated more during the past 50 years than during the preceding 2,000 years.

In areas where the atmosphere contains ammonia gas (NH_3), some of the sulfuric acid droplets can react with the ammonia to form solid particles of ammonium sulfate: $H_2SO_4 + 2NH_3 \rightarrow (NH_4)_2SO_4$. These particles fall to the earth as *dry deposition* or dissolve in rainwater and make up another component of acid deposition. Small particles of these sulfates by themselves are not harmful to humans and provide plants with an essential nutrient. However, droplets of sulfuric acid in the atmosphere can become attached to these particles and inhaled. There is some evidence that this combination of sulfuric acid droplets and ammonium sulfate particles can interact synergistically to cause more harm to the lungs than either pollutant acting alone.

A variety of types of particulate matter is also found in industrial smog. Dust, soot, and other forms enter the atmosphere from active volcanoes, forest fires, dust storms, sea spray, and other natural sources. Particulate matter is also added to the atmosphere as a result of human activities such as the clearing of land for agriculture and urbanization and the burning of wood and fossil fuels in stoves, industrial and power plant boilers, and motor vehicles.

Unlike most major air pollutants, particulate matter consists of many different chemicals that form particles and droplets of widely varying sizes (Figure 21-3) and with diverse health effects and average residence times in the atmosphere. *Large particles*, with diameters greater than 10 micrometers (about 0.000039 inch) make up most of the natural emissions from dust storms and active volcanoes. They normally remain in the troposphere only a day or two before either falling to earth or being washed out of the atmosphere by precipitation. They are normally not considered to be major threats to human health because they are prevented from entering the lungs by hairs in the nostrils.

Medium-sized particles, with diameters between 1 and 10 micrometers, tend to remain suspended in the air longer. Many of the particles are found in coal dust and in fly ash produced by coal-

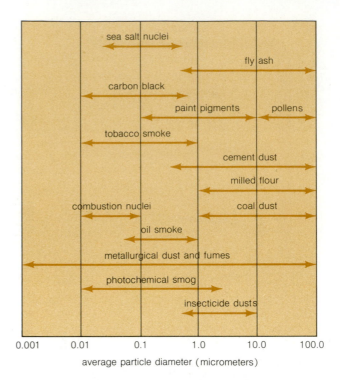

Figure 21-3 Suspended particulate matter is found in a wide variety of types and sizes.

burning electric power and industrial plants. They tend to remain in the atmosphere longer than large particles. Most medium-sized particles rarely pose a serious threat to human health because they are captured by the mucous lining in the upper airway passage and carried to the throat and swallowed.

The most serious threats to human health are posed by some *fine particles*, with diameters less than 1 micrometer (Figure 21-3). Hazardous types of fine particles remain suspended in the troposphere for 1 to 2 weeks and in the stratosphere for 1 to 5 years before they clump together to form large particles, which settle to the earth or are washed out of the atmosphere. These particles can be hazardous to human health because they are small enough to penetrate deeply into the lungs and may bring with them toxic or cancer-causing pollutants that have adsorbed on their surfaces. They are also a major factor in reducing visibility and possibly in altering global weather and climate.

Formation of Photochemical Smog Two gaseous substances involved in the formation of photochemical smog are colorless nitric oxide and yellow-brown nitrogen dioxide. About 80 percent of the nitric oxide in the atmosphere comes from natural sources, mostly from bacterial action and a smaller amount from lightning and forest fires. Because this natural production is spread out over much of the earth's surface and is diluted in the troposphere, clean, dry air normally has no more than trace levels of nitric oxide.

Nitric oxide also enters the atmosphere as a result of the combustion of fossil fuels. At normal atmospheric temperatures the nitrogen gas and oxygen gas making up most of the atmosphere do not react with each other. At the high temperatures inside an internal combustion engine or fossil-fuel-burning power plant, however, they react to produce nitric oxide, which is exhausted into the atmosphere: $N_2 + O_2 \rightarrow 2NO$. Because these human sources of nitric oxide are concentrated in urban and industrial areas, this gas can build up to harmful levels near such areas.

Once in the atmosphere, nitric oxide reacts with oxygen gas to form nitrogen dioxide, a yellowish-brown gas with a pungent, choking odor: $2NO + O_2 \rightarrow 2NO_2$. The resulting nitrogen dioxide gas is responsible for the brownish haze that hangs over many cities during sunny afternoons. This gas can aggravate respiratory disease, irritate the lungs and eyes, decrease atmospheric visibility, and contribute to heart, lung, liver, and kidney damage.

Typically nitrogen dioxide remains in the atmosphere for about 3 days. Small amounts of this gas react with water vapor in the atmosphere to form nitric acid and nitric oxide: $3NO_2 + H_2O \rightarrow 2HNO_3 + NO$. Some nitric acid is washed out of the atmosphere as another component of acid deposition. Some acid can also react with ammonia gas in the atmosphere to form particles of ammonium nitrate, which eventually fall to earth or are washed out of the atmosphere by rainfall and make up ammonium nitrate, another component of acid deposition: $HNO_3 + NH_3 \rightarrow NH_4NO_3$.

Any chemical reaction activated by light is called a **photochemical reaction**. When nitrogen dioxide is exposed to ultraviolet radiation from sunlight, it undergoes a photochemical reaction that converts it to nitric oxide and highly reactive atomic oxygen (O): $NO_2 \rightarrow NO + O$. The atomic oxygen then reacts with gaseous oxygen in the atmosphere to produce the photochemical oxidant *ozone*: $O_2 + O \rightarrow O_3$. Through a complex sequence of reactions, atomic oxygen also reacts with volatile gaseous hydrocarbons (evaporated into the atmosphere from spilled or partially burned gasoline) to form various *peroxyacyl nitrates* (PANs). PANs also react with atmospheric hydrocarbons to produce *aldehydes* such as formaldehyde, which burn the eyes like tear gas. This complex mixture of secondary pollutants makes up the **photochemical smog** that hangs over many cities such as Los Angeles and Salt Lake City (see the photo on p. 263). A greatly simplified summary of the formation of photochemical smog is given in Figure 21-4. Mere traces of ozone, PANs, and aldehydes in this type of smog can cause burning of the eyes, irritation of the respiratory tract, and damage to crops and trees.

Figure 21-5 shows a characteristic pattern of the variation of concentrations of nitric oxide, nitrogen dioxide, and ozone throughout a typical smoggy day in Los Angeles. Early in the morning when commuter traffic begins to build, atmospheric concentrations of nitric oxide and hydrocarbons start to rise. As some of the NO is converted to NO_2, its concentration rises. Then under the influence of sunlight, NO_2 concentrations decrease as this gas is converted to O atoms. The highly reactive O atoms combine with O_2 and hydrocarbons to produce rising levels of ozone and aldehydes, which peak near midday (Figure 21-5).

Climate, Topography, and Air Pollution The frequency and the severity of smog in a given area depend on climate, topography, heating practices, traffic, and the density of population and industry. Normally, solar radiation heats air near the ground during the day and causes it to expand and rise. Air from surrounding high-pressure areas moves into the low-pressure area created when the hot air rises. This mixing of the air dilutes and carries pollutants into higher levels of the troposphere and helps prevent pollutants from reaching dangerous levels near the ground (Figure 21-6, left).

But under some climatic conditions, often related to topography, a layer of dense cool air is trapped beneath a layer of light warm air in a particular urban area or valley. This is called a **thermal inversion** or temperature inversion (Figure 21-6, right, and Figure 21-7, right). In effect, a lid is clamped over the region, trapping pollutants normally removed by upward air currents; if the thermal inversion is prolonged, air pollutants can accumulate to harmful and even lethal levels.

Thermal inversions can happen more often and last longer over towns or cities that lie in valleys surrounded by mountains or are near a coast. Put several million people with an almost equal number of automobiles together in a subtropical climate with light winds. Then put mountains on three sides and the ocean on the other and you have Los Angeles, and the ideal recipe for photochemical smog. During the day, when ocean breezes predominate, cool air from the ocean flows into the Los Angeles valley, and after sunset when land breezes predominate, cool air flows from the mountains into the valley. These two processes create almost daily inversions in the Los Angeles basin.

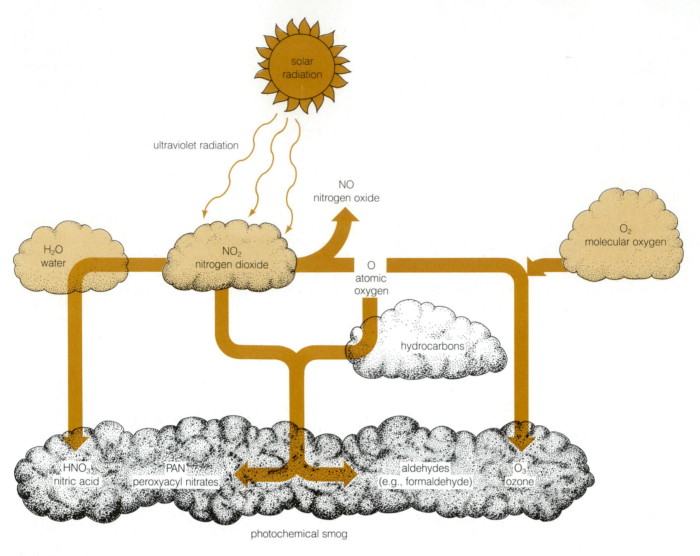

Figure 21-4 Greatly simplified scheme of the formation of photochemical smog.

Usually, thermal inversions last for only a few hours, but occasionally a high-pressure air mass stalls over an area for days. Prolonged thermal inversions, especially during cold weather when more fossil fuels are burned to keep buildings warm, have caused most air pollution disasters where hundreds to thousands of people have been killed or injured.

21-3 Effects of Air Pollution on Property, Plants, Animals, and Human Health

Damage to Property, Plants, and Animals Marble statues and building materials such as limestone, marble, mortar, and slate are discolored and attacked by sulfuric acid formed from emissions of sulfur oxides and by nitric acid formed from emissions of nitrogen oxides. The cost of architectural damage in the United States from air pollution is

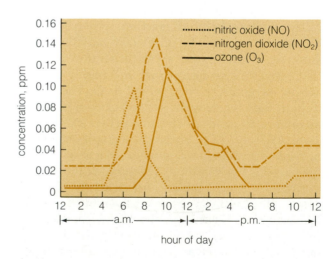

Figure 21-5 Variation of atmospheric concentrations of nitric oxide, nitrogen dioxide, and ozone with time of day: Los Angeles, July 19, 1965. (Source: National Air Pollution Control Administration.)

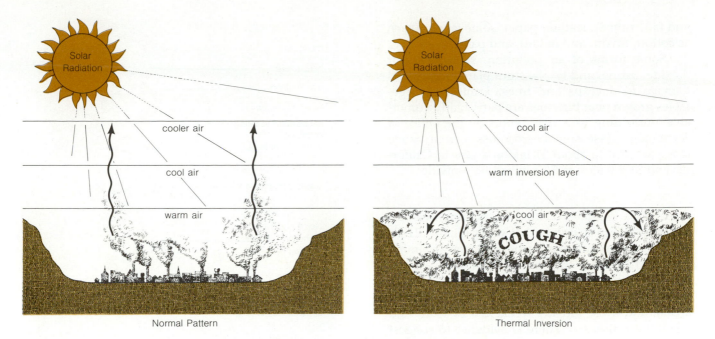

Normal Pattern

Thermal Inversion

Figure 21-6 Thermal inversion traps pollutants in a layer of cool air that cannot rise to carry the pollutants away.

estimated to be at least $2 billion a year and perhaps as high as $10 billion a year. Atmospheric fall-out of soot and grit also soils statues, buildings, cars, and clothing—costing billions of dollars each year for cleaning and maintenance.

Sulfuric acid, sulfur dioxide, nitrogen oxides, nitric acid, and some particulates also greatly accelerate the corrosion of metals, especially steel, iron, and zinc, causing billions of dollars of damage each year. Sulfuric acid and ozone also attack

Figure 21-7 Two faces of New York City. The almost clear view was photographed on a Saturday afternoon (November 26, 1966). The effect of more cars in the city and a thermal inversion is shown in the right-hand photograph, taken the previous day.

and fade rubber, leather, paper, some fabrics (such as cotton, rayon, and nylon), and paint.

Some forms of air pollution, such as sulfur dioxide, ozone, and PANs, stunt plant growth and damage food crops and trees. Fruits and vegetables grown near big cities are particularly vulnerable. According to a 1982 report by the Office of Technology Assessment, crop losses from ozone alone amount to about $1 billion a year in California and $1.9 billion to $4.5 billion nationwide.

Damage to Human Health   Air pollution can affect humans in a number of ways (Figure 21-8). Decades of research have produced overwhelming statistical evidence that *high levels* of some air pollutants are particularly harmful to the very young, the old, the poor (who often live in highly polluted areas), and those already weakened by heart and respiratory diseases. There is evidence to suggest that air pollution emitted by the burning of fossil fuels in the United States contributes to the premature death of at least 53,000 people each year.

Despite massive statistical evidence that air pollution kills and harms people, it is nearly impossible to establish that a particular pollutant causes a particular disease or death because of: **(1)** the number and variety of air pollutants, **(2)** the difficulty in detecting pollutants that may cause harm at extremely low concentrations, **(3)** synergistic interactions of pollutants (Section 6-2), **(4)** the difficulty of isolating single harmful factors when people are exposed to so many potentially harmful chemicals over several decades, **(5)** the unreliability of records of disease and death, **(6)** multiple causes and lengthy incubation times of diseases like emphysema, chronic bronchitis, lung cancer, and heart disease (Chapter 20), and **(7)** problems associated with extrapolating test data on laboratory animals to humans.

Largely because of these difficulties, many people are misled when they hear the statement "Science has not proven absolutely that smoking or exposure to a certain level of a particular air pollutant has killed anyone." Like "Cats are not elephants," such a statement is true but meaningless. Science can be used to disprove hypotheses, but it never has proved anything absolutely and never will. Science does not establish absolute truths but only a degree of probability or confidence in the validity of an idea, usually based on statistical or circumstantial evidence.

To understand the health effects of air pollution, look at what happens to the air we inhale (Figure 21-9). Each breath swirls down the trachea, which divides into two big bronchial tubes that enter the lungs. These tubes divide and subdivide

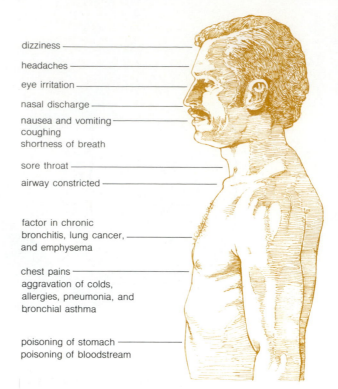

dizziness

headaches

eye irritation

nasal discharge

nausea and vomiting
coughing
shortness of breath

sore throat

airway constricted

factor in chronic
bronchitis, lung cancer,
and emphysema

chest pains
aggravation of colds,
allergies, pneumonia, and
bronchial asthma

poisoning of stomach
poisoning of bloodstream

Figure 21-8   Some possible effects of air pollution on the human body.

into many small ducts, or bronchiole tubes. At the end of the bronchiole tubes are about 500 million bubblelike air sacs, called **alveoli**, lying like clusters of tiny grapes inside the lungs. Oxygen in the air passes through the walls of the alveoli and combines with hemoglobin in the blood. At the same time carbon dioxide passes from the blood back through the alveolar walls into the lungs for exhaling.

The harmful effects of carbon monoxide arise when it combines chemically with the oxygen-carrying hemoglobin found in red blood cells. Carbon monoxide from smoking and from automobile exhausts reacts with hemoglobin in red blood cells about 220 times more rapidly than the oxygen gas we inhale. Thus, it ties up much of the hemoglobin in the blood and forces the heart to pump harder to supply enough oxygen. This aggravates cardiovascular disease and eventually causes enlargement of the heart.

Breathing carbon monoxide can: **(1)** impair judgment and reflexes and cause slight headache and fatigue with exposure for 1 hour or more to moderate concentrations (50 to 100 ppm), **(2)** cause severe headache, dizziness, coma, and irreversible brain cell destruction with exposure for 2 hours or more to a concentration of 250 ppm, and **(3)** cause unconsciousness with exposure to 750 ppm for 1 hour and death after 3 to 4 hours of such exposure.

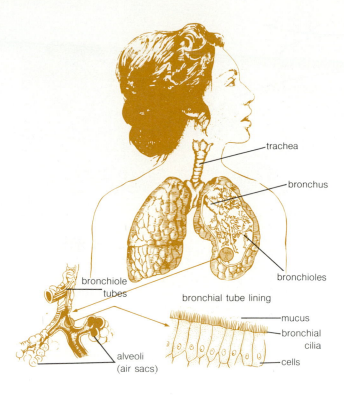

trachea

bronchus

bronchioles

bronchiole tubes

bronchial tube lining

mucus

bronchial cilia

cells

alveoli (air sacs)

Figure 21-9 The human respiratory system.

Each puff of inhaled cigarette smoke contains 200 to 400 ppm of carbon monoxide, which ties up at least 5 percent of a heavy smoker's hemoglobin during smoking. People are not exposed to high concentrations of carbon monoxide in traffic jams (25 to 115 ppm) and on crowded freeways (10 to 75 ppm). However, such relatively low exposure levels can impair judgment and slow reflexes enough to be an indirect cause of many automobile accidents—especially for drivers who smoke. The increased incidence of heart disease in heavy smokers and in taxi drivers, automobile mechanics, traffic policemen, and others exposed daily to moderate concentrations of carbon monoxide is believed to be partially caused by the increased heart size necessary for pumping sufficient oxygen throughout the body.

Humans have several defenses against dirty air: **(1)** hairs in the nose filter out large particles; **(2)** small amounts of mucus secreted constantly in the upper respiratory tract wash out or dissolve irritants and trap particles from the air; and **(3)** hundreds of thousands of **cilia**, tiny mucus-coated hairs in the upper respiratory tract, continually wave back and forth to eject mucus and foreign matter. Smoking and exposure to excessive levels of air pollutants like ozone, sulfur dioxide, and nitrogen dioxide, and some types of particulate matter apparently destroy, stiffen, or slow the cilia and thus make them less effective. As a result, bacteria

and tiny particles can penetrate the alveoli, increasing the chances of respiratory infection and lung cancer.

If the lungs become irritated, mucus flows more freely to help remove the irritants. A coughing mechanism then expels the dirty air and some of the contaminated mucus. Excessive cigarette smoking and exposure to high levels of pollutants such as sulfur dioxide and sulfuric acid can trigger so much mucus flow that air passages become blocked, causing more coughing. As muscles surrounding the bronchial tubes weaken from prolonged coughing, more mucus accumulates, and breathing becomes progressively more difficult. If this cycle persists, it indicates **chronic bronchitis**—a persistent inflammation of the mucous membranes of the trachea and bronchi that now affects one out of every five American men between the ages of 40 and 60.

Pulmonary emphysema, which results when an individual cannot expel most of the air from the alveoli lining the lungs, is usually accompanied by chronic bronchitis. Irritation by cigarette smoke and other forms of air pollution can cause the bronchial tubes to close up. The trapped air may expand and fuse clusters of alveoli together. The air sacs then lose their ability to expand and contract and may even tear. As a result, the lungs become enlarged and less efficient. As more alveoli are damaged, the bronchial tubes tend to collapse and breathing gets harder. Walking becomes painful and running impossible. After a period of years, breathing efficiency may become so low that the victim dies of suffocation or heart failure.

Pulmonary emphysema affects about 1.5 million Americans (half of them under age 65) and is the fastest-growing cause of death in the United States. This disease is incurable and basically untreatable. It is caused and aggravated by a number of factors, including smoking, air pollution, and heredity. About 25 percent of emphysema cases seem to be due to a hereditary condition characterized by the absence of a protein that gives the air sacs their elasticity. A test has been devised to detect this genetic defect, which predisposes its carriers to contracting emphysema, especially if they smoke, or live or work in a polluted area. Such persons should have the test made.

Lung cancer is the abnormal, runaway growth of cells in the mucous membranes of the bronchial passages. Some air pollutants can cause lung cancer directly, and others can disrupt the action of the cilia. If the cilia and the mucus do not remove carcinogenic pollutants, lung cancer is more likely. Although smoking—a form of personal air pollution—is considered to be the number one cause, lung cancer has also been linked to inhalation of

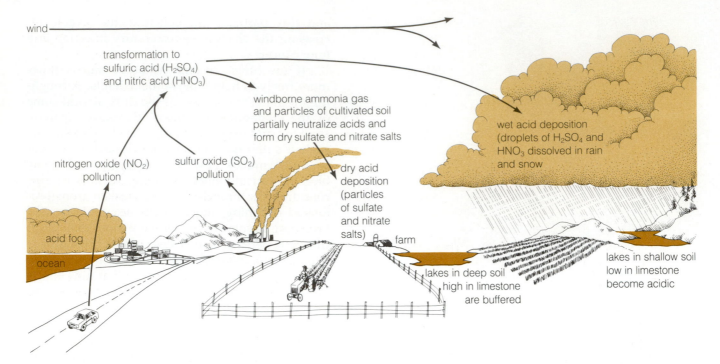

wind

transformation to
sulfuric acid (H₂SO₄)
and nitric acid (HNO₃)

windborne ammonia gas
and particles of cultivated soil
partially neutralize acids and
form dry sulfate and nitrate salts

wet acid deposition
(droplets of H₂SO₄ and
HNO₃ dissolved in rain
and snow

nitrogen oxide (NO₂)
pollution

sulfur oxide (SO₂)
pollution

dry acid
deposition
(particles
of sulfate
and nitrate
salts)

farm

acid fog

ocean

lakes in deep soil
high in limestone
are buffered

lakes in shallow soil
low in limestone
become acidic

Figure 21-10   Acid deposition and acid fog.

other air pollutants including: **(1)** some radioactive isotopes, such as plutonium-239 (Section 18-1), **(2)** polynuclear aromatic hydrocarbons such as 3,4-benzopyrene, found in cigarette and other types of smoke, **(3)** automobile exhaust, and **(4)** particulate matter—especially particles of asbestos, beryllium, arsenic, cadmium, and nickel (Section 24-5).

21-4   Effects of Air Pollution on Ecosystems: Acid Fog and Acid Deposition

Components of Acid Fog and Acid Deposition Droplets of sulfuric acid formed in the air from emissions of sulfur dioxide from coal-burning power plants, factories, and metal smelters and droplets of nitric acid formed in the air from nitrogen oxides emitted by automobiles and smokestacks are the major components of **acid fog**. This fog, along with minute concentrations of ozone in the atmosphere, can damage the leaves and needles of trees growing at high elevations. These acid-containing droplets can also dissolve in rain and snow and fall out of the atmosphere in a process called *wet deposition*, which can damage tree roots and kill aquatic life when it runs off into nearby lakes and streams. Particles of sulfate and nitrate salts can also be dissolved in rainwater and be removed by wet deposition or simply fall out of the atmosphere as particles in a process called *dry deposition*. The combined dry and wet deposition

of these secondary pollutants onto the surface of the earth is known as **acid deposition** (Figure 21-10), commonly and inaccurately referred to as *acid rain* or *acid precipitation*.

Relative levels of acidity and basicity of water solutions of substances are commonly expressed in terms of **pH**. The pH scale typically runs from 0 to 14. As shown in Figure 21-11, a water solution with a pH value of 7 is considered *neutral*; one with a pH value below 7 is *acidic*; and one with a pH value above 7 is *basic*. The lower the pH value, the higher the acidity. Because the pH scale is logarithmic, each whole-number decrease in pH represents a *tenfold increase* in acidity. Thus, a solution with a pH of 3 is 10 times more acidic than one with a pH of 4, 100 times more acidic than one with a pH of 5, and 1,000 times more acidic than one with a pH of 6.

Pure distilled water not exposed to the air is neutral (pH = 7). Normal rainwater, however, is slightly acidic (pH ≈ 5.6) because of the formation of a weak solution of carbonic acid (H₂CO₃) when rainwater dissolves some of the naturally occurring carbon dioxide in the atmosphere. This slight acidity of natural rainfall helps water dissolve soil minerals for use by plant and animal life. But when human activities cause the acidity of rain and snow to increase beyond this natural level, harmful effects occur to trees and aquatic life. Acid fog can have a pH typically ranging from about 2 to 3.5—which is 128 to 4,000 times more acidic than normal rainwater. Acid deposition falling in most of

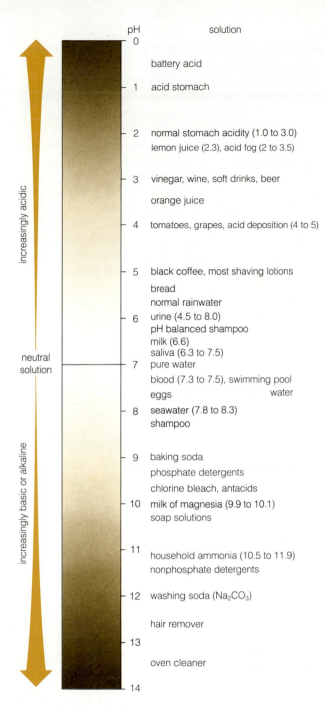

pH	solution

increasingly acidic

- 0
- battery acid
- 1 — acid stomach
- 2 — normal stomach acidity (1.0 to 3.0)
  lemon juice (2.3), acid fog (2 to 3.5)
- 3 — vinegar, wine, soft drinks, beer
  orange juice
- 4 — tomatoes, grapes, acid deposition (4 to 5)
- 5 — black coffee, most shaving lotions
  bread
  normal rainwater
- 6 — urine (4.5 to 8.0)
  pH balanced shampoo
  milk (6.6)
  saliva (6.3 to 7.5)

neutral solution

- 7 — pure water
  blood (7.3 to 7.5), swimming pool
  eggs   water
- 8 — seawater (7.8 to 8.3)
  shampoo

increasingly basic or alkaline

- 9 — baking soda
  phosphate detergents
  chlorine bleach, antacids
- 10 — milk of magnesia (9.9 to 10.1)
  soap solutions
- 11 — household ammonia (10.5 to 11.9)
  nonphosphate detergents
- 12 — washing soda (Na_2CO_3)
  hair remover
- 13 — oven cleaner
- 14

Figure 21-11 The pH scale used to measure acidity and basicity or alkalinity of water solutions. Values shown for common aqueous solutions are approximate.

the eastern United States and some regions of the West, southeastern Canada, and most of western Europe has pH values 4 to 40 times more acidic than normal rainwater.

Effects of Acid Fog and Acid Deposition Signs of damage to forests and lakes from excess acidity have been reported in many areas throughout the world, including much of the eastern United States, southeastern Canada, Brazil, South Africa,

much of central and northern Europe, Poland, Czechoslovakia, the Soviet Union, and China. Damage to crops, fish, wildlife, and materials alone from acid fog and acid deposition is estimated conservatively to be at least $2.4 billion a year in the United states and $50 billion to $60 billion a year in western Europe.

According to a 1983 study by the National Academy of Sciences, acid fog and acid deposition can **(1)** destroy most forms of aquatic life when large quantities of acidic rain or snow run off into lakes and slow moving streams, **(2)** kill or stunt the growth of fish and other forms of aquatic life by leaching toxic aluminum salts from the soil so that they run off into nearby lakes and streams, **(3)** convert fairly harmless mercury deposits in lake bottom sediments to highly toxic methyl mercury (Section 24-5), **(4)** kill some of the microorganisms in the soil responsible for breaking down organic matter and recycling nitrogen and carbon through food webs (Section 5-3), **(5)** inhibit seed germination, injure leaves of trees and other plants by leaching calcium and magnesium nutrients and breaking down a waxy protective coating on leaves that helps prevent damage from diseases and pests, **(6)** damage trees and reduce timber production when roots take up toxic compounds of aluminum, cadmium, copper, iron, zinc, and other salts leached out of the soil, **(7)** damage trees and reduce timber production by leaching vital plant nutrients such as calcium, magnesium, and potassium from the soil, **(8)** damage and decrease yields of food crops like soybeans, spinach, and alfalfa unless farmers neutralize excessive soil acidity by applying lime or crushed limestone, **(9)** damage marble and limestone monuments and buildings and corrode exposed metal culverts, roofs, bridges, and expressway support beams that are not adequately painted and maintained, and **(10)** leach toxic metal compounds into drinking water supplies from soils and water pipes. According to a study by the Brookhaven National Laboratory, acid fog and deposition may also play a key role in the premature deaths of 50,000 Americans a year.

Damage to trees, especially conifers such as spruce and fir, is most severe at high altitudes because the mountaintops are bathed in fogs that are typically 5 to 10 times more acidic than the acids deposited on the mountaintop soils. Measurements taken from Maine to North Carolina have shown a 40 percent reduction in growth for 31 tree species found at high elevations. Throughout Scandinavia, forests are stunted by acid fog and acid precipitation, most of which apparently comes from tall smokestacks in the industrial regions of England and West Germany. By 1983, 34 percent of the forests in West Germany had been damaged by a combination of ozone, acid fog, and acid deposition.

The effect of acid deposition on lakes, ponds, streams, and rivers varies widely in different locations because of differences in soil types and vegetation of the watersheds involved. The soils in some watersheds contain limestone ($CaCO_3$), dolomite ($MgCO_3$), and other alkaline (basic) substances that can react with and neutralize the acids before they run off into nearby lakes and streams. But poor, thin soils, such as those lying over granite and some types of sandstone, are already acidic and have little ability to neutralize additional acids. Such soils are found throughout much of Scandinavia and southeastern Canada (especially in the provinces of Quebec and Ontario) and in the United States throughout large portions of New England, the Great Smoky Mountains, northern Wisconsin and Minnesota, the Pacific Northwest, and the Colorado Rockies. Most of the soils in the Midwest, where a large part of the sulfur dioxide and nitrogen dioxide emissions is produced, are alkaline enough to prevent serious damage.

By 1982, an estimated 3,000 lakes and 40,000 kilometers (25,000 miles) of streams in the eastern states had been altered by excess acidity. An additional 6,000 lakes and 56,000 kilometers (35,000 miles) of streams elsewhere in the East are also threatened by acid deposition. Public health officials in New York State advise children and pregnant women not to eat *any* fish caught in the state's lakes because of possible contamination with aluminum and/or mercury leached into the lakes from nearby soil or released from bottom sediments by increased acidity. Canada estimates that at least 50 percent of its acid deposition comes from the United States and that up to 1,400 of its water bodies have been so seriously affected by acid deposition that they are devoid of fish. At least 6,000 of Sweden's lakes contain no fish because of excess acidity.

Controlling Acid Fog and Acid Deposition So far most of the emphasis has been placed on reducing emissions of sulfur dioxide by at least 50 percent, using methods discussed in Section 21-7. However, there is increasing evidence that effective control of excessive acidity in the air as well as plant-damaging ozone will require a similar reduction in emissions of nitrogen oxides from power plants, industrial plants, and motor vehicles, as discussed in Section 21-8.

The coal and automobile industries have fought the enactment of more stringent air pollution controls in the United States because they contend that: **(1)** evidence used to show that acidity levels are increasing in intensity and geographic

scope is suspect because it is not based on adequate measurements; **(2)** the extent of damage from acid fog and acid deposition is a matter of considerable scientific dispute and needs much additional research before any effective plan of action can be drawn up and implemented; **(3)** there is no reliable evidence linking emissions of sulfur and nitrogen oxides from specific midwestern power and industrial plants to the alleged increases in the acidity of precipitation in the Northeast and Canada; **(4)** achieving a 50 percent reduction in sulfur dioxide emissions by installing pollution control equipment to remove sulfur dioxide from the smokestack emissions of coal-burning power plants would cost an estimated $10 billion a year for many years and would cause sharp rises in electricity bills; **(5)** requiring only coal-burning power and industrial plants to remove at least 50 percent of the sulfur dioxide and nitrogen oxides from their smokestack emissions will probably not solve the problem because there is considerable evidence that much of the damage, especially to trees, is caused by exposure to nitric acid and nitrate salts produced mostly by emissions of nitrogen oxides from motor vehicles and from ozone produced when nitrogen oxides react with hydrocarbons and other chemicals in the atmosphere; and **(6)** further reduction of emissions of nitrogen oxides from motor vehicles would increase manufacturing costs and the cost of motor vehicles to consumers.

Since 1981 the Reagan administration has used these arguments as reasons for not developing a policy and committing funds for decreasing the damage from acid fog and acid deposition until further research establishes the extent of damage and the causes of this complex problem. However, a number of air pollution scientists believe that indications are adequate that reducing the present levels of sulfur dioxide and nitrogen oxides emitted by power plants, industrial plants, and motor vehicles by at least 50 percent would significantly reduce the damage of these pollutants to human health and to forest and aquatic ecosystems.

21-5   Effects of Air Pollution on the Ozone Layer and Global Climate

Protecting the Ozone Layer The *stratosphere* contains an average concentration of a few parts per million of *ozone*, a pale blue poisonous gas. This gas is concentrated in the upper portion of the stratosphere, called the **ozone layer**, located about 20 to 35 kilometers (12 to 22 miles) above the earth's surface. The ozone gas in this layer is formed when oxygen molecules in the stratosphere interact with ultraviolet radiation from the sun in a complex sequence of reactions that can be summarized by the overall reaction: $3O_2$ + ultraviolet radiation → $2O_3$.

The importance of the ozone layer lies in the fact that its molecules of ozone interact with incoming radiation from the sun and filter out more than 99% of the ultraviolet radiation from the sun that could harm most plant and animal life on earth. In recent years a number of scientists have become concerned that the average concentration of ozone in the ozone layer could be decreased by: **(1)** large-scale nuclear war (which would destroy most of the ozone layer), **(2)** direct injection of nitrous oxide (N_2O) from the exhaust of supersonic airplanes flying in the stratosphere, and **(3)** the movement from the troposphere to the stratosphere of chlorofluorocarbons (CFCs, also known as Freons), such as CCl_2F_2, released from aerosol spray cans and from discarded or leaking refrigeration or air conditioning equipment. When these chemicals come under the influence of high-energy ultraviolet radiation in the stratosphere, they can be converted to highly reactive species capable of reacting with ozone and decreasing its concentration in the ozone layer.

Estimates of the effects of various chemicals on the ozone layer are based primarily on theoretical mathematical models of at least 192 chemical reactions, many of them poorly understood, which are known to be taking place in the stratosphere. According to the latest models, each 1 percent decrease in ozone would increase the amount of ultraviolet radiation reaching the earth's surface by 1 to 3 percent. Direct measurements cannot be used to confirm or deny such models because the concentration of ozone in various parts of the stratosphere normally varies as much as 5 percent a year. Decreasing the concentration of ozone in the stratosphere could: **(1)** increase the number of skin cancer cases by 2 to 5 percent for each 1 percent decrease in ozone concentration, **(2)** increase the number of cases of severe sunburn in unprotected light-skinned people, **(3)** damage many species of land plants and some aquatic species and possibly decrease the yields of some important food crops such as corn, rice, and wheat, and **(4)** cause unpredictable changes in world climatic patterns.

There is general agreement that large-scale nuclear war would probably destroy most of the ozone layer. Recent research has indicated that nitrous oxide exhausted by supersonic jet planes flying in the stratosphere poses a much less serious threat to the ozone layer than originally projected. A 1984 study by the National Academy of Sciences projected a 2 to 4 percent decrease in total ozone in the ozone layer over the next 100 years if CFCs continue to be produced at 1977 rates and a 10 to 12 percent reduction if usage should rise sharply above the 1977 level.

Since 1978, the United States, Canada, Sweden, Norway, and Denmark banned the use of CFCs in aerosol spray cans, but these substances are still widely used as refrigerating agents and as foaming agents for various polymers. Chemists quickly developed substitutes for CFCs as propellants in aerosol spray cans, but so far other compounds are not as efficient and nontoxic as CFCs in refrigeration and air conditioning units.

Global Warming from Carbon Dioxide Emissions: The Greenhouse Effect Because carbon dioxide is a normal constituent of the atmosphere as a result of the carbon and oxygen cycles (Section 5-3), it is not commonly considered to be an air pollutant. However, extremely large amounts of fossil fuels are being burned, and this human intervention into the carbon and oxygen cycles is producing carbon dioxide faster than these natural cycles can adjust to it.

As a result, the average concentration of carbon dioxide in the atmosphere has been increasing since 1860 (when the world began a shift to increased reliance on coal as a fuel) and at a more rapid rate since 1958 (primarily because of the increased combustion of gasoline, fuel oil, and natural gas). There is concern that this could lead to a gradual increase in the average temperature of the earth's atmosphere, which in turn could affect global climate patterns.

About 70 percent of the incoming radiant energy from the sun passes through the atmosphere and strikes the surface of the earth much like light passing through glass. This incoming radiation is absorbed by the land and water and radiated back toward space as longer wavelength infrared (IR) radiation, or heat energy (Figure 4-10). But not all this infrared heat makes it back into space. Some is absorbed primarily by carbon dioxide gas and water vapor in the atmosphere and reradiated toward the earth's surface to warm the lower atmosphere. This warming action is commonly called the **greenhouse effect**, because the atmosphere acts similarly to the glass in a greenhouse or a car window, which allows visible light to enter but hinders the escape of longer wavelength infrared heat.* Without the greenhouse effect, the earth would be a cold and lifeless planet.

Because of the large amount of water vapor in the atmosphere and its relatively rapid rate of recycling in the water or hydrologic cycle (Section

---

*The term *greenhouse effect* is misleading, despite widespread use. The temperature in a greenhouse or closed car rises on a sunny day because the glass in a greenhouse or car window **(1)** allows visible light to enter but hinders the escape of longer wavelength infrared heat, and **(2)** also keeps the warmed air from blowing away.

5-3), human activities can apparently have relatively little effect on the average concentration of water vapor in the earth's atmosphere. However, because each 908 kg (1 ton) of fossil fuel that is burned completely releases about 2.7×10^3 kg (3 tons) of carbon dioxide into the atmosphere, a number of atmospheric scientists are concerned that the combustion of fossil fuels can increase the average levels of carbon dioxide in the earth's atmosphere.

A 1983 report by the National Academy of Sciences indicated a general consensus among 70 atmospheric scientists that based on present computer models of atmospheric processes, a doubling of the 1980 carbon dioxide levels in the atmosphere would raise the average global temperature of the atmosphere between 1.5 and 4.5°C (2.7 and 8.1°F) and 2 to 3 times this temperature increase at the earth's polar regions. A 1985 model of the atmosphere suggested that global warming from CO_2 buildup may be only half as great as these earlier projections because the denser and wetter clouds containing more CO_2 should reflect more sunlight into space. However, other recent studies indicate that dozens of other gases such as chlorofluorocarbons found in trace amounts in the atmosphere could produce a global warming at least as great as that caused by carbon dioxide alone.

At first glance a warmer average climate might seem desirable, resulting in lower heating bills, longer growing seasons, and possible increases in crop productivity in some parts of the world. Unfortunately, however, there could be serious consequences in other parts of the world. Two possible harmful effects of global warming are: (1) changes in the distribution of rainfall and snowfall over much of the earth, requiring that the world's major food-growing regions such as those in much of the United States be shifted northward to areas of Canada and the Soviet Union, where the soils tend to be poorer and less productive, and (2) some melting of glaciers and icefields in polar regions, with a projected rise in the average sea levels of about 2.4 m (8 ft) by 2100—thus possibly causing flooding of coastal cities and industrial areas.

Suggestions for preventing this problem from reaching crisis levels include: (1) reducing the use of fossil fuels (especially coal) over the next 50 years and relying more on either a combination of energy conservation and renewable energy from the sun, wind, flowing water, and geothermal energy, or increased use of nuclear power, should this source of energy become economically feasible and politically and environmentally acceptable, (2) using scrubbers to remove carbon dioxide from the smokestack emissions of coal-burning power and industrial plants and the exhaust from vehicles—a

technically feasible but expensive solution, (3) planting trees worldwide to reduce the greenhouse effect by increasing the uptake of carbon dioxide from the atmosphere through photosynthesis as well as reducing the harmful effects of the increased rate of deforestation that is presently taking place (Section 12-5), (4) using soil conservation measures (Section 9-4) to reduce soil erosion, which releases carbon dioxide into the atmosphere.

Because it will be difficult to get the world's nations to reduce the use of fossil fuels, plant trees, and take steps to reduce soil erosion enough to prevent significant rises in atmospheric carbon dioxide levels, some scientists suggest that we prepare for the effects of long-term global warming through: (1) breeding plant strains that need less water and can thrive in water too salty for ordinary crops, (2) improving the efficiency of irrigation so that less water is wasted (Section 10-6), (3) erecting levees to protect coastal areas from flooding, as the Dutch have done for hundreds of years, (4) using zoning ordinances to prohibit new construction on undeveloped, low-lying coastal areas, and (5) storing up several years' supply of food worldwide as insurance against climate changes that could shift the world's major food-growing regions and decrease overall crop yields.

Global Cooling from Particulate Matter   Some atmospheric scientists worry that the average atmospheric temperature might fall instead of rising because of the large-scale injection of particles of solids and liquids into the atmosphere from a combination of clearing land for agriculture and urbanization, and through smokestack, chimney, and automobile emissions, as well as from neutral processes such as volcanic eruptions, dust storms, and forest fires. Particles, especially those in the upper atmosphere, can reflect some of the incoming solar radiation back into space, reducing the amount of solar heat reaching the earth's surface, thus tending to cool the atmosphere. A cooling of only a few degrees could trigger a new ice age.

However, particles from volcanic eruptions and human activities can either raise or lower atmospheric temperature depending on their size, composition, reflective properties, and altitude, and the reflectivity of the earth's surface. Fine particles that either rise or are injected into the stratosphere (for example, in aircraft emissions) cause heating of the air there by absorbing some of the incoming radiation. In the troposphere, particles over a dark surface (such as the ocean) normally cool the lower atmosphere because they reflect and reradiate incoming solar radiation into space. But over much of the land, which is lighter than the oceans, particles can absorb heat energy and radiate some of it back toward the earth, thus warming the lower atmosphere.

The overall effects on global climate of particles from natural and human sources are complex and still poorly understood. Estimates of the human production of atmospheric dust range from 5 to 50 percent of the total annual input. It is argued that the climatic effects of most particles from human activities are primarily local because they end up in the troposphere, where they remain for only a few days before falling out or being washed out by rain.

21-6 Principles of Pollution Control

Internal and External Costs To understand possible approaches for controlling air and water pollution and land disruption, we need to distinguish between what economists call *internal costs* and *external costs*. In making or using anything there are **internal costs**. For example, the price one pays for a new car reflects the costs of construction and operation of the factory, raw materials and labor, marketing expenses, and shipping costs, as well as automobile company and dealer profits.

There are also economic and health side effects not directly associated with the act of production. These **external costs** are passed on to someone else, usually the general public. For example, the price one pays for a car does not include the external costs resulting from the land disruption and air and water pollution caused by mining and processing the manufacturing materials, the noise and air pollution caused by driving the vehicle, and the losses of productivity, good health, and life due to automobile accidents. Thus, the *true cost* of any item or service is its internal cost plus the hidden external costs.

Our present economic system makes it profitable to pollute because many of the present and future external costs from producing and using various goods and services are not included in the initial price of the items. Thus, the initial cost of an item or service does not provide the purchaser with accurate information about its true long-term cost. As a way out of this dilemma, economists have proposed *internalizing the external costs*, so that the initial cost of an item or service reflects as closely as possible its true cost.

Setting Pollution Control Standards *The basic issues* in pollution control involve *deciding how much pollution is too much and how much we are willing to*

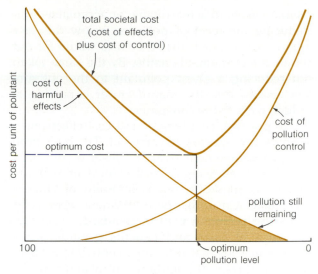

Figure 21-12 Cost-effectiveness analysis involves comparing the costs from the harmful effects of pollution with the cost of pollution control to minimize the total costs of pollution control and still reduce pollution to an acceptable level. The shaded area shows that some harmful effects remain, but removing these residual damages would make the costs of pollution control too high.

pay to reduce pollution to levels acceptable by the majority of the people. Finding out the level of exposure to a particular pollutant that is harmful to most individuals in a particular species is costly, time-consuming, and difficult. Because of the complex and mostly unknown effects and interactions of pollutants, we probably will never be able to set exact standards or determine specific pollutant levels that cause certain diseases or ecological effects. However, just because we can't determine allowable pollutant levels perfectly does not mean that we shouldn't establish tentative levels that can be adjusted as we get better information.

To determine the allowed level of pollution that will protect most of the public and still not force industries out of business, experts on both sides of the controversy often attempt to compare pollution control *costs* with the *benefits* that may occur from pollution control. This procedure is called **cost-benefit analysis.** Industry spokespersons emphasize the pollution control costs, which are fairly easy to document, whereas environmentalists and health officials emphasize the pollution control benefits, which are often harder to document and put a price tag on.

Next **cost-effectiveness analysis** is used to determine how many dollars will be needed to achieve various levels of pollution control. As illustrated in Figure 21-12, a goal of cost-effectiveness analysis could be to minimize the total costs of pollution control and still reduce harmful environmental effects to a reasonable or acceptable level. Reducing the level of pollution below this level might cost so much that the costs would far outweigh the economic benefits. By the same token, not reducing a given pollutant to the optimum level means that the manufacturer is passing on hidden costs to the consumer.

In practice, cost-benefit and cost-effectiveness analyses are not easy to carry out. The basic problem is disagreement over the level of risk associated with allowing a certain level of pollution that is economically, psychologically, or ethically acceptable. Critics of these economic approaches to setting pollution control standards argue that estimates of costs and benefits generally come from industries that have an incentive to exaggerate present costs of pollution control and to downgrade the economic value of a benefit in the future. An even more serious problem is that many important things cannot be reduced to dollars and cents. For example, some of the costs of air pollution, such as extra laundry bills, house repainting, and ruined crops are fairly easy to estimate. But it is difficult, if not impossible, to put a price tag on human life, clean air and water, whooping cranes, a clear sky, beautiful scenery, and the ability of natural ecosystems to degrade and recycle our wastes, because such quantifications involve political, social, and ethical issues as well as economic ones. Environmentalists contend that ethics and common sense—not economic analysis—should be the overriding factors.

Air Pollution Control Laws in the United States
Dissatisfaction with the automobile industry's progress and rising pressure from those concerned about the effects of air pollution on human health led to the passage of several air pollution control laws (Appendix 3). Indeed, the Clean Air Act of 1970 is considered to be one of the strongest pieces of environmental legislation passed in the United States.

The goal of the Clean Air Act of 1970 was to have air "safe enough to protect the public's health" by May 31, 1975. To accomplish this goal, the Environmental Protection Agency was required to set primary and secondary *national ambient air quality standards* for major primary air pollutants, as listed in Appendix 6. Each standard specifies the maximum allowable level for a certain air pollutant averaged over a given period, beyond which harmful health effects may occur. A deadline was set for attainment of each primary standard, with enforcement left to the states. Although these laws allow no consideration of the costs required to meet the primary standards, there is con-

tinuing pressure by industry to allow the use of cost-benefit economic analysis in setting the value for primary standards.

For automobiles Congress went even further and set standards itself, hoping to force the automobile companies to speed research on emission reduction and alternative engines. The act required a 90 percent reduction in emissions of hydrocarbons and nitrogen oxides from 1970 levels for all 1975 automobiles, and a 90 percent reduction in emissions of nitrogen oxides from 1971 levels for 1976 models. By 1976, however, these standards had not been met because the EPA had granted a series of 1-year extensions to the automobile industry.

After 7 years of pressure from all sides, Congress amended its 1970 legislation by passing the Clean Air Act of 1977. Despite vigorous industry pressure, the principles of the 1970 act were reaffirmed. But the dates set for compliance with standards in the 1970 act were postponed. The 1977 amendments also established *prevention of significant deterioration* regulations, partly to protect parts of the country where the air is cleaner than that required by one or more national ambient air quality standards and partly to prevent industries from relocating facilities from exceptionally dirty areas to clean areas.

Between 1975 and 1983 air pollution control laws in the United States were a major factor in reducing average levels of sulfur dioxide by 36 percent, carbon monoxide by 33 percent, ozone by 17 percent, lead by 67 percent, and particulate matter by 20 percent. Despite this important progress, there is far to go. In 1985 Congress was carrying out an investigation to determine why since 1971 the EPA had regulated only five of the 600 potentially dangerous air pollutants it is empowered to control.

Methods of Pollution Control
Three general methods used to control pollution once a given control standard has been adopted are: **(1)** *input or preventive control*, designed to prevent a pollutant from being formed or to sharply reduce its formation, **(2)** *throughput control*, in which the amount of pollution produced in a process is reduced by altering the process or reducing the rate at which it is carried out, and **(3)** *output control*, in which a pollutant is removed after it is produced, either just before or after it enters the ecosphere. Table 21-3 gives examples of each of these approaches. Because output control methods tend to be expensive and difficult, it is usually easier and cheaper in the long run to rely more on input and throughput methods for pollution control.

Table 21-3 Approaches to Pollution Control

Input Control	Throughput Control	Output Control
Prevent or reduce the amount of pollutant from reaching the atmosphere or a body of water (e.g., use soil conservation techniques to reduce dust blown into the air and sediment washed into aquatic systems).	Reduce the rate of throughput: that is, slow down production and consumption (e.g., reduce consumption by price increases, pollution taxes, economic incentives, or, as a last resort, rationing).	Remove the pollutant or dilute it at the emission source (exhaust pipe, smokestack, or sewage line).
Select inputs that contain or produce little if any of the pollutant (e.g., use natural gas, coal gasification, or low-sulfur oil for electric power plants).	Find and promote shifts to substitute products or services that are less harmful (e.g., emphasize mass transit and paratransit rather than cars in cities and use reusable soft drink bottles instead of cans).	Remove the pollutant or lower its concentration (usually harder and more expensive because the pollutant is dispersed—the second energy law again).
Remove the pollutant before using the input (e.g., remove the sulfur from coal and oil).	Alter or replace a process to generate less or none of the pollutant (e.g., develop a car engine or a paper-making process that produces less pollution).	Convert the pollutant to a less harmful form (e.g., convert very toxic methyl mercury to less harmful inorganic forms of mercury, as discussed in Section 24-5).
Improve the natural ability of an ecosystem to dilute or degrade a pollutant (e.g., add oxygen or air to a river or lake to increase its ability to degrade oxygen-demanding wastes).	Make the process more efficient so that less energy and matter are wasted and less pollution is produced (e.g., convert aluminum ore to aluminum metal with a process requiring less electricity).	Choose the time and place of discharge to minimize damage (e.g., use tall smokestacks to disperse air pollutants at high levels where they may be dispersed more effectively, or stagger work hours to reduce air pollution by motor vehicles). This approach does not reduce the total pollution load, but it can spread it out so that harmful levels may not be exceeded.

21-7 Control of Sulfur Dioxide and Particulate Matter in Industrial Smog

Control of Sulfur Oxides Emissions The major methods for controlling emissions of sulfur oxides can be divided into input and output approaches, as summarized in the accompanying box.

Control of Particulate Matter Emissions Major input and output methods for controlling emissions of particulate matter are summarized in the box on page 303.

A combination of most of these methods reduced total particulate matter emissions in the United States by about 65 percent between 1970 and 1983. Despite this significant progress, the to-

---

Methods for Controlling Sulfur Oxides Emissions

Input Approaches

1. *Reduce population growth (Chapter 8) and the wasteful use of energy (Section 19-7). Because at least 43 percent of the energy used each year in the U.S. is wasted unnecessarily, many energy experts argue that reducing waste is the cheapest, easiest, and most effective approach.*

2. *Shift from fossil fuels to a mix of other energy sources such as nuclear, solar, wind, hydropower, or geothermal energy (Chapters 17, 18, and 19). Even if half the electrical power used in the United States is produced by such alternative energy sources by the year 2000, existing and projected coal-burning plants will still emit massive quantities of sulfur oxides unless other input and output methods are used to reduce these emissions.*

3. *Convert solid coal to a gaseous fuel or a liquid fuel to remove most of its sulfur impurities and to re-*

duce emissions of sulfur oxides from burning solid coal (Section 17-3). Economic feasibility is still unknown, the net useful energy yield is low, and the solid waste produced must be disposed of safely (Section 24-2).

4. *Shift to low-sulfur coal (containing less than 1 percent sulfur). Major U.S. supplies of low-sulfur coal are located west of the Mississippi, far from major eastern power plants and industrial centers to which the coal would have to be transported, at high cost and with the use of much energy. In addition, western coal tends to have a lower energy value per unit of weight than high-sulfur eastern coal, and boilers in a number of older power plants could not burn low-sulfur coal without expensive modifications.*

5. *Remove sulfur from the fuel before burning. Existing physical and chemical processes can remove 20 to 40 percent of the sulfur before burning, but can increase fuel costs by 25 to*

50 percent depending on the method used and the amount of sulfur removed.

Output Approaches

1. *Require the use of scrubbers, fluidized-bed combustion, limestone injection, or some other method to remove sulfur oxides during combustion or from smokestack exhaust gases.* A scrubbing technology that removes as much as 90 percent of the sulfur dioxide from smokestack emissions is *flue gas desulfurization*. Stack gases pass through a chamber containing a slurry of limestone ($CaCO_3$) and water. The slurry absorbs the sulfur dioxide and converts it to calcium sulfate ($CaSO_4$), which can safely be used as landfill, or to calcium sulfite ($CaSO_3$), a gooey substance that can harm groundwater (Section 22-5) and is difficult to dispose of. Such scrubbers, however, can cost as much as $300 million per large installation. From 90 to 98 percent of the sulfur dioxide and essentially all the nitrogen oxides produced during the combustion of coal can be removed by using *fluidized-bed combustion*, in which a mixture of pulverized coal and limestone burns while air is blown through it (Figure 17-5). This method should be widely available by 1990, but it costs about $300 million to $400 million per large installation. In a newer *limestone injection* scrubbing technology, limestone is injected directly into the furnace of the power plant while the coal is burning and combines with sulfur compounds before they are converted to sulfur dioxide to produce calcium sulfate. This process is cheaper (up to about $125 million per large installation) than flue gas desulfurization and fluidized-bed combustion but removes only about 50 to 60 percent of the sulfur dioxide. A new titanium dioxide (TiO_2) process that can remove as much as 99.7 percent of sulfur dioxide emissions and 99.6 percent of particulate matter emissions is being tested. Since 1980, the EPA has required all new coal-burning power plants to use scrubbers that remove at least 70 percent of the sulfur dioxide from smokestack emissions. However, this standard does not apply to existing coal-burning power plants or to oil-burning plants being converted to burn coal. Each of these older coal-burning plants emits almost seven times as much sulfur dioxide a year as a new plant. Reducing acid deposition by 50 percent probably would require that these older plants be retrofitted with scrubbers or fluidized-bed combustion. According to the EPA, requiring efficient air pollution control devices on all fossil-fuel-burning plants would increase the consumer's cost of electricity by 15 to 20 percent—far less than the external health and materials costs to consumers from not controlling these emissions.

2. *Discharge emissions from smokestacks tall enough to pierce the thermal inversion layer.* This widely used approach is favored by industry because it is cheaper than removing pollutants from stack gases. It is opposed, however, by most environmentalists. This method can decrease concentrations of sulfur oxides and particulate matter in areas near the power or industrial plant, but it can lead to increased levels of these pollutants and secondary pollutants making up acid fog and acid deposition in distant areas. In 1985 the Supreme Court ordered the EPA to require pollution controls on all of the 650 tall stacks in the United States after attempts by the EPA and the Reagan administration to have such orders by four federal courts overturned.

3. *Use intermittent emission control.* That is, shut down a plant or switch to low-sulfur fuels during adverse climate conditions, especially during hot, wet months when electricity use is high and acid deposition is apparently heaviest. At other times emissions through tall smokestacks or burning of high-sulfur coal would be allowed. This method is favored by industry but opposed by the EPA and by most environmentalists for the same reasons discussed for output method 2.

4. *Add a tax on each unit of sulfur dioxide emitted, to reduce emissions and to encourage development of more efficient and cost-effective methods of emissions control* (see Section 21-6).

5. *Add lime or ground limestone to soil and acidified lakes to neutralize acidity.* This procedure is favored over the installation of pollution control devices by utilities and coal-burning industries. Liming, however is expensive ($100 to $150 per acre), and must be repeated; and the costs must be borne by individual landowners or by taxpayers through increased state and local taxes. Environmentalists call this a "Band-aid" approach to acid deposition and say that it fails to address the real problem.

Methods for Controlling Emissions of Particulate Matter

Input Approaches

1. *Reduce population growth (Chapter 8) and energy waste by power plants and automobiles. This is the most effective way to reduce the amount of dangerous fine particles formed in the atmosphere (Figure 21-3).*

2. *Shift from fossil fuels to other energy sources such as nuclear, solar, wind, and geothermal energy to produce electricity (Chapters 16, 17, and 18).*

3. *Burn liquefied or gasified coal instead of solid coal (Section 17-3).*

4. *Discourage automobile use and encourage a shift to mass transit and paratransit (Section 14-3).*

5. *Shift to less polluting automobile engines and fuels (Section 21-8).*

Output Approaches

1. *Remove particulates from stack exhaust gases.* This is the most widely used method in electric power and industrial plants. Several methods are in use, including: **(a)** *electrostatic precipitators* (Figures 21-13*a* and 21-14), which remove up to 99.5 percent of the total mass of particulate matter (but not most fine particles) by means of an electrostatic field that charges the particles so that they can be attracted to a series of electrodes and removed from exhaust gas, **(b)** *baghouse filters* (Figure 21-13*b*), which can remove up to 99.9 percent of the particles (including most fine particles) as exhaust gas is passed through fiber bags in a large housing, **(c)** *cyclone separators* (Figure 21-13*c*), which remove 50 to 90 percent of the large particles (but very few of the medium-sized and fine particles) by swirling

exhaust gas through a funnel-shaped chamber in which particles collect through centrifugal force, and **(d)** *wet scrubbers* (Figure 21-13*d*), which remove up to 99.5 percent of the particles (but not most fine particles) and 80 to 95 percent of the sulfur dioxide. Except for baghouse filters, none of these methods removes many of the more hazardous fine particles, and all methods produce hazardous solid wastes, or sludge, that must be disposed of safely (Section 24-6). Except for cyclone separators, all methods are expensive, and none removes particles formed as secondary pollutants in the atmosphere.

2. *Discharge emissions from smokestacks tall enough to pierce the thermal inversion layer (see comments in the box for sulfur oxides control methods).*

3. *Use intermittent emission control (see comments in the box for sulfur oxides control methods).*

4. *Use emission control devices on automobiles to remove particles from exhausts or to reduce particle formation through more complete combustion (Section 21-8).*

---

tal amount of more hazardous fine particles emitted into the atmosphere has increased since 1970 and is expected to increase even more in the future because: **(1)** it is extremely difficult and expensive to remove fine particles from smokestacks and automobile exhaust; **(2)** most fine particles are secondary pollutants formed in the atmosphere by the reaction of various chemicals and thus are not removed by pollution control devices for smokestacks and automobiles; **(3)** there is increasing reliance on coal to produce electricity; and **(4)** there are more vehicles than ever on the road.

21-8 Control of Photochemical Smog: The Automobile Problem

Methods for Controlling Pollution from Automobiles Controlling emissions from 164 million mobile sources in the United States is more difficult politically and economically than controlling emissions from several thousand stationary sources. This explains in part why emissions of nitrogen dioxide in the United States rose by 10 percent between 1970 and 1983.

Figure 21-13 Four commonly used methods for removing particulates from the exhaust gases of electric power and industrial plants.

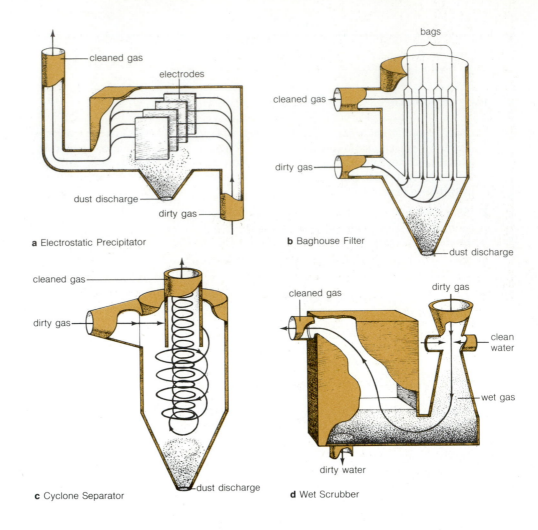

a Electrostatic Precipitator

b Baghouse Filter

c Cyclone Separator

d Wet Scrubber

Eastman Kodak Company

Figure 21-14 The effectiveness of an electrostatic precipitator in reducing particulate emissions. A stack with the precipitator turned off (left) and with the precipitator operating (right). Although this method can remove up to 99.5 percent of the total mass of particulates in exhaust gases, it does not remove very many of the invisible and more harmful fine particles.

The major methods for controlling emissions from mobile sources can be divided into input and output approaches, as summarized in the accompanying box.

Methods for Controlling Emissions from Automobiles and Other Mobile Sources

Input Approaches

1. *Reduce automobile use* by increasing existing taxes on gasoline, horsepower, and car weight, banning cars in downtown areas, and modifying life-styles and designing cities to require less automobile use (Section 14-4).

2. *Develop mass transit and paratransit systems* (Section 14-3).

3. *Use a cleaner burning fuel for the internal combustion engine.* Examples include natural gas, fuel cells (batteries), and hydrogen gas.

4. *Improve fuel efficiency* by reducing the size, weight, wind resistance, and power of cars and by improving the energy efficiency of transmissions, air conditioning, and other accessories.

5. *Modify the internal combustion engine to lower emissions and improve gas mileage and develop less-polluting engines.* Such modifications include carburetor adjustments, fuel injection, and diesel and stratified-charge engines.

An Output Approach

1. *Control emissions from the internal combustion engine more strictly*, for example, by the use of afterburners and catalytic converters.

Air pollution can no longer be addressed as simply a local urban problem.

Erik P. Eckholm

Discussion Topics

1. Why is it so difficult to establish an air quality standard for a specific air pollutant? Does this mean that such standards shouldn't be established?

2. Distinguish between photochemical smog and industrial smog in terms of major pollutants and sources, major human health effects, time when worst episodes occur, and methods of control.

3. Evaluate the pros and cons of the statement "Since we have not proven absolutely that anyone has died or suffered serious disease from nitrogen oxides, automobile manufacturers should not be required to meet the federal air pollution standards."

4. Rising oil and natural gas prices and environmental concerns over nuclear power plants could force the United States to depend more on coal, its most plentiful fossil fuel, for electric power. Comment on this in terms of air pollution. Would you favor a return to coal instead of increased use of nuclear power? Why?

5. How should we deal with the problem of air pollution from fine particulate matter? Why is this one of our more serious pollution problems?

6. Should chlorofluorocarbons be banned from use in refrigeration and air conditioning units? Why or why not?

7. Debate the idea that we should set up a world food bank to store several years' supply of food as insurance against climate change. How would you decide who gets this food in times of need?

8. What good and bad effects would internalizing the external costs of pollution have on the U.S. economy? Do you favor doing this? How might it affect your life-style? The life-style of the poor? If possible, have an economist discuss these problems with your class.

9. What obligations, if any, concerning the environment do we have to future generations. Try to list the major beneficial and harmful aspects of the environment that were passed on to you during the past 50 years by the two preceding generations.

10. Do you favor or oppose requiring a reduction by 50 percent between 1986 and 1996 in average levels of sulfur dioxide and nitrogen oxides emitted by

fossil-fuel-burning electric power and industrial plants and average nitrogen oxide levels emitted by motor vehicles in the United States? Defend your position. What would be the effect on your life-style of such a policy aimed at protecting human health and decreasing the damage to forests and aquatic ecosystems from acid fog and acid deposition?

11. Why is controlling air pollution generally a more difficult political problem than controlling water pollution? Give several reasons.

12. Simulate an air pollution hearing at which automobile manufacturers request a 3-year delay in meeting the air pollution standards set for the coming year. Assign three members of the class as members of a decision-making board and other members as the president of an automobile manufacturing company, two lawyers for that company, two government attorneys representing the EPA, the chief engineer for an automobile manufacturer, a public health official, and two citizens (one opposing and one favoring the proposal). Have a class discussion of the implications of the final ruling of the board.

13. Use the literature to determine what changes, if any, have been made since 1984 in the Clean Air Acts of 1970 and 1976. Which changes do you believe strengthened the act? Which ones weakened the act? Explain your position.

Water Pollution

22-1 Types, Sources, and Effects of Water Pollution

Major Water Pollutants **Water pollution** occurs when a substance or condition (such as heat) so degrades the quality of a body of water that the water fails to meet specified water quality standards or cannot be used for a specific purpose. Water that is too polluted to drink may be satisfactory for industrial use. Water too polluted for swimming may not be too polluted for fishing. Water too polluted for fishing may still be suitable for sailing or for generating electrical power.

Table 22-1 summarizes the sources, effects, and methods for controlling eight major types of water pollutants. In terms of quantity, sediments or water-suspended solids from soil erosion constitute the largest source of water pollution in the United States as well as throughout most of the world (Section 9-3).

Some Indicators of Water Quality Three indicators of water quality are the concentration of *dissolved oxygen (DO)*, the *biological oxygen demand (BOD)*, and the *fecal coliform bacteria count*. Most animals and plants that live in water need oxygen to carry out aerobic respiration (Section 5-3). The **dissolved oxygen content**, or DO content, is the amount of oxygen gas dissolved in a given quantity of water at a particular temperature and atmospheric pressure. At a typical temperature of 20°C (68°F) and normal atmospheric pressure, the maximum concentration of dissolved oxygen is 9 parts of oxygen per million parts of water (9 ppm).

Bacteria in water use dissolved oxygen to break down organic wastes. The amount of dissolved oxygen required for such bacterial decomposition is determined by measuring the **biological oxygen demand (BOD)** in terms of the parts per million (or milligrams per liter) of dissolved oxygen consumed over 5 days at 20°C (68°F) and normal atmospheric pressure. If a water system is overloaded with oxygen-demanding wastes, bacterial activity can reduce the dissolved oxygen content to levels so low that some aquatic species die. Water is considered seriously polluted when the BOD causes the dissolved oxygen content to fall below 5 ppm and gravely polluted if the DO level falls below 4 ppm.

A good indicator of drinking and swimming water quality is the number of colonies of **fecal coliform bacteria** present in a 100-milliliter (0.40 cup) sample of water. Coliform bacteria are intestinal microorganisms found in human and animal wastes. Although most coliform bacteria themselves are not harmful, their presence in water indicates the possible presence of other bacteria from untreated human and animal waste that can cause typhoid fever, cholera, dysentery, viral hepatitis, encephalitis, and other waterborne bacterial diseases (Chapter 20). Usually several samples are taken, and water is considered safe to drink by EPA standards when the arithmetic mean of all samples does not exceed 1 coliform bacterial colony per 100 milliliters of water, with no single sample having a count higher than 4 colonies per 100 milliliters. The EPA-recommended level for water safe for swimming is "not more than 200 colonies per 100 milliliters," although some cities and states allow swimming at higher levels. In 1981 about 35 percent of all samples taken from U.S. waters exceeded the EPA fecal coliform bacteria standard for safe swimming.

Point and Nonpoint Sources Some wastes are discharged from identifiable locations called **point sources**. They include: **(1)** pipes dumping untreated sewage from cities and industries into waterways and oceans, **(2)** sewage treatment plants that remove some but not all pollutants

Table 22-1 Major Water Pollutants

Pollutant	Sources	Effects	Control Methods
Oxygen-demanding wastes	Natural runoff from land; human sewage; animal wastes; decaying plant life; industrial wastes (from oil refineries, paper mills, food processing, etc.); urban storm runoff	Decomposition by oxygen-consuming bacteria depletes dissolved oxygen in water; fish die or migrate away; plant life destroyed; foul odors; poisoned livestock	Treat wastewater; minimize agricultural runoff
Disease-causing agents	Domestic sewage; animal wastes	Outbreaks of waterborne diseases, such as typhoid, infectious hepatitis, cholera, and dysentery (Chapter 20); infected livestock	Treat wastewater; minimize agricultural runoff; establish a dual water supply and waste disposal system
Inorganic chemicals and minerals			
Acids	Mine drainage; industrial wastes; acid deposition (Section 21-4)	Kills some organisms; increases solubility of some harmful minerals	Seal mines; treat wastewater; reduce atmospheric emissions of sulfur and nitrogen oxides (Chapter 21)
Salts	Natural runoff from land; irrigation; mining; industrial wastes; oil fields; urban storm runoff; deicing of roads with salts	Kills freshwater organisms; causes salinity buildup in soil; makes water unfit for domestic use, irrigation, and many industrial uses	Treat wastewater; reclaim mined land; use drip irrigation; ban brine effluents from oil fields
Lead	Leaded gasoline; pesticides; smelting of lead (see Section 24-5)	Toxic to many organisms, including humans	Ban leaded gasoline and pesticides; treat wastewater
Mercury	Natural evaporation and dissolving; industrial wastes; fungicides	Highly toxic to humans (especially methyl mercury)	Treat wastewater; ban unessential uses (Section 24-5)
Plant nutrients (phosphates and nitrates)	Natural runoff from land; agricultural runoff; mining; domestic sewage; industrial wastes; inadequate wastewater treatment; food-processing industries; phosphates in detergents	Algal blooms and excessive aquatic growth; kills fish and upsets aquatic ecosystems; eutrophication; possibly toxic to infants and livestock; foul odors	Advanced treatment of industrial, domestic, and food-processing wastes; recycle sewage and animal wastes to land; minimize soil erosion
Sediments	Natural erosion, poor soil conservation; runoff from agricultural, mining, forestry, and construction activities	Major source of pollution (700 times solid sewage discharge); fills in waterways, harbors, and reservoirs; reduces shellfish and fish populations; reduces ability of water to assimilate oxygen-demanding wastes	More extensive soil conservation practices (Section 9-4)
Radioactive substances	Natural sources (rocks and soils); uranium mining and processing; nuclear power generation; nuclear weapons testing	Cancer; genetic defects (see Section 18-1)	Ban or reduce use of nuclear power plants and weapons testing; more strict control over processing, shipping, and use of nuclear fuels and wastes (see Section 18-2)
Heat	Cooling water from industrial and electric power plants	Decreases solubility of oxygen in water; can kill some fish; increases susceptibility of some aquatic organisms to parasites, disease, and chemical toxins; changes composition of and disrupts aquatic ecosystems	Decrease energy use and waste; return heated water to ponds or canals or transfer waste heat to the air; use to heat homes, buildings, and greenhouses
Organic chemicals			
Oil and grease	Machine and automobile wastes; pipeline breaks; offshore oil well blowouts; natural ocean seepages; tanker spills and cleaning operation	Potential disruption of ecosystems; economic, recreational, and aesthetic damage to coasts, fish, and waterfowl; taste and odor problems	Strictly regulate oil drilling, transportation, and storage; collect and reprocess oil and grease from service stations and industry; develop means to contain and mop up spills

Table 22-1 Major Water Pollutants *(continued)*

Pollutant	Sources	Effects	Control Methods
Pesticides and herbicides	Agriculture; forestry; mosquito control	Toxic or harmful to some fish, shellfish, predatory birds, and mammals; concentrates in human fat; some compounds toxic to humans; possible birth and genetic defects and cancer (see Section 20-3)	Reduce use; ban harmful chemicals; switch to biological and ecological control of insects (Section 23-6)
Plastics	Homes and industries	Kills fish; effects mostly unknown	Ban dumping, encourage recycling of plastics; reduce use in packaging
Detergents (phosphates)	Homes and industries	Encourages growth of algae and aquatic weeds; kills fish and causes foul odors as dissolved oxygen is depleted	Ban use of phosphate detergents in crucial areas; treat wastewater (see Section 22-7)
Chlorine compounds	Water disinfection with chlorine; paper and other industries (bleaching)	Sometimes fatal to plankton and fish; foul tastes and odors; possible cancer in humans	Treat wastewater; use ozone for disinfection and activated charcoal to remove synthetic organic compounds

(Section 22-7), **(3)** combined storm and sewer lines that are overloaded during storms, **(4)** animal feedlots, and **(5)** offshore oil well blowouts and oil tanker accidents.

Nonpoint sources are spread out and difficult to identify and control. They yield the diffuse discharge of wastes from **(1)** runoff of sediment from forest fires, construction, farming, and most logging, **(2)** runoff of chemical fertilizers, pesticides, and saline irrigation water from croplands, **(3)** urban storm water runoff, **(4)** drainage of acids, minerals, and sediments from active and abandoned mines, **(5)** atmospheric precipitation and deposition of acids produced mostly by coal-burning electric power and industrial plants, and **(6)** untraceable spills or dumping of oil from tankers at sea and various hazardous materials from industries (Chapter 24). Nonpoint source pollution is now recognized as a major and growing problem.

22-2   Drinking Water Quality

For much of the world's population in LDCs, the major water pollution problem is drinking water contaminated with bacteria and viruses that cause sickness and death. By contrast, in the MDCs, purification of drinking water has led to a sharp drop in the incidence of waterborne diseases. Some scientists, however, are concerned about actual and potential contamination of many U.S. drinking water supplies with some of the 63,000 inorganic and organic chemicals in commercial use. For example, more than 700 synthetic organic chemicals have been found in various drinking water supplies in the United States.

Studies indicate some of these synthetic organic chemicals can cause kidney disorders, birth defects, and various types of cancer in laboratory test animals. Most of these chemicals are discharged directly or accidentally into surface water systems or into underground aquifers used as sources of drinking water. Studies have indicated, however, that a few of these chemicals may be formed by chemical reactions between substances discharged into bodies of water or used to disinfect water. For example, evidence indicates that potentially harmful trihalomethane compounds, such as chloroform ($CHCl_3$), may be formed when the chlorine used to kill bacteria in drinking water combines with natural organic matter in untreated water or with other synthetic organic chemicals discharged into some rivers.

Under provisions of the Safe Drinking Water Act of 1974, the Environmental Protection Agency is required to establish national drinking water standards, called maximum contaminant levels, to protect public health. By early 1985 the EPA had set maximum contaminant levels for only 20 potential drinking water pollutants, as shown in Appendix 7. According to the EPA, 20 percent of the nation's public water systems did not meet one or more of the existing federal drinking water standards in 1983. Amendments added to the Safe Drinking Water Act in 1985 require the EPA to set standards for 14 of the most potentially harmful synthetic organic chemicals by 1986, and to set standards for 50 other contaminants no later than 1988.

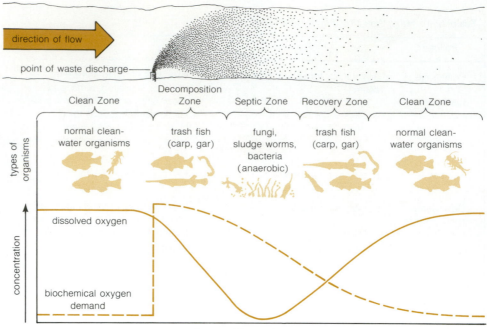

Figure 22-1 The oxygen sag curve (solid) versus oxygen demand (dashes). Depending on flow rates and the amount of pollutants, rivers recover from oxygen-demanding wastes and heat if given enough time.

direction of flow

point of waste discharge

| Clean Zone | Decomposition Zone | Septic Zone | Recovery Zone | Clean Zone |

types of organisms

normal clean-water organisms   trash fish (carp, gar)   fungi, sludge worms, bacteria (anaerobic)   trash fish (carp, gar)   normal clean-water organisms

concentration

dissolved oxygen

biochemical oxygen demand

time or distance downstream

Wells for millions of individual homes in suburban and rural areas are not required to meet federal safe drinking water standards—primarily because of the cost (at least $1000) of testing each well regularly and because of political problems associated with verifying individual compliance. A study conducted for the EPA by Cornell University scientists in 1982 indicated that 39 million rural area residents—two out of three of all rural Americans—were drinking water from private wells that did not meet one or more federal water quality standards.

Contaminated wells and concern about possible contamination of public drinking water supplies has led about 1 out of every 20 Americans to drink bottled water at an average cost of $1 a gallon (3.8 liters). Bottled water is regulated by the Food and Drug Administration (FDA) based on standards that are required to be equivalent to EPA standards for drinking water. However, the FDA does little testing of bottled water and generally relies on tests by the bottled water industry. Furthermore, there is no assurance that such water won't contain synthetic organic chemicals and other chemicals not regulated by EPA standards.

Most activated charcoal filter units for attachment under home sinks remove most synthetic organic chemicals if properly maintained. However, they are not effective in removing bacteria, viruses, and toxic metals. Filter units that contain "bacteriocides" prevent bacteria from building up *in the filter only*—not in the water.

22-3   Surface Water Pollution of Rivers

River Characteristics and Pollution If not overloaded, flowing rivers can dilute many wastes and can renew their supply of dissolved oxygen gas through turbulence and other exposure to the atmosphere. As a result, most rivers recover fairly rapidly from excess heat, oxygen-demanding wastes, and other rapidly degradable pollutants (Figure 22-1)—if not overloaded. Just below the area where large quantities of oxygen-demanding wastes are added, DO levels in the water drop sharply. Further downstream, however, the DO content returns to its normal level. The depth and width of the *oxygen sag curve*, shown in Figure 22-1, and the time and distance a river takes to recover, depend on a river's volume and flow rate and the volume of incoming oxygen-demanding wastes. Slowly degradable and nondegradable pollutants, however, are not eliminated by dilution and natural breakdown processes in rivers and must be prevented from entering them.

Along many rivers, water for drinking is removed *upstream* from a town, and industrial and sewage wastes are discharged *downstream*. This pattern is usually repeated hundreds to thousands of times. Because the rivers normally do not have enough time to recover before receiving the next load of wastes, pollution tends to intensify closer to the ocean. If each town were to withdraw its drinking water downstream rather than upstream,

Table 22-2 Undesirable and Desirable Effects of Heat Added to Water

Undesirable Effects	Desirable Effects
Thermal shock (the sudden death of thermally sensitive aquatic life due to sharp changes in temperature). When a power plant first opens, the sudden injection of hot water can kill some existing species. When the plant shuts down for repairs, the sudden temperature drop could kill the new heat-resistant species that have moved in, especially in winter.	Longer commercial fishing season and increased catches when desirable warm-water species are attracted to heated water areas.
Increased susceptibility of aquatic organisms to parasites, disease, and toxic chemicals.	Reduction in winter ice cover.
Disruption of fish migration patterns.	Increased recreational use because of the warming of very cold bodies of water.
Lowered dissolved oxygen concentrations at a time when the higher water temperature raises organisms' oxygen requirements.	Use of warm water for aquaculture to cultivate catfish, shrimp, lobsters, carp, oysters, and other species eaten by humans.
Fewer eggs and fewer surviving young for thermally sensitive species.	
Reduction of diversity of species by elimination of thermally sensitive organisms.	Use of heated water to heat buildings and greenhouses, provide hot water, remove snow, desalt ocean and brackish water, and provide low-temperature heat for some industrial processes. These uses, however, require that fossil fuel and nuclear power plants be built near urban areas, increasing the risks from air pollution and accidental releases of radioactivity.
Shifts in species composition. (This may be beneficial or harmful, depending on the new species that thrive in heated water. Undesirable slime and blue-green algae thrive in heated water.)	
Disruption of food webs by loss of one or several key species, especially plankton, at lower levels of the food webs.	
Delay of spring and fall lake turnover.	
Mutilation or killing of small organisms and fish sucked through power plant intake pumps, pipes, and heat-exchange condensers.	
Fish kills and other ecological damage caused by the chlorine, copper sulfate, or other chemicals used to keep bacteria and other microbes from fouling the water-cooling pipes in power plants.	

river water quality would improve dramatically!

A 1982 report by the United Nations Environmental Program reported that between 1972 and 1982 more U.S. rivers were moderately polluted than in the preceding decade, but fewer rivers were highly polluted. Perhaps the most spectacular river cleanup has occurred in Great Britain. In the 1950s the River Thames was little more than a flowing sewer. But after 25 years of effort, $200 million of British taxpayers' money, and millions more spent by industry, the Thames has made a remarkable recovery. By 1978 dissolved oxygen levels had risen almost 30 percent, the river was supporting an increasing population of almost 95 species of fish, including pollution-sensitive salmon, commercial fishing was thriving, and many species of waterfowl and wading birds had returned to their former feeding grounds.

Thermal Pollution of Rivers and Lakes Almost half of all water withdrawn in the United States each year is for cooling electric power plants. The cheapest and easiest method is to withdraw cool water from a nearby body of water and return the heated water to the same body of water. About 98 percent of this cooling water is returned to rivers, lakes, and estuaries without incurring significant contamination by chemicals. However, this water contains huge quantities of heat that can have ecological effects on the bodies of water.

Although average temperature increases for the entire body of water may not be large, most of the hot water is discharged near the ecologically vulnerable shoreline, where mature fish spawn and young fish spend their first few weeks. One or several power plants may use a given body of water without serious damage, thus giving the misleading impression that others can be built. Then just one more plant can exceed the aquatic system's threshold level (Section 6-2) and cause serious ecological damage to plant and animal life.

There is controversy over the seriousness of this problem. Some scientists view waste heat as a potential water pollutant that can damage and disrupt aquatic ecosystems, while others talk about using heated water for beneficial purposes and speak of *thermal enrichment* rather than *thermal pollution*, as summarized in Table 22-2. Despite talk about thermal enrichment, most existing and pro-

Figure 22-2 Cooling towers for the Rancho Seco nuclear power plant near Sacramento, California. Compare the size of the towers with the power plant and automobiles. Each tower is more than 120 meters (400 feet) high and could hold a baseball field in its base.

posed sites for electric power plants prevent the beneficial use of their enormous outputs of hot water. Pumping water for long distances to croplands, aquaculture ponds, and buildings requires energy (and money), and much of the heat is lost during transport.

Most analysts argue that the best course is to reduce thermal water pollution as much as possible by using one or a combination of the following approaches: **(1)** reducing the need for additional power plants by using and wasting less energy (Section 19-7), **(2)** limiting the number of power and industrial plants allowed to discharge heated water into a given body of water, **(3)** minimizing damage by returning heated water away from the fragile shore zone, **(4)** using wet cooling towers (Figure 22-2) to inject waste heat into the atmosphere by spraying heated water up through a draft of air and allowing it to be cooled by evaporation, **(5)** using dry cooling towers to pump heated water through tubes, with the heat transferred to the atmosphere by conduction, and **(6)** dissipating the heat into the atmosphere by evaporation in shallow cooling ponds or canals by pumping hot water into one end and withdrawing cooler water for reuse from the other end.

Most new power plants use wet cooling towers. This approach has several disadvantages, including a large daily input of cooling water (to replace water lost to the atmosphere by evaporation), visual pollution from the gigantic cooling towers (Figure 22-2), high construction costs (about $100 million per tower for a 1,000-megawatt plant), high operating costs, and excessive fog and mist in nearby areas. Dry towers are seldom used because they cost two to four times more to build than wet towers. Cooling ponds and canals are useful where enough affordable land is available (about 1,000 acres for a 1,000-megawatt plant).

22-4 Surface Water Pollution of Lakes

Lake Characteristics Lakes consist of three distinct zones (Figure 22-3): **(1)** the **littoral zone** near the shore, in which rooted aquatic plants are found, **(2)** the **limnetic zone** (or open water surface layer) through which sunlight can penetrate, which is dominated by tiny floating phytoplankton (which use sunlight to carry out photosynthesis), and free-swimming fish, and **(3)** the **profundal zone** of deep water, which is not penetrated by sunlight and is inhabited by bottom-dwelling aquatic organisms.

The flushing time for lakes is 1 to 100 years, compared to weeks for rivers (Table 10-1). As a result of this relatively slow flow rate, lakes are more susceptible than rivers to biological magnification of persistent and nondegradable chemicals. Contamination with oil, sludge, pesticides, metals, and other toxic substances that can kill fish and destroy bottom life are major pollution problems in lakes.

Natural and Cultural Eutrophication of Lakes
As part of its natural aging process, a lake receives nutrients such as phosphorus (as PO_4^{3-}) and nitrogen (as NO_3^-) by precipitation and by drainage from the surrounding land basin, from bottom sediments, and from organisms living in the lake. This natural erosion and runoff of plant nutrients from the land into lakes is called **eutrophication**. A lake with a low supply of plant nutrients is called an **oligotrophic lake**. Such lakes tend to have crystal-clear water, low algae populations, and high DO concentrations; they contain fish species like smallmouth bass and lake trout.

A lake with a large or excessive supply of plant nutrients like phosphates and nitrates is called a **eutrophic lake**. Such lakes tend to have a large

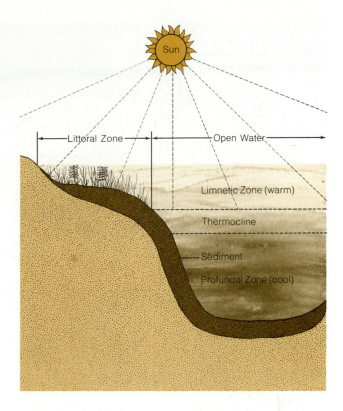

Figure 22-3 The major zones of a lake.

blue-green algae population, low DO content, and fish species like bullhead and carp. Most lakes become more eutrophic with time, as plant nutrients flow in from surrounding land. Depending on local climate and geologic conditions, the natural eutrophication of a lake typically takes thousands to millions of years.

One of the most widespread pollution problems of lakes—particularly shallow ones near urban or agricultural centers—is the acceleration of **natural eutrophication** from the increased flow of plant nutrients into a lake as a result of human activities. This stepped-up process, known as **cultural eutrophication**, can produce in a few decades the same effect as natural eutrophication. Cultural eutrophication is caused by excessive addition of phosphates and nitrates from effluents from sewage treatment plants, runoff of fertilizers and animal wastes, and soil erosion from construction, mining, and poor land use (Figure 22-4).

When a lake is overloaded with phosphate and nitrate plant nutrients, rooted plants like water chestnuts and water hyacinths in the littoral zone and green and blue-green algae in the limnetic zone undergo population explosions until they cover much of the lake's surface. Some algae, particularly blue-green algae, make the water taste and smell bad.

Through photosynthesis these blooms contribute oxygen to the layer of a lake during daylight

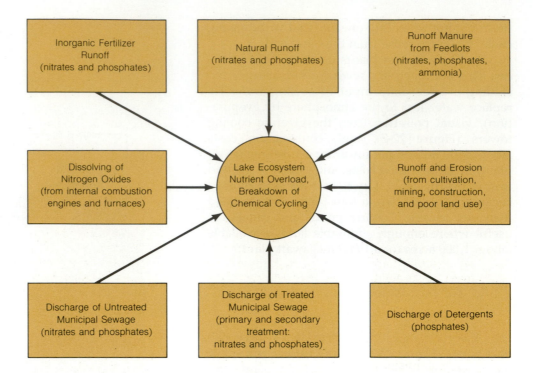

Figure 22-4 Nutrient overload, or cultural eutrophication, from human activities can disrupt the chemical cycles in a lake.

Inorganic Fertilizer Runoff (nitrates and phosphates)

Natural Runoff (nitrates and phosphates)

Runoff Manure from Feedlots (nitrates, phosphates, ammonia)

Dissolving of Nitrogen Oxides (from internal combustion engines and furnaces)

Lake Ecosystem Nutrient Overload, Breakdown of Chemical Cycling

Runoff and Erosion (from cultivation, mining, construction, and poor land use)

Discharge of Untreated Municipal Sewage (nitrates and phosphates)

Discharge of Treated Municipal Sewage (primary and secondary treatment: nitrates and phosphates)

Discharge of Detergents (phosphates)

hours. However, when they die, fall to the bottom, and are decomposed, the decomposing bacteria deplete the lake's lower layer of dissolved oxygen. Trout, whitefish, and many other deep-water species of fish die of oxygen starvation, while species like perch and carp, which need less oxygen and live in the upper layers, thrive. The actual number of fish may even increase, but there will be *fewer of the kinds* of fish that most humans prefer to catch and eat.

In shallow lakes or in the shore zone in deep lakes, the dead algae can also exhaust dissolved oxygen and kill fish in the surface layer of water. If excess nutrients continue to flow into a lake, the bottom water becomes foul and almost devoid of animals as anaerobic (non-oxygen-requiring) bacteria increase in numbers and produce smelly hydrogen sulfide and other chemicals.

A National Eutrophication Survey conducted by the EPA estimated that in 1975 one-third of the 100,000 medium to large lakes in the United States and about 85 percent of the larger lakes near major population centers were suffering from some degree of cultural eutrophication. For decades billions of gallons of untreated sewage along with industrial and agricultural wastes were discharged into the Great Lakes, causing cultural eutrophication, killing fish, contaminating water supplies, and forcing the closure of many bathing beaches. Although all five of the Great Lakes have been affected, the impact on Lake Erie has been particularly intense. In the 1960s there were persistent rumors that Lake Erie was dying. Actually much of it was overnourished and too alive with plant

life such as blue-green algae from cultural eutrophication. Lake Erie's pollution problems are intensified because it is fairly shallow and is located near major population centers in the United States and Canada. Lake Ontario's small size and shallowness also make it susceptible to cultural eutrophication.

The Great Lakes still receive large inputs of wastes, especially phosphates and toxic chemicals. However, the rate of cultural eutrophication has been slowed, and even reversed in some areas, by pollution control measures taken by the United States and Canada since 1972. As a result, most swimming beaches in Lake Erie that had been shut down for 30 years are again open and crowded with people during the summer. Commercial fishing is also making a comeback. Continued pollution from toxic chemicals, however, has caused health officials to warn pregnant women and women of childbearing age not to eat any fish from Lake Ontario. Others are advised to eat no more than one fish meal a month. Residents of Toronto are drinking more bottled water after the discovery of more than 50 potentially hazardous chemicals in the city's drinking water supply, which is drawn from Lake Ontario.

Controlling Cultural Eutrophication There are two basic categories of methods for controlling cultural eutrophication: *output approaches* (which treat the symptoms) and *input approaches* (which treat the causes), as summarized in the accompanying box.

Methods for Controlling Cultural Eutrophication

Output Approaches

1. Bypass the lake by diverting wastewater to fast-moving streams or to the ocean. This is not possible in most places. Where this approach is possible, it may transfer the problem from a lake to a nearby estuary system.

2. Dredge lake sediments to remove excess nutrient buildup. This is impractical in large, deep lakes and not very effective in shallow lakes. Dredging often reduces local water quality, and the dredged material must go somewhere—usually into the ocean.

3. Remove or harvest excess weeds, debris, and rough fish. With present technology this is difficult and expensive in large lakes.

4. Control nuisance plant growth with herbicides and algicides. The direct addition of such toxic substances as well as magnification in food chains may upset the ecosystem (Chapter 23).

Input Approaches

1. Remove phosphates from sewage treatment plant effluents before they reach the lake. This eliminates most point sources by removing phosphates from industrial and municipal sewage.

2. Recycle valuable phosphate and nitrogen nutrients removed from sewage treatment plant effluents and recycle them to the land as fertilizers rather than allowing them to overfertilize lakes and estuaries. This will require improved technologies for transporting treated sewage to agricultural and forest areas and for preventing toxic metal buildup in the soil. The basic problem here is economic, not technological, because synthetic fertilizer is so cheap. Some suggest taxing commercial fertilizer to make recycling of such organic fertilizer more competitive.

3. Ban or set low limits (less than 4 percent) on phosphates in detergents, which would decrease phosphate input by 30 to 70 percent.

4. Control land use near watershed areas (Section 14-4) and use sound soil conservation practices (Section 9-4) to reduce the runoff of inorganic fertilizers, manure from animal feedlots, and erosion from cultivation, mining, construction, and other human activities that disturb the soil.

The Detergent Controversy   Until 1945 soap was used to wash clothes in the United States. But in areas with hard water (which contains ions of calcium, magnesium, or iron), soap leaves a greasy, grayish film—the soap curd. Around 1945 chemists invented synthetic detergents, which contain chemicals such as sodium tripolyphosphate to prevent the oily film from forming. Manufacturers advertised the new products heavily, and now detergents have almost completely replaced soap for washing clothes.

Much of the phosphate in detergents is released into lakes directly or indirectly in the effluents discharged from most sewage treatment plants, which do not remove phosphates. By 1983 detergents accounted for 30 to 70 percent of the phosphorus in wastewater in regions that had not banned phosphate detergents.

For lakes where phosphorus is the limiting factor in the prevention of excessive eutrophication, the best approach to minimizing cultural eutrophication is to drastically reduce the input of phosphates. But there is disagreement over whether this should be done by banning or limiting phosphates in detergents or by removing phosphates from wastewater at sewage treatment plants. The detergent industry favors the second plan, and city governments and consumer groups favor the first. Banning or limiting phosphates in detergents is faster and cheaper than upgrading sewage plants and could cut phosphate inputs into lakes where phosphorus is the limiting factor by 30 to 70 percent in a single stroke.

Detergent industry officials, however, contend that in most areas a phosphate ban alone would not eliminate phosphate-related problems because phosphate detergents contribute only a small amount of the phosphorus that finds its way in aquatic systems from natural runoff, human wastes, and fertilizer runoff. They argue that removal of phosphates at sewage treatment plants would eliminate about 50 to 80 percent of the input.

By 1985, Michigan, Wisconsin, Vermont, Indiana, New York, Minnesota, and Canada had banned or limited the amount of phosphate in phosphate-containing detergents, household cleaners, and water conditioners. Bans have also been in effect in Chicago, Miami, Akron, and other cities.

Another approach is to return to the use of soap. Approximately 60 percent of the U.S. population lives in areas where soap works well because the water is soft or only slightly hard. People who live in areas with moderate to very hard water (check with local water officials) could use soap with a water softener, such as washing soda. Clothes will not be quite as bright because modern automatic washers are not designed for soap, even with added water softeners. Even so, soap will work if you are not hooked on having your clothes sparkling white.

22-5 Groundwater Pollution

A Growing Threat One of the major water pollution problems of the 1980s is the growing contamination of the underground aquifers (Figure 10-1) that provide drinking water for one out of two Americans and 95 percent of those in rural areas. The EPA estimated that by early 1985 only about 1 to 2 percent of the nation's usable groundwater supplies was polluted. But even this small percentage affects more than 5 million people, and officials are finding that the rate of groundwater contamination is increasing—especially near many of the 75 percent of major U.S. cities that depend on groundwater for most of their supply. Furthermore, according to a 1984 report by the Office of Technology Assessment, the EPA's contamination figure is probably low because there has been no uniform or comprehensive testing of the nation's groundwater resources, and only 18 of the reported 700 different substances found in groundwater were covered by federal water quality standards by early 1985. Serious contamination of groundwater has been reported in 34 states and is suspected in another 6 states.

In Florida, where 9 out of 10 people rely on groundwater for their water supplies, contamination had reached crisis proportions by 1984. Between 1978 and 1983 more than 2,800 public and private drinking water wells in 20 states were closed because of contamination. Between 1945 and 1980 there were more than 31,000 reported

cases of illness caused by contamination of groundwater by viruses, bacteria, or parasites and 57 cases from groundwater contaminated with toxic organic chemicals. A 1984 study by researchers from Harvard University's School of Public Health of 3,000 households in Woburn, Massachusetts, where some residents drank contaminated groundwater between 1963 and 1979, revealed a consistent pattern linking contaminated wells and high incidences of childhood leukemia, stillbirths, infant deaths, and birth defects such as cleft palate and Down's syndrome.

Sources of Groundwater Contamination Groundwater can be contaminated from a number of sources (Figure 22-5). These include: (1) sewer and pipeline leaks, (2) accidental chemical or oil spills, (3) leaks from the more than 300 deep underground waste injection wells and 493,000 relatively shallow wells used to dispose of industrial and oil field wastes (at least 5,000 new waste disposal wells are dug each year), (4) leaks of gasoline and other hazardous chemicals from an estimated 100,000 of the nation's 3.5 million aging underground storage tanks (most of which are unprotected from corrosion and have an expected life span of about 15 years), (5) leaks from at least 22,000 abandoned hazardous waste disposal sites, 1,500 active hazardous waste landfills, and 15,000 active municipal landfills that are unlined and located above or near aquifers (Chapter 24), (6) the discharge of sewage and household wastes into the soil each year from the nation's 20 million septic tanks, one-third of which are estimated to be working improperly, (7) runoff of animal wastes from feedlots (bacteria and nitrates), (8) runoff of salts spread on icy roads, (9) runoff of fertilizers (nitrates) and pesticides, and (10) leaks from at least 31,000 of the estimated 205,000 industrial, mining, municipal, and agricultural surface water impoundments (lagoons, pits, and ponds) used for the storage, treatment, and disposal of liquid wastes.

Control of Groundwater Pollution Groundwater pollution is more difficult to control than surface water pollution for several reasons: (1) Some bacteria and most suspended solid pollutants are removed as contaminated surface water percolates through the soil into aquifers, but the effectiveness of this process varies with the type of soil, and septic tanks can discharge bacteria and viruses directly into some groundwater supplies; (2) bacterial degradation of oxygen-demanding wastes reaching aquifers does not occur readily because of

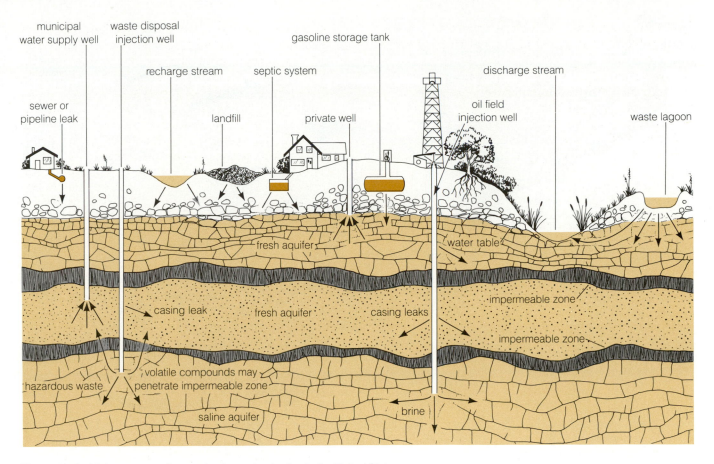

Figure 22-5 Major sources of groundwater contamination in the United States.

a lack of dissolved oxygen and sufficient microorganisms in groundwater; **(3)** soil is not effective in filtering out viruses and most synthetic organic chemicals; **(4)** because contaminants are not effectively biologically degraded, pollution levels in groundwater are typically many times higher than the levels found in surface water; **(5)** the rate of movement of most groundwater is so slow (typically about 30 centimeters or 1 foot a day) that there is relatively little dispersion and dilution of wastes; **(6)** because pollutants entering groundwater tend to spread in a plume whose shape and rate of movement are difficult to monitor and predict, groundwater may be heavily contaminated in one place and safe for drinking only a few hundred feet away; **(7)** locating and monitoring groundwater pollution is expensive (up to $10,000 per monitoring well), and usually many monitoring wells must be sunk to determine the area of contamination; **(8)** groundwater pollution is essentially irreversible because most aquifers are recharged very slowly; **(9)** pumping out an aquifer, cleaning it up, and then pumping the water back into the ground costs $5 million to $10 million for a single aquifer; and **(10)** in the United States there are only about

3,500 technically trained people in the groundwater field, most of them new to the business.

Because of these difficulties, it is generally recognized that preventing contamination in the first place is the only effective long-range way to protect groundwater resources. Despite the seriousness of this threat to drinking water supplies, by 1985 there was no single federal law designed to protect groundwater supplies. Some aspects of groundwater pollution, however, can be controlled by using parts of existing water pollution control laws (Section 22-8).

In 1984, after what many environmentalists considered years of unnecessary delay, the Environmental Protection Agency proposed a national groundwater strategy. This plan places primary responsibility for protecting groundwater supplies on the states. The EPA suggests that each state classify its groundwater resources into one of three categories. *Class I*, consisting of irreplaceable sources of drinking water and ecologically vital sources, should be given the highest protection. *Class II* includes all other groundwater currently in use or potentially available for drinking water or for other beneficial uses. *Class III* groundwater is

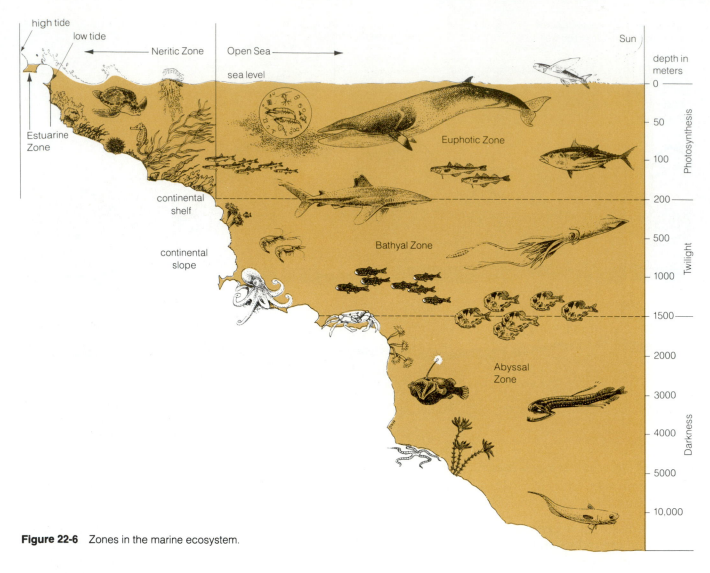

Figure 22-6 Zones in the marine ecosystem.

not a potential source of drinking water and is of limited beneficial use, usually because it is already too contaminated. Environmentalists have called the plan a "non-strategy" and have criticized the federal government for passing this serious national problem on to already economically overburdened states with little or no economic assistance.

22-6 Ocean and Estuarine Zone Pollution

The Marine Ecosystem The ocean can be divided into several zones (Figure 22-6). The **neritic zone** includes the **estuarine zone**, containing estuaries and coastal wetlands, and extends out to the edge of the continental shelf. **Estuaries** are thin, fragile zones along coastlines where freshwater streams and rivers meet and mix with salty oceans. **Coastal wetlands** are normally wet or flooded shallow shelves that extend back from the freshwater–salt-water interface. They consist of a complex maze of marshes (see inside back cover), bays, lagoons, tidal flats, and mangrove swamps.

The estuarine zone, representing less than 10 percent of the total ocean area, contains 90 percent of all sea life. Sunlight can penetrate the waters in this shallow zone, allowing photosynthesis to occur among its vast floating population of phytoplankton, the grass of the sea. These plants support the zooplankton and bottom-feeding invertebrates like shellfish, which support larger fish and add to our food supply. Strings of *barrier islands* are also found in the neritic zone along some coastal areas.

Moving out from the continental shelf, we enter the **open sea** or the *oceanic zone*, making up about 90 percent of the total ocean area. Because there are few nutrients in the open sea, it supports relatively little life compared with the neritic zone.

The open sea is divided into three vertical zones. The layer through which light can penetrate is called the **euphotic zone**; this is the zone in

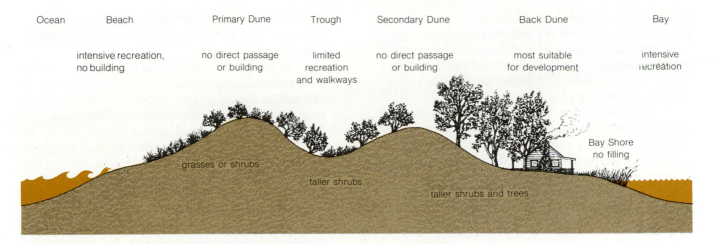

Ocean	Beach	Primary Dune	Trough	Secondary Dune	Back Dune	Bay
intensive recreation, no building	no direct passage or building	limited recreation and walkways		no direct passage or building	most suitable for development	intensive recreation

grasses or shrubs

taller shrubs

taller shrubs and trees

Bay Shore
no filling

Figure 22-7 Primary and secondary dunes offer natural protection from flooding. Ideally, construction and development should be allowed only behind the second strip of dunes, with walkways built over the dunes to the beach.

which photosynthesis can occur. Below the euphotic zone is the **bathyal zone**, or *dark open water* zone, in which many of the larger marine organisms like tuna and whale cruise. Finally, we encounter the **abyssal zone** of the ocean, consisting of deep water and the benthos (ocean bottom).

The Ultimate Sink

The oceans are the ultimate sink for natural and human wastes. Water used and contaminated in homes, factories, and farms flows into rivers, which eventually empty into the ocean. In addition, wastes are loaded on barges and dumped directly into the ocean. The major pollution problems of the oceans occur around its edges—the estuaries, wetlands, bays and harbors, and the inland seas like the Baltic and Mediterranean near large cities, industrial centers, and the mouths of polluted rivers.

Reports of the death of the oceans in the late 1960s and 1970s were exaggerated. Fortunately, the vastness of ocean waters and their constant mixing dilute and disperse many types of waste to harmless levels. Other wastes are broken down and recycled by natural chemical cycles (Section 5-3) in ocean ecosystems. Marine life has also proved to be more resilient than some scientists expected.

Although the ocean can dilute and break down large amounts of sewage and some types of industrial waste—especially in its deep-water areas—its capacity to do so has limits. The sheer magnitude of discharges, especially near the coasts, can overload an ocean's natural purifying systems. In addition, these natural processes cannot readily degrade many of the plastics, pesticides, and other synthetic chemicals created by human ingenuity. For example, recent studies by marine wildlife researchers indicate that 1 to 2 million seabirds and more than 100,000 marine mammals, including whales, seals, dolphins, and manatees, die each year as a result of plastic pollution.

Importance of Estuarine Zones

According to the economic land-use ethic (Section 12-1), estuaries and coastal wetlands are desolate, mosquito-infested, worthless lands that should be drained, dredged, filled in, and built on. Nothing could be further from the truth. From 60 to 80 percent of the commercially valuable species of saltwater fish, mollusks, and shellfish depend on the estuarine zone at some stage of life for food, spawning grounds, or nurseries for their young. Thus, filling in or damaging the productivity of estuarine zones destroys a major source of protein and affects the livelihood of millions of people involved in the $12 billion a year U.S. commercial and recreational fishing industry. In addition, estuarine zones feed and shelter wading marsh birds (herons, egrets, ibis, cranes), birds of prey (ospreys, marsh hawks, bald eagles), and other waterfowl (cormorants, pelicans, grebes, loons). Also, migratory birds (ducks, geese, snipes) rest in estuarine zones.

Estuarine zones and adjoining sand dunes are also important natural flood control devices, absorbing damaging waves caused by violent storms and serving as a giant sponge to absorb floodwaters before they can reach human coastal habitats on higher ground. Under natural conditions, the area behind most beaches is protected by two sets of sand dunes held together by sea oats and other grasses and shrubs (Figure 22-7). For free and effective flood protection, buildings should be placed behind these primary and secondary dunes, with walkways built over both dunes to the beach. When coastal developers remove these protective dunes or build behind the first set of dunes, minor hurricanes and sea storms can sweep away the cottages, homes, and buildings. Some people call such events "natural" disasters and insist on insurance payments and loans so that they can build again and wait for the next disaster.

If not overloaded, estuaries and coastal wetlands provide another important, free service by removing large amounts of pollutants from coastal waters. It is estimated that 0.004 square kilometer (1 acre) of tidal estuary substitutes for a $75,000 waste treatment plant and is worth a total of $83,000 when sport and food fish production is included.

Stresses on U.S. Estuarine Zones   The fate of the coastal environment directly or indirectly affects every U.S. citizen. By 1985, almost two out of three Americans lived in counties bordering the nation's shoreline (including the Great Lakes). Nine of the nation's largest cities, about 40 percent of the manufacturing plants, and two out of every three nuclear and coal-fired electric power plants are located in coastal counties.

Because of these multiple uses and stresses, more than 50 percent of the coastal estuaries in the continental United States have been destroyed or damaged, primarily by dredging and filling. According to a 1983 EPA study, the Chesapeake Bay, the nation's largest estuary, is an ecosystem in decline from pollution by toxic chemicals and excessive inputs of nitrogen and phosphorus plant nutrients, and the region could soon lose its $750 million a year seafood industry. In contrast, the once heavily polluted Delaware Bay was making a comeback by 1984 after three decades of cleanup efforts by federal, state, and local authorities, including improved sewage treatment plants and tougher controls on industrial effluents. Pollution control is made easier because the Delaware Bay flushes itself out in about 100 days, about three to four times faster than the Chesapeake Bay.

Coastal Zone Management   Fortunately, about two-thirds of the nation's estuaries and wetlands remain undeveloped. But trying to protect these lands while still allowing reasonable use is a difficult task because: **(1)** more than 70 percent of the nation's coastline (excluding Alaska) is privately owned; **(2)** plans for protecting and using one estuarine system may not apply to another; **(3)** even when an ecologically sound plan exists, there is tremendous pressure to use estuarine areas primarily for economic purposes; and **(4)** protection plans are hampered because of the conflicting goals of the many different coastal municipalities, counties, and states sharing the use of estuarine zones.

The National Coastal Zone Management Act of 1972 and amendments added in 1980 provide federal aid to the 35 coastal states and territories to help them develop voluntary, comprehensive programs for protecting and managing coastlines. Although this law is an important step, it has been hindered by inadequate funding and by the voluntary nature of plans drawn up and approved for funding. By 1984, 30 of the 37 eligible states and territories had federally approved coastal management plans, and most others were in advanced stages of program development. But many of the voluntary state plans are vague and do not provide sufficient enforcement authority to ensure adequate protection of coastal lands. California and North Carolina are considered to have the strongest programs, but there are continuing efforts by developers and other interests to have these programs weakened.

Ocean Dumping   Barges and ships dump heavily contaminated sewage sludge, industrial wastes, and dredged material at designated sites near the Atlantic, Gulf, and Pacific coasts. One of the most intensely used shallow-water ocean dumping sites is the New York Bight, 19 kilometers (12 miles) off the New York–New Jersey coast near the mouth of the Hudson River. An estimated 95 percent of all U.S. ocean dumping occurs in this area. A 105-square-kilometer (40-square-mile) area of the ocean bottom in the New York Bight is covered with a black toxic sludge that severely alters animal communities living on the sea bottom and is slowly moving toward the shore.

Since the 1960s some industrial wastes have also been dumped at a deep-ocean site 170 kilometers (106 miles) east of New York City. There scientists have found little degradation of marine life. Because of the greater depth of the water and nearness to Gulf Stream currents, this site apparently has a much greater capacity to dilute and disperse wastes and may soon replace the New York Bight as the major U.S. ocean dumping site.

By 1975, 54 nations, including all major maritime powers, had agreed to stop dumping high-level radioactive wastes, biological and chemical warfare agents, various kinds of oil, some pesticides, durable plastics, mercury, and cadmium into the ocean. The agreement does not include wastes related to offshore mineral exploration and development of seabed mineral resources, wastes dumped into rivers and lakes that empty into the oceans, and arsenic, lead, fluorides, cyanides,

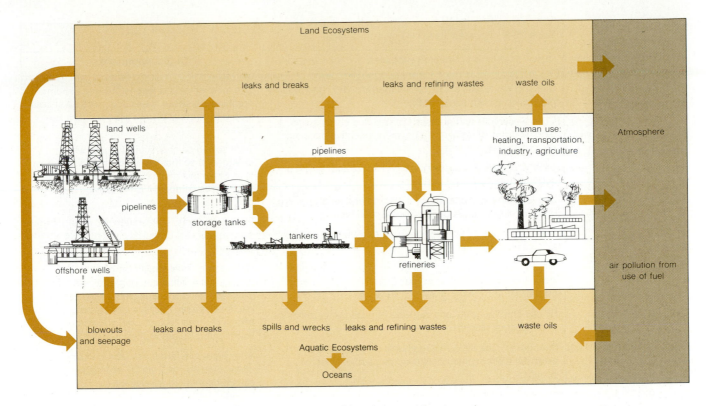

Figure 22-8 Major sources of oil pollution of the hydrosphere, the lithosphere, and the atmosphere.

zinc, and several other toxic substances (Chapter 24).

Because of this international ocean dumping agreement and the U.S. Ocean Dumping Act of 1972, the volume of industrial wastes dumped into U.S. ocean waters was almost halved between 1972 and 1981. However the amount of dredged materials, which are not regulated, increased almost fourfold between 1977 and 1983.

Some scientists argue that while the deep ocean may be better equipped than land to dilute sewage and some forms of industrial waste to harmless levels, it should not be used for the dumping of slowly degradable or nondegradable pollutants like dioxin, polychlorinated biphenyls (PCBs), some radioactive isotopes and pesticides, mercury compounds, and other persistent substances that can reach human food sources by being biologically magnified in ocean food chains. Unfortunately, many of these materials are mixed in with some types of sludge and industrial waste. Marine scientists point out that it is difficult—if not almost impossible—to determine the threshold levels at which various ocean pollutants could cause serious harm to marine life and to humans who depend on ocean food sources.

Ocean Oil Pollution Each year more than 6 billion kilograms (7 million tons) of oil and petroleum products is added to the oceans (Figure 22-8). Ac-

cording to a 1983 report by the Department of the Interior, about 15 percent of the annual input into U.S. ocean waters comes from natural seeps of crude petroleum from deposits below the ocean bottom. Human activities account for the remaining 85 percent, which is added in the forms of *crude petroleum* (oil as it comes out of the ground) and *refined petroleum* (obtained by distillation and chemical processing of crude petroleum). River and urban runoff, mostly from the disposal of lubricating oil from machines and automobile crankcases, accounts for about 44 percent of the annual input; oil tanker accidents and routine discharges of oil from tankers during loading, unloading, and cleaning, 20 percent; and ruptures or blowouts of offshore oil wells about 5 percent.

Tanker accidents and blowouts, however, could become a more important source of ocean pollution in the future, as more oil is drilled and transported to meet world energy demands. On June 3, 1979, a well being drilled in the Caribbean Sea by Pemex, the Mexican national petroleum company, suffered a blowout. Before it was capped 295 days later, it had poured some 531 million liters (140 million gallons) of oil valued at $100 million into the sea. The largest tanker accident to date was the breakup of the supertanker *Amoco Cadiz* in 1978. This vessel released more than 254 million liters (67 million gallons) of oil valued at $24 million, polluting 322 kilometers (200 miles) of European coastline. In 1984 a federal district judge

Table 22-3 Approaches to Oil Pollution Control

Input Approaches	Output Approaches
Use and waste less oil per capita (Section 19-7) and reduce population growth (Chapter 8).	Use mechanical barriers to prevent oil from reaching the shore, then vacuum oil up or soak it up with straw or pillows filled with chicken feathers.
Collect used oils and greases from service stations and other sources (possibly by a payment incentive plan) and reprocess them for reuse.	Treat spilled oil chemically (usually with detergents) so that it will disperse, dissolve, or sink. Since this method can kill more marine life than the oil does, it is not favored by ecologists.
Strictly regulate the building, maintenance, and routing of supertankers and superports.	
Use load-on-top procedures for loading and emptying all oil tankers (already done on 80 percent of all tankers).	Ship oil in a solid state, much like a gel, so that it can be picked up quickly and easily if an accident occurs.
Build supertankers with double hulls, to reduce chances of a spill and to separate oil cargo from ballast water.	Develop bacterial strains (by genetic recombination) that can degrade compounds in oil faster and more efficiently than natural bacterial strains. Possible ecological side effects of these "superbugs" should be investigated before widespread use.
Strictly enforce safety and disposal regulations for offshore wells and international agreements prohibiting discharge of oily ballast and cleaning water from tanks.	
Strictly enforce safety and disposal regulations for refineries and industrial plants.	Add oil-soluble ferrofluids (iron-containing material) to the spill, which will enable electromagnets to remove the oil.
Strengthen international agreements on oil spills and establish a strong international control authority for the oceans.	

ruled that Standard Oil Company of Indiana (Amoco), which operated the tanker, had been negligent "with respect to the design, operation, maintenance, repair, and crew training" of the tanker and was liable for most of the damages (up to $2 billion) caused by the spill.

There is considerable dispute, uncertainty, and conflicting evidence concerning the short-term and long-term effects of oil on ocean ecosystems. The effects of oil spills are difficult to predict because they depend on a number of factors, including the type of oil spilled (crude or refined), the amount spilled, the distance of the spill from the shore, the time of year, the weather, the tidal currents, and wave action.

Crude oil and refined oil are collections of hundreds of substances with widely differing properties. Low-boiling, aromatic hydrocarbons are the primary cause of immediate kills of a number of aquatic organisms, especially in their larval forms. Fortunately, most of these toxic chemicals evaporate into the atmosphere within a day or two. Some other chemicals remain on the water surface and form floating, tarlike globs that can be as big as tennis balls, while other chemicals sink to the ocean bottom. A number of these chemicals are degraded by marine microorganisms, but this natural process is slow (especially in cold Arctic and Antarctic waters), requires a large amount of dissolved oxygen, and tends to be least effective on some of the most toxic petroleum chemicals.

Some marine birds, especially diving birds, die when oil interferes with their normal body processes or destroys the natural insulating properties of their feathers. Some oil components find their way into the fatty tissues of some fish and shellfish, making the fish unfit for human consumption because of their oily taste. Some petroleum chemicals can cause subtle changes in the behavioral patterns of aquatic organisms. For example, lobsters and some fish may lose their abilities to locate food, avoid injury, escape enemies, find a habitat, communicate, migrate, and reproduce. Floating oil slicks can also concentrate oil-soluble compounds like DDT and other pesticides.

There is some evidence that spills of crude oil may cause less damage than those of refined oil. Studies of the effects of several crude oil spills between 1969 and 1978 reveal that most forms of marine life recovered nearly completely within 3 years. In contrast, spills of oil—especially refined oil—near shore or in estuarine zones, where sea life is most abundant, have much more damaging and long-lasting effects. For example, damage to estuarine zone species from the spill of refined oil at West Falmouth, Massachusetts, in 1969, was still being detected 10 years later. The major input approaches (used to keep oil from reaching the ocean) and output approaches (to remove or minimize its effects once it gets there) are summarized in Table 22-3.

22-7 Approaches to Water Pollution Control

Soil Conservation Because sediment from soil erosion is the single largest source of water pollution, soil conservation is the most important approach for reducing sedimentation in streams and maintaining the fertility of the soil. Methods for reducing soil erosion are discussed in Section 9-4.

Sewage Treatment Two major approaches for dealing with the liquid wastes of civilization are: **(1)** dumping them into the nearest waterway, and **(2)** purifying them to varying degrees in septic tanks, lagoons, or sewage treatment plants. In many LDCs the first approach is used, causing widespread infection from waterborne diseases (Chapter 20).

By contrast, in MDCs most waterborne wastes from homes, businesses, and factories and from storm runoff flow through a network of sewer pipes to sewage treatment plants. Some areas have separate lines for sewage and storm water runoff. In many areas, however, lines for these two sources are combined because it is cheaper. During almost any rain the total volume of wastewater and storm runoff in such combined systems is too large to be handled completely by the sewage treatment plant system. When this occurs, some untreated wastes overflow into rivers and streams.

When sewage reaches a treatment plant, it can undergo various levels of treatment, or purification, depending on the sophistication of the plant and the degree of purity desired. Figure 22-9 and the accompanying box summarize primary, secondary, and tertiary sewage treatment methods.

The liquid wastes of about 71 percent of the U.S. population and those of 87,000 industries

---

Waste Treatment Methods

Primary treatment is a *mechanical* process that uses screens to filter out debris like sticks, stones, and rags, and a sedimentation tank, where suspended solids settle out as **sludge** (Figure 22-9). These two operations remove about 60 percent of the solid material but only one-third of the oxygen-demanding wastes. Chemicals are sometimes added to speed up the settling of suspended solids in a process called *flocculation*.

Secondary treatment is a *biological* process that uses bacteria to break down wastes. This removes up to 90 percent of the oxygen-demanding wastes by using either **(1)** *trickling filters*, where sewage is degraded by bacteria as it seeps through a bed of stones, or **(2)** an *activated sludge process*, where sewage is aerated with air or pure oxygen to aid bacterial degradation. The water from the trickling filter or aeration basin is then sent to a sedimentation tank, where most suspended solids settle out as sludge. Primary plus secondary treatment still *leaves* 10 to 15 percent of the oxygen-demanding wastes, 10 percent of the suspended solids, 50 percent of the nitrogen (mostly as nitrates), 70 percent of the phosphorus (mostly as phosphates), 30 percent of most toxic metal compounds, 30 percent of most synthetic organic compounds, and essentially all the long-lived radioactive isotopes and dissolved and persistent organic substances like some pesticides.

Tertiary treatment refers to a series of specialized chemical and physical processes used to reduce the quantity of one or more of the pollutants remaining after primary and secondary treatment. Tertiary treatment is rarely used because many methods are still in the experimental stage and it is expensive (twice as costly to build the plant and up to four times the operating costs of primary plus secondary treatment). Three forms of tertiary treatment used in some places are *precipitation*, for removing suspended solids and phosphates, *adsorption*, in which activated carbon is used to remove dissolved organic compounds, and *electrodialysis or reverse osmosis*, for reducing levels of dissolved organic and inorganic substances.

Disinfection is carried on as a part of all three forms of sewage treatment to remove water coloration and to kill disease-carrying bacteria and some (but not all) viruses. Chlorine is the most widely used disinfectant. A newer method uses ultrasonic energy to break down wastes mechanically, and other disinfectants, such as ozone or chlorine dioxide, to kill bacteria. This approach, however, is more expensive than chlorination. A less costly approach, where local supplies are heavily contaminated with trihalomethanes and other synthetic organic chemicals, is to filter the water or wait until organic materials have settled; then chlorine is added and the water is filtered with activated charcoal.

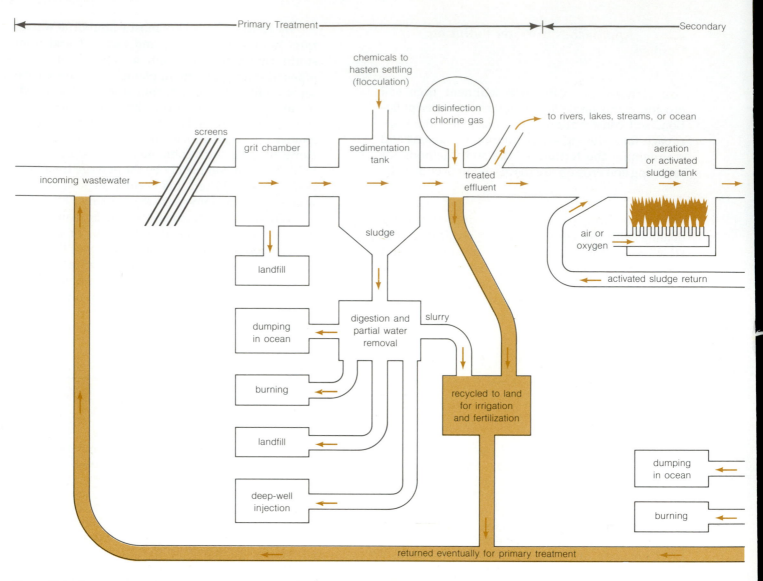

Figure 22-9 Primary, secondary, and tertiary sewage treatment. The shaded areas show the recycling of nutrients to the land, an ecological approach to dealing with these nutrients that is used little so far.

were treated in about 15,000 municipal sewage treatment plants. The wastes of two-thirds of these individuals received secondary sewage treatment, 5 percent received some form of tertiary treatment, and the remainder received only primary treatment. The liquid wastes of the 29 percent of the mostly suburban and rural U.S. population not served by sewage treatment plants were either degraded in cesspools and septic tanks or discharged directly into waterways.

Ecological Waste Management and Recycling
Building secondary sewage treatment plants throughout the United States is an important step in water pollution control, especially in reducing oxygen-demanding wastes, suspended solids, and bacterial contamination. But there are problems associated with this approach. When bacteria de-

grade oxygen-demanding wastes, the resulting effluents are rich in phosphates and nitrates. Allowing these nutrient-rich effluents to flow into lakes and slow-moving rivers can overload these systems, triggering algae blooms and oxygen depletion (Figure 22-4).

Many scientists argue that instead of overloading aquatic systems with phosphate- and nitrate-rich sewage effluents, these plant nutrients should be returned as fertilizer to the land (forests, parks, and croplands) or to aquaculture ponds. Such an ecological approach to waste management mimics nature by recycling plant nutrient wastes to the land, as shown in the shaded areas of Figure 22-9.

There is concern, however, that this approach could allow bacteria, viruses, toxic metal compounds, and hazardous synthetic organic chemicals to build up in the soil, food crops, and fish and shellfish grown by aquaculture, and to contami-

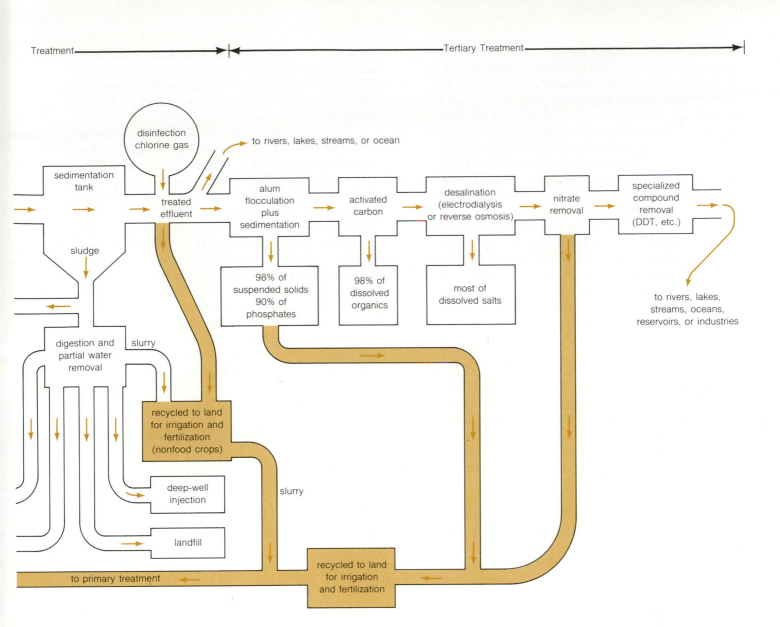

nate groundwater aquifers. But others argue that these problems could be controlled by removing certain pollutants, banning the use of highly contaminated liquid wastewater and sewage sludge, and using wastewater and sludge for noncrop purposes on forest lands and surface-mined lands. Liquid effluent from sewage treatment plants is already being used successfully as fertilizer in some cities near farming areas, forest lands, and estuaries. Since this effluent has already been treated, contamination from bacteria, viruses, and toxic metals is not a major problem.

At the personal level, you should never dispose of products containing harmful chemicals by pouring them down house or street drains or flushing them down the toilet. Examples of products harmful to aquatic systems that are not usually removed by sewage treatment include: **(1)** waste oil drained from automobiles (put it in a con-

tainer and take to a local service station, where it will be turned in for recycling), **(2)** antifreeze (pour it onto a porous surface like gravel away from water supplies), **(3)** pesticides and herbicides, **(4)** paints, lacquers, thinners, brush cleaners, wood preservatives, and turpentine, and **(5)** household cleaners containing organic solvents. Call your local health or water treatment department for information about proper disposal of wastes in categories 3, 4, and 5.

22-8 Water Pollution Control Laws in the United States

Two major water pollution control laws in the United States are the Federal Water Pollution Act of 1972 and the Clean Water Act of 1977. As amended in 1981, these two acts (1) require the EPA to establish a system of national effluent standards for major water pollutants, (2) require all municipalities to use secondary sewage treatment by 1988, (3) set interim goals of making all U.S. waters safe for fishing and swimming by 1983 and eliminating the discharge of 129 priority toxic pollutants (zero discharge) into U.S. waters by 1985, (4) allow the point source discharge of pollutants into U.S. waterways only with a permit from the EPA or an EPA-approved state agency, (5) require all industries to use the *best practicable technology* for control of conventional pollutants by 1984 and nonconventional pollutants by 1987, with the EPA allowed to grant waivers in cases where costs outweigh benefits, (6) require industries to use the best practicable technology for control of 129 priority toxic pollutants by 1984 (no waivers allowed) and to use the more stringent *best available technology* that is reasonable and affordable to control these pollutants by 1987, and (7) authorized the expenditure of up to $52.5 billion between 1972 and 1986 for federal grants to aid states and localities to construct new secondary sewage plants, upgrade existing ones, or use innovative waste treatment technologies like spraying effluents on land for fertilizer and reclaiming and reusing wastewater.

By 1984, 94 percent of the industries affected had met the goal of using the best practicable technology for control of conventional pollutants. But progress toward achieving the best practicable technology control of nonconventional and toxic pollutants by 1984 was far behind schedule. Some 37 states were unable to meet the 1983 goal of fishable and swimmable waters, mainly because of their inability to control nonpoint pollution from agricultural and urban runoff—a type of pollution that by 1983 had received only 1 percent of all federal water pollution funds.

The sewage plant construction program has had numerous problems. By 1983, at least 7,000 waste treatment plants had not met the 1977 goal of secondary sewage treatment. Despite $37 billion in federal grants since 1972, only about 4,000 of 19,000 funded projects had been completed by 1983 because of fraud, overbuilding, bureaucratic and construction delays, and after 1981, a lack of sufficient funds.

As a result of water pollution laws, the total amount of industrial pollutants entering U.S. waterways was reduced by one-half between 1972 and 1980. Existing laws, if funded and enforced, should lead to greater improvement in the future and could save about $12.3 billion annually in estimated damages from water pollution. However, a 1984 study by the General Accounting Office showed that most of the 33 industries studied violated their waste discharge permits issued by the EPA and accused the EPA of doing little to enforce compliance of existing water pollution control laws.

The Clean Water Act of 1977 was due for renewal in 1984. In 1983 the Reagan administration made proposals that would delay industry pollution control deadlines, loosen permit review procedures, and sharply reduce federal grants for sewage plant construction. However, environmentalists and many health scientists proposed that the act be strengthened by (1) including goals and standards for preventing groundwater contamination, (2) requiring more stringent monitoring and control of nonpoint sources of water pollution, and (3) requiring the EPA to establish a comprehensive monitoring program for the nation's surface and underground waters and funding it adequately to accomplish this goal. By mid-1985 Congress was still debating proposed changes in this important act.

The reason we have water pollution is not basically the paper or pulp mills. It is, rather, the social side of humans—our unwillingness to support reform government, to place into office the best qualified candidates, to keep in office the best talent, and to see to it that legislation both evolves from and inspires wise social planning with a human orientation.

Stewart L. Udall

Discussion Topics

1. How would you control (a) nondegradable pollutants, (b) slowly degradable (persistent) pollutants, and (c) rapidly degradable (nonpersistent) pollutants?

2. Give examples of a point pollution source and a nonpoint pollution source in the water and in the air, and explain how you would control pollutants from these sources.

3. Explain why "dilution is not always the solution to water pollution," and relate your explanation to the second law of energy (Section 3-3). Cite examples and conditions for which this solution is and is not applicable.

4. Explain how a river can cleanse itself of oxygen-demanding wastes. Under what conditions can this natural cleansing system fail?

5. Give your reasons for agreeing or disagreeing with the idea that we should deliberately dump most of our wastes in the ocean. It is a vast sink for diluting and mixing, and if it becomes polluted, we can

get food from other sources. Let the ocean go as a living system so that we can live.

6. Should all dumping of wastes in the ocean be banned? If so, where would you put these wastes? What exceptions, if any, would you permit? Under what circumstances? Explain why banning ocean dumping alone will not stop ocean pollution.

7. Should the United States (or any other coastal nation) ban all offshore oil wells? Why or why not?

What might be the consequences of this restriction for the nation? For foreign policy? For security? For your town? For you? What might be the consequences of not doing this?

8. Explain why building expensive primary and secondary sewage treatment plants, although necessary, may hasten the deterioration of water quality through cultural eutrophication. What are some solutions to this dilemma?

23

Pesticides and Pest Control

A **pest** is any unwanted organism that directly or indirectly interferes with human activity. Since 1945 vast fields planted with one or a few crops, as well as home gardens and lawns, have been blanketed with a variety of synthetic chemicals called **pesticides** (or *biocides*) to kill organisms that humans consider to be undesirable. The substances in this arsenal kill unwanted insects (*insecticides*), plants (*herbicides*), rodents (*rodenticides*), fungi (*fungicides*), and other organisms.

Pesticides can improve crop yields and help control populations of disease organisms transmitted by insects, rodents, and other organisms. However, there is considerable evidence that their widespread use can have harmful effects on wildlife, ecosystem structure and function, and human health. Their overuse can even lead to increased crop loss and resurgence of diseases and pests they are supposed to help control. In 1962 Rachel Carson's book *The Silent Spring* dramatized the potential dangers of pesticides to wildlife and humans and set off a controversy between environmentalists and pesticide industry officials that is still going on. A number of pesticides, such as DDT, have been banned for most uses in the United States. However, the pesticide industry maintains that the evidence on which such bans were based is inconclusive and that the benefits of using DDT and other pesticides outweigh their harmful effects. Is this so, and are there alternatives to relying almost exclusively on synthetic

Figure 23-1 Throughout recorded history locusts have devoured wild and cultivated plants used to feed the world's human population.

FAO photo by Jean Manuel

pesticides to control pests? In this chapter we look at these important questions by concentrating on the two most widely used pesticide types: insecticides and herbicides.

23-1 Chemicals Against Pests: A Brief History

Natural Control of Pests in Diverse Ecosystems

Only about 10,000 or 1 percent of the at least 1 million cataloged insect species have become pests by human standards. Only about 100 of these insects cause about 90 percent of the damage to food crops. Indeed, most insects, fungi, rodents, and soil microorganisms are helpful in bringing about natural cycling of chemicals essential to life (Section 5-3) and in pollinating plant species. Furthermore, in a diverse, relatively undisturbed natural ecosystem, populations of natural insect predators, disease organisms, and parasites usually keep populations of potentially harmful species from reaching levels high enough to cause significant economic loss of food crops or livestock. In addition, through natural selection many plants and insects contain or give off chemicals that tend to protect them from certain predators, parasites, and diseases.

Why the Need for Pest Control Has Increased

Throughout history, locusts (Figure 23-1) have ravaged wild and cultivated plants that humans eat. Over the past 200 years, however, an increasing number of insects and other pests have become major threats to crops used to feed the world's mushrooming population. The major reason for this disquieting development is that large areas of diverse ecosystems, containing small populations of many species, have been replaced with greatly simplified agricultural ecosystems and lawns containing large populations of one or only a few plant species. In such biologically simplified ecosystems, organisms whose populations were held to acceptable levels by natural regulatory mechanisms in more diverse ecosystems can grow in number and achieve *pest* status. As a result, humans have had to spend an increasing amount of time, energy, and money to control pests in crop fields, lawns, and other simplified ecosystems.

The Ideal Pest Control Method

The ideal method of pest control: **(1)** would kill only the target pest, **(2)** would have no short- or long-term health effects on nontarget organisms, including humans, **(3)**

would be nonpersistent and would break down into harmless chemicals in a relatively short time, **(4)** would not result in genetic resistance (see Section 23-4), and **(5)** would be less costly than the economic losses resulting from not using pest control. Unfortunately, no known pest control method meets all these criteria.

First-Generation Pesticides

The most widely used approach has been to spray fields with synthetic pesticides. Before 1940 there were only a few dozen pesticides on the market. Many of these *first-generation* commercial pesticides were *nonpersistent organic compounds* made or extracted from insect poisons found naturally in plants. For example, pyrethrum, an organic compound made from dried chrysanthemum flowers, was used as an insecticide by the Chinese 2,000 years ago and is still in use today. Caffeine is also an excellent insecticide that can be used to control tobacco hornworms, mealworms, milkweed bugs, and mosquito larvae. Other insecticides derived from natural plant sources include nicotine (as nicotine sulfate) from tobacco, rotenone from the tropical derris plant, and garlic oil and lemon oil, which can be used against fleas, mosquito larvae, houseflies, and other insects.

A second type of *first-generation* commercial pesticides in use before 1940 consisted of *persistent inorganic compounds* made from toxic metals such as arsenic, lead, and mercury. Most of these compounds are no longer used because they are highly toxic to humans and other animals (Section 24-5), contaminate the soil for 100 years or more, and tend to accumulate in soils to the point of inhibiting plant growth.

Second-Generation Pesticides

The major revolution in insect pest control occurred in 1939 when it was discovered that DDT (short for dichlorodiphenyltrichloroethane), a chemical known since 1874, was a potent insecticide. Since 1945 chemists have developed a large number of such synthetic organic chemicals for killing insects, weeds, rodents, and other pests. The chemicals in this array are known as *second-generation pesticides*. DDT and many related second-generation pesticides have been widely used because they are: **(1)** easy and cheap to produce, **(2)** *broad-spectrum* chemicals that can kill a wide variety of pest organisms, **(3)** *persistent* chemicals that can kill target pests for a long time (because they are not broken down readily in the environment), and **(4)** insoluble in water (hence are not dissolved and washed away by rain or irrigation water).

Worldwide, about 227 million kilograms (2.50 million tons) of second-generation pesticides is used each year—amounting to an average of about 0.45 kilogram (1 pound) of pesticides for each person on earth. About 45 percent of these pesticides is used in Europe, 34 percent in North America, and the remaining 21 percent mostly in LDCs. It is projected that global use of pesticides will more than double between 1975 and 2000, with usage increasing four- to sixfold in LDCs.

In the United States about 1,200 different chemicals are mixed to make 40,000 pesticide formulations. Total pesticide use in the United States use almost tripled between 1964 and 1984, with pesticide sales increasing from $40 million to more than $4 billion between 1940 and 1984. Of the pesticides used on U.S. crop fields, about 51 percent are herbicides, 35 percent are insecticides, and most of the remaining 14 percent are fungicides. Today about 12 percent of all U.S. cropland (including pastures) is treated with insecticides, 34 percent with herbicides, and 2 percent with fungicides. About two-thirds of all *insecticides* applied each year in the United States are used on only two crops: cotton (50 percent) and corn (17 percent). The major applications of *fungicides* are to fruit crops (60 percent) and vegetables (26 percent).

At least 75 percent of the *herbicides* applied in the United States are used to control the growth of weeds on cultivated land, with 71 percent of all herbicides applied to only two crops: corn (53 percent) and soybeans (21 percent). By reducing the need to use mechanical cultivation to control weeds, the use of herbicides also reduces soil erosion (Sections 9-3 and 9-4), the consumption of fossil fuels (Section 19-7), and human labor costs; water conservation is another result. The remaining major uses of herbicides include spraying them to reduce or eliminate: **(1)** plant growth near roadways and pipelines and under electric powerlines, **(2)** unwanted trees and other plants in tree farms and in national forests, **(3)** brush and poisonous plants on rangelands, **(4)** aquatic plants that clog waterways, and **(5)** plants such as ragweed, poison oak, and poison ivy that can harm susceptible humans.

Since 1970 the use of herbicides in the United States has increased more rapidly than any other type of pesticide. Their use is expected to increase even more as farmers switch to minimum-till and no-till farming (Section 9-4), which require, respectively, 13 and 17 percent more herbicides per area of land than conventional tillage farming.

In the United States about one-third of all pesticides are used on home gardens, trees, and lawns. An EPA study showed that 92 percent of all U.S. households use one or more types of pesticide to control insects, weeds, fungi, or rodents. This study also showed that the average application of pesticides per unit of land area by homeowners is much higher than that by farmers and that in 1980 more than 250,000 Americans became sick because of pesticides used in the home.

23-2 Major Types and Properties of Insecticides and Herbicides

Insecticides Table 23-1 compares some of the properties of three major groups of complex synthetic organic compounds used as insecticides: *chlorinated hydrocarbons*, which contain several atoms of chlorine per molecule, *organophosphates*, which contain at least one atom of phosphorus per molecule, and *carbamates*, which contain at least one atom of nitrogen per molecule. A fourth group of insecticides consists of the organic compounds known as *pyrethroids*. These substances are extracted from the pyrethrum flower and either used directly or modified chemically. Pyrethroids are particularly useful because of their specificity for certain insects, their nontoxic effects on birds and mammals, and their relatively rapid breakdown in the environment. A fifth group consists of *inorganic compounds made from toxic metals such as arsenic, lead, and mercury* (Section 24-5). Because most of these insecticides are highly toxic to wildlife and humans and remain active in the environment for 100 years or more, they are rarely used today.

Most chemical insecticides are *broad-spectrum* nerve poisons that kill most of the target and nontarget insects in the sprayed area by disrupting their nervous systems. One important property of any pesticide is its *persistence*, the length of time it remains active in killing insects and weeds. Table 23-1 shows that chlorinated hydrocarbons are *persistent pesticides* that remain active for 2 to 15 years. Persistence varies, however, with different climate conditions. For example, it may take 10 to 15 years for half the amount of DDT applied in the United States to break down, whereas in the tropics half of it may break down in 6 months or less.

Most organophosphate and carbamate insecticides remain active for a few hours to several months before decomposing to harmless products and are called *nonpersistent insecticides*. Although most organophosphate pesticides break down fairly rapidly in the environment, *most are much more toxic to birds, humans, and other mammals than chlorinated hydrocarbon pesticides*. Furthermore, to compensate for this fairly rapid breakdown, farm-

Table 23-1 Major Types of Insecticides

Type	Examples	Action on Insects	Persistence
Chlorinated hydrocarbons	DDT, DDE, DDD, aldrin, dieldrin, endrin, heptachlor, toxaphene, lindane, chlordane, kepone, mirex	Nerve poisons that cause convulsions, paralysis, and death	High (2 to 15 years)
Organophosphates	Malathion, parathion, Azodrin, Phosdrin, methyl parathion, Diazinon, TEPP, DDVP	Nerve poisons that inactivate the enzyme that transmits nerve impulses	Low to moderate (normally 1 to 12 weeks but some can last several years)
Carbamates	Carbaryl (Sevin), Zineb, maneb, Baygon, Zectran, Temik, Matacil	Nerve poisons	Usually low (days to weeks)

ers usually apply nonpersistent insecticides at regular intervals to help ensure more effective insect control. As a result, these chemicals are often present in the environment almost continuously, like persistent pesticides.

Herbicides More than 180 different types of synthetic organic herbicide are used in the United States. These herbicides can be placed into three classes based on their effect on plants: contact herbicides, systemic herbicides, and soil sterilants. *Contact herbicides* kill plant foliage within a few days through direct contact. *Triazines*, which kill plants by interfering with photosynthesis, are one class of triazines. The triazine known as atrazine is one of the most widely used herbicides.

Systemic herbicides are absorbed by the foliage or roots and then transferred through the entire plant. One class of systemic herbicides are *phenoxy compounds*, such as 2,4-D (short for 2,4-dichlorophenoxyacetic acid), 2,4,5-T (short for 2,4,5-trichlorophenoxyacetic acid), and silvex. When absorbed, these herbicides create excess hormones that stimulate uncontrollable growth. The plant literally grows itself to death because it cannot provide enough nutrients to keep up with the greatly accelerated growth rate. A second class of systemic herbicides are *substituted ureas*, such as diuron, norea, and fenuron. These water-soluble compounds, which can be absorbed quickly by plant roots, accumulate in leaves, where they kill the plant by inhibiting photosynthesis. *Soil sterilant herbicides* kill plants by killing soil microorganisms essential to plant growth for 48 hours to 2 years. Examples include Treflan, Dymid, Dowpon, and Sutan. Most of these herbicides also act as systemic herbicides.

Most herbicides do not remain active for a long time and are classified as *nonpersistent* pesticides.

However, the triazine herbicides atrazine and prometone, and several soil sterilants, are fairly persistent.

23-3 The Case for Insecticides and Herbicides

Using Insecticides to Control Disease During World War II, DDT was sprayed directly on the bodies of soldiers and war refugees to control body lice, which spread typhus. Because of this prophylactic measure, fewer soldiers died of typhus than from battle wounds for the first time in history. The World Health Organization also used DDT and related second-generation pesticides to help control the number of cases and the spread of insect-transmitted diseases such as malaria (transmitted by the *Anopheles* mosquito), bubonic plague (rat fleas), typhus (body lice and fleas), sleeping sickness (tsetse fly), and Chagas' disease (kissing bugs) (Chapter 20).

Thanks largely to DDT, dieldrin, and several other chlorinated hydrocarbon insecticides (Table 23-1), more than 1 billion people have been freed from the risk of malaria, and the lives of at least 7 million people have been saved since 1947. Thus, *DDT and other insecticides probably have saved more lives than any other chemicals synthesized since humans have inhabited the earth.*

Although DDT and several other chlorinated hydrocarbon insecticides deserve their reputation as givers of life, they are no longer effective in many parts of the world. By early 1985, the World Health Organization reported that 51 of the 60 malaria-carrying species of mosquitoes had become genetically resistant to DDT and one or more of the other chlorinated hydrocarbon insecticides widely used to control the disease. As a result, between 1970 and 1984 there was a thirty- to fortyfold increase in malaria in a number of countries where it had been almost eradicated. Despite the increasing ineffectiveness of DDT and other insecticides, the World Health Organization points out that a ban on DDT and its substitutes, where they are still useful, would lead to large increases in disease, human suffering, and death.

Using Insecticides and Herbicides to Increase Food Supplies *Each year pests and disease consume or destroy about 45 percent of the world's food supply:* 33 percent of this loss occurs before harvest and 12 percent after harvest. Even in the United States, which uses vast amounts of pesticides and has sophisticated food storage and transportation networks, the total loss due to pests and disease each year is estimated to be 42 percent of potential production—33 percent before harvest and 9 percent after. This leads to annual losses of about $20 billion.

It is estimated that for each dollar invested in the United States for pesticide control, about $4 is returned in increased crop yields, though some studies halve the benefits per dollar invested. Without the use of insecticides and herbicides in the United States, the Department of Agriculture and the Office of Technology Assessment estimate that total annual production of crops, livestock, and forests would drop by 25 to 30 percent and food prices could rise by 50 to 75 percent. However, according to entomologist David Pimentel, a total pesticide ban in the United States would increase preharvest losses from pests and crop diseases by only 9 percent (from 33 percent to 45 percent) and would cause no serious food shortages because production of foods eaten by humans would decrease by only about 5 percent.

There are alternatives to relying on synthetic insecticides and herbicides to control insects and weeds, as discussed in Section 23-6. But some argue that synthetic pesticides have several advantages over other approaches: **(1)** a variety of insecticides and herbicides are available to control most insect pests quickly and at a reasonable cost; **(2)** when genetic resistance occurs in pest insects and weeds, farmers can usually switch to other pesticides; **(3)** the pesticides have a relatively long shelf life and are easily shipped and applied; **(4)** an increasing number of narrow-spectrum, nonpersistent pesticides have been developed to control specific insects, weeds, and diseases without the widespread killing of other nonpest organisms; **(5)** not all pesticides have been shown to have harmful side effects; and **(6)** when properly handled and applied, these substances are safe.

23-4 The Case Against Insecticides and Herbicides

Do the Benefits Outweigh the Risks? In view of the benefits that pesticides can offer, you may wonder why anyone would question their use. There is considerable evidence, however, that the widespread use of pesticides can cause undesirable and harmful effects in ecosystems and in nonpest organisms, including humans. The major problems associated with the widespread use of insecticides are: **(1)** development of genetic resistance against pesticides by target species, **(2)** killing of natural enemies of target pest species, **(3)** creation of new pests, **(4)** global mobility of persistent insecticides, **(5)** biological magnification of persistent insecticides in food chains and food webs, **(6)** threats to wildlife, and **(7)** threats to human health.

Because of these problems, some analysts question the long-term value of pesticides in increasing food supplies and in controlling diseases spread by insects. For example, even though insecticide use in the United States increased tenfold between 1940 and 1980, crop losses from insects almost doubled (from 7.1 percent to 13 percent) during the same period. In addition, diseases such as malaria have made a dramatic comeback because most strains of malaria-carrying mosquitoes have become resistant to DDT and other widely used persistent pesticides (Section 20-2). Let's look in more detail at the major problems associated with pesticides.

Development of Genetic Resistance *The most serious drawback to using chemicals to control pests is that most pest species, especially insects, can develop genetic resistance to any chemical poison.* When an area is sprayed with a pesticide, most of the pest organisms are killed. However, due to random mutations and natural selection (Section 6-2), a few individuals in a given population of a particular species survive because they have genes that make them resistant or immune to a particular pesticide.

Because most pest species—especially insects and disease organisms—have short generation times, a few surviving organisms can reproduce a large number of more resistant offspring in a short time. For example, the boll weevil (Figure 23-2), a major cotton pest, can produce a new generation every 21 days. When populations of offspring of resistant parents are repeatedly sprayed with the same pesticide, each succeeding generation contains a higher percentage of resistant individuals. Thus the widespread use of any chemical to control a rapidly producing insect pest species typi-

Figure 23-2 About 35 percent of the pesticides used in the United States are deployed against the cotton boll weevil.

U.S. Department of Agriculture

cally becomes ineffective within about 5 years—even sooner in the hot, wet tropics, where insects and disease organisms can adapt quickly to new environmental conditions.

Genetic resistance develops more rapidly in most insect pest species than in their predators. This occurs because predators, which are higher in the food chain, normally have smaller populations and reproduce at a lower rate than the herbivore pests on which they feed. Weeds and plant diseases also develop genetic resistance, but not as quickly as most insects.

Genetic resistance to various pesticides increased more than fourfold between 1960 and 1984. Worldwide, by 1984, at least 447 species of insects, 50 species of fungi, most species of rats, and at least 50 species of weeds had strains resistant to one or more synthetic pesticides. For example, cotton, which receives half the insecticides used in the United States, now has more than 25 pesticide-resistant insect pests, including boll weevils, bollworms, budworms, and spider mites. Because half of all pesticides applied worldwide are herbicides, genetic resistance in weeds is expected to increase significantly.

There are alternatives to the increased use of pesticides to keep populations of pest species below levels that cause economic loss (see Section 23-6). However, farmers receive at least 90 percent of their information about pest control from advertisements and brochures from pesticide manufacturers and from sales representatives of these companies. Thus, it is not surprising that when genetic resistance develops, pesticide company sales representatives usually recommend more frequent applications, stronger doses, or switching to a different formula to keep the resistant species under control, rather than suggesting alternative

methods that reduce their commissions and the profits of the companies they work for.

This narrow chemical approach can harness farmers on a *pesticide treadmill* in which the cost of using pesticides increases while their effectiveness decreases as genetic resistance develops ever more rapidly. Eventually insecticide costs can exceed the economic loss that would result from not using these chemicals.

Killing of Natural Pest Enemies Most pesticides are *broad-spectrum* poisons that kill not only the target pest species but also a number of natural predators and parasites that may have been keeping the pest species at reasonably low levels. Without sufficient natural enemies, and with lots of food available, a rapidly reproducing insect pest species can make a strong comeback a few days or weeks after being initially controlled. This revival of the pest population requires the use of more pesticides and is another factor that can place a farmer on a pesticide treadmill.

For example, in California's San Joaquin Valley, farmers sprayed cotton crops with heavy dosages of the organophosphate insecticide Azodrin to control the cotton bollworm. After only three sprayings, so many natural predators had been killed that the cotton bollworm was able to destroy 20 percent of the cotton crop. Independent research scientists at the University of California estimated that if no pesticides had been used, only about 5 percent of the cotton crop would have been destroyed by the bollworms.

Creation of New Pests The repeated use of broad-spectrum pesticides can also create new pests and convert minor pests to major pests—the reverse of what pesticides are supposed to do. This occurs when pesticides kill off natural predators that normally keep populations of minor pests such as parasites and insects (especially mites) under control. This phenomenon contributes to the pesticide treadmill by creating a need for additional pesticides. For example, the spider mite feeds on the chlorophyll found in the leaves and evergreen needles of many trees found in western forests. In a normal forest ecosystem it is only a minor pest because its population is kept low by natural predators such as ladybugs and other predator mites. When a major pest, the spruce budworm, began devastating western forests, the EPA allowed the U.S. Forest Service to spray these areas with DDT. While controlling the spruce budworm, however, the DDT killed off the natural enemies of the spider mite, which then became a major pest that caused more damage than the spruce budworm.

A combination of genetic resistance, killing of natural predators, and the rise of new pest species can lead to the collapse of a profitable agricultural operation, as happened to the cotton industry in northeastern Mexico in the 1960s. In the 1950s the boll weevil became resistant to chlorinated insecticides such as DDT and carbamate insecticides. Cotton farmers switched to organophosphates, but their use wiped out natural enemies of the cotton bollworm, a minor pest, and elevated it to major pest status. For a while, higher doses of organophosphates helped control the boll weevil and the bollworm but greatly increased production costs. Then in the mid-1960s the bollworm developed genetic resistance to the major organophosphate insecticide being used (methyl parathion). By that time some growers were treating their fields 15 to 18 times each growing season, and despite great costs were still suffering major crop losses from the increasingly resistant bollworms. Profits dropped so sharply that the multimillion-dollar cotton industry disappeared from the area, causing a severe economic depression in northeastern Mexico.

Global Mobility and Biological Magnification of Persistent Pesticides Once applied, DDT and other persistent pesticides can be dispersed far beyond the site of application, as shown in Figure 23-3. This *global mobility* of DDT and related pesticides occurs because of their persistence and because droplets of sprayed and evaporated pesticides can be transported by the wind and deposited in distant areas by rain, snow, and particulate matter to which they become attached. These persistent chemicals also wash into streams and eventually into the oceans, where currents move them long distances. Migratory birds and fish that feed on species containing pesticide residues also help disperse such chemicals to distant areas. Persistent pesticides can also gradually seep downward into groundwater aquifers and contaminate drinking water supplies (Section 22-5).

During the 1950s and 1960s scientists began detecting trace amounts of DDT and related persistent pesticides in the fatty tissue of living creatures throughout the world: Arctic seals, Antarctic penguins, and Eskimos located far from sprayed agricultural areas were found to be affected. Because of the chemicals' persistence, mobility, high solubility in fats, and relatively low solubility in water, concentrations of DDT and other related persistent pesticides can be *biologically magnified* in food chains and webs to levels hundreds to millions of times higher than those in the soil or water, as shown in Figures 5-6 and 23-3. The high levels of pesticides stored in the fatty tissues of organisms feeding on contaminated organisms lower in the food chain can kill many forms of wildlife outright or interfere with their ability to produce live offspring, as discussed later in this section.

Because of biological magnification, everyone born after 1950 has carried since birth several parts per million of such persistent compounds as DDT and nonpesticide polychlorinated biphenyls (PCBs) in their fatty tissues. During the 1960s the milk of many nursing mothers was found to contain levels of DDT higher than those allowed in cow's milk by the Food and Drug Administration. It is too early to tell what harmful effects, if any, will result from the long-term exposure of humans throughout the world to low levels of such substances.

Biological magnification, with its resulting potential or actual effects on wildlife and humans, was the main reason for the ban on DDT for all but emergency uses in the United States since 1972. In 1971 Americans carried an average of about 8 ppm of DDT in their bodies; by 1980, only 8 years after the partial ban, the average level had dropped to about 4 ppm.

Threats to Wildlife Marine organisms, especially shellfish, can be killed by minute concentrations of chlorinated hydrocarbon pesticides. Trout build up a layer of fat during summer months for use as a source of energy in winter. If this fat is contaminated with DDT washed into streams from crop fields, the fish can be killed by the large levels of DDT released into the bloodstream when the fat is used as a food source in winter. The young offspring of some species are often particularly sus-

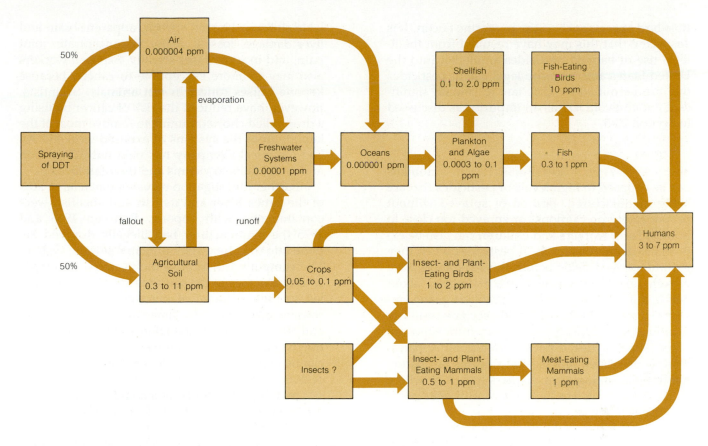

Figure 23-3 The movement and biological magnification of DDT in the ecosphere.

ceptible to pesticides. For example, adult mink remain healthy after consuming DDT-contaminated fish, but about 80 percent of their newborn infants die within a few days.

Honeybees, which pollinate crops that provide one-third of the food consumed in the United States, are extremely susceptible to pesticide poisoning. Each year an estimated 20 percent of all honeybee colonies in the United States are killed by pesticides and another 15 percent of the colonies are damaged—causing annual losses of at least $135 million from the reduced pollination of vital crops.

In the 1950s there was a drastic decline in populations of robins and other species of songbirds in forests and areas sprayed with DDT and other chlorinated hydrocarbon pesticides. Research showed that robins were accumulating lethal levels of DDT when they fed on earthworms that had consumed the leaves of elm trees sprayed with the pesticide to kill the bark beetle, which carries the fungus of Dutch elm disease. The strange springtime silence of such forests led Rachel Carson to choose the famous title for her book warning of the harmful effects of pesticides.

DDT stored in fatty tissue does slowly break down to related persistent compounds such as DDD and DDE, and perhaps various PCBs, which

are also produced synthetically (Section 24-5). To many species, however, these compounds are as harmful—in some cases more harmful—as DDT. During the 1950s and 1960s there were drastic declines in populations of fish-eating birds such as the osprey, cormorant, brown pelican, and bald eagle. There were also sharp population decreases of predatory birds such as the peregrine falcon, prairie falcon, sparrow hawk, and Bermuda petrel, which help control populations of rabbits, ground squirrels, and other crop-damaging small mammals.

Research has shown that many of these population declines occurred because the DDE produced by the breakdown of high levels of DDT accumulated in the bodies of the affected bird species and reduced the deposition of calcium in the shells of their eggs. The resulting thin-shelled eggs are so fragile that many of them break, and the unborn chicks die before they can hatch normally. Since the U.S. ban on DDT in 1972, DDT, DDD, and DDE residues have dropped in populations of most of these bird species. As a result, populations of species such as ospreys, brown pelicans, and prairie hawks have made a comeback.

By 1980, however, it was discovered that in some areas levels of DDT and other banned pesticides were beginning to rise again in some susceptible bird species such as the peregrine falcon. It is suspected that this may have resulted from the illegal use of banned pesticides smuggled into the United States and from the legal use of pesticides that are permitted by U.S. law to contain significant amounts of DDT as an impurity, as discussed in Section 23-5.

Short-Term Threats to Human Health Humans can be exposed to high levels of pesticides that are being manufactured, poured or sprayed without proper safety precautions, or sprayed too close to homes. Farm workers and children can also be exposed to dangerous levels of pesticides when they go into sprayed fields before the substances have broken down. By conservative estimates, about half a million farm workers, pesticide plant employees, and children worldwide become seriously ill, and about 20,000 die each year, from exposure to toxic insecticides, especially organophosphates. Insecticide-related illnesses and deaths are particularly high among farm workers in LDCs, where educational levels are low and control over pesticide use is often lax.

Since the ban of DDT and related persistent pesticides in the 1970s, U.S. farmers have switched to less persistent organophosphate and carbamate insecticides (Table 23-1)—some of which are more toxic to wildlife and humans than the chlorinated hydrocarbons they replaced. Humans exposed to high levels of organophosphates can suffer from severe convulsions, tremors, paralysis, coma, and death. Exposure to low levels of such compounds can cause confusion, cramps, vomiting, diarrhea, difficulty in breathing, and headaches. In the United States insecticides cause an estimated 45,000 illnesses and 200 deaths each year. Some observers believe that the number of pesticide-related illnesses among farm workers in the United States and throughout the world is greatly underestimated because of poor records, lack of doctors and reporting in rural areas, and faulty diagnosis.

Deaths and injuries can also occur from the manufacture of pesticides. In 1975 state officials found that 70 of the 150 employees in a pesticide manufacturing plant in Hopewell, Virginia, had been poisoned by exposure to high levels of *kepone* (chlorodecone), a persistent chlorinated hydrocarbon pesticide used as an ant and roach poison. In the plant kepone dust filled the air, covered equipment, and was even found in the employees' lunch area. Some of the workers also brought kepone dust home on their clothes and contaminated family members. The plant, which was associated with Allied Chemical Company, was shut down in 1975, and 29 workers were hospitalized with uncontrol-

lable shaking, blurred speech, apparent brain and liver damage, loss of ability to concentrate, joint pain, and in some cases sterility. Exposed workers also may be more susceptible to cancer, because kepone causes cancer in test animals. Scientists, however, have reduced the risk of cancer by using a drug called cholestryamine to remove most of the kepone from the systems of exposed workers. Allied Chemical Company has paid out $13 million in damage suits to victims and their families.

Further investigation revealed that a large area of the James River and its fish and shellfish were contaminated with kepone. Between 1966 and 1975 the manufacturer had illegally dumped kepone into the Hopewell municipal sewage system. The compound disrupted the bacterial decomposition processes in the sewage treatment plant and led to the discharge of untreated, kepone-laden sewage into the nearby James River. Between 1975 and 1980 more than 160 kilometers (100 miles) of the river and its tributaries were closed to commercial fishing, resulting in a loss of jobs and millions of dollars.

In 1984 the world's worst industrial accident occurred at a Union Carbide pesticide plant located in Bhopal, India. More than 2,500 people were killed when highly toxic methyl isocyanate gas, used in the manufacture of pesticides, leaked from a storage tank. At least 20,000 (perhaps as many as 100,000) people were left with blindness, sterility, kidney and liver infections, tuberculosis, brain damage, and other disorders that can lead to premature death. This tragedy might have been prevented by the expenditure of perhaps no more than a million dollars to ensure more adequate plant safety. This incident has aroused concern over the safety of the 11,583 chemical plants located in the United States.

In the 1960s a controversy began over the possible health effects from the use of the herbicide 2,4,5-T. Between 1962 and 1970, Agent Orange, a 50-50 mixture of 2,4-D and 2,4,5-T, was sprayed to defoliate swamps and forests in South Vietnam to prevent guerrilla ambushes, discourage the movement of troops and supplies through demilitarized zones, clear areas around military camps, and destroy crops that could feed North Vietnamese soldiers. This campaign of biological warfare destroyed vast areas of farmland, more than half the mangrove forests of South Vietnam, and about 5 percent of that nation's hardwood forests—a supply of wood worth at least $500 million that could have lasted the country for 30 years.

In 1965 and 1966 a study commissioned by the National Cancer Institute found that low levels of 2,4,5-T caused high rates of birth defects in laboratory animals. This report was not released to the public until 1969. Because of the resulting pressure from environmentalists and health officials, however, the Vietnam defoliation program was halted in 1970. Investigations revealed that the birth defects in laboratory animals were probably caused by a highly toxic dioxin called TCDD, apparently formed in minute quantities as an unavoidable contaminant during the manufacture of 2,4,5-T. Soil and industrial wastes contaminated with TCDD have also been deposited in numerous sites in the United States, and the compound is considered a serious hazardous waste, as discussed in Section 24-5.

In the late 1970s as many as 40,000 previously healthy Vietnam veterans began experiencing a variety of medical disorders including dizziness, blurred vision, insomnia, fits of uncontrollable rage, nausea, chloracne on large areas of their skin, and depression. An abnormally high percentage of these veterans fathered infants who were aborted prematurely, born dead, or had multiple birth defects. Other veterans had higher than expected incidences of leukemia, lymphoma, and rare testicular cancer.

By 1980, more than 1,200 Vietnam War veterans had filed claims with the Veterans Administration for disabilities allegedly caused by exposure to Agent Orange. The VA and the manufacturers of the chemical, however, continue to deny any connection between these medical disorders and Agent Orange and attribute the problems to the development of post-Vietnam stress syndrome. A long-term VA study of the health effects of exposure to Agent Orange is not expected to be completed until 1987–1990. In 1984 the companies making Agent Orange settled out of court a claim brought by Vietnam veterans, without admitting any guilt or connection between the disorders and the use of the herbicidal agent.

Long-Term Threats to Human Health Many scientists are concerned about the possible effects on humans of long-term, low-level exposure to DDT and other persistent pesticides. Such possible long-term effects, if any, won't be known for at least several decades, because the oldest people who have carried these chemicals in their bodies since conception were only 40 years old in 1985. The results of this long-term worldwide experiment, with human beings involuntarily playing the role of guinea pigs, may never be known, because it is almost impossible to determine that a specific chemical such as DDT caused a particular cancer or other harmful effect.

However, some disturbing but inconclusive evidence has emerged. DDT, aldrin, dieldrin, hep-

tachlor, mirex, endrin, and 19 other pesticides have all been found to cause cancer in test animals, especially liver cancer in mice. In addition, autopsies have shown that the bodies of people who died from various cancers, cirrhosis of the liver, hypertension, cerebral hemorrhage, and softening of the brain contained fairly high levels of DDT or its breakdown products DDD and DDE. Recent National Academy of Sciences studies indicate that of the 1,400 different chemicals used in registered pesticides in the United States, up to 25 percent (350 chemicals) may cause cancer in people, and 66 percent (924 chemicals) have not been adequately tested for their potential to produce cancer, genetic changes, and birth defects.

23-5 Pesticide Bans in the United States

Pesticide Laws Because of the potentially harmful effects of pesticides on wildlife and humans, as just discussed, Congress passed the Federal Insecticide, Fungicide, and Rodenticide Act in 1972. This act, which was amended in 1975 and 1978, requires that all commercially available pesticides be registered with the Environmental Protection Agency. Using information provided by the pesticide manufacturer, the EPA may refuse to approve the use of a pesticide or may classify it for general or restricted use. The EPA can also cancel or suspend the use of a pesticide already on the market if new evidence of harmful side effects on wildlife or humans is received.

Insecticide and Herbicide Bans Since 1972 the EPA has banned the use except for emergency situations of DDT and several other persistent chlorinated hydrocarbon pesticides, including aldrin, dieldrin, heptachlor, lindane, chlordane, and toxaphene. These pesticides were banned: (1) because their persistence and ability to be concentrated to high levels as they pass through food chains and webs threatens some forms of wildlife, and (2) because laboratory tests showed that they caused birth defects, some types of cancer, and neurological disorders in laboratory test animals. In 1974 the EPA permitted the U.S. Forest Service to use DDT on a short-term, emergency basis to help prevent the tussock moth from destroying valuable stands of coniferous trees in the Northwest.

In 1976 mirex was withdrawn from use by its manufacturer before an EPA ban was imposed because it was shown that the substance accumulates in human tissue, causes cancer in test animals, and had contaminated the Great Lakes. Similarly, kepone was withdrawn from the market in 1976 before it was officially banned as a result of the contamination incident described in Section 23-4. Large quantities of DDT and other chlorinated hydrocarbon pesticides are still used in Africa, South America, and Asia, where their use has not been banned.

Until 1983 ethylene dibromide (EDB) was widely used by the citrus and grain industries to fumigate soil, stored grain, and grain milling machinery. Since 1974, studies have shown that relatively low doses of EDB, either absorbed through the skin or ingested, caused stomach cancer and genetic mutations in all test animals in a relatively short period of time. Despite this knowledge, the EPA waited until 1983 to ban the use of EDB as a pesticide and to establish maximum permissible residue levels in foods already treated with EDB. This action was triggered by the discovery of alarming levels of EDB in groundwater supplies in fruit-growing areas of Florida, Texas, Hawaii, California, and Arizona, and in grain products, including some cake and muffin mixes on grocery store shelves. The food industry and the FDA have proposed that exposure of certain foods to low levels of radiation be used to replace the use of EDB, as discussed in Section 23-6. By 1985 the EPA had reviewed the potential health effects of only 70 of the 1,200 chemicals such as EDB used to formulate 10,000 different pesticide products in the United States.

The related herbicides silvex and 2,4,5-T were banned in 1979, except to control weeds on rice, sugar cane, and rangeland, because the traces of dioxin found in them may cause miscarriages and birth defects in humans. All uses of these two herbicides were banned in 1985. However, manufacturers of these herbicides contend that the bans were based on inconclusive evidence and continue their fight to have the decisions overturned in court or through political pressure on the Environmental Protection Agency.

Export of Banned Pesticides In response to a slowdown in the rate of growth of pesticide use in the United States since 1981, the U.S. chemical industry has increased exports of pesticides or their basic ingredients to other countries (mostly LDCs) where they have not been banned. In most LDCs more than half, and in some cases up to 70 percent, of these pesticides are applied to crops such as cotton, coffee, cocoa, and bananas destined for export to Europe, Japan, and the United States. A 1979 report by the General Accounting Office concluded that a large proportion of food and natural fiber imported into the United States from LDCs may

contain unsafe residues of pesticides that have been banned here.

In 1978 Congress amended the Federal Insecticide, Fungicide, and Rodenticide Act to require foreign purchasers of such pesticides to acknowledge their awareness that these chemicals are not approved for use in the United States. The EPA then notifies the government of the importing country that unregistered products are being shipped. However, many of these shipments arrive and are used long before this information reaches the appropriate official with the power to refuse to accept the cargo. In 1983 the United Nations and the Food and Agricultural Organization passed a resolution that calls for (1) stringent restrictions on the export of products whose use has been banned or severely restricted in the exporting nation and (2) the widespread publication of an easily understandable list of such products. The United States was the only nation opposing this resolution.

Proponents of exports of banned or unregistered pesticides, drugs, and other chemicals argue that the countries receiving the exports should be free to use these chemicals if they so desire and that if U.S. companies do not sell these substances to them, someone else will.

Has DDT Really Been Banned? A 1983 study showed that 44 percent of the fruits and vegetables grown in California contained higher than expected residues of 19 different pesticides, including DDT and other banned substances. Investigators suspect that this is due to a combination of two factors: (1) the illegal smuggling of relatively large amounts of DDT and some other banned insecticides into the United States from Mexico, for use mostly in western states, and (2) a loophole in U.S. pesticide control legislation that allows the sale in the United States of up to 50 insecticides that can contain as much as 15 percent by weight DDT. This DDT is allowed because it is classified by the EPA as an "unintentional impurity" that occurs when certain insecticides are manufactured. For example, dicofol (usually sold under the trade name Kelthane) contains as much as 15 percent DDT. Although dicofol is not manufactured in the United States, more than 910 thousand kilograms (2 million pounds) is imported each year and used to control mites on cotton and citrus crops. The application to register the use of dicofol in the United States, which stated plainly that DDT was used in its manufacture, was submitted to the EPA in 1972, only a few days after the agency had banned the general use of DDT in this country. In early 1985 the EPA began a review of its long-standing policy of allowing significant percentages of DDT and other banned pesticides to be allowed as impurities in pesticides used in the United States.

23-6 Alternative Methods of Insect Control

Possible Alternatives to Insecticides Major alternatives that can reduce and in some cases eliminate the need to use conventional pesticides to control populations of insect pests include: **(1)** planting crops at various times and places and in certain ways (cultural control), **(2)** encouraging or introducing natural predators, parasites, and diseases (biological control), **(3)** using radiation or chemicals to sterilize adult insect males, **(4)** developing new varieties of crops and livestock animals resistant to various pest insects, parasites, and disease, **(5)** using sound, light, and sex attractants to trap insects or confuse male insects so they can't find mates, **(6)** using synthetic or extracted chemical insect hormones to disrupt the life cycle of insects, to prevent them from reaching maturity or to render them unable to reproduce, **(7)** irradiating foods to kill pests and increase shelf life, **(8)** changing the attitudes of consumers and farmers about pests and about the cosmetic appearance of fruits and vegetables, and **(9)** using a carefully planned variety of cultural, biological, and chemical methods in proper sequence and timing (integrated pest management). In the remainder of this section we examine each of these approaches in more detail.

Cultural Control For centuries farmers have modified the agricultural environment to create an environment that is unfavorable to pest populations. Examples of such *cultural control* of insect populations include: **(1)** rotating or altering the crops grown in a given field from year to year to prevent pest species (especially nonmigrating organisms living in the soil) that feed on a particular crop from establishing a permanent habitat in the field, **(2)** planting two or more different crops in alternate rows or strips in the same field (for example, such *intercropping* of peanuts and corn can reduce crop losses from corn borers by up to 80 percent, primarily because the peanut plants serve as a habitat for insects that prey on the corn borer); **(3)** adjusting planting times either to ensure that most of the pest population starves to death before the crop is available or to favor natural predators over the pests, **(4)** removing stalks and debris that serve as breeding places for insects, **(5)** planting hedgerows and alternating rows of different crops as barriers to insect invasion, **(6)** planting a low-valued trap crop such as alfalfa next to cotton to attract pests such as the lygus bug away from the more highly valued crop, **(7)** using scarecrows and noisemakers to frighten away crop-eating birds, **(8)** preventing invasions of lawns and pastures by crab grass and unwanted weeds by never cutting the grass lower than 7.6 centimeters (3 inches), **(9)** using customs inspections and quarantines to hinder or prevent the spread of pests, and **(10)** growing crops in areas where certain major pests do not exist.

Unfortunately, to increase short-term profits, large numbers of farmers in MDCs such as the United States have abandoned many of these cultural control methods. In addition, the mechanized vehicles and large irrigating systems that are widely used in modern industrialized agriculture are best suited to large fields of single crops and thus discourage the use of practices such as intercropping to reduce pest infestations.

Biological Control For decades, agricultural scientists have been using *biological control* to regulate the populations of some insects, weeds, rodents, and other pests by reintroducing once effective natural predators, parasites, and pathogens (disease-causing bacteria and viruses) or by importing new ones. Worldwide there have been more than 300 successful biological control projects, especially in China and the Soviet Union. Examples include using: **(1)** a species of ladybug to control the cottony-cushion-scale insect that attacked citrus orchards in California, **(2)** ladybugs and praying mantises (Figure 23-4) to control aphids, **(3)** purple martin birds to control mosquitoes, **(4)** a bacterial agent (*Bacillus thuringiensis*), sold commercially as a dry powder, to control many strains of leaf-eating caterpillars, mosquitoes, and gypsy moths, **(5)** milky spore disease, obtainable in garden stores, to control Japanese beetles, **(6)** virus sprays to help control the Douglas fir tussock moth, the gypsy moth, and the cotton bollworm, and **(7)** tiny parasitic wasps to control various crop-eating moths.

Biological control has several major advantages, including: **(1)** normally, the regulation of only the target species, **(2)** persistence of the control, **(3)** avoidance of the creation of new pests, and **(4)** ability of the predator species to undergo natu-

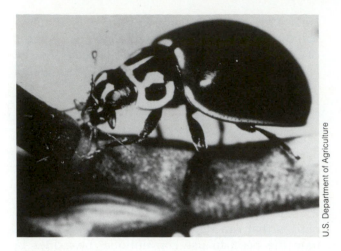

U.S. Department of Agriculture

Figure 23-4 The praying mantis (left) and the ladybug (right, eating an aphid) are insects used to control insect pests.

ral selection along with the target pest species so that genetic resistance is minimized.

Using biological control, however, is not simple and requires expensive and painstaking efforts by highly trained scientists. Obstacles to widespread use of biological control agents include difficulties in **(1)** mass production, **(2)** getting the agents established in fields, **(3)** making them work consistently in fields characterized by wide variations in temperature and moisture, **(4)** protecting them from being killed by pesticides sprayed in nearby fields, **(5)** ensuring that the agents themselves don't become pests and that they are not harmful to humans, livestock, and wildlife, and **(6)** obtaining research funds (such serious efforts to find and establish biological controls have been carried out for less than 10 percent of the known agricultural pests). Some pest organisms can also develop genetic resistance to viruses and bacterial agents used for biological control. In addition, farmers find that pesticides are faster acting and simpler to apply than biological agents.

Genetic Control by Sterilization   Males of an insect species can be raised in the laboratory, sterilized by radiation or certain chemicals, and then released in an infested area to mate unsuccessfully with fertile wild females. If sterile males outnumber fertile males by 10 to 1, pest species in a given area can be eradicated in about four generations, provided reinfestation does not occur. This *sterile male technique* works best if: **(1)** the females mate only once; **(2)** the infested area is isolated so that it can't be periodically reinfested with new nonsterilized males; and **(3)** the insect pest population has already been reduced to a fairly low level by weather, pesticides, or other factors.

This technique has been used to eradicate the

oriental fruit fly in Guam and the Mediterranean fruit fly, nicknamed the *medfly*, on the isle of Capri. A serious outbreak of the medfly, an imported pest of fruits and vegetables, could cause a loss of billions of dollars a year in California and Florida, devastating the citrus fruit and vegetable industries in these states. In 1981 the medfly appeared in California, but sterilized male medflies were imported from Hawaii and used along with traps containing malathion to control the infestation. This approach worked well in two infested areas but was only partially successful elsewhere because of inadequate funding and because some of the males released were not sterile—thus compounding the problem. The medfly, however, did not adapt well to the California environment, and populations never expanded to highly destructive levels. The Department of Agriculture, however, threatened to place a quarantine on all California produce. This action, coupled with nervous reactions from many of the state's produce growers, forced the governor to order massive aerial spraying of pesticides, which apparently eradicated the medfly.

A major livestock pest found in South America, Central America, Mexico, and the southeastern and southwestern United States is the screwworm fly. This metallic blue-green insect, about two to three times the size of the common housefly, deposits its eggs in open wounds of warm-blooded animals, such as cattle and deer. Within a few hours the eggs hatch into parasitic larva that feed on the flesh of the host animal. A severe infestation of this pest can kill a mature steer within 10 days. The Department of Agriculture used the sterile-male approach to essentially eliminate the screwworm fly from the southeastern states between 1962 and 1971. In 1972, however, the pest made a dramatic comeback, infesting

100,000 cattle and causing serious losses until 1976, when a new strain of the males was developed, sterilized, and released to bring the situation under temporary control. For some unknown reason, males in the original strain of flies raised and sterilized in captivity were no longer able to mate with wild females. Scientists have speculated that they may have lost their resistance to certain natural predators or disease organisms present in the natural environment or were unable to fly as far and as fast as fertile wild males. This extremely expensive, taxpayer-supported government program has saved the cattle industry about $120 million a year since 1977. But to prevent resurgences of this pest, new strains of the sterile male flies will probably have to be developed, sterilized, and released every few years.

Major problems with this approach include: **(1)** ensuring that sterile males are not overwhelmed numerically by nonsterile males, **(2)** having sufficient knowledge of the mating times and behavior of each target insect, **(3)** the possibility that laboratory-produced strains of sterile males will not be as sexually active as normal wild males, **(4)** preventing reinfestation with new nonsterilized males, and **(5)** high costs.

Genetic Control by Breeding Resistant Crops and Animals

For many decades agricultural scientists have used artificial selection, cross-breeding, and genetic engineering to develop new varieties of plants and animals resistant to certain insects, fungi, and diseases. For example, scientists have developed a number of wheat strains that are resistant to the Hessian fly, a major wheat pest that was accidentally introduced to the United States in the straw bedding of German mercenaries during the Revolutionary War. During the past 35 years the Hessian fly has adapted and overcome resistant varieties of wheat, but so far scientists have been able to develop new resistant strains of wheat.

Breeding new resistant strains of crops and animals, however, is expensive and can require from 10 to 20 years of painstaking work by highly trained scientists. The new strains must also produce high yields. Furthermore, insect pests and plant diseases can adapt and develop new strains that can attack the once resistant varieties, forcing scientists to continually develop new resistant strains.

Chemical Control Using Natural Sex Attractants and Hormones

Various *pheromones*, which serve as insect sex attractants, and *hormones*, which can be used to disrupt the natural life cycles of insects, can be used to control populations of insect pest species. Some observers believe that these two new types of chemical agent, sometimes called *third-generation pesticides*, may eventually replace or sharply reduce the use of less desirable second-generation pesticides currently in use.

In many insect species, when a virgin female is ready to mate she releases a minute amount (typically about 0.00000001 gram) of a species-specific chemical sex attractant called a *pheromone*. Males of the species up to a half-mile away can detect the chemical with their antennal receptors and follow the scent upwind to its source. Pheromones extracted from an insect pest species or synthesized in the laboratory can be used to lure pests into traps containing toxic chemicals. An infested area can also be sprayed with the appropriate pheromone or covered with millions of tiny cardboard squares impregnated with the substance so that most of the males become confused and are unable to find a mate because they detect the smell of virgin females almost everywhere.

Pheromones come fairly close to being the ideal pesticide. They are: **(1)** highly specific (because the pheromone of one species will affect only that species), **(2)** effective in extremely minute concentrations, **(3)** biodegradable (typically breaking down within a week), **(4)** especially effective in reducing or eliminating pest populations that are already at low levels, **(5)** not poisonous to animals (thus having no effect on wildlife and not creating new pests), and **(6)** have relatively little chance of causing genetic resistance. Pheromones are now commercially available for use against 25 major pests, including the pink bollworm, cotton boll weevil, cabbage looper, Japanese beetle, bark beetle, tobacco budworm, and oriental fruit fly.

However, there are difficulties in **(1)** identifying and isolating the specific sex attractant for each pest species, **(2)** ascertaining the mating behavior of the target insect, and **(3)** coping with periodic reinfestation from surrounding areas. The major problem is lack of availability. To produce a pheromone, scientists use elaborate and expensive chemical procedures to isolate a few milligrams of a specific pheromone from hundreds of pounds of insects. If the chemical structure of these complex molecules can be determined, however, some of them can be produced synthetically in the laboratory, often using cheaper techniques.

Hormones are chemicals produced by cells in organisms that travel through the bloodstream and control various aspects of an organism's growth and development. As shown in Figure 23-5, each step in the life cycle of a typical insect is regulated by the release of *juvenile hormones (JH)* and *molting*

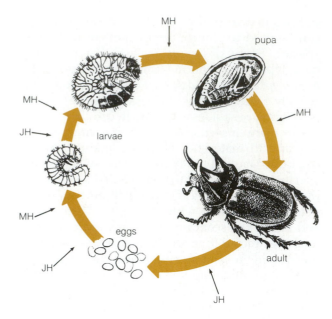

Figure 23-5 For normal growth, development, and reproduction, certain juvenile hormones (JH) and molting hormones (MH) must be present at genetically determined stages in the typical life cycle of an insect. If applied at the right time, synthetic hormones can be used to disrupt the life cycle of insect pests.

hormones (MH). For example, an insect in its larval stage continuously produces a particular juvenile hormone to keep it at this relatively undeveloped stage until it is time to move to the next step in the life cycle. If insect larvae are sprayed with the appropriate JH at a certain point in the life cycle, they will not mature into the adult stage and thus will not be able to mate. Other juvenile hormones or molting hormones applied at certain stages in an insect's life cycle can produce abnormalities that cause the insect to die before it can reach maturity and reproduce (Figure 23-6).

The first commercially marketed hormone was used successfully to control three species of floodwater mosquitoes, including one strain of *Anopheles*, which transmits malaria in Central and South America. Insect hormones have the same important advantages as pheromones, except that they can sometimes affect natural predators of the target insect species and other nonpest species. These hormones also: **(1)** take weeks rather than minutes to kill, **(2)** are often ineffective with a large infestation, **(3)** sometimes break down chemically in the environment before they can act, **(4)** must be applied at the right time in the life cycle of a target insect, as shown in Figure 23-5, and **(5)** are difficult and costly to isolate and produce.

Irradiation of Foods In 1985 the food industry and the FDA proposed the use of low levels of radiation to **(1)** replace fumigants such as EDB, **(2)** extend the shelf life of foods such as fresh fish, ground beef, and fresh fruits, berries, and vegetables, **(3)** destroy organisms that cause trichinosis from pork, salmonellosis from poultry, and deadly botulism from cured meats (thus reducing the need for nitrate and nitrite preservatives, which can cause cancer in test animals). In this process, the food is exposed to X rays, streams of electrons, or gamma rays from radioactive isotopes such as cobalt-60 or cesium-137. A food does not become radioactive when it is radiated, just as diagnostic X rays do not make your body radioactive.

Opponents of this proposal question the safety of eating irradiated foods. They argue that: **(1)** several studies have shown damage to organs in animals fed irradiated foods; **(2)** no tests have been carried out to determine whether eating irradiated foods will cause detrimental health effects to turn up after 30 to 40 years; **(3)** nonradioactive substances, whose potential short- and long-term harmful effects have not been adequately tested, are formed when foods are irradiated; and **(4)** high-dose irradiation can cause measurable loss of vitamins in some foods.

Food industry representatives maintain that: **(1)** the allegations of damage to organs in animals fed irradiated foods are based on only a few flawed studies, and a 1978 review of more than 1,000 studies concluded that there is no confirmed evidence of any danger from eating irradiated foods; **(2)** since World War II the testing of irradiated foods by the army, their use in 25 other countries, including Canada, Japan, France, Italy, and the U.S.S.R., and studies of U.S. astronauts and cancer patients whose diet is made up totally of irradiated foods, have shown no harmful effects from eating such foods; **(3)** radiation is no more—and in many instances is less—destructive of vitamins than conventional food processing; and **(4)** irradiation of foods, on balance, is likely to reduce health hazards to humans by decreasing the use of some potentially damaging pesticides and food additives.

Opponents argue that if the FDA approves the irradiation of foods in the United States, it should require that these foods be labeled clearly as such. The food industry, with support from the FDA, opposes such labeling, fearing that consumers might not buy products identified as irradiated because many would not understand that the foods contain no radioactive substances.

Changing the Attitudes of Consumers and Farmers Three human attitudes tend to increase the use of pesticides and lock us into the pesticide treadmill. First, many people believe that the only

U.S. Department of Agriculture

Figure 23-6 Some chemical hormones can keep insects from maturing completely, thus making it impossible for them to reproduce. Compare the normal mealworm (left) with one that failed to develop an adult abdomen after being sprayed with a synthetic hormone (right).

good bug is a dead bug. Second, most consumers insist on buying only perfect, unblemished fruits and vegetables, even though a few holes or frayed leaves do not significantly affect the taste, nutrition, or shelf life of such produce. Third, most farmers and consumers want pests to be controlled instantly but have not seriously considered the long-term effects of such a demand. This last attitude, which is encouraged by the pesticide industry, leads to the mistaken belief that there are no alternatives to the widespread use of pesticides and that without them diseases such as malaria

would be rampant and crop losses would be great enough to cause mass starvation. Educating farmers and consumers to change these three attitudes would go a long way toward reducing the use of pesticides.

Integrated Pest Management An increasing number of experts believe that in many cases our almost exclusively pesticide-based approach can be replaced with a carefully designed ecological approach—called *integrated pest management (IPM)*. In

In IPM each crop and its major pests are considered as an ecological system, and a control program is developed that uses a variety of biological, chemical, and cultural methods in proper sequence and timing. The overall aim is not eradication but keeping pest populations just below the level of economic loss. Fields are carefully monitored to check whether pests have reached an economically damaging level. When such a level is reached, farmers first use biological and cultural controls. Pesticides are applied only when absolutely necessary, and in small amounts, with different chemicals being used to retard development of genetic resistance.

Over the past 30 years more than three dozen IPM programs have been used successfully. These experiments have shown that a properly designed IPM program can reduce preharvest pest-induced crop losses by 50 percent, reduce pesticide use and pesticide control costs by 50 to 75 percent, reduce fertilizer and irrigation needs, and at the same time increase yields and reduce costs. By 1979 IPM was being used on more than a third of all the cotton acreage in 14 major cotton-producing states. Its major advantages include: (1) minimal use of pesticides, (2) long-term control costs cheaper than those from using conventional pesticides, (3) reduction of the hazards from the killing of natural predators, creation of new pests, and biological magnification in food chains associated with the use of many pesticides, and (4) decreased soil erosion.

Some of the major problems are as follows: (1) IPM is complex and requires expert knowledge about each pest–crop situation; (2) it is slower acting and more labor intensive than the use of conventional pesticides; (3) methods developed for a given crop in one area may not be applicable to another area with slightly different growing conditions; (4) although IPM's long-term costs are typically lower than those of using conventional pesticides, initial costs may be higher; and (5) highly trained IPM consultants must charge for their services, whereas pesticide sales representatives offer farmers free advice.

The banning of most pesticides and the gradual switch to IPM is difficult—if not impossible—because of the high costs involved and because of political pressure from powerful agricultural chemical companies that see little profit in such methods. As a result, IPM will have to be developed mostly by greatly increased financial and technical support from federal and state agencies.

Getting Off the Pesticide Treadmill In addition to changing the three attitudes discussed earlier that lead to increased and unnecessary use of pesticides, some environmentalists have urged that the use of pesticides be reduced and their safety improved by: (1) educating and encouraging farmers to cut the number of applications by applying pesticides sparingly and only at the most effective time, (2) having the Department of Agriculture license a corps of pest management advisers trained in entomology, microbiology, and the ecology of pest management (these professionals would be fully independent of pesticide manufacturers, and financially supported by local, state, or federal governments or from dues paid by cooperative organizations of farmers and consumers), (3) requiring that pesticides be used by farmers and other individuals only through prescriptions issued by federally licensed pest management advisers, (4) providing government subsidies and perhaps government-backed crop-loss insurance to farmers who use IPM or other approved alternatives to the widespread use of pesticides, (5) requiring chemical companies to have all new and existing pesticides tested by licensed laboratories independent of the pesticide industry, to determine whether they cause cancers, birth defects, or genetic mutations in test animals, (6) streamlining the EPA's lengthy administrative procedures for canceling or delaying registration of pesticides it finds unsafe, (7) lowering the requirements for registering natural pest control pheromones and hormones (which have few, if any, of the harmful side effects associated with the use of most second-generation pesticides), (8) allowing citizens access to safety data on any registered or proposed pesticide product and more participation in the regulatory process, (9) requiring labels for all produce, indicating any pesticide used in production, and lowering the rigid cosmetic standards that exclude most organically grown produce from many grocery stores, (10) prohibiting the export of pesticides and pesticide ingredients that either are not registered for use in the United States or have been banned here, (11) much more strictly enforcing existing regulations banning the import of food and fiber items that contain pesticide residues above specified levels, and (12) increasing the funding for research on biological controls, natural insect pheromones and hormones, and effective IPM strategies for all major pests.

We need to recognize that pest control is basically an ecological, not a chemical problem.

Robert L. Rudd

Discussion Topics

1. Should the United States abandon or sharply decrease the use of pesticides? Explain. What might be the consequences for LDCs? For the United States? For you?

2. Explain how the use of insecticides can actually increase the number of insect pest problems and threaten some carnivores, including humans?

3. How does genetic resistance to a particular pesticide occur? What major advantages do insects have over humans in this respect?

4. Debate the following resolution: Because DDT and the other banned chlorinated hydrocarbon pesticides pose no demonstrable threats to human health and have probably saved more lives than any other chemicals in human history, they should again be approved for use in the United States.

5. Do you agree or disagree with the 12 suggestions made at the end of Section 23-6 for reducing the use of pesticides? In each case, defend your position and try to have agricultural scientists or pesticide company representatives give their position on each suggestion.

24

Solid Waste and Hazardous Wastes

The shift from a throwaway society to a recycling one can help restore a broad-based gain in living standards.

Lester R. Brown and Edward C. Wolf

24-1 Solid Waste Production in the United States

What Is Solid Waste?   Any useless, unwanted, or discarded material that is not a liquid or a gas is classified as **solid waste**. It is yesterday's newspaper and junk mail, today's dinner table scraps, raked leaves and grass clippings, nonreturnable bottles and cans, worn-out appliances and furniture, abandoned cars, animal manure, crop residues, food-processing wastes, sewage sludge, fly ash, mining and industrial wastes, and an array of other cast-off materials. Some solid waste that is hazardous to human health must be detoxified or isolated from the environment. However, most of the things we throw away should be regarded not as solid waste but as *wasted solids*, which should be reused or recycled.

Amount and Sources of Solid Waste   Each American directly or indirectly produces 19,300 kilograms (21 tons) of solid waste each year, or 53 kilograms (115 pounds) a day. Direct per capita production of solid waste amounts to about 0.7 kilogram (1.5 pounds) a day, or 256 kilograms (548 pounds) a year.

Most solid waste in the United States is produced *indirectly* by agricultural, mining, and industrial activities. Animal, crop, and forest wastes from agricultural activities make up 56 percent of the total. About 34 percent of all solid waste in the United States consist of piles of rock, dirt, sand, and slag left behind from the mining and processing of energy resources and nonfuel mineral resources.

The total amount of industrial solid waste produced is small compared with solid waste from agricultural and mining activities, with industry accounting for only about 6 percent of the waste produced each year. Much of this is scrap metal, plastics, slag, paper, sludge from sewage treatment plants, and fly ash from electrical power plants. The last two categories, sludge and fly ash, are increasing rapidly because of stricter laws controlling the pollution of water (Section 22-8) and air (Section 21-6). Fly ash wastes will increase dramatically in coming years if more coal is used to produce electricity (Section 17-3).

Urban solid waste produced by homes and businesses in or near urban areas amounts to about 4 percent of the total produced each year in the United States. Because this solid waste is concentrated in highly populated areas, it must be removed quickly and efficiently. Because solid wastes from agricultural, mining, and industrial activities are discussed elsewhere in this book, the next two sections of this chapter are devoted to possible solutions to the problem of urban solid waste.

24-2 Disposal of Urban Solid Waste: Dump, Bury, or Burn?

What happens to the trash and garbage that is left on the curb for pickup in the United States? Most citizens and businesses don't care where their trash is disposed of as long as they don't have to smell it, see it, or pay too much to have it taken away. Based on this "out of sight, out of mind" principle, local sanitation departments or privately owned services collect the wastes and dump them on the land, bury them, or burn them. In 1984 $6.5 billion was spent to collect and dispose of urban solid waste in the United States. Between 1975 and 1984, average collection and disposal costs have at least doubled and are continuing to rise as many urban areas run out of places to dispose of their refuse.

Table 24-1 Comparison of Methods for Solid Waste Disposal and Resource Recovery*

Methods	Advantages	Disadvantages
Littering	Easy	Unsightly; very expensive to clean up,† wastes resources
Open dump	Easy to manage; low initial investment and operating costs; can be put into operation in a short period of time; can receive all kinds of wastes.	Unsightly; breeds disease-carrying pests; foul odors; causes air pollution when wastes are burned; can contaminate groundwater and surface water through leaching and runoff; ecologically valuable marshes and wetlands may be erroneously considered "useless" and filled; wastes resources; difficult to site because of public opposition
Sanitary landfill	Easy to manage; relatively low initial investment and operating costs; can be put into operation in a short period of time; if properly designed and operated, can minimize pest, aesthetic, disease, air pollution, and water pollution problems; methane gas produced by waste decomposition can be used as a fuel; can receive all kinds of wastes; can be used to reclaim and enhance the value of submarginal land§	Can degenerate into an open dump if not properly designed and managed; requires large amount of land; difficult to site because of citizen opposition and rising land prices‡; wastes resources; leaching may cause water pollution; filled land may settle; production of methane gas from decomposing wastes can create a fire or explosion hazard; obtaining adequate cover material may be difficult; hauling waste to distant sites is costly and energy inefficient
Incineration	Removes odors and disease-carrying organic matter; reduces volume of wastes by at least 80%; extends life of landfills; requires little land; can produce some income from salvaged metals and glass and use of waste heat to heat nearby buildings	High initial investment; high operating costs; frequent and costly maintenance and repairs; requires skilled operators; resulting residue and fly ash must be disposed of; causes air pollution unless very costly controls are installed; fine-particle air pollution (Section 21-1) even with pollution control; wastes some resources
Composting	Converts organic waste to soil conditioner that can be sold for use on land; moderate operating costs; most disease-causing bacteria destroyed	Can be used only for organic wastes; wastes must be separated; limited U.S. market for resulting soil conditioner; American solid waste poorly suited for composting because of low organic waste content
Resource recovery plant	High public acceptance; if designed and operated properly, produces little air and water pollution; reduces waste of resources; extends life of landfills; can produce income from salvaged metals, glass, and other materials and from sale of recovered energy for heating nearby buildings; may be easier to site than landfill or conventional incinerator	High initial investment; high operating costs; technology for many operations not fully proven; requires markets for recovered materials or energy produced; costly maintenance and repair; requires skilled operators; can cause air pollution if not properly controlled; profitable only with high volume of waste; discourages low-technology, sustainable earth approach emphasizing reuse and decreased use and waste of resources (Table 15-1)

*The costs for all methods listed vary widely with location, have been rising rapidly and do not include the costs of land, plant construction, and waste collection. Typical collection costs in urban areas range from $50 to $80 per metric ton of waste.
†Litter is so widely dispersed that labor and operating costs for collection are extremely high (the second energy law again).
‡By 1983 over two-thirds of the cities in the United States had run out of landfill sites.
§In Virginia Beach, Virginia, a landfill known as Mount Trashmore is used as an amphitheater; near Chicago, a landfill is used as a ski slope.

Currently about 82 percent of the collected urban solid waste is deposited in landfills, 8 percent is burned in municipal incinerators, 10 percent is recycled, and a tiny fraction is composted. Composting occurs when organic matter is broken down in the presence of oxygen by aerobic (oxygen-needing) bacteria to produce a humuslike end product, *compost*, which can be used as a soil conditioner. Table 24-1 lists the major advantages and disadvantages of these methods for solid waste disposal and resource recovery.

Four methods used for the land disposal of solid waste are: open dumps, landfills, sanitary landfills, and secured landfills. An **open dump** is a land disposal site where solid and liquid wastes are deposited and left uncovered, with little or no regard for control of scavengers or for aesthetic,

disease, air pollution, and water pollution problems. A **landfill** is a slightly upgraded version of the open dump. This type of land waste disposal site is normally located with little, if any, regard for possible pollution of groundwater and surface water due to runoff and leaching; waste is covered intermittently with a layer of earth.

A **sanitary landfill** (Figure 24-1) is a land waste disposal site that is located to minimize water pollution from runoff and leaching; waste is spread in thin layers, compacted, and covered with a fresh layer of soil each day to minimize pests and aesthetic, disease, air pollution, and water pollution problems. A **secured landfill** is a site for the containerized burial and storage of hazardous solid and liquid wastes; the site has restricted access and is continually monitored. The landfill is located

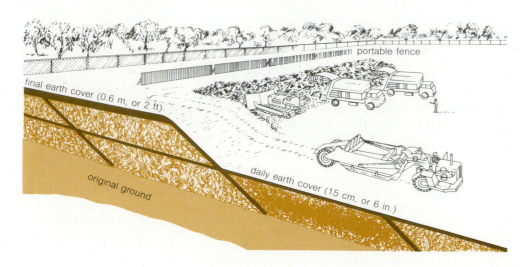

Figure 24-1 A sanitary landfill. Wastes are spread in a thin layer and then compacted with a bulldozer. A scraper (foreground) covers the wastes with a fresh layer of soil at the end of each day. Portable fences are used to catch and hold debris blown by the wind.

portable fence

final earth cover (0.6 m, or 2 ft)

original ground

daily earth cover (15 cm, or 6 in.)

above geologic strata that should prevent wastes from leaching into groundwater, as discussed further in Section 24-6.

The Resource Conservation and Recovery Act of 1976 required that all open dumps be closed or upgraded to landfills by 1983 and banned the creation of new open dumps. However, this law does not clearly distinguish between landfills and sanitary landfills. As a result, many of the 15,000 sites designated as sanitary landfills are really landfills. Because of their unfavorable location, many of these landfills cause water pollution problems, especially groundwater contamination (Section 22-5).

24-3   Resource Recovery from Solid Waste

The High-Technology Approach There is disagreement over whether most resources should be recycled by a *centralized, high-technology approach* or by a *decentralized, low-technology approach*. In the high-technology approach, large, centralized *resource recovery plants* ideally would shred and automatically separate mixed urban wastes to recover glass, iron, aluminum, and other valuable materials, which would be sold to manufacturing industries for recycling. The remaining paper, plastics, and other combustible wastes would be incinerated to produce steam, hot water, or electricity, which could be used in municipal facilities or sold to nearby buildings and manufacturing plants. The incinerator residue, including particulates removed to prevent air pollution, could be used as landfill to reclaim damaged land or processed into cinder blocks, bricks, or other building materials.

By 1984, more than 37 large-scale resource recovery plants had been built and 17 others were under construction in the United States. Although

a few of these plants separate and recover some iron, aluminum, and glass for recycling, most are sophisticated incinerators used to produce energy by burning trash. Denmark uses incinerator plants to convert about 60 percent of its burnable wastes to energy, Switzerland 40 percent, and the Netherlands and Sweden each 30 percent.

Unlike their European counterparts, most resource recovery plants in the United States have been a bitter disappointment. They have been expensive to build ($50 million to $500 million per plant) and have suffered from delays, breakdowns, high operating and maintenance costs, lack of enough waste each day to make them economical to operate, and continuing financial losses. Even though 55 percent of the cost of these facilities was subsidized by federal funds, several have gone bankrupt and have been abandoned. One notable exception is a steam-producing plant in Saugus, Massachusetts, which has a steady and reliable supply of refuse and was built using tried-and-true European technology. Even so, it took the plant 4 years to make a profit—the only U.S. resource recovery plant operating in the black by 1984. Some cities talk about a new wave of resource recovery plants based on the Saugus model. Such efforts, however, may be hindered by drastic cutbacks in federal funds since 1981 and by citizens campaigning against having landfills or resource recovery plants nearby.

The Low-Technology Approach Most waste materials recovered in the United States are recycled in a *low-technology approach* involving source separation. In this simpler, small-scale approach, homes and businesses place waste materials such as glass, paper, metals, and food scraps into separate containers. Compartmentalized city collection trucks, private haulers, or voluntary recycling organizations pick up the segregated wastes, clean them up if necessary, and sell them to scrap dealers, compost plants, and manufacturers. Studies have shown that this source separation takes only 16 minutes a week for the average American family.

By 1984, more than 260 U.S. cities had curbside pickup of separated wastes. There were also more than 3,000 municipal or community-based recycling centers operating in the United States, together recycling 12 million tons of trash for an annual cash return exceeding $380 million. If 75 percent of all municipal waste were burned in high-technology energy recovery plants, this would provide only 1 percent of the nation's annual energy use. By contrast, a comprehensive low-technology recycling program could save 5 percent of the annual U.S. energy use—more than the energy generated by all U.S. nuclear power plants—at perhaps one-hundredth of the capital and operating costs.

Mixing wastes and sending them off to landfills or resource recovery plants also hinders the recycling of paper and encourages the use of throwaway cans and bottles. Indeed, proponents of large-scale resource recovery plants in the United States have opposed federal and state laws to ban or discourage the use of nonreturnable cans and bottles and to encourage the recycling of paper, because removal of these materials from mixed urban refuse could make such plants unprofitable.

Aluminum Cans Versus Returnable Bottles for Beverages Throwing away an aluminum beverage container wastes as much energy as filling such a can with gasoline and pouring out half. At least one-fourth of all U.S. aluminum production goes into packaging, half of which is for beverage containers. Although about 54 percent of the aluminum beverage cans made and used in the United States are recycled at more than 2,500 recycling centers set up by the aluminum industry, this percentage could be increased to at least 90 percent by: (1) *banning nonreturnable beverage containers*, as has been done in Denmark, and requiring that all beverages be supplied in standardized bottle sizes refillable by any bottler, or (2) discouraging the use of throwaways by requiring a deposit on each throwaway beverage bottle and can.

By 1984, the second approach had been adopted in Sweden, Norway, the Netherlands, several provinces in Canada, and in nine states in the United States: Oregon (in 1972), Vermont (in 1973), Maine (in 1978), Michigan (in 1979), Iowa (in 1979), Connecticut (in 1980), Delaware (in 1982), Massachusetts (in 1983), and New York (in 1983).

In the United States these bottle laws were opposed by a powerful and well-funded lobby of steel and aluminum companies, metalworkers' unions, supermarket chains, and most major brewers and soft drink bottlers. Expensive ad campaigns financed by these groups have helped prevent passage of such laws in a number of states as well as a national beverage container deposit bill. This has happened even though surveys show that 73 percent of the Americans polled would support such legislation.

Environmental Protection Agency and the General Accounting Office studies estimate that a national container deposit law would: (1) reduce roadside beverage container litter by 60 to 70 percent, (2) reduce urban solid waste by 1 percent and thus save $25 million to $50 million a year in waste disposal costs, (3) reduce the use of bauxite ore to make aluminum by 53 to 74 percent and the use of iron ore by 45 to 83 percent, (4) reduce air, water, and solid waste pollution from the beverage industry by 44 to 86 percent, (5) save energy equivalent to that needed to provide the annual electrical needs for 2 million to 7.7 million homes, (6) produce a net increase of 80,000 to 100,000 jobs, and (7) save consumers at least $1 billion annually.

The approach favored by steel, aluminum, and glass companies, and others involved in the sale of beverage containers, is to enact litter recycling laws. This type of law levies a tax on industries whose products pose a potential threat as litter or landfill clutter. Revenues from the tax are used to establish and maintain statewide recycling centers. By 1983, at least eight states had this type of law: Washington, Virginia, Ohio, Nebraska, Louisiana, California, Arkansas, and Tennessee. Environmentalists contend, however, that this approach attacks the *symptoms* of solid waste disposal problems, not the underlying causes.

Recycling Paper: A Case Study Although paper can be recycled at a fairly high rate, only about 25 percent of the world's waste paper is now recycled. This percentage represents 50 percent in Mexico, 45 percent in Japan, 43 percent in the Netherlands, 35 percent in West Germany, 34 percent in Sweden, 26 percent in the United States, and 18 percent in Canada. Mexico, Japan, and the Netherlands have high paper recycling rates primarily because they are sparsely forested. In Sweden, the separation of waste paper from all garbage in homes, shops, and offices has been required by law since 1980. A number of analysts believe that with sufficient economic incentives and laws, at least half the world's waste paper could be recycled by the end of the century.

Each American uses directly or indirectly an average of about 275 kilograms (600 pounds) of paper per year—about 8 times the world average and about 40 times that in LDCs. Every Sunday edition of the *New York Times* consumes about 0.60 square kilometer (150 acres) of forest. Almost three-fourths of the U.S. paper production ends up in the trash, with waste paper making up about half the volume of urban solid waste produced each year. During World War II the United States recycled 40 to 45 percent of its waste paper when paper drives and recycling were national priorities.

In addition to saving trees and land, recycling paper saves about 30 to 55 percent of the energy needed to produce paper from virgin pulpwood and can reduce air pollution from pulp mills by about 95 percent. If half the discarded paper were recycled, the United States would save enough energy to provide 10 million people with electrical power each year.

Having individual homes and businesses separate out paper for recycling is an important key to increased recycling. Otherwise the paper becomes so contaminated with other trash that waste paper dealers will not buy it. Such source separation is feasible primarily for newspapers from homes, corrugated boxes from commercial and industrial establishments, and printing and writing papers from offices. Slick paper magazines, magazine sections, and advertising supplements, however, cause contamination problems and must not be included. By 1984, more than 150 U.S. cities required residences and businesses to separate out newspapers and cardboard for pickup and recycling, and at least 700 American companies and organizations were separating and selling computer cards and high-grade office waste paper for recycling.

Factors hindering waste paper recycling in the United States include: (1) federal tax subsidies and other financial incentives that make it cheaper to produce paper from trees than from recycling, (2) fluctuating prices because of variations in demand, and (3) the increased burning of paper in some cities in resource recovery plants to produce energy. Furthermore, the capital gains of producers of virgin timber are taxed at only 30 percent, whereas the paper recycling industry pays the normal corporate rate of 48 percent. This means that virgin timber producers are getting more than $100 million a year in tax breaks, to motivate them to grow more pulp trees.

A requirement at all levels of government that a certain percentage of all paper purchased be made from recycled fibers would help create a fixed demand and make it more profitable to recycle paper. The Resource Conservation and Recovery Act requires all federal agencies to buy products composed of the highest percentage of recycled materials practicable, but this law has largely been ignored. By contrast, since 1977 Maryland has complied with a similar law and has increased the amount of paper purchased by the state as recycled stock to 25 percent.

The Low-Waste Society: Going Beyond Disposal and Recycling A number of analysts believe that a fundamental goal of any national resource and solid waste management program should be to waste fewer resources by a combination of (1) reduced resource use per product (smaller cars and thinner walled containers), (2) reduced resource use per person (fewer cars per family), (3) increased product lifetimes (longer lasting cars, tires, and appliances), and (4) increased resource reuse by substituting packaging that can be reused in its original form, such as glass bottles, which can be refilled, for throwaway items such as aluminum beverage cans. Some examples of this approach are found in Table 15-1.

24-4 Types, Sources, and Effects of Hazardous Wastes

What Is Hazardous Waste? Any discarded material that may pose a substantial threat or potential hazard to human health or the environment when improperly handled is called a **hazardous waste**. These wastes, which are growing at a rate of 3 percent a year, include a variety of toxic, ignitable, corrosive, or dangerously reactive substances. Examples include acids, cyanides, and pesticides; solvents from dry cleaners; compounds of lead, mercury, arsenic, and cadmium; soil contaminated with toxic PCBs and dioxin; infectious wastes from hospitals and research laboratories; improperly treated sewage sludge; obsolete explosives, herbicides, and nerve gas stockpiled and waiting disposal by the Department of Defense, and low- and high-level radioactive materials (Section 18-2).

Since almost any substance can cause a harmful effect at a high enough level, regulators have great difficulty in classifying individual chemicals as hazardous. Another problem is that a substance normally considered to be fairly harmless may interact synergistically with one or more chemicals in the environment to produce one or more highly toxic substances.

Volume, Sources, and Effects of Hazardous Wastes Since 1950 an estimated 5 trillion kilograms (6 billion tons) of hazardous wastes have been deposited in or on the land throughout the United States. Many of the wastes have been improperly disposed of and represent chemical time bombs, which greatly increase the risk of exposure to substances that can cause cancer, birth defects, miscarriages, blood diseases, nervous disorders, and damage to human kidneys, liver, and genes. Each year more than 250 billion kilograms (275 million tons) of new hazardous wastes are produced in the United States by approximately 750,000 different producers. In 1984 this amounted to an average of 1,070 kilograms (2,350 pounds) of hazardous waste for each inhabitant of the United States—enough to fill the New Orleans Superdome from floor to ceiling twice a day. The actual amount of hazardous wastes may be much larger, because a 1984 study by the National Academy of Sciences revealed that only about 20 percent of the almost 49,000 different chemicals in commercial use have been subjected to extensive toxicity testing, and one-third of these chemicals have never been tested at all for toxicity.

About 71 percent of the hazardous wastes produced in the United States come from the chemical and petroleum industries, about 22 percent from metal-related industries, and the remaining 7 percent from a variety of other sources. Although all states produce hazardous wastes, about 65 percent of the volume is produced, in decreasing order, by Texas, Ohio, Pennsylvania, Louisiana, Michigan, Indiana, Illinois, Tennessee, West Virginia, and California.

The EPA estimates that only about 15 percent of the hazardous wastes produced in the United States are disposed of in an environmentally sound manner. About 70 percent are buried in unsafe landfills or unlined lagoons and ponds. Another 15 percent are burned, injected into deep wells, dumped at sea, and illegally dumped (into municipal landfills, rivers, sewer drains, wells, empty lots and fields, old quarries, abandoned mines), or spread along the roadsides. In Tennessee illegal dumpers have sent freight cars loaded with hazardous wastes to fictitious addresses, C.O.D. Toxic wastes are sprayed on ordinary dirt roads. Trash and garbage collectors are bribed to dispose of toxic wastes in landfills not designed to handle such substances. Toxic wastes are also illegally mixed with heating oil and burned in the boilers of schools, hospitals, and office and apartment buildings, resulting in the release of some of these toxic chemicals into the air. Law enforcement officials warn that illegal dumping is becoming more frequent as waste generators and haulers try to cut costs. Some officials have also warned that the lure of large profits and generally lax law enforcement have led to increased involvement of organized crime in every aspect of the hazardous waste disposal industry all over the United States.

A study by the state of California in 1983 revealed that hazardous wastes produced mostly during the manufacture of computer chips were leaking from 36 of the 49 underground storage tanks in one area. These wastes had contaminated surrounding soil, pockets of groundwater, and some private and community water wells in such "Silicon Valley" communities as Santa Clara, Mountain View, Sunnyvale, and San Jose. A 1980 survey by the California Department of Industrial Relations also found that workers in the microelectronics industry had more than three times as many illnesses per 100 workers as those in general manufacturing.

In Bullitt County, Kentucky, an area known as the "Valley of the Drums" contains 100,000 barrels of highly toxic chemicals. All the EPA can do is prevent the chemicals inside the rusting drums from leaking into a nearby stream and periodically check the area for groundwater contamination.

In Elizabeth City, New Jersey, a fire broke out in 1980 at a dump for highly explosive wastes used by the bankrupt Chemical Control Corporation. A serious incident was averted when winds blew the toxic clouds away from heavily populated areas.

Transporting hazardous wastes, mostly by truck and train, is another area of increasing concern. The EPA and the Department of Transportation estimate that each year there are about 16,000 transportation incidents involving spills of hazardous materials (not all of them wastes), resulting in about 20 deaths, 600 injuries, and at least $10 million in property damage.

Hazardous wastes can threaten human health from **(1)** direct exposure of workers in some industries, transportation accidents, and children playing near dump sites or swimming in contaminated streams, **(2)** long-term or essentially permanent contamination of groundwater from leaching and direct injection wells (Section 22-5), **(3)** contamination of surface water by runoff from hazardous waste disposal and storage landfills and other areas and overflow of hazardous waste treatment lagoons, **(4)** biological magnification of some materials in food chains and food webs from improper disposal, and **(5)** air pollution from open burning or poorly controlled incineration.

Health effects from exposure to hazardous chemicals can be *acute*, such as skin and eye irritation, disease, dizziness, nausea, blurred vision, tremors, headaches, chemical skin burns, blindness, and death, or *chronic*, such as various types of cancer (Section 20-3), stillbirths, birth defects, sterility, allergies, and heart, liver, and lung damage. Improper management and disposal of hazardous wastes can also cause fish kills, livestock losses, loss of vital habitat for fish and wildlife, soil contamination, cultural eutrophication of lakes (Section 22-4), and depletion of microorganisms that are important in nutrient cycling and nitrogen fixation (Section 5-3).

24-5 Some Examples of Hazardous Wastes: Dioxin, PCBs, and Toxic Metals

Dioxin In 1981 it was discovered that the soil in Times Beach, Missouri, a suburb of St. Louis, was contaminated with oil containing dioxin that had been sprayed on dirt roads to control dust. The problem was uncovered when birds and horses exposed to dioxin-contaminated soil began dying. In 1983 the EPA bought out the entire town at a cost of $36.7 million and had to relocate 2,200 people. Twenty-six other sites in Missouri are known to be contaminated with dioxin, and 75 more are suspected. Significant levels of dioxin have also been found in some rivers in Michigan, in some fish taken from the Great Lakes, and in flooded basements of homes near the Love Canal in Niagara Falls (Section 6-3).

Technically, the term *dioxin* refers to a family of 75 different organic compounds containing carbon, hydrogen, and chlorine. One form in particular, 2,3,7,8-tetrachlorodibenzo-paradioxin (2,3,7,8-TCCD, often referred to as TCCD) has been shown to be extremely toxic and to cause liver cancer, birth defects, and death in laboratory animals at extremely low levels. TCCD persists in the environment, especially in soil and in the human body, and can apparently be biologically magnified to higher levels in food chains and webs. As discussed in Section 23-4, this dioxin frequently appears as an unavoidable trace contaminant formed as a by-product in the chemical reactions used to make the herbicides silvex and 2,4,5-T, and Agent Orange (a 50–50 mixture of 2,4-D and 2,4,5-T).

TCCD and several other dioxins may also be formed in trace amounts during the high-temperature combustion of various organic compounds in incinerators and other combustion processes. In 1981 the EPA concluded that the quantity of TCCD and other dioxins released into the atmosphere during combustion processes is

small and diffuse enough to be relatively harmless, though this finding is disputed by some scientists. The major potential threat comes from larger quantities of TCCD present primarily in industrial dump sites, many of them abandoned.

Although the toxicity of 2,3,7,8-TCCD varies with the test animals exposed, in guinea pigs it is 170,000 times more toxic than cyanide. Workers and others exposed to TCCD in accidents have complained of headaches, weight and hair loss, liver disorders, irritability, insomnia, nerve damage in the arms and legs, loss of sex drive, and chloracne (a severe, painful, and often disfiguring form of acne). Although there have not been enough cases to permit statistical analysis, there is some evidence that humans exposed to chemicals contaminated with TCCD may have an abnormally high incidence of soft-tissue carcinoma, an extremely rare form of cancer that strikes muscles, fat, and nerve cells. The drug cholestryamine, used to remove kepone from the fatty tissues of workers exposed to this pesticide (Section 23-4), may also be useful in purging TCCD from the bodies of victims of dioxin poisoning.

Some studies have also suggested that TCCD can cause birth defects, but other studies have failed to confirm this finding. In 1983 the EPA began a 4-year program to identify and clean known and suspected sites believed to contain 80 to 95 percent of the TCCD produced in the United States.

Polychlorinated Biphenyls (PCBs) Since 1966 scientists have found widespread contamination from a widely used group of toxic, oily synthetic organic chemicals known as *polychlorinated biphenyls* (*PCBs*). PCBs are mixtures of about 70 different but closely related chlorinated hydrocarbon compounds that are made, like DDT, of carbon, hydrogen, and chlorine. PCBs, which were manufactured in the United States from 1929 to 1979, are still widely used as insulating and cooling fluids in electrical transformers and capacitors. Until 1979 PCBs were also used in the production of plastics, paints, rubber, adhesives, sealants, printing inks, carbonless copy paper, waxes, and pesticide extenders, and for dust control on roads.

There was little concern about PCBs until 1968, when some 1,300 Japanese came down with chloracne and suffered liver and kidney damage after they had eaten rice oil accidentally contaminated with PCBs that had leaked from a heat exchanger. Although precise data are not available, statistical analysis suggests that victims of this accident may suffer from an abnormally high incidence of stomach and liver cancer. Other studies of different occupational groups exposed to PCBs have led to inconsistent and conflicting results, primarily because of the difficulty in determining the amount and types of exposure of individuals to various mixtures of PCBs.

Like DDT, PCBs are insoluble in water, soluble in fats, and very resistant to biological and chemical degradation. Thus, they have the ideal properties for persistence and magnification in food chains and webs. Like DDT, PCBs have been found in Antarctic penguins, in predatory birds at the top of food chains, and in the milk of nursing mothers. The EPA estimated that by 1976 at least 98 percent of all Americans had detectable levels of PCBs in their fatty tissues, but that only 8 percent of the population had concentrations above the level (3 ppm) believed to cause adverse health effects. In 1974 the U.S. chemical industry voluntarily stopped producing PCBs for all uses except closed systems such as electrical transformers. In 1981, two years after most uses of PCBs in the United States had been banned by the EPA, a nationwide study showed that only 1 percent of the U.S. population had concentrations of PCBs greater than 3 ppm.

As with DDT (Section 23-4), the long-term health effects on humans exposed to low levels of PCBs are unknown. But large doses of PCBs have produced liver cancer, kidney damage, weight loss, and reproductive disorders in laboratory animals.

Even if a total ban on all uses of PCBs in the United States is instituted, as proposed by the EPA in 1983, sealed electrical transformers and capacitors used by electric utility companies still contain some 341 million kilograms (375,000 tons) of PCBs. Each year PCBs are released into the environment when some of these transformers and capacitors leak, catch fire, or explode. Since 1980 the EPA has required that all material containing PCBs be labeled and disposed of only at EPA-approved, secured landfills or high-temperature incinerators.

The EPA estimates that an additional 132 million kilograms (145,000 tons) of PCBs has been disposed of in dumps and landfills, many of them subject to leaking and leaching. It is estimated that an additional 68 million kilograms (75,000 tons) is dispersed throughout the United States, much of it dumped illegally on roadsides and other areas.

Cadmium Cadmium is widely used in electroplating metals to prevent corrosion, in plastics and paints, and in nickel–cadmium batteries. It is also a contaminant in phosphate fertilizers.

Japanese physicians reported in 1955 that *itai-itai byo*, or "ouch-ouch disease," was related to exposure to high levels of cadmium, although some scientists dispute this claim. Ouch-ouch disease causes excruciating pain in the joints and slowly weakens the bones when the calcium in bones is replaced by cadmium. Even standing, coughing, or sneezing can break bones, and the disease can be fatal. Between 1955 and 1968 several hundred cases and at least 100 deaths were reported in northern Japan, mostly among people who ate rice and soybeans grown in fields contaminated by nearby mines and cadmium-using industries.

The real threat to most of us, however, may be exposure to low levels of cadmium obtained from air, water, and diet over long periods. About 4 to 8 percent of the cadmium we eat or drink stays in the body, whereas from 40 to 50 percent of the cadmium we inhale remains. For this reason, tobacco smoke is the major source of cadmium for most people. Pack-a-day smokers carry about twice as much cadmium in their bodies as nonsmokers. People in the same room with smokers are also exposed to cadmium and other dangerous chemicals present in cigarette smoke.

The second most important source of cadmium is leafy vegetables, which pick it up from agricultural soils polluted primarily from atmospheric fallout of cadmium, sludge application, and commercial phosphate fertilizers. Cadmium is also found in small amounts in soft water. Adequate and balanced dietary intake of iron, zinc, calcium, copper, protein, and vitamins D and C can reduce the absorption of cadmium from food and water.

Because half the cadmium entering the human body remains there for an average of 200 days, cadmium gradually accumulates in our bodies—primarily in the liver and kidneys. Several studies in test animals and humans have shown that levels of cadmium near or slightly above the average daily intake of 50 to 70 micrograms can lead to high blood pressure (hypertension) and atherosclerosis. Sweden banned most products containing cadmium in 1980.

Lead Some lead enters the ecosystem from natural sources, but most comes from human activities. Lead has been used in pottery, batteries, solder, plumbing, cooking vessels, pesticides, and household paints. It is emitted into the atmosphere by lead smelters and by the burning of coal and leaded gasoline. As a result, lead levels have been increasing throughout the world since 800 B.C. (when humans began mining and using lead), even in remote areas such as Greenland. The sharpest rise in lead levels in the environment occurred between 1940 and 1967, when many automobiles began burning leaded gasoline, which contains tetraethyl lead as an antiknock additive.

In the United States 90 percent of the tiny particles of lead found in the air we inhale comes from the burning of leaded gasoline in cars, and another 5 percent from industrial smelters. Airborne lead can also enter some of the food we ingest when lead particles from vehicle exhausts settle over agricultural and grazing areas, especially those near highways. Some of this is absorbed by the plant, but much of it can be removed by careful washing. Another source of lead in food is the solder used to seal the seams on cans of food. For example, freshly opened cans of grapefruit and orange juice with lead-soldered seams exceed the EPA standard for lead in drinking water. To be safe, some scientists have proposed that the FDA ban the use of soldered cans for food. The remaining lead we ingest comes from drinking water contaminated primarily by atmospheric fallout of lead particles. Once lead enters the blood, about 10 percent is excreted from the body and the rest is stored in the bone.

Children are particularly vulnerable to lead poisoning because their bodies absorb lead more readily than those of adults. A 1981 survey by the National Center for Health Statistics revealed that one out of every 25 American preschool children and one out of every five inner-city black preschool children had dangerously high levels of lead in the blood. About 200 American children die each year from lead poisoning. Another 12,000 to 16,000 children each year are treated for lead poisoning and survive. About 30 percent of those who survive suffer from palsy, partial paralysis, blindness, and mental retardation. Data from a government study of more than 20,000 people, conducted between 1976 and 1980, indicated that even low levels of lead in the blood of children and adults can lead to high blood pressure.

Most cases of lead poisoning involve children between the ages of 1 and 3 years who have *pica*. This abnormal craving for unnatural foods, including dirt, paper, putty, and plastic may result from a diet deficiency. Children with pica used to eat paint chips from houses painted with lead-containing paint, but these paints were banned for interior use in the United States in 1940.

Companies making lead additives for gasoline and other lead-using industries note the absence of concrete evidence that airborne lead harms humans. But experiments with animals have indicated that low levels of lead can increase animal death rates at all ages by as much as one-quarter.

As a result, a number of prominent scientists have urged that lead additives be banned from gasoline and food cans and that smelting, battery, and other lead-using industries be required to control lead emissions into the air and water.

In 1985, after 13 years of delay and legal challenges by the gasoline industry, the EPA announced final rules for reducing the lead in gasoline by 91 percent in 1986. Between 1986 and 1992, the EPA says that this reduction will (1) reduce the blood pressure of 1.8 million people between the ages of 40 and 59, (2) cause 172,000 fewer children to have unacceptably high levels of lead in their blood, (3) raise the IQ scores of 42,000 children under age 7 by an average of 2.2 points, (4) save $1 billion a year, mostly from reduced health problems attributable to lead in gasoline, and (5) raise the cost of producing all gasoline by an average of 2 cents per gallon. The EPA has proposed that all lead be banned from gasoline by 1988, but this proposal is being strongly opposed by makers of lead gasoline additives.

Mercury Get a tooth filled, flip a silent light switch, install an automatic furnace or air conditioner, use fluorescent lights, and you are depending on mercury—a chemical that has been used in various forms for more than 27 centuries. It is also used in some paints, floor waxes, and furniture polishes, in antibacterial and antimildew agents, in medicines and in fungicides for seeds, and in making plastics, paper, clothing, and camera film.

Human input of mercury into the air and water has been increasing, especially because of the burning of coal (which contains mercury as a trace contaminant) and by deliberate and accidental discharges into rivers, streams, and lakes. These inputs, however, are relatively small compared to natural inputs of mercury into the environment due to vaporization from the earth's crust and from the vast amounts naturally stored as bottom sediments in the ocean. It is dangerous to consume large amounts of pike, tuna, swordfish, and other large ocean species that contain high levels of mercury. But most, if not all, of this mercury comes from natural sources, and the danger has probably always been present. Although the human input into the ocean is insignificant, some lakes, rivers, bays, and estuaries near mercury-using industries are being threatened.

Elemental mercury (Hg) is not a dangerous poison unless vaporized and inhaled directly over a fairly long period. While preparing fillings, for example, dentists and dental workers can be exposed to high levels of mercury vapor because of

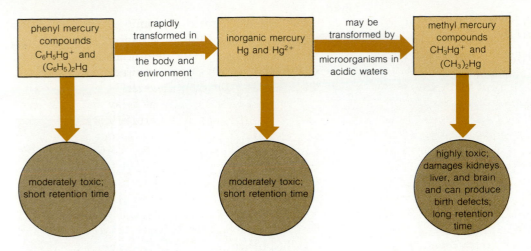

Figure 24-2 Some chemical forms of mercury and how they may be transformed.

leaky containers, spills, and poor ventilation. This condition can cause neurologic problems, such as anxiety or depression, and even death.

The major threat from mercury is an extremely toxic organic mercury compound known as methyl mercury (CH_3Hg^+). It stays in the body more than 10 times longer than metallic mercury, can attack the central nervous system, kidneys, liver, and brain tissue, and can cause birth defects. The major forms of mercury and the ways they are transformed from one form to another are summarized in Figure 24-2.

Between 1953 and 1960, 52 people died and 150 suffered serious brain and nerve damage from methyl mercury discharged into Minamata Bay, Japan, by a nearby chemical plant. Most of the victims in this seaside village area ate fish contaminated with methyl mercury three times a day. In 1969 a New Mexico farm laborer fed seed grain treated with methyl mercury to his hogs. After he and his family ate meat from these animals, three of the children became severely crippled. A fourth child, poisoned in his mother's womb, was born blind and later was found to be mentally retarded. In Iraq in 1972, a large shipment of seed grain treated with methyl mercury was distributed to villagers, who fed it to animals and used it to bake bread. It was reported that 459 people died and estimated that 6,530 were injured.

In 1969 two Swedish scientists found that anaerobic (non-oxygen-requiring) bacteria dwelling in bottom mud could convert relatively harmless elemental mercury and inorganic mercury (Hg^{2+}) salts into highly toxic methyl mercury (Figure 24-2). Fortunately, most waters apparently are not acidic enough to enable this transformation. However, we still know far too little about the complex chemistry of mercury in living systems, and the increases in acidity of many lakes from acid deposition (Sections 21-4 and 22-4) may aggravate this problem.

24-6 Control and Management of Hazardous Wastes

Methods for Dealing with Hazardous Wastes

The major options for dealing with hazardous wastes include: **(1)** long-term storage in secured landfills, surface impoundments such as lagoons, underground geologic formations (primarily for radioactive wastes), and injection into underground wells, **(2)** conversion to less hazardous or nonhazardous materials by chemical or physical processes, spreading wastes on the land where they can decompose biologically, or incineration on land or at sea using especially designed incinerator ships, and **(3)** reduction or elimination of waste materials through modification of industrial processes and recycling and reuse.

Land Disposal of Hazardous Wastes So far the primary method of handling about 95 percent of the regulated hazardous wastes in the United States has been land disposal—by putting the waste into surface impoundments or landfills (secured and unsecured) or injecting them into deep underground wells. This approach is relatively cheap and until recently was not regulated. However, increased regulation, higher land costs, and vigorous opposition by citizens who want hazardous wastes disposed of but not near them have made land disposal a less attractive option. In most communities, NMBY (Not in My Backyard) has become the political battle cry of citizens opposing waste dumps.

Since 1976 the Resource Conservation and Recovery Act has required that any landfill used for the storage of hazardous wastes be a secure landfill, that operators of such landfills show financial responsibility for up to $10 million in damages due to accidents, and that the facilities be monitored

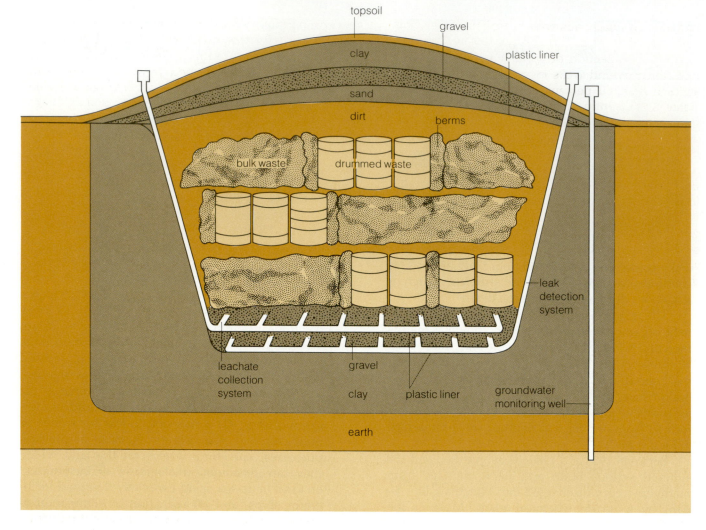

Figure 24-3 A secure landfill for the long-term storage of hazardous wastes.

for at least 30 years to minimize the chance of hazardous wastes escaping into the environment. Figure 24-3 shows a typical *secure landfill*. Ideally a secure landfill is sited in thick natural clay deposits, isolated from surface or subsurface water supplies, not expected to be subjected to flooding, earthquakes, or other disruptions, and so located that the transport of leachate to an underground water source is unlikely.

In addition, such a landfill has: **(1)** two plastic liners about 30 times the thickness of a plastic trash bag to help prevent leakage, **(2)** a bottom layer of gravel and perforated pipes to collect the leachate that inevitably seeps to the bottom of the pit and should be pumped to the surface for treatment, **(3)** drainage ditches to prevent flooding from surface water, **(4)** a cover consisting of another plastic liner plus layers of sand, gravel, and packed clay shaped to divert rainwater, and **(5)** a monitoring well system to check for chemicals leaking into groundwater. For even greater security, materials placed in the landfill can be solidified to reduce their volume and encased in cement, asphalt, glass, or organic polymers to decrease the chance of escape into the air or water.

In evaluating hazardous waste control methods in 1983, the Office of Technology Assessment stated: "All land disposal methods will eventually fail." The only question is, when. Sooner or later any secure landfill will leak from tears in the plastic liners caused by bulldozers or freezing temperatures, leachate disintegration of the liner, crushing of leachate collection pipes by the weight of the waste, clogging of the perforations in leachate collection pipes by debris, or disruption of the protective cover by erosion, new construction on the site, or subsidence. This finding is supported by a National Academy of Sciences report that concluded that land disposal should be the method of last resort.

In West Germany 60 percent of the hazardous wastes produced are detoxified rather than being dumped. In Denmark, where 98 percent of the nation's drinking water comes from groundwater, virtually no hazardous wastes are disposed of on land.

Because injection of liquid hazardous wastes into underground wells is relatively inexpensive, an estimated 60 percent of all liquid hazardous wastes is currently disposed of this way, and the number of new wells is increasing by 20 percent a year. Deep-well injection is now permitted in about 20 states, with many other states under local pressure to allow this form of disposal. However, unless injection wells are located in carefully selected and monitored geologically stable areas, well below groundwater supplies, there is the danger of essentially irreversible groundwater contamination (Section 22-5). The disposal of radioactive wastes in deep geologic formations is discussed in Section 18-2.

Conversion to Less Hazardous or Nonhazardous Materials Hazardous organic compounds that contain little or no toxic metal compounds, volatile materials, and persistent organic compounds can be detoxified biologically by landfarming. In *landfarming* wastes are applied onto or beneath surface soil and mixed to expose the contaminated material to oxygen; then microorganisms and nutrients are added as needed to ensure biological decomposition. This method is particularly useful for sewage, petroleum refinery, and paper mill sludges applied to forestlands and for reclamation of surface-mined land.

Other biological treatment and decomposition processes include the use of composting, trickling filters, activated sludge, and aerated lagoons (Section 22-7). There are also experiments using mutant bacteria produced by genetic manipulation to detoxify specific waste materials. But critics of this approach worry that these "superbugs" may get out of control and destroy other material before it has a chance to become waste.

Physical processes for detoxifying wastes include: **(1)** neutralization of acidic or alkaline wastes, **(2)** oxidation or reduction by chemical reactions into different substances, **(3)** removal of toxic metals and other compounds by precipitation or absorption (by a chemical such as activated carbon), and **(4)** selective removal of ions in ion exchangers by passing wastewater over a packed bed of resin.

Hazardous wastes can also be decomposed by *incineration* in the presence of oxygen at high temperatures. The Netherlands incinerates about half its hazardous waste, and the EPA estimates that about 60 percent of all U.S. hazardous wastes could be incinerated. Incineration has a number of advantages: **(1)** it can detoxify complex organic compounds by breaking them down into harmless gases; **(2)** it is potentially the safest method of disposal for most types of hazardous wastes; **(3)** it is the most effective method of waste disposal and can destroy at least 99.99 percent of organic waste material such as pesticides, solvents, and PCBs; **(4)** it greatly reduces the volume of waste (in the form of ash) to be disposed of; **(5)** large land areas are not required; and **(6)** the heat produced can be converted to steam and used for electricity or heating nearby buildings.

But there are also some disadvantages: **(1)** incineration is the most expensive method (helping explain why only about 3 percent of the hazardous wastes in the United States is burned); **(2)** the ash left over, whether toxic or nontoxic, must be disposed of; **(3)** not all hazardous wastes are combustible; and **(4)** the gaseous and particulate combustion products emitted by the incinerator can be hazardous to health if not controlled. According to the EPA only about a third of the industrial hazardous wastes incinerated in the United States is burned in environmentally acceptable facilities.

Because most citizens object to living near incinerators, interest has been growing in incinerating liquid hazardous wastes at sea in specially designed ships. This approach is about one-third cheaper than on-land incineration and minimizes the dangers to people from an accident. Since 1977 there have been several experimental burns of U.S. hazardous wastes in incinerator ships in the Pacific Ocean and the Gulf of Mexico. Though measurements indicated that the test burns were successful, several scientists contend that the measurements were inadequate and in some cases invalid.

Environmentalists have opposed the burning of wastes at sea, fearing that chemical spills as a result of accidents from human error, fog, storms, or reefs, or residue from the incomplete destruction of toxic wastes could threaten marine life. In addition, they suspect that some companies would take shortcuts to save money or would cover up accidents at sea, far away from scrutiny. They also point out that at-sea incineration works only for liquid waste and thus is not appropriate for most of the hazardous wastes generated in the United States. Despite these concerns, in 1985 the EPA proposed rules for incinerating liquid hazardous wastes at sea.

Recycling and Reuse In Europe, waste exchanges or clearinghouses are used to transfer about one-third of the waste produced by one firm to another firm that can use it as raw material. By early 1985 at least 30 regional waste exchanges in the United States were transferring about 10 percent of the listed wastes, and this fraction could be increased significantly in the future. For example, the Minnesota Mining and Manufacturing Company sells ammonium sulfate, a corrosive by-product of videotape manufacture, to fertilizer makers.

However, only about 4 percent of the toxic materials produced by U.S. industry is recycled or reused. Despite the enormous potential for recycling and waste trading, since 1979 the EPA has spent almost no money and has assigned only one person to promote this approach.

Federal Legislation and Control of Hazardous Wastes The Toxic Substances Control Act of 1976 and the Resource Conservation and Recovery Act (RCRA) of 1976, as amended in 1984, require the EPA to identify hazardous wastes, set standards for their management, and issue guidelines and provide some financial aid to establish state programs for managing such wastes. The RCRA also requires all firms that store, treat, or dispose of more than 100 kilograms (220 pounds) of hazardous wastes per month to apply to the EPA for a permit.

To reduce illegal dumping, hazardous waste producers granted disposal permits by the EPA must use a "cradle-to-grave" manifest system; that is, they must keep track of hazardous wastes from point of origin to point of disposal at licensed and inspected facilities. EPA administrators, however, point out that this requirement is almost impossible to enforce because the EPA and state regulatory agencies do not have enough personnel to review the paper trails of more than 150,000 hazardous waste generators and 15,000 haulers each year—let alone verify them and prosecute offenders.

Environmentalists argue that the Resource Conservation and Recovery acts of 1976 and 1984 have several serious loopholes and difficulties. These include: **(1)** the slowness of the EPA to inventory and identify chemicals not presently on the fairly short hazardous wastes list, **(2)** inadequate sampling and testing procedures for use by waste producers to determine whether their wastes are classified as hazardous under federal guidelines, **(3)** exclusion from the regulations of at least 4.5 billion kilograms (5 million tons) of hazardous waste discharged down sewers as domestic sewage, including most of the metal-finishing industry's highly toxic wastes, **(4)** exemption from regulation of recycled or reused hazardous wastes (such as the recycled dioxin-contaminated oil

sprayed on roads in Times Beach), **(5)** the absence of a requirement for states to regulate all the hazardous wastes identified by the EPA, which makes states with weaker programs likely choices by industry as dumping grounds for certain wastes, and **(6)** the absence of a requirement that thousands of older but still active waste sites be upgraded to the safety standards required for new disposal sites. In addition, since federal money for inspections ended in 1982, only a few states regularly monitor city and county landfills to determine whether leachate is percolating into groundwater.

In 1980 Congress passed the Comprehensive Environmental Response, Compensation and Liability Act, known as the Superfund program. The Superfund itself consisted of $1.6 billion, to be used by the EPA between 1980 and 1985, to clean up abandoned or inactive hazardous waste dump sites. The EPA is authorized to collect fines and sue the owners later (if they can be found) to recover up to three times the cleanup costs, but by early 1985 only about $7 million had been collected by the EPA. About 87 percent of the cleanup funds comes from taxes on certain chemical and petrochemical industries and the remainder from general federal tax revenues.

This legislation also requires states to provide 10 percent of the cleanup costs for sites located on private property and 50 percent of the costs for those on public land. The EPA cannot use any money from the Superfund for cleanup unless the states make their contributions. By 1985 only eight states had such cleanup funds in their budgets.

By mid-1985 the EPA had identified almost 20,000 potentially hazardous waste sites in the United States and estimated that the total could reach 22,000. By mid-1985, 840 of these sites (many of them used by the Department of Defense) were included on a national priorities list as especially hazardous and eligible for cleanup using federal funds, provided sufficient state funds were made available. The EPA estimates that the total number of especially hazardous sites on the priorities list eventually may run as high as 2,200, with the cleanup at some sites costing hundreds of millions of dollars. By mid-1985, emergency action had removed the immediate threat at 328 of the 840 priority sites. But only 20 sites had been cleaned up completely enough to be removed from the list of priority sites.

Drastic EPA budget cuts since 1981 have made it difficult to implement the Superfund legislation. In 1983 critics charged the EPA with letting some noncomplying firms off too easily and settling for superficial cleanups. Investigations by Congress in 1983 of Superfund mismanagement and alleged inside deals with some industries generating hazardous wastes led to the firing of the director of the program and the resignation of the head of the EPA.

The EPA estimates that the $1.6 billion provided in the Superfund for use through 1985 will probably clean up no more than 100 abandoned dump sites—only a fraction of 786 to 2,200 priority sites the EPA estimates will need cleanup, long-term monitoring, or both. The estimated cost for cleaning up all the present priority hazardous waste dump sites and those that are projected to be added to the priority list ranges from $12 billion to $260 billion, with a 1985 study by the Office of Technology estimating a cost as high as $100 billion. A 1983 study by the Office of Technology Assessment concluded that the Superfund program might be ineffective in the long term because wastes are simply moved from one burial site to another, and leakage eventually will occur.

Waste is a human concept. In nature nothing is wasted, for everything is part of a continuous cycle. Even the death of a creature provides nutrients that will eventually be reincorporated in the chain of life.

Denis Hayes

Discussion Topics

1. List the advantages and disadvantages of each of the following methods for disposal of solid waste: **(a)** open dumping, **(b)** sanitary landfill, **(c)** incineration, and **(d)** composting.

2. How is solid waste collected and disposed of in your community? Are the land waste disposal sites true sanitary landfills or merely landfills? Does the community have any secured landfill sites for disposal of hazardous wastes?

3. List the advantages and disadvantages of the high-technology (resource recovery plant) and the low-technology (source separation) approaches to recycling. Would you favor requiring all households and businesses to separate recyclable materials? Defend your answer.

4. Investigate to determine whether **(a)** your college and your city have recycling programs, **(b)** your college and your local government require that a certain fraction of all paper purchases contain recycled fiber, **(c)** teachers in your college and in local schools expect everyone to write on both sides of paper, **(d)** your college sells soft drinks in throwaway cans or bottles, and **(e)** your state has or is contemplating a law requiring deposits on all beverage containers.

5. Give your reasons for agreeing or disagreeing with each of the following proposals for dealing with solid waste in the United States.
   a. Remove all subsidies, preferential transportation charges, depletion allowances, and tax breaks for primary materials industries.
   b. Provide tax breaks and incentives for recycling industries and for all manufacturers who use recycled materials.
   c. Pass a law requiring deposits on all beverage containers, phasing gradually to a complete ban on all nonreturnable beverage containers.
   d. Give tax breaks on comparable products that use smaller amounts of resources per unit or last longer.
   e. Standardize package sizes for each item and add a tax for excess packaging.
   f. Allow taxpayers to deduct the cost of appliance repairs from their taxable income to discourage disposal of such items.
   g. Require all households and businesses to separate trash into recyclable components.
   h. Require that all products be labeled to show the amount and type of recycled materials.
   i. Require local, state, and federal agencies to buy materials composed of the highest available percentage of recycled materials.

6. Would you oppose the location of a secured landfill for the storage of hazardous wastes on property near your home? If you oppose such a site, what alternatives would you suggest? Would you oppose the location of an incinerator for the decomposition of hazardous wastes near your home? Why or why not?

7. Criticize the following statements:
   a. Because the average lead level in blood tends to be below the level for classical lead poisoning, there is no serious problem.
   b. Because there is no absolute proof that lead in gasoline has killed any Americans, there is no need to reduce or phase out the use of lead compounds in gasoline.
   c. Because human activities are not the major source of mercury pollution in the ocean, there is no cause for concern.
   d. We should no longer eat tuna and swordfish.
   e. Since metallic mercury is not highly toxic to humans, we have little to fear.

8. Give your reasons for agreeing or disagreeing with each of the following proposals for dealing with hazardous wastes in the United States.
   a. Burn all liquid hazardous wastes in federally approved at-sea incinerator ships.
   b. Reduce the use of hazardous wastes and encourage recycling and reuse of such materials by

levying a tax or fee on producers for each quantity of such wastes generated.

c. Ban or impose a higher fee on the land disposal of high-priority hazardous wastes as a means of encouraging recycling, reuse, treatment, and destruction technologies for dealing with hazardous wastes.

d. Provide low-interest loans, tax breaks, and other financial incentives for industries producing hazardous wastes to encourage the recycling, reuse, treatment, destruction, and reduced production of such wastes.

Epilogue

Achieving a Sustainable Earth Society

When there is no dream, the people perish.
Proverbs 29:18

The frontier or throwaway mentality sees the earth as a place of unlimited room and resources, where ever-increasing production, consumption, and technology inevitably lead to a better life for everyone. If we pollute one area, we merely move to another or eliminate or control the pollution through technology. This mentality represents an attempt to dominate nature.

Many environmentalists argue that over the next 50 years or so we must change from our present *throwaway* or *frontier rules* to a new set of *sustainable earth* or *conserver rules* designed to maintain the earth's vital life-support systems not only for our own sustenance, well-being, and security, but also to fulfill our obligation to pass this vital legacy along to future generations.

In contrast to the old or new frontier mind-set, a sustainable earth or conserver mentality sees that the earth is a place of limited room and resources and that ever-increasing production and consumption can put severe stress on the natural processes that renew and maintain the air, water, and soil on which we depend. Sustaining the earth calls for cooperating with nature, rather than blindly attempting to dominate it.

Achieving a sustainable earth world view involves working our way through four levels of environmental awareness summarized in the box on page 362.

But the sustainable earth view by itself is not enough. It is unrealistic to expect poor people living at the margin of existence to think about the long-term survival of the planet. When people need to burn wood to keep from freezing, they will cut down trees. When their livestock face starvation, they will overgraze grasslands. For this reason, analysts argue that an equally important element in the transition to a sustainable earth society is generous and effective aid from the MDCs to LDCs that are in need of grants, loans, and technical advice. According to these observers, such aid should help the poorer nations to become more self-reliant rather than causing them to become increasingly dependent on the MDCs for goods and services.

We can read and talk about environmental problems, but finally it comes down to what you and I are willing to do individually and collectively. We can begin at the individual level and work outward by joining with others to amplify our actions. This is the way the world is changed. Envision the world as made up of all kinds of cycles and flows in a beautiful and diverse web of interrelationships and a kaleidoscope of patterns and rhythms whose very complexity and multitude of potentials remind us that cooperation, honesty, humility, and love must be the guidelines for our behavior toward one another and the earth.

It is not too late. There is time to deal with the complex environmental problems we face if enough of us really care. It's not up to "them," it's up to "us." Don't wait.

The main ingredients of an environmental ethic are caring about the planet and all of its inhabitants, allowing unselfishness to control the immediate self-interest that harms others, and living each day so as to leave the lightest possible footprints on the planet.
Robert Cahn

Four Levels of Environmental Awareness

First Level: Pollution

Discovering the symptoms. At this level, we must point out and try to stop irresponsible acts of pollution by individuals and large and small organizations, and we must resist being duped by slick corporate advertising. But we must at the same time change our own life-styles. We have all been drilling holes in the bottom of the boat. Arguing over who is drilling the biggest hole only diverts us from working together to keep the boat from sinking. The problem of remaining at the pollution level is that individuals and industries see their own impacts as too tiny to matter, not realizing that millions of individual impacts acting together can overload our life-support systems. Remaining at the pollution awareness level also leads people to see the crisis as a problem comparable to a "moon shot," and to look for a quick technological solution: Have technology fix us up, send me the bill at the end of the month, but don't ask me to change my way of living.

Second Level: Neo-Malthusian Overpopulation

Recognizing that the cause of pollution is not just people but their level of consumption and the environmental impact of various types of production. At the second level the answers seem obvious. We must simultaneously reduce both the number of people in MDCs and LDCs and wasteful consumption of matter and energy resources—especially in the MDCs, which, with less than 30 percent of the world's population, account for about 90 percent of environmental pollution worldwide.

Third Level: Spaceship Earth (Shallow Ecology)

Becoming aware that population and resource use will not be controlled until a reasonable number of leaders and citizens recognize that protecting and preserving the life-support systems that sustain all life must be our primary aim. The goal at this level is to use technology, economics, and conventional politics to control population growth, pollution, and resource depletion to prevent ecological overload. Some argue that the spaceship earth level is a sophisticated expression of our arrogance toward nature—the idea that through technology we can control nature and create artificial environments to avoid environmental overload. They point out that using this approach eventually will pose a dire threat to individual human freedom because to protect the life-support systems that are necessary in space, a centralized authority (ground control) rigidly controls astronauts' lives. Instead of novelty, spontaneity, joy, and freedom, the spaceship model is based on cultural homogenization, social regimentation, artificiality, monotony, and gadgetry. In addition, it is argued that this approach can cause environmental overload in the long run because it is based on the false idea that we have essentially complete understanding of how nature works.

Fourth Level: Sustainable Earth (Deep Ecology)

Recognizing five fundamental conditions. (1) everyone and every living species is interconnected; (2) the role of humans is not to rule and control nature but to work with nature and selectively control relatively small parts of nature on the basis of ecological understanding; (3) because the living organisms on earth and their interactions are so diverse and complex that we will never fully understand them, attempts at excessive control will sooner or later backfire (Section 6-3); (4) our major goal should be to preserve the ecological integrity, stability, and diversity of the life-support systems for all living organisms; and (5) since all living species by virtue of their existence have an inalienable right to life in their natural environments, the forces of biological evolution, not human technological control, should determine which species live or die.

Appendix 1

Periodicals

The following journals will aid the intelligent citizen in keeping well informed and up to date on environmental problems. Those marked with an asterisk are recommended as basic reading. Subscription prices, which tend to change, are not given.

American Forests, American Forestry Association, 1319 18th St., N.W., Washington, DC 20036. Popular treatment, "seeks to promote an enlightened public appreciation of natural resources."

Audubon, National Audubon Society, 950 Third Ave., New York, NY 10022. Conservationist viewpoint; covers more than bird-watching. Good popularizer of environmental concerns; well-produced, sophisticated graphics.

BioScience, published monthly by the American Institute of Biological Sciences, 1401 Wilson Blvd., Arlington, VA 22209. Official publication of AIBS; gives major coverage to biological aspects of the environment, including population. Style ranges from semipopular to technical. Features and news sections attentive to legislative and governmental issues.

Bulletin of the Atomic Scientists, 935 East 60th St., Chicago, IL 60637. In recent years has increased coverage of environmental issues, particularly in relation to nuclear power and nuclear testing and fallout.

Catalyst for Environmental/Energy, 274 Madison Ave., New York, NY 10016. High-level, popular treatment; substantial articles on all aspects of environment, including population control. Reviews books and films suited to environmental education.

Ceres, Food and Agricultural Organization of the United Nations (FAO), UNIPUB, Inc., 650 First Avenue, P.O. Box 433, New York, NY 10016. Contains articles on the population–food problem.

The CoEvolution Quarterly, P.O. Box 428, Sausalito, CA 94965. Covers a wide range of environmental and self-sufficiency topics. Also publishes *The New Whole Earth Catalog* (1980).

Conservation Foundation Letter, The Conservation Foundation, 1717 Massachusetts Ave., N.W., Washington, DC 20036. Usually 12 pages long. Good summaries of key issues.

Conservation News, National Wildlife Foundation, 1412 16th St., N.W., Washington, DC 20036. Good coverage of wildlife issues.

Design and Environment, 355 Lexington Ave., New York, NY 10017. Useful for architects, engineers, and city planners.

Earth Shelter Digest, 479 Fort Road, St. Paul, MN 55102. Gives the latest information on earth-sheltered (underground) housing.

The Ecologist, Ecosystems Ltd., 73 Molesworth St., Wadebridge, Cornway PL27 7DS, United Kingdom. Wide range of articles on environmental issues from an international viewpoint.

Ecology, Ecological Society of America, Dr. Ralph E. Good, Business Manager, Department of Biology, Rutgers University, Camden, NJ 08102. Good source of information on more technical ecology research.

Ecology Law Quarterly, University of California, Boalt Hall School of Law, Berkeley, CA 94720. Good treatment of latest developments in environmental law.

Ekistics, Athens Center of Ekistics, 24 Strat Syndesmou, Athens 136, Greece. Reviews the problems and science of human settlements. Reflects ideas of such planners as Constantine Doxiadis, the late R. Buckminster Fuller, and the late John McHale.

Environment, Heldref Publications, 4000 Albemarle St., N.W., Washington, DC 20016. Seeks to put environmental information before the public. Excellent in-depth articles on key issues.

Environmental Abstracts, Environment Information Center, Inc., 48 West 38th St., New York, NY 10018. Compilation of environmental abstracts; basic bibliographic tool. Too expensive for individual subscription but should be available in your library.

Environmental Action, Room 731, 1346 Connecticut Ave., N.W., Washington, DC 20036. Political orientation. Excellent coverage of environmental issues from legal, political, and social action viewpoints.

The Environmental Professional, Editorial Office, Department of Geography, University of Iowa, Iowa City, IA 52242. Excellent discussion of environmental issues.

Environmental Science & Technology, American Chemical Society, 1155 16th St., N.W., Washington, DC 20036. Emphasis on water, air, and solid waste chemistry. Basic reference for keeping up to date on technological developments.

EPA Journal, Environmental Protection Agency. Order from Government Printing Office, Washington, DC 20402. Broad coverage of environmental issues and updates on EPA activities.

Family Planning Perspectives, Planned Parenthood–World Population, Editorial Offices, 666 Fifth Ave., New York, NY 10019. Excellent coverage of population issues and latest information on birth control methods.

The Futurist, World Future Society, P.O. Box 19285, Twentieth Street Station, Washington, DC 20036. Covers wide range of societal problems, including environmental, population, and food issues. A fascinating and readable journal.

Impact of Science on Society, UNESCO, 317 East 34th St., New York, NY 10016. Essays on the social consequences of science and technology.

Journal of the Air Pollution Control Association, 4400 5th Ave., Pittsburgh, PA 15213. Technical research articles.

Journal of the American Public Health Association, 1015 18th St., N.W., Washington, DC 20036. Some coverage of environmental health issues.

Journal of Environmental Education, Heldref Publications, 4000 Albemarle St., N.W., Suite 504, Washington, DC 20016. Good for teachers.

Journal of Environmental Health, National Environmental Health Association, 1600 Pennsylvania Ave., Denver, CO 80203. Good coverage of technical research.

Journal of the Water Pollution Control Federation, 2626 Pennsylvania Ave., N.W., Washington, DC 20037. Technical research articles.

Journal of Wildlife Management, Wildlife Society, Suite 611, 7101 Wisconsin Ave., N.W., Washington, DC 20014. Good coverage of basic issues and information.

Living Wilderness, The Wilderness Society, 1901 Pennsylvania Ave., N.W., Washington, DC 20006. Strong statement of "wild areas" viewpoint.

Mother Earth News, P.O. Box 70, Hendersonville, NC 28739. Superb articles on organic farming, alternative energy systems, and alternative life-styles.

National Parks and Conservation Magazine, National Parks and Conservation Association, 1701 18th St., N.W., Washington, DC 20009. Good coverage of parks and wildlife issues.

National Wildlife, National Wildlife Federation, 1412 16th St., N.W., Washington, DC 20036. Good summaries of issues with wildlife emphasis. Action oriented, with a "Washington report."

Natural History, American Museum of Natural History, Central Park West at 79th St., New York, NY 10024. Popular; wide school and library circulation. Regularly concerned with environment.

Nature, 711 National Press Building, Washington, DC 20045. British equivalent to *Science*; enjoys outstanding reputation.

New Scientist, 128 Long Acre, London, WC 2, England. Excellent general science journal with extensive coverage of environmental issues.

Not Man Apart, Friends of the Earth, 1245 Spear Street, San Francisco, CA 94105. Excellent capsule summaries and a few in-depth articles on national and international environmental issues.

Organic Gardening & Farming Magazine, Rodale Press, Inc., 33 E. Minor St., Emmaus, PA 18049. The best guide to organic gardening.

Pollution Abstracts, Data Courier, Inc., 620 S. 5th St., Louisville, KY 40202. Basic bibliographic tool. Too expensive for individual subscription but should be available in your library.

Population and Vital Statistics Report, UN Publications Sales Section, New York, NY 10017. Latest world figures.

Population Bulletin, Population Reference Bureau, 2213 M St., N.W., Washington, DC 20037. Nontechnical articles on population issues. Highly recommended.

Population Bulletin, UN Publications Sales Section, New York, NY 10017. Statistical summaries. English and French editions.

Resources, Resources for the Future, Inc., 1755 Massachusetts Ave., N.W., Washington, DC 20036. Free on request; summarizes information and research on natural resources.

Science, American Association for the Advancement of Science, 1515 Massachusetts Ave., N.W., Washington, DC 20036. A basic resource. Probably the single best source for key environmental articles.

Science News, Science Service, Inc., 1719 N St., N.W., Washington, DC 20036. Good popular summaries of scientific developments, including environmental topics.

Scientific American, 415 Madison Ave., New York, NY 10017. Outstanding journal for the intelligent citizen who wants to keep up with science. Many general articles on environment and ecology.

The Sierra Club Bulletin, 530 Bush St., San Francisco, CA 94108. Excellent coverage of a wide range of environmental problems and of citizen action. Beautiful photographs.

Technology Review, Room E219-430, Massachusetts Institute of Technology, Cambridge, MA 02139. Not specialized or always technical, but addressed to a sophisticated audience. In recent years has devoted more than half its pages to environmentally related material; also strong on issues of science policy.

UNESCO Courier, UNESCO Publications Center, 317 East 34th St., New York, NY 10016. A magazine for the general reader; frequently attentive to environmental issues.

Worldwatch Papers, Worldwatch Institute, 1776 Massachusetts Ave., N.W., Washington, DC 20036. A series of reports designed to serve as an early warning system on major environmental problems. Worldwatch also publishes an annual *State of the Environment* in book form. Highly recommended.

Appendix 2

Units of Measurement

Length
Metric
1 kilometer (km) = 1,000 meters (m)
1 meter (m) = 100 centimeters (cm)
1 meter (m) = 1,000 millimeters (mm)
1 centimeter (cm) = 0.01 meter (m)
1 millimeter (mm) = 0.001 meter (m)
English
1 foot (ft) = 12 inches (in.)
1 yard (yd) = 3 feet (ft)
1 mile (mi) = 5,280 feet (ft)
Metric-English
1 kilometer (km) = 0.621 mile (mi)
1 meter (m) = 39.4 inches (in.)
1 inch (in.) = 2.54 centimeters (cm)
1 foot (ft) = 0.305 meter (m)
1 yard (yd) = 0.914 meter (m)

Area
Metric
1 square kilometer (km^2) = 1,000,000 square meters (m^2)
1 square meter (m^2) = 1,000,000 square millimeters (mm^2)
1 hectare (ha) = 10,000 square meters (m^2)
1 hectare (ha) = 0.01 square kilometer (km^2)
English
1 square foot (ft^2) = 144 square inches (in.^{2})
1 square yard (yd^2) = 9 square feet (ft^2)
1 square mile (mi^2) = 27,880,000 square feet (ft^2)
1 acre (ac) = 43,560 square feet (ft^2)
Metric-English
1 hectare (ha) = 2.471 acres (ac)
1 square kilometer (km^2) = 0.386 square mile (mi^2)
1 square meter (m^2) = 1.196 square yards (yd^2)
1 square meter (m^2) = 10.76 square feet (ft^2)
1 square centimeter (cm^2) = 0.155 square inch (in.^{2})

Volume
Metric
1 cubic kilometer (km^3) = 1,000,000 cubic meters (m^3)
1 cubic meter (m^3) = 1,000,000 cubic centimeters (cm^3)
1 liter (L) = 1,000 milliliters (mL) = 1,000 cubic centimeters (cm^3)
1 milliliter (mL) = 0.001 liter (L)
1 milliliter (mL) = 1 cubic centimeter (cm^3)
English
1 gallon (gal) = 4 quarts (qt)
1 quart (qt) = 2 pints (pt)
Metric-English
1 liter (L) = 0.265 gallon (gal)
1 liter (L) = 1.06 quarts (qt)
1 liter (L) = 0.0353 cubic foot (ft^3)
1 cubic meter (m^3) = 35.3 cubic feet (ft^3)
1 cubic meter (m^3) = 0.765 cubic yard (yd^3)
1 cubic kilometer (km^3) = 0.24 cubic mile (mi^3)
1 barrel (bbl) = 159 liters (L)
1 barrel (bbl) = 42 U.S. gallons (gal)

Mass
Metric
1 kilogram (kg) = 1,000 grams (g)
1 gram (g) = 1,000 milligrams (mg)
1 gram (g) = 1,000,000 micrograms (μg)
1 milligram (mg) = 0.001 gram (g)
1 microgram (μg) = 0.000001 gram (g)
1 metric ton (mt) = 1,000 kilograms (kg)
English
1 ton (t) = 2,000 pounds (lb)
1 pound (lb) = 16 ounces (oz)
Metric-English
1 metric ton = 2,200 pounds (lb)
1 kilogram (kg) = 2.20 pounds (lb)
1 pound (lb) = 454 grams (g)
1 gram (g) = 0.035 ounce (oz)

Energy and Power
Metric
1 kilojoule (kJ) = 1,000 joules (J)
1 kilocalorie (kcal) = 1,000 calories (cal)
1 calorie (cal) = 4.184 joules (J)
Metric-English
1 kilojoule (kJ) = 0.949 British thermal unit (Btu)
1 kilojoule (kJ) = 0.000278 kilowatt-hour (kW-h)
1 kilocalorie (kcal) = 3.97 British thermal units (Btu)
1 kilocalorie (kcal) = 0.00116 kilowatt-hour (kW-h)
1 kilowatt-hour (kW-h) = 860 kilocalories (kcal)
1 kilowatt-hour (kW-h) = 3,400 British thermal units (Btu)
1 quad (Q) = 1,050,000,000,000,000 kilojoules (kJ)
1 quad (Q) = 2,930,000,000,000 kilowatt-hours (kW-h)
Approximate crude oil equivalent
1 barrel (bbl) crude oil = 6,000,000 kilojoules (kJ)
1 barrel (bbl) crude oil = 2,000,000 kilocalories (kcal)
1 barrel (bbl) crude oil = 6,000,000 British thermal units (Btu)
1 barrel (bbl) crude oil = 2,000 kilowatt-hours (kW-h)
Approximate natural gas equivalent
1 cubic foot (ft^3) natural gas = 1,000 kilojoules (kJ)
1 cubic foot (ft^3) natural gas = 260 kilocalories (kcal)
1 cubic foot (ft^3) natural gas = 1,000 British thermal units (Btu)
1 cubic foot (ft^3) natural gas = 0.3 kilowatt-hour (kW-h)
Approximate hard coal equivalent
1 ton (t) coal = 20,000,000 kilojoules (kJ)
1 ton (t) coal = 6,000,000 kilocalories (kcal)
1 ton (t) coal = 20,000,000 British thermal units (Btu)
1 ton (t) coal = 6,000 kilowatt-hours (kW-h)

Temperature Conversions
Fahrenheit (°F) to Celsius (°C):

$$°C = \frac{(°F - 32.0)}{1.80}$$

Celsius (°C) to Fahrenheit (°F):

$$°F = (°C \times 1.80) + 32.0$$

Appendix 3

Major U.S. Environmental Legislation

General

National Environmental Policy Act of 1969 (NEPA)

Energy

National Energy Acts of 1978 and 1980

Water Quality

Federal Water Pollution Control Act of 1972

Ocean Dumping Act of 1972

Safe Drinking Water Act of 1974

Toxic Substances Control Act of 1976

Clean Water Act of 1977

Air Quality

Clean Air Act of 1965

Clean Air Act of 1970

Clean Air Act of 1977

Noise Control

Noise Control Act of 1972

Quiet Communities Act of 1978

Resources and Solid Waste Management

Solid Waste Disposal Act of 1965

Resource Recovery Act of 1970

Resource Conservation and Recovery Act of 1976

Wildlife

Species Conservation Act of 1966

Federal Insecticide, Fungicide, and Rodenticide Control Act of 1972

Marine Protection, Research, and Sanctuaries Act of 1972

Endangered Species Act of 1973

Land Use

Multiple Use Sustained Yield Act of 1960

Wilderness Act of 1964

Wild and Scenic River Act of 1968

National Coastal Zone Management Acts of 1972 and 1980

Forest Reserves Management Act of 1974

Forest Reserves Management Act of 1976

National Forest Management Act of 1976

Surface Mining Control and Reclamation Act of 1977

Endangered American Wildnerness Act of 1978

Alaskan Land-Use Bill of 1980

Appendix 4

Ways to Save Water

In the bathroom (65 percent of residential water use)

1. Begin with the toilet, which accounts for about 40 percent of all water used in a home. To reduce the amount of water used per flush, put a plastic container with its top cut off and weighted with a few stones into each toilet tank or buy (for about $10) and insert a toilet dam made of plastic and rubber; bricks also work but tend to disintegrate and gum up the water. If every toilet in the United States had such a device, about 19 million liters (5 million gallons) of water would be saved each day. Flush only when necessary, and don't use your toilet for a wastebasket or an ashtray. For new houses, install water-saving toilets or, where health codes permit, waterless or composting toilets.

2. Take short showers rather than baths (which use more water than a 5-minute shower), or fill the bathtub to the minimal water level. Use water-saving shower flow restrictors (which cost less than a dollar and can be installed in a few minutes by almost anyone), and shower by wetting down, turning off the water while you soap up, and then rinsing off.

3. Check for toilet, shower, and sink leaks frequently and repair them promptly.

4. Don't keep the water running when brushing your teeth, shaving, or washing.

In the laundry room (15 percent of residential water use)

1. Wash only when you have a full load, using the short cycle and filling the machine to the lowest possible water level.

2. When buying a new washer, get one that uses the least amount of water and fills up to different levels for loads of different sizes.

3. Check for leaks frequently and repair all leaks promptly.

In the kitchen (10 percent of residential water use for drinking and cooking)

1. Use an automatic dishwasher only when you have a full load, using the short cycle and letting the dishes air-dry overnight to save energy.

2. When washing dishes by hand, don't let the faucet run. Instead use one filled dishpan for washing and another for rinsing.

3. Keep a jug of cold water in the refrigerator rather than running drinkable water from a tap until it gets cold.

4. While waiting for faucet water to get hot, catch the cool water in a pan and use it for cooking or to water plants.

5. Check for sink and dishwasher leaks frequently and repair them promptly.

6. Try not to use a garbage disposal or water-softening system (both are major water users).

Outdoors (10 percent of residential use but higher in arid areas)

1. If you own a car, wash it less frequently.

2. Wash vehicles from a bucket of soapy water; use the hose only for rinsing.

3. Sweep walks and driveways instead of hosing them off.

4. Water lawns and gardens in the early morning or in the evening, not in the heat of midday and not when it is windy. Better yet, landscape with native plants instead of grass or plants that need watering.

5. Do your utmost to conserve water in summer and during hot periods, when stream flows are low and demands are great.

Appendix 5

Ways to Save Energy

Home Heating

1. Have the local power company or a qualified energy auditor inspect your home (sometimes free but usually costs $10 to $20) to determine the best ways to improve energy efficiency (15 to 20% savings, depending on what is done).

2. Turn thermostat down to 18°C (65°F) during the day and 13°C (55°F) at night, use a humidifier to provide comfort at lower temperatures, wear sweaters or insulated underwear indoors, and use sleeping bags or extra nonelectric blankets at night [saves 1.6%/°C (3%/°F)].

3. Close off and do not heat closets and unused rooms (variable savings).

4. Add extra insulation in ceilings, walls, and floors, and insulate heating ducts (20 to 50% savings).

5. Caulk and weatherstrip leaky windows, doors, and electric receptacles (10 to 30% savings).

6. Cover windows with insulated draperies, quilts, or shutters at night and during cloudy days; open the covers for windows exposed to the sun during the day, and keep the windows clean to ensure maximum solar gain (5 to 25% savings).

7. Have furnace cleaned and tuned at least once a year, clean furnace air intake filters at least every 2 weeks and replace as needed, and clean thermostat once a year to remove dust (10 to 20% savings).

8. Do not use a conventional fireplace for heating unless the doors and heating ducts in the room where it is located are covered and a window closest to the fireplace is cracked about 1 cm (0.5 in.) to reduce overall loss of heat from the house (10 to 20% savings).

9. Add storm windows and doors or double-airlock entry doors (10 to 15% savings).

10. When fireplace is not in use, cover opening with a glass screen or insulated cover and close dampers (8% savings).

11. Do not block heating duct outlets with drapes, furniture, or other items (variable savings).

12. Use kitchen and bathroom exhaust fans to remove house heat only when absolutely necessary, since they remove house heat (variable savings).

13. After taking a bath or washing dishes, let the hot water stand in the tub or sink until it cools to room temperature so that heat is added to house rather than going down the drain.

Home Cooling (variable savings)

1. Add extra insulation and make house airtight (see items 4 and 5 for home heating).

2. Use natural breezes as much as possible by opening windows and doors.

3. Shade windows from direct sunlight by using awnings, overhangs, leaf-shedding (deciduous) trees, vines, or insulated drapes or shutters with light colors to reflect sunlight.

4. Use a manual or automatic clock to raise minimum temperature for an air-conditioned house to 20°C (78°F) during the day and a few degrees higher at night, and wear light clothing indoors.

5. Close insulated drapes, quilts, or shutters on the sunny side of the house during the day and open them at night.

6. Use energy-efficient ceiling fans, small room fans, or whole-house window or attic fans to eliminate or reduce the need for air conditioning.

7. Open windows at night to bring in cooler night air, close them during hot days, keep closet doors closed, and close off rooms on the nonsunny side of the house during the day to keep them several degrees cooler.

8. Close off doors and air conditioning ducts to unused rooms.

9. In warm climates use light-colored roofing and outside paint to reflect incoming solar radiation.

10. During baths or showers, close bathroom doors and use exhaust fan or open bathroom window to prevent transfer of heat and humid air to rest of house.

11. Try to schedule heat- and moisture-producing activities such as bathing, mopping, ironing, and washing and drying clothes for the coolest part of the day; cover pans when cooking.

12. Turn off pilot light on gas stove (saves $20 a year).

13. To reduce heat load and electricity, use lower lighting levels, turn off lights, television sets, and other appliances as much as possible, switch to more efficient screw-in fluorescent light bulbs, use low-wattage (4 or 7 watt) night lights to light hallways, bathrooms, and other rooms when brighter lighting is not needed, and use dimmer switches and lights focused on desks and other areas to provide lighting only as needed.

Hot Water (savings variable)

1. Insulate hot water pipes and hot water heater (saves at least $60 a year).

2. Repair all leaky faucets—especially those providing hot water.

3. Install aerators on all sink faucets and flow restrictors on showerheads to reduce overall water use and hot water use.

4. Wash dishes no more than once a day in a double sink or two pans (one for washing and one for rinsing), without running water except to fill the sinks or pans; or use a dishwasher only with a full load and disconnect drying cycle to let dishes air dry.

5. Machine-wash clothes in cold or warm water using only a full load, or adjust water level for small loads. Hang

clothes on outside line during summer instead of using dryer.

6. Adjust the thermostat on hot water heater to 49°C (120°F) (15% savings) or no higher than 60°C (140°F) if you have an automatic dishwasher.

7. Once or twice a year remove sediment (which reduces efficiency) by hooking up a hose to drain pipe on hot water heater.

8. Turn off hot water heater when house will be empty for a few days.

9. Take 3- to 5-minute showers instead of baths. After getting wet, turn off water, soap down, and then turn water on again for rinsing.

10. Do not let water run while bathing, shaving, brushing teeth, or washing dishes.

11. When purchasing a new hot water heater try to buy a tankless instant gas-fired heater or an active or passive solar water heater.

12. See if your utility company provides a lower rate allowing them to install a meter with a timer that cuts off your hot water heater for a few hours a day when the utility's electric load is highest.

Appliances

1. When purchasing a new stove, refrigerator, air conditioner, dishwasher, furnace, hot water heater, or other appliance, buy the most energy-efficient model available to save energy and money on a lifetime-cost basis.

2. When possible, use a microwave oven instead of a conventional oven, especially during the summer.

3. Choose the smallest refrigerator or freezer that will meet your needs and keep it full but not overfull. Select a chest freezer rather than an upright model to prevent unnecessary loss of cool air when the door is opened.

4. Keep the condenser coils at the back of refrigerators and freezers clean and unobstructed to improve cooling efficiency.

5. If possible, do not locate a refrigerator or freezer near a stove or other source of heat.

6. Thaw frozen foods before cooking.

7. Don't place hot foods in the refrigerator.

Transportation

1. Walk or use a bicycle or moped as much as possible, especially for trips less than 8 kilometers (5 miles) (80 to 100% savings).

2. Use a carpool, mass transit, or paratransit as much as possible (50% or more savings).

3. Plan carefully to eliminate unnecessary trips (up to 50% savings).

4. Purchase the most energy-efficient vehicle available (30 to 70% savings).

5. Keep engine well tuned, and clean or replace air filter regularly (20 to 50% savings).

6. Obey speed limits (20% or more savings).

7. Drive smoothly (no jerks, fast starts, or excessive braking), and turn off engine if car must be idled longer than a minute (15 to 20% savings).

8. Purchase car without air conditioner or use air conditioning sparingly (9 to 20% savings).

9. Keep tires inflated to proper pressure (2 to 5% savings).

Appendix 6

National Ambient Air Quality Standards for the United States in 1985

Pollutant	Averaging Time	Primary Standard Levels [micrograms (μg) or milligrams (mg) per cubic meter (m³) and parts per million (ppm)]	Secondary Standard Levels [micrograms (μg) or milligrams (mg) per cubic meter (m³) and parts per million (ppm)]
Particulate matter	Annual (geometric mean)	75 μg/m³	60 μg/m³
	24 hours*	260 μg/m³	150 μg/m³
Sulfur oxides	Annual (arithmetic mean)	80 μg/m³ (0.03 ppm)	—
	24 hours*	365 μg/m³ (0.14 ppm)	—
	3 hours*		1300 μg/m³ (0.5 ppm)
Carbon monoxides	8 hours*	10,000 μg/m³ (9 ppm)	10,000 μg/m³ (9 ppm)
	1 hour	40 mg/m³* (35 ppm)	40 mg/m³* (35 ppm)
Nitrogen dioxide	Annual (arithmetic mean)	100 μg/m³ (0.05 ppm)	100 μg/m³ (0.05 ppm)
Ozone	1 hour	240 μg/m³ (0.12 ppm)	240 μg/m³ (0.12 ppm)
Hydrocarbons (nonmethane)[†]	3 hours (6 to 9 a.m.)	160 μg/m³ (0.24 ppm)	160 μg/m³ (0.24 ppm)
Lead	3 months	1.5 μg/m³	1.5 μg/m³

Source: Environmental Protection Agency.
*Not to be exceeded more than once a year.
[†]A non-health-related standard used as a guide for ozone control.

Federal Drinking Water Standards for the United States in 1985

Contaminant	Maximum Contaminant Level (MCL)	Possible Effects of Exceeding MCL	Contaminant	Maximum Contaminant Level (MCL)	Possible Effects of Exceeding MCL
Microbiological			**Synthetic Organic Chemicals (SOCs)**		
Coliform bacteria	No more than 4 colonies per 100 milliliters, with the arithmetic mean of all samples from a source not to exceed 1 colony per 100 milliliters	Waterborne infectious diseases	Endrin (pesticide)	0.2 ppb	Carcinogenic, toxic
			Lindane (pesticide)	4 ppb	Suspected carcinogen
			Methoxychlor (pesticide)	100 ppb	Toxic
			Toxaphene (pesticide)	5 ppb	Toxic
Inorganic Chemicals					
Arsenic	50 parts per billion (ppb)*	Carcinogenic, toxic	2,4,-D (herbicide)	100 ppb	Suspected carcinogen
Barium	1,000 ppb	Toxic	2,4,5-T or Silvex (herbicide)	10 ppb	Suspected carcinogen
Cadmium	10 ppb	Carcinogenic, toxic, biologically magnified			
			Trihalomethanes (e.g., chloroform)	100 ppb	Some are carcinogenic
Chromium	50 ppb	Carcinogenic, toxic, causes ulcers	**Radioactive Substances**		
Fluoride	1,400–2,400 ppb	Causes fluorosis	Naturally occuring:		
Lead	50 ppb	Toxic, biologically magnified	Radium-226 and radium-228	5 picocuries (pCi) per liter†	Tissue and genetic damage
Mercury	2 ppb	Toxic	Gross alpha-particle Activity	15 pCi per liter	Tissue and genetic damage
Nitrates	10,000 ppb	Toxic, especially in infants			
Selenium	10 ppb	Toxic, linked with dental caries	Human sources of beta-particle and proton activity	4 millirems per year‡	Tissue and genetic damage
Silver	50 ppb	Carcinogenic, toxic			

*One part per billion is equivalent to 0.001 milligram per liter.
†The picocurie is a standard unit of measurement of the total amount of radioactivity.
‡A millirem is 0.001 rem. The rem is a measure of a dose of radioactivity, weighted to reflect potential biological damage to the human body.

Readings

A more detailed list of readings is found in a related text: Miller, G. Tyler, Jr. 1985. *Living in the Environment*, 4th ed. Belmont, Calif.: Wadsworth.

Chapter 1   Population, Resources, and Pollution

Brown, Lester R., et al. 1984 and 1985. *State of the World*. New York: W. W. Norton. Superb overviews of environmental problems and their interrelationships published annually since 1984 by the Worldwatch Institute, Washington, D.C.

Conservation Foundation. 1984. *State of the Environment 1984*. Washington, D.C.: Conservation Foundation. Excellent overview.

Council on Environmental Quality and U.S. Department of State. 1980. *The Global 2000 Report to the President*, Vols. 1–3. Washington, D.C.: Government Printing Office. Outstanding summary of global population, resource, and pollution problems with projections to the year 2000.

Eckholm, Erik P. 1982. *Down to Earth: Environment and Human Needs*. New York: W. W. Norton. Superb summary of global environmental problems and progress being made on these problems.

Meadows, Donella, et al. 1982. *Groping in the Dark: The First Decade of Global Modeling*. New York: John Wiley. Superb summary and evaluation of the successes and failures of various global models of the interactions of population, pollution, and resource use.

Myers, Norman, ed. 1984. *Gaia: An Atlas of Planet Management*. Garden City, N.Y.: Anchor Press/Doubleday. Superb overview of the planet's resources and environmental problems with outstanding illustrations.

Simon, Julian L. 1981. *The Ultimate Resource*. Princeton, N.J.: Princeton University Press. Readable presentation of the cornucopian position that economic growth and technology will solve the world's population, resource, and pollution problems. The author makes some good points but bases most of his arguments on the false mathematical idea that all resource supplies are infinite because anything that is infinitely divisible is infinite in quantity—equivalent to saying that since you can divide the money in your bank account into an infinite number of fractions of a cent, then you have an infinite amount of money.

Simon, Julian L., and Herman Kahn, eds. 1984. *The Resourceful Earth*. New York: Basil Blackwell. Useful collection of articles by a series of experts who in general support the cornucopian position. The editors' strong cornucopian position, however, is not fully supported by the articles in this volume.

Chapter 2   Human Impact on the Earth

Conservation Foundation. 1982. *State of the Environment 1982*. Washington, DC: Conservation Foundation. Excellent summary. Documents attempts of the Reagan administration to undo much of the progress made on environmental problems during the 1970s.

Graham, Frank. 1971. *Man's Dominion: The Story of Conservation in America*. New York: M. Evans. Outstanding history of conservation.

Lash, Jonathan, et al. 1984. *A Season of Spoils*. New York: Pantheon. Well-researched book documenting antienvironmental policies during the first years of the Reagan administration.

Leopold, Aldo. 1949. *A Sand County Almanac*. New York: Oxford University Press. An environmental classic describing Leopold's ecological land-use ethic.

Marsh, George Perkins. 1864. *Man and Nature*. New York: Charles Scribner. An environmental classic considered to be one of the greatest American works on the environment.

Nash, Roderick. 1968. *The American Environment: Readings in the History of Conservation*. Reading, Mass.: Addison-Wesley. Excellent collection of articles.

Vig, Norman J., and Michael E. Craft. 1984. *Environmental Policy in the 1980s*. Washington, D.C.: Congressional Quarterly Press. Useful summary of Reagan's environmental policy.

Chapter 3   Some Matter and Energy Laws

Bent, Henry A. 1971. "Haste Makes Waste: Pollution and Entropy." *Chemistry*, vol. 44, 6–15. Excellent and very readable account of the relationship between entropy (disorder) and the environmental crisis.

Christensen, John W. 1981. *Energy, Resources, and Environment*. Dubuque, Iowa: Kendall/Hunt. Excellent overview for use at the high school level.

Fowler, John M. 1984. *Energy and the Environment*, 2nd ed. New York: McGraw-Hill. Excellent overview.

Rifkin, Jeremy. 1980. *Entropy: A New World View*. New York: Viking Press. Superb nontechnical description of the need to develop a sustainable earth society based on the second law of thermodynamics.

Chapter 4   Ecosystem Structure: What Is an Ecosystem?

Colinvaux, Paul A. 1978. *Why Big Fierce Animals Are Rare*. Princeton, N.J.: Princeton University Press. Fascinating and very readable description of major ecological principles. Highly recommended.

Ehrlich, Paul R., Anne H. Ehrlich, and John P. Holdren. 1977. *Ecoscience: Population, Resources and Environment*. San Francisco: W. H. Freeman. Superb, more detailed text at a higher level.

Kormondy, Edward J. 1984. *Concepts of Ecology*, 3rd ed. Englewood Cliffs, N.J.: Prentice-Hall. First-rate introduction at a slightly higher level.

Odum, Eugene P. 1983. *Basic Ecology*. Philadelphia: W. B. Saunders. Outstanding textbook by a prominent ecologist.

Ricklefs, Robert E. 1976. *The Economy of Nature*. Portland, Ore.: Chiron Press. Beautifully written introduction to ecology at a slightly higher level.

Smith, Robert L. 1980. *Ecology and Field Biology*, 3rd ed. New York: Harper & Row.

Outstanding basic text using the ecosystem approach.

Watt, Kenneth E. F. 1982. *Understanding the Environment*. Newton, Mass.: Allyn & Bacon. Excellent introduction at a somewhat higher level.

Chapter 5 Ecosystem Function: How Do Ecosystems Work?

See the Readings for Chapter 4.

Chapter 6 Changes in Ecosystems: What Can Happen to Ecosystems?

See also the Readings for Chapter 4.

Adams, Ruth, and Susan Culen, eds. 1982. *The Final Epidemic: Physicians and Scientists on Nuclear War*. Chicago: University of Chicago Press. Excellent summary by physicians and scientists of the effects of nuclear war.

Ehrlich, Paul R. 1980. "Variety Is the Key to Life." *Technology Review*, March/April. Important article showing the need for preserving diversity

Ehrlich, Paul R., et al. 1984. *The Cold and the Dark: The World After Nuclear War*. New York: W. W. Norton. Superb overview of ecological and health effects of nuclear war.

Epstein, Samuel S., et al. 1982. *Hazardous Waste in America*. San Francisco: Sierra Club Books. Detailed study of hazardous waste problems, with excellent description of the Love Canal episode.

Harwell, Mark A. 1984. *Nuclear Winter*. New York: Springer-Verlag. Excellent description of ecological effects of nuclear war.

Odum, Eugene P. 1969. "The Strategy of Ecosystem Development." *Science*, vol. 164, 262–270. Excellent summary of succession.

Woodwell, G. M. 1970. "Effects of Pollution on the Structure and Physiology of Ecosystems." *Science*, vol. 168, 429–433. Excellent analysis.

Chapter 7 Human Population Dynamics

Bouvier, Leon F. 1980. "America's Baby Boom Generation: The Fateful Bulge." *Population Bulletin*, April, pp. 1–35. Superb discussion of the implications of the baby boom for American society.

Bouvier, Leon F. 1984. "Planet Earth 1984–2034: A Demographic Vision." *Population Bulletin*, vol. 39, no. 1, 1–39. Outstanding overview of future population trends and problems.

Commission on Population Growth and the American Future. 1972. *Population and the American Future*. Washington, D.C.: Government Printing Office. Also available in paperback (Signet, New American Library). Important historical document showing the need for controlling U.S. population growth.

Dickenson, J. P., et al. 1983. *A Geography of the Third World*. New York: Methuen. Excellent overview.

Haupt, Arthur, and Thomas T. Kane. 1978. *The Population Handbook*. Washington, D.C.: Population Reference Bureau. Superb introduction to demographic terms and concepts.

Population Reference Bureau. Annual. *World Population Data Sheet*. Washington, D.C.: Population Reference Bureau. This concise annual summary is the source for most of the population data used in this book.

Population Reference Bureau. 1982. "U.S. Population: Where We Are; Where We're Going." *Population Bulletin*, vol. 37, no. 2. Excellent summary.

Weller, Robert, and Leon Bouvier. 1981. *Population: Demography and Policy*. New York: St. Martin's Press. Excellent college text at a somewhat higher level.

Chapter 8 Human Population Control

See also the Readings for Chapter 7.

Bouvier, Leon F. 1981. *The Impact of Immigration on U.S. Population Size*. Washington, D.C.: Population Reference Bureau. Excellent analysis.

Brown, Lester T. 1981. *Building a Sustainable Society*. New York: W. W. Norton. An outstanding discussion of the need for population control and the conservation of renewable and nonrenewable resources.

Connery, John. 1977. *Abortion: The Development of the Roman Catholic Perspective*. Chicago: Loyola University Press. Excellent overview.

Crewdson, John. 1983. *The Tarnished Door*. New York: New York Times Books. Excel-

lent overview of U.S. immigration policies and problems.

Hardin, Garrett. 1982. *Naked Emperors, Essays of a Taboo Stalker*. San Francisco: William Kaufman. Excellent series of thought-provoking essays on a variety of subjects including overpopulation and abortion.

Jaffe, Frederick S., et al. 1980. *Abortion Politics*. New York: Alan Guttmacher Institute. Excellent balanced approach.

Keyfitz, Nathan. 1984. "The Population of China." *Scientific American*. vol. 250, no. 2, 38–47. Excellent overview.

Newland, Kathleen. 1977. *Women and Population Growth: Choice Beyond Childbearing*. Washington, D.C.: Worldwatch Institute. Superb discussion of women's roles.

Teitelbaum, Michael S. 1975. "Relevance of Demographic Transition Theory for Developing Countries." *Science*, vol. 188, 420–425. Excellent summary of why the demographic transition may or may not work for today's LDCs.

Tien, H. Yuan. 1983. "China: Demographic Billionaire." *Population Bulletin*, vol. 38, no. 2, 1–42. Excellent summary of China's efforts to control its population.

Chapter 9 Soil Resources

Battie, Sandra S. 1983. *Soil Erosion: Crisis in America's Croplands*? Washington, D.C.: Conservation Foundation. Useful and objective analysis.

Beasley, R. P. 1972. *Erosion and Sediment Pollution Control*. Ames: Iowa State University Press. Excellent discussion of soil conservation.

Brown, Lester R., and Edward C. Wolf. 1984. *Soil Erosion: Quiet Crisis in the World Economy*. Washington, D.C.: Worldwatch Institute. Superb overview.

Dale, Tom, and V. G. Carter. 1955. *Topsoil and Civilization*. Norman: University of Oklahoma Press. Classic work describing soil abuse throughout history.

Sophen, C. D., and J. V. Baird. 1982. *Soils and Soil Management*. Reston, Va.: Reston Publishing Co. Excellent introductory text that is easy to read but scientifically sound.

Chapter 10 Water Resources

Ashworth, William. 1982. *Nor Any Drop To Drink*. New York: Summit Books. Outstanding overview of the water crisis in the United States.

Cousteau, Jacques-Yves, et al. 1981. *The Cousteau Almanac: An Inventory of Life on Our Water Planet*. Garden City, N.Y.: Doubleday. Superb source of information.

Postel, Sandra. 1984. *Water: Rethinking Management in an Age of Scarcity*. Washington, D.C.: Worldwatch Institute. Superb overview.

Pringle, Laurence. 1982. *Water—The Next Great Resource Battle*. New York: Macmillan. Superb overview of the water crisis in the United States.

Sheaffer, John, and Leonard Stevens. 1983. *Future Water*. New York: William Morrow. Excellent overview of U.S. water resource problems with suggested solutions.

Stokes, Bruce. 1983. "Water Shortages: The Next Energy Crisis." *The Futurist*, April, pp. 38–47. Excellent overview.

U.S. Geological Survey. 1984. *Estimated Use of Water in the United States in 1980*. Washington, D.C.: Government Printing Office. Useful source of data.

U.S. Water Resources Council. 1979. *The Nation's Water Resources, 1975–2000*. Vols. 1–4. Washington, D.C.: Government Printing Office. Excellent source of data.

Chapter 11   Food Resources and World Hunger

Battie, Sandra S., and Robert G. Healy. 1983. "The Future of American Agriculture." *Scientific American*, vol. 248, no. 2, 44–53. Superb overview.

Bourlag, Norman E. 1983. "Contributions of Conventional Plant Breeding to Food Production." *Science*, vol. 219, 689–693. Excellent analysis of the green revolution by the father of the movement.

Brown, Lester R., et al. 1984 and 1985. *State of the World*. New York: W. W. Norton. Excellent summaries of some world food problems and proposed solutions.

Crosson, Pierre R. 1984. "Agricultural Land: Will There Be Enough?" *Environment*, vol. 26, no. 7, 17–20, 40–45. Excellent analysis.

Eckholm, Erik P. 1976. *Losing Ground: Environmental Stresses and World Food Prospects*. New York: W. W. Norton. Outstanding survey of environmental problems associated with agriculture throughout the world.

Lappé, Francis M., and Joseph Collins. 1977. *Food First*. Boston: Houghton Mifflin. Provocative discussion of world food problems. Also see an analysis of these proposals by Peter Huessy. *The Food First Debate* (San Francisco: Institute for Food and Development Policy, 1978).

Mollison, Bill, and David Holmgren. 1978. *Permaculture One*. Tasmania, Australia: Tagari Books. Excellent summary of this ecological approach to sustainable agriculture.

Murphy, Elaine M. 1984. *Food and Population: A Global Concern*. Washington, D.C.: Population Reference Bureau. Excellent overview.

Pimentel, David, and Marcia Pimentel. 1979. *Food, Energy, and Society*. New York: John Wiley. Outstanding discussion of food problems and possible solutions, with emphasis on energy use and food production.

Reichert, Walt. 1982. "Agriculture's Diminishing Diversity." *Environment*, vol. 24, no. 9, 6–11 and 39–43. Excellent summary of threats from loss of genetic diversity.

Todd, Nancy J., and John Todd. 1984. *Bioshelters, Ocean Arks, City Farming: Ecology as a Basis for Design*. San Francisco: Sierra Club Books. Outstanding description of ecological approaches to producing food.

U.S. Department of Agriculture. 1980. *Report and Recommendations on Organic Farming*. Washington, D.C.: Department of Agriculture. Excellent summary of research.

Chapter 12   Land Resources: Wilderness, Parks, Forests, and Rangelands

Brown, Lester R., et al. 1984 and 1985. *State of the World*. New York: W. W. Norton. Excellent overview of the status of the world's forests and ways to protect them.

Clawson, Marion. 1983. *The Federal Lands Revisited*. Washington, D.C.: Resources for the Future. Excellent survey of federal land use with suggestions for future policies.

Connally, Eugenia, ed. 1982. *National Parks in Crisis*. Washington, D.C.: National Parks and Conservation Association. Useful collection of articles on problems facing national parks, with rec-

ommendations for future policies and actions.

Daniel, T. W., et al. 1979. *Principles of Silviculture*. New York: McGraw-Hill. Excellent standard text.

Eckholm, Erik, et al. 1984. *Fuelwood: The Energy Crisis That Won't Go Away*. Washington, D.C.: Earthscan. Superb overview.

Hardin, Garrett. 1968. "The Tragedy of the Commons." *Science*, vol. 162, 1243–1248. Classic environmental article describing how land and other resources are abused when they are shared by everyone.

Minckler, Leon S. 1980. *Woodland Ecology*, 2nd ed. Syracuse, N.Y.: Syracuse University Press. Superb and readable introduction to ecological management of forests.

Myers, Norman. 1984. *The Primary Source: Tropical Forests and Our Future*. New York: W. W. Norton. Superb analysis by an expert.

Nash, Roderick. 1982. *Wilderness and the American Mind*. 3rd ed. New Haven, Conn.: Yale University Press. Outstanding book on American attitudes toward wilderness and conservation.

National Park Service. 1980. *The State of the Parks—1980*. Washington, D.C.: Department of the Interior. Excellent overview.

Smith, D. M. 1982. *The Practice of Silviculture*. New York: John Wiley. Excellent standard text.

U.S. Office of Technology Assessment. 1984. *Technologies to Sustain Tropical Forest Resources*. Washington, D.C.: Government Printing Office. Excellent analysis.

Chapter 13 Wildlife Resources

Eckholm, Erik P. 1978. *Disappearing Species: The Social Challenge*. Washington, D.C.: Worldwatch Institute. Superb overview of the need for wildlife conservation.

Ehrlich, Paul, and Anne Ehrlich. 1981. *Extinction*. New York: Random House. One of the best treatments of the value of wildlife and the causes of extinction, with suggestions for preventing extinction.

Elton, Charles S. 1958. *The Ecology of Invasions by Plants and Animals*. London: Methuen. An environmental classic on species invasions and introductions.

Koopowitz, Harold, and Hilary Kaye. 1983. *Plant Extinctions: A Global Crisis*. Washington, D.C.: Stone Wall Press. Up-to-date summary of this problem with suggestions for preventing plant extinctions.

Livingston, John. 1981. *The Fallacy of Wildlife Conservation*. London: McClelland & Stewart. Thought-provoking critique of the idea that wildlife resources should be managed by wildlife experts to benefit humans.

Myers, Norman. 1983. *A Wealth of Wild Species: Storehouse for Human Welfare*. Boulder, Colo.: Westview Press. Superb presentation of the value of wild species to humans.

Reagan, Tom, and P. Singer. 1976. *Animal Rights and Human Obligation*. Englewood Cliffs, N.J.: Prentice-Hall. Excellent discussion of this controversial issue.

Chapter 14   Urban Land Use and Land-Use Planning

Brown, Lester R., et al. 1980. *Running on Empty: The Future of the Automobile in an Oil-Short World*. New York: W. W. Norton. Excellent analysis.

Choate, Pat, and Susan Walter. 1981. *America in Ruins: Beyond the Public Works Pork Barrel*. Washington, D.C.: Council on State Planning Agencies. Superb discussion of the deterioration of America's physical plant and suggestions for correcting the problem.

Dantzig, George B., and Thomas L. Saaty. 1973. *Compact City: A Plan for a Livable Environment*. San Francisco: W. H. Freeman. Outstanding analysis of urban design for a more ecologically sound and self-reliant city.

McHarg, Ian L. 1969. *Design with Nature*. Garden City, N.Y.: Natural History Press. A beautifully written and illustrated description of an ecological approach to land-use planning. Also available in paperback from Doubleday.

Meyer, John R., and Jose A. Gomez-Ibanez. 1981. *Autos, Transit, and Cities*. Cambridge, Mass.: Harvard University Press. Excellent analysis of problems and possible solutions.

Morris, David. 1982. *Energy and the Transformation of Urban America*. San Francisco: Sierra Club Books. Excellent discussion of how urban areas, particularly small cities, can take steps toward increased energy efficiency and resource self-reliance.

National Academy of Sciences. 1983. *Future Directions of Urban Public Transportation*. Washington, D.C.: National Academy of Sciences. Excellent analysis.

Odum, Eugene P. 1969. "The Strategy of Ecosystem Development." *Science*, vol. 164, 262–270. Classic article on ecological principles and land use.

Whyte, William H. 1980. *The Social Life of Small Urban Spaces*. Washington, D.C.: Conservation Foundation. Superb analysis of how small urban open spaces should be designed for maximum use by people. These research findings are also discussed in the NOVA television program "City Spaces, Human Places," orginally broadcast on PBS November 29, 1981, and available for classroom use.

Chapter 15   Nonrenewable Mineral Resources

Barnet, Richard J. 1980. *The Lean Years: Politics in an Age of Scarcity*. New York: Simon & Schuster. Superb discussion of the politics and economics of resource use and increasing scarcity.

Barnett, Harold J. 1967. "The Myth of Our Vanishing Resources." *Transactions—Social Sciences & Modern Society*, June, pp. 7–10. Classic article summarizing the cornucopian view of natural resources. Compare with the article by Preston Cloud (1975).

Brown, Lester, R., et al. 1984. *State of the World 1984*. New York: W. W. Norton. See Chapter 6 for excellent overview of recycling.

Cloud, Preston E., Jr. 1975. "Mineral Resources Today and Tomorrow." In William W. Murdoch, ed., *Environment: Resources, Pollution and Society*, 2nd ed. Sunderland, Mass.: Sinauer. Superb summary of the neo-Malthusian view. Compare with the works by Barnett (1967), Simon (1981), and Smith (1979).

Leontief, Wassily, et al. 1983. *The Future of Nonfuel Minerals in the U.S. and World Economy: 1980–2030*. Lexington Mass.: Lexington (Heath). Excellent analysis.

Park, Charles F., Jr. 1975. *Earthbound: Minerals, Energy, and Man's Future*. San Francisco: W. H. Freeman. Superb overview emphasizing the neo-Malthusian view.

Purcell, Arthur H. 1980. *The Waste Watchers: A Citizen's Handbook for Conserving Energy*. Garden City, N.Y.: Anchor Press/

Doubleday. Superb guide for achieving a low-waste society.

Simon, Julian L. 1981. *The Ultimate Resource*. Princeton, N.J.: Princeton University Press. Effective presentation of the cornucopian position. Simon makes some good points but many arguments are based on the false ideas that resource supplies are infinite and that energy can be recycled.

Smith, V. Kerry. 1979. *Scarcity and Growth Reconsidered*. Baltimore: Johns Hopkins University Press. Excellent analysis of the cornucopian view of world resource supplies.

Ward, Barbara. 1979. *Progress for a Small Planet*. New York: W. W. Norton. Superb discussion of the need for a new international world economic order.

Chapter 16   Energy Resources: Types, Use, and Concepts

See also the Readings for Chapters 2, 17, 18, and 19.

American Physical Society. 1975. *Efficient Use of Energy*. New York: American Institute of Physics. Summary of energy waste and opportunities for energy conservation based on second-law energy efficiencies in the United States.

Brown, Lester R., et al. 1984 and 1985. *State of the World*. New York: W. W. Norton. Excellent overviews of oil use, nuclear power, and renewable energy.

Clark, Wilson and Jake Page. 1981. *Energy, Vulnerability, and War: Alternatives for America*. New York: W. W. Norton. Excellent analysis of vulnerability of centralized U.S. energy system to electromagnetic pulses from nuclear attack and to cutoffs of imported oil. Based on a study carried out by energy expert Clark for the Department of Defense.

Colorado Energy Research Institute. 1976. *Net Energy Analysis: An Energy Balance Study of Fossil Fuel Resources*. Golden, Colo.: Colorado Energy Research Institute.

Farallones Institute. 1979. *The Integral Urban House: Self-Reliant Living in the City*. San Francisco: Sierra Club Books. Excellent discussion of how to survive in the city.

Flavin, Christopher. 1980. *Energy and Architecture: The Solar and Conservation Potential*. Washington, D.C.: Worldwatch Institute. Excellent summary.

Flavin, Christopher. 1984. *Electricity's Future: The Shift to Efficiency and Small-Scale Power*. Washington, D.C.: Worldwatch Institute. Excellent analysis.

Lovins, Amory B. 1977. *Soft Energy Paths*. Cambridge, Mass.: Ballinger. Superb analysis of energy alternatives. For pros and cons of the soft energy path, see Hugh Nash, ed., *The Energy Controversy: Soft Path Questions and Answers* (San Francisco: Friends of the Earth, 1979).

Lovins, Amory B., and L. Hunter Lovins. 1982. *Brittle Power: Energy Strategy for National Security*. Andover, Mass.: Brick House. Superb analysis of how the present U.S. centralized energy system is highly vulnerable to disruption and how this situation could be corrected by following the soft energy path.

Sant, Roger W., and Dennis W. Bakke. 1983. *Creating Energy Abundance*. New York: McGraw-Hill. Excellent summary of why saving energy saves money.

Solar Energy Research Institute. 1981. *A New Prosperity: Building a Sustainable Energy Future*. Andover, Mass.: Brick House. Outstanding study showing how more efficient use of energy and greatly expanded use of nonrenewable energy sources could lead to a 25 percent reduction in U.S. energy consumption and virtually eliminate oil imports.

Chapter 17   Nonrenewable Energy Resources: Fossil Fuels and Geothermal Energy

Also see the Readings for Chapter 16.

Ackerman, Bruce A., and William T. Hassler. 1981. *Clean Coal/Dirty Air*. New Haven, Conn.: Yale University Press. Excellent discussion of air pollution from coal-burning power plants with suggestions for improvement.

Allar, Bruce. 1984. "No More Coal-Smoked Skies?" *Environment*, vol. 26, no. 2, 25–30. Excellent summary of fluidized-bed combustion of coal.

Leon, George de Lucenay. 1982. *Energy Forever: Power for Today and Tomorrow*. New York: Arco Publishing. Excellent nontechnical overview of nonrenewable and renewable energy alternatives.

Perry, Harry. 1983. "Coal in the United States: A Status Report." *Science*, vol. 222, no. 4622, 377–394. Excellent overview.

Shahinpoor, Mohsen. 1982. "Making Oil from Sand." *Technology Review*, February–March, pp. 49–54. Excellent summary of oil sands.

U.S. Department of Energy. 1980. *Geothermal Energy and Our Environment*. Washington, D.C.: Department of Energy. Superb summary.

Chapter 18   Nonrenewable Energy Resources: Nuclear Power

Also see Readings for Chapter 16.

Brown, Lester R., et al. 1984. *State of the World 1984*. New York: W. W. Norton. See Chapter 7 for excellent overview of the economics of nuclear power.

Cohen, Bernard L. 1983. *Before It's Too Late: A Scientist's Case for Nuclear Power*. New York: Plenum Press. Probably the best available case for nuclear power by an expert.

Ford, Daniel F. 1983. *Three Mile Island: Thirty Minutes to Meltdown*. New York: Penguin Books. Excellent overview by a nuclear power expert.

Gofman, John W. 1981. *Radiation and Human Health*. San Francisco: Sierra Club Books. An expert's detailed and controversial evaluation of the health effects of exposure to low-level radiation.

Harding, Jim. 1984. "Lights Dim for Nuclear Power." *Not Man Apart*, April, pp. 21–22. Excellent summary of economics of nuclear power in the United States and elsewhere.

Hippenheimer, T. A. 1984. *The Man-Made Sun: The Quest for Fusion Power*. Boston: Little, Brown. Excellent overview.

Kaku, Michio, and Jennifer Trainer. 1982. *Nuclear Power: Both Sides*. New York: W. W. Norton. Useful collection of pro and con essays.

Kemeny, John G. 1980. "Saving American Democracy: The Lessons of Three Mile Island." *Technology Review*, June–July, pp. 65–75. Excellent analysis by the head of the presidential panel that investigated this nuclear accident.

Komanoff, Charles. 1981. *Power Plant Cost Escalation*. New York: Van Nostrand Reinhold. Detailed analysis by an expert of the unfavorable economics of nuclear power.

Kulcinski, G. L., et al. 1979. "Energy for the Long Run: Fission or Fusion?" *American Scientist*, vol. 67, 78–89. Superb evaluation of nuclear fusion.

League of Women Voters Education Fund. 1982. *A Nuclear Power Primer: Issues for Citizens*. Washington, D.C.: League of Women Voters. Outstanding, readable, balanced summary.

League of Women Voters Education Fund. 1980. *A Nuclear Waste Primer*. Washington D.C.: League of Women Voters. Excellent readable, balanced summary.

Lidsky, Lawrence M. 1983. "The Trouble with Fusion." *Technology Review*, October, pp. 32–44. Suberb evaluation by one of the world's most prominent nuclear fusion scientists.

McCracken, Samuel. 1982. *The War Against the Atom*. New York: Basic Books. Excellent defense of nuclear power.

O'Banion, Kerry. 1981. "Long-Term Nuclear Options." *Environmental Science & Technology*, vol. 15, no. 10, 1130–1136. Excellent comparison of the environmental effects of nuclear fission breeder reactors and nuclear fusion reactors.

Upton, Arthur C. 1982. "The Biological Effects of Low-Level Ionizing Radiation." *Scientific American*, vol. 246, no. 2, 41–49. Excellent overview.

Weinberg, Alvin M. 1980. "Is Nuclear Energy Necessary?" *Bulletin of the Atomic Scientists*, March, pp. 31–35. Excellent case for nuclear power.

Chapter 19 Renewable Energy Resources

Also see Readings for Chapter 16.

Brown, Lester R., et al. 1984 and 1985. *State of the World*. New York: W. W. Norton. Excellent overviews of renewable energy resources.

Butti, Ken, and John Perlin. *A Golden Thread: 2500 Years of Solar Architecture and Technology*. New York: Cheshire Books. Superb overview.

Center for Science in the Public Interest. 1977. *99 Ways to a Simple Lifestyle*. Garden City, N.Y.: Doubleday. Superb summary of how you can conserve matter and energy.

Deudney, Daniel, and Christopher Flavin. 1983. *Renewable Energy: The Power to Choose*. 1983. New York: W. W. Norton. Excellent overview of renewable energy resources.

Finneran, Kevin. 1983. "Solar Technology: A Whether Report." *Technology Review*, April, pp. 48–59. Excellent overview.

Holdren, John P. 1982. "Energy Hazards: What to Measure, What to Compare."

Technology Review, April, pp. 34–75. Excellent discussion by an expert of how to evaluate risks of various energy options.

Kendall, Henry, and Steven Nadis. 1980. *Energy Strategies: Toward a Solar Future*. Cambridge, Mass.: Ballinger. Splendid analysis of energy alternatives.

Mazria, Edward. 1979. *The Passive Solar Energy Book: A Complete Guide to Passive Solar Home, Greenhouse, and Building Design*. Emmaus, Pa.: Rodale Press. One of the best available books on passive solar energy design.

Medsker, Larry. 1982. *Side Effects of Renewable Energy Resources*. New York: National Audubon Society. Excellent summary of environmental effects.

Pimentel, David, et al. 1984. "Environmental and Social Costs of Biomass Energy." *BioScience*, February, pp. 89-93. Excellent overview.

Pryde, Philip R. 1983. *Nonconventional Energy Resources*. New York: Wiley-Interscience. Excellent overview with emphasis on nonrenewable energy resources.

Chapter 20
The Environment and Human Health: Disease, Food Additives, and Noise

Beattie, Edward J. 1980. *Toward the Conquest of Cancer*. New York: Crown Publishers. Superb analysis showing how 40 to 50 percent of cancers can be prevented and about 50 percent can be cured.

Chandler, William U. 1984. *Improving World Health: A Least Cost Strategy*. Washington, D.C.: Worldwatch Institute. Superb analysis.

Davis, Devra L. 1981. "Cancer in the Workplace: The Case for Prevention." *Environment*, vol. 23, no. 6, 25–37. Excellent overview.

Eckholm, Erik P. 1977. *The Picture of Health: Environmental Sources of Disease*. New York: W. W. Norton. Superb overview of environmental health problems throughout the world.

Efron, Edith. 1984. *The Apocalyptics: Cancer and the Big Lie*. New York: Simon & Schuster. Attack by a freelance writer accusing environmentalists of overstating the effects of human-produced chemicals as a cause of cancer.

Epstein, Samuel S. 1979. *The Politics of Cancer*. Garden City, N.Y.: Anchor/Doubleday. Excellent overview of cancer by an expert, with emphasis on occupational exposure to carcinogens.

Gorman, James. 1979. *Hazards to Your Health: The Problem of Environmental Disease*. New York: New York Academy of Sciences. Superb overview.

Kessler, David A. 1984. "Food Safety: Revising the Statute." *Science*, vol. 223, 1034–1040. Excellent analysis of whether food-safety laws should be changed.

National Academy of Sciences. 1981. *Effects on Human Health from Long-Term Exposures to Noise*. Washington, D.C.: National Academy of Sciences. Excellent overview of research on the effects of noise.

National Academy of Sciences. 1982. *Diet, Nutrition, and Cancer*. Washington, D.C.: National Academy Press. Authoritative review of possible relationships between diet and cancer.

Reif, Arnold E. 1981. "The Causes of Cancer." *American Scientist*, vol. 69, 437–447. Excellent overview.

Whelan, Elizabeth M., and Frederick J. Stare. 1983. *The 100% Natural, Purely Organic, Cholesterol-Free, Megavitamin, Low-Carbohydrate Nutrition Hoax*. New York: Atheneum Press. Presentation heavily weighted toward the food industry; debunks concern over the dangers of food additives and pesticides and derides people who grow and eat health foods and use vitamin supplements.

Winter, Ruth A. 1978. *A Consumer's Dictionary of Food Additives*. New York: Crown Publishers. Excellent guide to additives.

Chapter 21
Air Pollution

Andrews, Richard N. L. 1981. "Will Benefit-Cost Analysis Reform Regulations?" *Environmental Science & Technology*, vol. 15, no. 9, 1016–1021. Excellent summary of cost-benefit analysis.

Boyle, Robert H., and R. Alexander Boyle. 1983. *Acid Rain*. New York: Schocken Books. Excellent overview.

Bryson, Reid A., and Thomas J. Murray. 1977. *Climates of Hunger: Mankind and the World's Changing Weather*. Madison: University of Wisconsin Press. Excellent summary of the position that the world may be cooling.

Council on Environmental Quality. 1981. *Global Energy Futures and the Carbon Dioxide Problem*. Washington, D.C.: Council on Environmental Quality. Superb overview.

Fennelly, Paul F. 1976. "The Origin and Influence of Airborne Particulates." *American Scientist*, vol. 64, 46–56. Superb overview.

Gold, Michael. 1980. "Indoor Air Pollution." *Science 80*, March–April, pp. 30–33. Excellent overview.

Idso, Sherwood B. 1982. *Carbon Dioxide: Friend or Foe?* Tempe, Ariz.: IBR Press. Challenges the consensus view that increased CO_2 levels will lead to a warming effect and the idea that if warming occurred, its harmful effects would outweigh its beneficial effects.

Lovins, Amory B., et al. 1981. *Least-Cost Energy: Solving the CO_2 Problem*. Andover, Mass.: Brick House. Excellent summary of how the problem could be minimized by a combination of energy conservation and a switch to solar, wind, hydro, biomass, and other forms of renewable energy.

Luoma, Jon R. 1984. *Troubled Skies, Troubled Waters: The Story of Acid Rain*. New York: Viking Press. Excellent overview.

National Academy of Sciences. 1983. *Acid Deposition: Atmospheric Processes in the United States*. Washington, D.C.: National Academy Press. Excellent overview and source of data.

National Academy of Sciences. 1983. *Changing Climate*. Washington, D.C.: National Academy Press. Excellent overview of the carbon dioxide problem.

Postel, Sandra. 1984. *Air Pollution, Acid Rain, and the Future of Forests*. Washington, D.C.: Worldwatch Institute. Superb overview.

Rose, David J., et al. 1984. "Reducing the Problem of Global Warming," *Technology Review*, May–June, pp. 49–58. Excellent overview.

Schneider, S. H., and R. S. Londer. 1984. *The Coevolution of Climate and Life*. San Francisco: Sierra Club Books. Excellent overview of possible effects of human activities on global climate.

Chapter 22
Water Pollution

Burmaster, David E. 1982. "The New Pollution: Groundwater Contamination." *Environment*, vol. 24, no. 2, 4–12, 33–36. Excellent overview.

Grundlach, Erich R., et al. 1983. "The Fate of *Amoco Cadiz* Oil." *Science*, vol. 221, 122–129. Scientific study of the recovery of most marine life within 3 years after this spill.

Lahey, William, and Michael Connor. 1983. "The Case for Ocean Waste Disposal." *Technology Review*, August–September, pp. 61–68. Excellent overview.

Marx, Wesley. 1981. *The Oceans: Our Last Resource*. San Francisco: Sierra Club Books. Excellent discussion of how to preserve the ocean's resources.

National Academy of Sciences. 1984. *Groundwater Contamination*. Washington, D.C.: National Academy Press. Excellent summary.

Simon, Anne W. 1984. *Neptune's Revenge: The Ocean of Tomorrow*. New York: Franklin Watts. Superb discussion of stresses on the oceans and ways to protect oceans from excessive abuse.

Woodwell, George M. 1977. "Recycling Sewage Through Plant Communities." *American Scientist*, vol. 65, 556–562. Excellent overview of this natural alternative to expensive waste treatment plants.

Chapter 23
Pesticides and Pest Control

Barrons, Keith C. 1981. *Are Pesticides Really Necessary?* Chicago: Regnery Gateway. Excellent presentation of both sides of the pesticide controversy, with emphasis on the benefits of pesticides.

Carson, Rachel. 1962. *Silent Spring*. Boston: Houghton Mifflin. An environmental classic that provided the first major warning about the dangerous side effects of pesticides.

Dunlap, Thomas R. 1981. *DDT: Scientists, Citizens, and Public Policy*. Princeton, N.J.: Princeton University Press. Excellent discussion of the history of the use of DDT and the problems that led to its banning in the United States.

Goldstein, Jerome. 1978. *The Least Is Best Pesticide Strategy*. Emmaus, Pa.: J.G. Press. Excellent discussion of integrated pest management.

Pimentel, David, et al. 1980. "Environmental and Social Costs of Pesticides: A Preliminary Assessment." *Oikos*, vol. 34, no. 2, 126–140. Superb overview.

van den Bosch, Robert. 1978. *The Pesticide Conspiracy*. Garden City, N.Y.: Doubleday. Pest management expert exposes political influence of pesticide companies in preventing widespread use of biological controls and integrated pest management.

van den Bosch, Robert, and Mary L. Flint. 1981. *Introduction to Integrated Pest Management*. New York: Plenum Press. Superb presentation.

Chapter 24   Solid Waste and Hazardous Wastes

Brown, Lester R., et al. 1984. *State of the World 1984*. New York: W. W. Norton. Chapter 6 is an excellent summary of recycling.

Brown, Michael. 1979. *Laying Waste: The Poisoning of America by Toxic Wastes*. New York: Pantheon. Critical attack with detailed discussion of the Love Canal disaster.

D'Itri, Patricia R., and Frank M. D'Itri. 1977. *Mercury Contamination*. New York: John Wiley. Superb summary.

Efron, Edith. 1984. *The Apocalyptics: Cancer and the Big Lie*. New York: Simon & Schuster. Contends that concern over hazardous wastes, pollutants, and other toxic and hazardous materials has been overblown and has little or no scientific foundation.

Environmental Protection Agency. 1977. *Fourth Report to Congress: Resource Recovery and Waste Reduction*. Washington, D.C.: Environmental Protection Agency. Superb analysis of solid waste disposal, high- and low-technology resource recovery, and resource conservation in the United States.

Epstein, Samuel S., et al. 1982. *Hazardous Waste in America*. San Francisco: Sierra Club Books. Superb analysis of the problems, along with suggested solutions.

Harrison, R. M., and D.P.H. Laxon. 1981. *Lead Pollution: Causes and Control*. London:

Chapman and Hall/Methuen. Excellent overview.

Hay, Alstair. 1982. *The Chemical Scythe: Lessons of 2,4,5-T and Dioxin*. New York: Plenum Press. Excellent overview.

Kriebel, David. 1981. "The Dioxins: Toxic and Still Troublesome." *Environment*, January–February, pp. 6–13. Excellent overview.

League of Women Voters Education Fund. 1981. *A Hazardous Waste Primer*. Washington, D.C.: League of Women Voters. Excellent overview.

Needleman, Herbert L. 1980. "Lead Exposure and Human Health: Recent Data on an Ancient Problem." *Technology Review*, March–April, pp. 39–45. Excellent overview.

U.S. Office of Technology Assessment. 1983. *Technologies and Management Strategies for Hazardous Waste Controls*. Washington, D.C.: Government Printing Office. Useful evaluation.

Epilogue   Achieving a Sustainable Earth Society

Cahn, Robert. 1978. *Footprints on the Planet: A Search for an Environmental Ethic*. New York: Universe Books. Outstanding book on environmental ethics and progress by a former member of the Council on Environmental Quality.

Callenbach, Ernest. 1975. *Ecotopia*. Berkeley, Calif.: Banyan Tree Books. Stirring vision of what the world could be like if we move to a sustainable earth society.

Devall, Bill and George Sessions. 1984. *Deep Ecology*. Salt Lake City, Utah: Earth First. Excellent collection of writings on deep ecology.

Fritsch, Albert J. 1980. *Environmental Ethics: Choices for Concerned Citizens*. New York: Anchor Books. Outstanding book.

Johnson, Warren. 1978. *Muddling Toward Frugality*. San Francisco: Sierra Club Books. Important and hopeful book showing how we might make it to a sustainable earth society.

Glossary

Abiotic   nonliving.

Abyssal zone   bottom zone of the ocean. compare *Bathyal zone, Euphotic zone.*

Acid deposition   combination of wet deposition from the atmosphere of droplets of sulfuric acid and nitric acid dissolved in rain and snow and dry deposition from the atmosphere of particles of sulfate and nitrate salts. These acids and salts are formed when water vapor in the air reacts with the air pollutants sulfur dioxide (SO_2) and nitrogen dioxide (NO_2).

Acid fog   droplets of sulfuric acid and nitric acid mixed with water vapor near the ground.

Acute disease   infectious disease (such as measles, whooping cough, typhoid fever) that normally lasts for a relatively short time before the victim either recovers or dies.

Aerobic organism   organism that requires oxygen to live.

Aerosols   liquid and solid particles suspended in air.

Age structure (age distribution)   number or percentage of persons at each age level in a population.

Age structure diagram   horizontal bar graph comparing the proportions of males and females in different age groups in the population.

Air pollution   air that contains one or more chemicals or possesses a physical condition like heat in high enough concentrations to harm humans, other animals, vegetation, or materials.

Albedo   fraction of the incident light that strikes a surface or body that is reflected from it. A measure of the reflectivity of the earth and its atmosphere.

Alga (algae)   simple one-celled or many-celled plant(s), usually aquatic, capable of carrying on photosynthesis.

Alpha particle   form of radiation consisting of a helium nucleus containing two protons and two neutrons with no electrons outside the nucleus.

Alveoli   tiny sacs at the end of the bronchiole tubes in the lungs, numbering in the hundreds of millions. Oxygen in the air passes through the alveolar walls to combine with hemoglobin in the blood, and carbon dioxide passes from the blood back through the alveolar walls for exhaling.

Ambient quality standard   maximum level of a specific pollutant allowed by the federal government in the air, water, soil, or food. May vary from region to region depending on conditions. compare *Emission standard.*

Amino acids   basic building block molecules of proteins. A long chain of certain amino acid molecules linked together chemically forms a specific protein molecule.

Anaerobic organism   organism that does not require oxygen to live.

Animal feedlot   confined area where hundreds or thousands of livestock animals are fattened for sale to slaughterhouses and meat processors.

Antagonistic effect   the interaction of two or more factors so that the net effect is less than that resulting from adding the independent effects of the factors. compare *Synergistic effect.*

Aquaculture   growing and harvesting of fish and shellfish in land-based ponds. compare *Mariculture.*

Aquifer   permeable layers of underground rock or sand that hold or transmit groundwater below the water table.

Area surface mining   type of mining in which minerals such as coal and phosphate are removed by cutting deep trenches in flat or rolling terrain.

Atmosphere   a region of gases and particulate matter extending above the earth's surface.

Atoms   extremely small particles that are the basic building blocks of all matter.

Autotrophic organism   organism that uses solar energy to photosynthesize organic food substances and other organic chemicals from carbon dioxide and water. compare *Heterotrophic organism.*

Background radiation   radiation in the environment from naturally radioactive materials and from cosmic rays entering the atmosphere.

Bacteria   smallest living organisms; with fungi, they comprise the decomposer level of the food chain.

Bathyal zone   middle or open-water zone in an ocean below the level of light penetration. compare *Abyssal zone, Euphotic zone.*

Beta particle   swiftly moving electron emitted by a radioactive substance. The isotopes strontium-90 and carbon-14 emit beta particles.

Biodegradable   capable of being broken down by bacteria into basic elements and compounds. Most organic wastes and paper are biodegradable.

Biofuels   gas or liquid fuels (such as ethyl alcohol) made from biomass (plants and trees).

Biogas   mixture of methane (CH_4) and carbon dioxide (CO_2) gases produced when anaerobic bacteria break down plants and organic wastes (such as manure).

Biogeochemical cycles   mechanisms by which chemicals such as carbon, oxygen, phosphorus, nitrogen, and water are moved through the ecosphere to be renewed over and over again. The three major cycle types are gaseous, sedimentary, and hydrologic.

Biological control   pest control that uses natural predators, parasites, or disease-causing bacteria and viruses (pathogens).

Biological magnification   buildup in concentration of a substance, such as DDT or some radioactive isotopes, in successively higher trophic levels of the food chain or web.

Biological oxygen demand (BOD)   amount of dissolved oxygen gas required for bacterial decomposition of organic wastes in water; usually expressed in terms of the parts per million (ppm) of dissolved oxygen consumed over 5 days at 20°C (68°F) and normal atmospheric pressure.

Biomass   total dry weight of all living organisms that can be supported at each trophic level in a food chain; total dry weight of all living organisms in a given area; plant and animal matter that can be used in any form as a source of energy.

Biome large terrestrial ecosystem characterized by distinctive types of plants and animals and maintained under the climatic conditions of the region.

Biosphere see *Ecosphere*.

Biotic living.

Biotic (reproductive) potential the maximum rate at which a population can reproduce, given unlimited resources and ideal environmental conditions.

Birth rate number of live births per 1,000 persons in the population at the midpoint of a given year.

Bitumen black, high-sulfur, tarlike heavy oil extracted from tar sands and upgraded to synthetic fuel oil.

Breeder reactor nuclear reactor that produces more nuclear fuel than it consumes by converting nonfissionable uranium-238 into fissionable plutonium-239.

Broad-spectrum pesticide chemical that kills organisms besides the target species.

Calorie amount of energy required to raise the temperature of 1 gram of water 1°C.

Cancer a group of more than 100 different diseases that strike people of all ages and races; characterized by uncontrolled growth of cells in body tissue.

Carcinogen a chemical or form of radiation (energy) that directly or indirectly causes a form of cancer.

Carnivore meat-eating organism.

Carrying capacity maximum population that a given ecosystem can support indefinitely under a given set of environmental conditions.

Cellular respiration see *Respiration*.

Channelization strengthening, widening, deepening, clearing, or lining of existing stream channels.

Chemical cycles see *Biogeochemical cycles*.

Chemical energy potential energy stored in the chemical bonds that hold together the atoms or ions in chemical compounds.

Chemical weathering attack and dissolving of parent rock by exposure to rainwater, surface water, oxygen and other gases in the atmosphere, and compounds secreted by organisms.

Chlorinated hydrocarbon insecticides synthetic, organic nervous system poisons containing chlorine, hydrogen, and carbon. Highly stable and fat soluble, they tend to be recycled through food chains, thereby affecting nontarget organisms. Members include DDT, aldrin, dieldrin, endrin, chlordane, heptachlor, toxaphene, and BHC.

Chronic bronchitis disease characterized by inflammation of the bronchial passages, excessive secretions of mucus, and recurrent coughing. It appears to be aggravated by air pollution, particularly sulfur oxides.

Chronic disease disease that lasts for a long time (often for life) and may **(1)** flare up periodically (malaria), **(2)** become progressively worse (cancer), or **(3)** disappear with advancing age (childhood asthma). Chronic diseases may be infectious (malaria, tuberculosis) or noninfectious (cardiovascular disorders, diabetes, hay fever).

Cilia tiny hairs lining the lungs that continually undulate and sweep foreign matter out of the lungs.

Clearcutting removing all trees from a given area in a single cutting.

Climate generalized weather at a given place on earth over a fairly long period.

Climax ecosystem (climax community) a relatively stable stage of ecological succession; a mature ecosystem with a diverse array of species and ecological niches, capable of using energy and cycling critical chemicals more efficiently than simpler, immature ecosystems.

Coal a solid, combustible organic material containing 55 to 90 percent carbon mixed with varying amounts of hydrogen, oxygen, nitrogen, and sulfur compounds.

Coal gasification process in which solid coal is converted to either low-heat-content industrial gas or high-heat-content synthetic natural gas (SNG).

Coal liquefaction process in which solid coal is converted to synthetic crude oil by the addition of hydrogen (hydrogenation).

Coastal wetlands shallow shelves that are normally wet or flooded and extend back from the freshwater–saltwater interface. They consist of a complex maze of marshes, bays, lagoons, tidal flats, and mangrove swamps.

Cogeneration the production of two useful forms of energy from the same process. In a factory, for example, steam produced for industrial processes or

space heating is run through turbines to generate electricity, which can be used by the industry or sold to power companies.

Combustion burning. Any very rapid chemical reaction in which heat and light are produced.

Commensalism symbiotic relationship between two different species in which one species benefits from the association while the other apparently is neither helped nor harmed. see *Mutualism, Parasitism, Symbiosis*.

Commons natural resources, especially land, reserved for common use. Many experts in environmental law also treat rivers, lakes, oceans, and the atmosphere as commons.

Community (natural) populations of different plants and animals living and interacting in a given area at a given time.

Competition two or more species in the same ecosystem attempting to use the same scarce resources.

Competitive exclusion principle no two species in the same ecosystem can occupy exactly the same ecological niche indefinitely.

Complete proteins animal proteins such as meat, fish, eggs, milk, and cheese that can provide humans with all eight essential amino acids.

Composting breakdown of organic matter in solid waste in the presence of oxygen by aerobic (oxygen-needing) bacteria to produce a humuslike end product, which can be used as a soil conditioner.

Compound substance composed of atoms or ions of two or more different elements held together in a fixed ratio by chemical bonds.

Concentration amount of a chemical or pollutant in a particular volume or weight of air, water, soil, or other medium.

Consumer organism that lives off other organisms. Generally divided into primary consumers (herbivores), secondary consumers (carnivores), and microconsumers (decomposers).

Contour farming plowing and planting along the sloping contours of land to reduce soil erosion and conserve water.

Contour surface mining mining by cutting out a series of contour bands on the side of a hill or mountain. Used primarily for coal. Usually the most destructive form of surface mining.

Cooling tower large tower used to transfer the heat in cooling water from a power or industrial plant to the atmosphere either by direct evaporation (wet, or evaporative, cooling tower) or by convection and conduction (dry cooling tower).

Cost-benefit analysis attempt to compare, for example, the pollution control costs and the costs (dollar and otherwise) of pollution damage with the benefits that may occur from successful application of pollution control technology. The goal is to minimize total costs, yet keep harmful effects to acceptable levels.

Cost-effectiveness analysis determination of how much it will cost to achieve a benefit from, say, pollution control and comparision of this amount to the cost of obtaining a higher or lower level of the benefit or using an alternative.

Crop rotation annual rotation of areas or strips planted with crops such as corn and tobacco that remove large amounts of nitrogen from the soil when harvested with legumes that add nitrogen to the soil or other crops such as oats, barley, or rye.

Crown fire intensely hot fire that can destroy all vegetation, kill wildlife, and accelerate erosion. compare *Ground fire*.

Crude birth rate see *Birth rate*.

Crude death rate see *Death rate*.

Crude oil see *Petroleum*.

Cultural eutrophication overnourishment of aquatic ecosystems with plant nutrients due to human activities such as agriculture, urbanization, and industrial discharge. see *Eutrophication*.

DDT dichlorodiphenyltrichloroethane, a chlorinated hydrocarbon that has been widely used as a pesticide.

Death rate number of deaths per 1,000 persons in the population at the midpoint of a given year.

Decibel (db) unit used to measure sound power or sound pressure.

Deciduous trees trees that lose their leaves during part of the year.

Decomposer bacterium or fungus that causes the chemical disintegration (rot or decay) of organic matter.

Degradable pollutant pollutant that can be decomposed, removed, or consumed and thus reduced to an acceptable level either by natural processes or by human-engineered processes. Pollutants that are broken down rapidly are called rapidly degraded, or nonpersistent, and those broken down slowly are called slowly degradable, or persistent. compare *Nondegradable pollutant*.

Demographic transition gradual change, supposedly brought about by economic development, from a condition of high birth and death rates to substantially lower birth and death rates for a given country or region.

Demography science of vital and social statistics of populations, such as numbers of births, deaths, and marriages, and number of reported cases of diseases.

Dependency load ratio of the number of old and young dependents in a population to the work force.

Depletion time time required to use up a certain fraction (usually 80 percent) of the known or estimated supply of a resource according to various assumed rates of use.

Desalination purification of salt or brackish water by removing the dissolved salts.

Desertification conversion of productive rangeland into desert usually through a combination of overgrazing, prolonged drought, and global climate change.

Detritus dead plant material, bodies of animals, and fecal matter.

Detritus food chain transfer of energy from one trophic level to another by decomposers. compare *Grazing food chain*.

Deuterium (D: hydrogen-2) isotope of the element hydrogen with a nucleus containing one proton and one neutron, thus having a mass number of 2. compare *Tritium*; see also *Heavy water*.

Disinfection sewage treatment to remove water coloration and to kill disease-carrying bacteria and some (but not all) viruses.

Dissolved oxygen (DO) content amount of oxygen gas (O_2) dissolved in a given quantity of water at a given temperature and atmospheric pressure. It is usually expressed as a concentration in parts per million (ppm) or as a percentage of saturation.

Diversity physical or biological complexity of a system. In many cases this property leads to ecosystem stability.

Doubling time time (usually years) that is necessary for a population to double in size.

Dredging surface mining of seabeds, primarily for sand and gravel.

Ecological efficiency (food chain efficiency) percent transfer of useful energy from one trophic level to another in a food chain.

Ecological equivalents species that occupy the same or similar ecological niches in similar ecosystems located in different parts of the world. For example, cattle in North America and kangaroos in Australia are both grassland grazers, hence are ecological equivalents.

Ecological niche description of a species' total structural and functional role in an ecosystem.

Ecological succession change in the structure and function of an ecosystem; repeated replacement of one kind of natural community of organisms with a different natural community over time. see *Primary succession, Secondary succession*.

Ecology study of the relationships of living organisms with each other and with their physical and biological environment; study of the structure and function of nature.

Ecosphere (biosphere) total of all the ecosystems on the planet, along with their interactions; the sphere of air, water, and land in which all life is found.

Ecosystem self-sustaining and self-regulating natural community of organisms interacting with one another and with their environment.

Effluent any substance, particularly a liquid, that enters the environment from a point source. Generally refers to wastewater from a sewage treatment or industrial plant.

Electromagnetic energy radiant energy that can move through a vacuum or through space as waves of oscillating electric and magnetic fields.

Electromagnetic spectrum span of electromagnetic energy ranging from short-wavelength gamma waves to long-wavelength radio waves.

Electron fundamental particle found moving around outside the nucleus of an atom. Each electron has one unit of negative charge (-1) and has extremely little mass.

Electrostatic precipitator device for removing particulate matter from smokestack emissions by causing the particles to become electrostatically charged and then attracting them to an oppositely charged plate, where they are removed from the air.

Element chemical such as iron (Fe), sodium (Na), carbon (C), nitrogen (N), and oxygen (O) whose distinctly different atoms serve as the 108 basic building blocks of all matter.

Emergency core-cooling system system designed to prevent meltdown if the core of a nuclear reactor overheats by instantaneous flooding of the core with large amounts of water.

Emigration   process of leaving one country to take up permanent residence in another.

Emission standard   maximum amount of a pollutant that is permitted by the federal government to be discharged from a single polluting source.

Emissivity   total amount of degraded heat energy flowing from the earth back into space.

Emphysema   see *Pulmonary emphysema*.

Endangered species   one having so few individual survivors that the species could soon become extinct in all or part of its region.

Energy   ability or capacity to do work or produce a change by pushing or pulling some form of matter.

Energy efficiency   the amount of useful energy produced by a source compared to the energy needed to obtain this amount of useful energy.

Energy flow   in ecology, the one-way transfer of energy through an ecosystem; more specifically, the way in which energy is converted and expended at each trophic level.

Energy pyramid   diagram representing the loss or degradation of useful energy at each step in a food chain. About 80 to 90 percent of the energy in each transfer is lost as waste heat, and the resulting shape of the energy levels is pyramidal.

Energy quality   ability of a form of energy to do useful work. High-quality energy (such as high-temperature heat, fossil fuels, and nuclear fuel) is concentrated, whereas low-quality energy (such as low-temperature heat) is dispersed or diluted.

Environment   aggregate of external conditions that influence the life of an individual organism or population.

Environmental resistance   all the limiting factors that act together to regulate the maximum allowable size, or carrying capacity, of a population.

EPA   the U.S. Environmental Protection Agency, the agency responsible for federal efforts to control air and water pollution, radiation and pesticide hazards, ecological research, and solid waste disposal.

Essential amino acid   one of the eight chemical building blocks for proteins that cannot be made in the human body and must be included in the diet for good health.

Estuarine zone   area near the coastline that consists of estuaries and coastal saltwater wetlands.

Estuary   thin zone along a coastline where freshwater system(s) and river(s) meet and mix with a salty ocean.

Euphotic zone   surface layer of an ocean, lake, or other body of water through which light can penetrate; thus, the zone of photosynthesis. compare *Abyssal zone, Bathyal zone.*

Eutrophic lake   lake with a large or excessive supply of plant nutrients (nitrates and phosphates).

Eutrophication (natural)   excess of plant nutrients from natural erosion and runoff from the land in an ecosystem supporting a large amount of aquatic life that can deplete the oxygen supply. see also *Cultural eutrophication.*

Evapotranspiration   combination of evaporation and transpiration of water into the atmosphere from living plants and soil.

Exponential growth   geometric growth by doubling; yields a J-shaped curve.

Exponential reserve index   estimated number of years until known world reserves of a nonrenewable resource will be 80 percent depleted if consumed at a rate increasing by a given percentage each year. compare *Static reserve index.*

External cost   cost of production or consumption that must be borne by society, not by the producer.

Extinction   complete disappearance of a species because of failure to adapt to environmental change.

Family planning   distribution of information and contraceptives to couples who have expressed an interest in limiting the number of children they have and/or in scheduling births.

Fecal coliform bacteria count   number of colonies of fecal coliform bacteria present in a 100-milliliter sample of water. Presence of coliform bacteria in water indicates possible presence of other harmful bacteria from untreated human and animal waste.

Feedback   signal sent back into a self-regulating system to induce some system response.

Feedback loop   return to the input of part of the output of a homeostatic system, which is then processed by the system.

Fermentation, anaerobic   process in which carbohydrates are converted in the

absence of oxygen to hydrocarbons (such as methane).

Fertilizer   substance that makes the land or soil capable of producing more vegetation or crops.

First energy law   see *First law of thermodynamics.*

First-law energy efficiency   ratio of the useful work or energy output of a device to the work or energy input that must be supplied to get the output. This ratio is normally multiplied by 100 so that the efficiency can be expressed as a percentage.

First law of thermodynamics (energy)   in any ordinary chemical or physical change, energy is neither created nor destroyed, but merely changed from one form to another. You can't get something for nothing; you can only break even; or, there is no free lunch.

Fissionable isotopes   isotopes that are capable of undergoing nuclear fission.

Floodplain   land next to a river that is covered by water when the river overflows its banks.

Fluidized-bed combustion   process using a flowing stream of hot air to suspend a mixture of powdered coal and limestone so that the coal burns more efficiently. In addition, the limestone removes 90 to 95 percent of the sulfur in the coal.

Fly ash   small, solid particles of ash and soot generated when coal, oil, or waste materials are burned.

Food additive   chemical deliberately added to a food, usually to enhance its color, flavor, shelf life, or nutritional characteristics.

Food chain   sequence of transfers of energy in the form of food from organisms in one trophic level to organism in another trophic level when one organism eats or decomposes another. see *Detritus food chain, Grazing food chain.*

Food contaminant   substance not deliberately added to food but usually resulting from poor sanitation, improper food processing or storage, or the careless use of compounds such as pesticides and radioactive materials.

Food web   complex, interlocking series of food chains.

Fossil fuel   buried deposits of decayed plants and animals that have been converted to crude oil, coal, natural gas, or heavy oils by exposure to heat and pressure in the earth's crust over hundreds of millions of years.

Freons   chlorofluorocarbon compounds

composed of atoms of carbon, chlorine, and fluorine.

Fungus organism without chlorophyll. The simpler forms are unicellular; the higher forms have branched filaments and complicated life cycles. Examples are molds, yeasts, and mushrooms.

Gamma rays high-energy electromagnetic waves with very short wavelengths, produced during the disintegration of some radioactive elements. Like X rays, they readily penetrate body tissues.

Gaseous cycle a biogeochemical cycle with the atmosphere as the primary reservoir. Examples include the oxygen and nitrogen cycles.

Gasohol vehicle fuel consisting of a mixture of gasoline and ethyl or methyl alcohol that typically contains 10 to 23 percent by volume alcohol.

Gene pool total genetic information possessed by a given reproducing population.

Genetic damage damage by radiation or chemicals to reproductive cells, resulting in mutations that can be passed on to future generations in the form of fetal and infant deaths and physical and mental disabilities.

Geometric growth see *Exponential growth.*

Geothermal energy heat energy produced when rocks lying below the earth's surface are heated to high temperature by energy from the decay of radioactive elements in the earth and from magma.

Grazing food chain transfer of energy in the form of food from one organism to another when green plants (producers) are eaten by plant eaters (herbivores), which in turn may be eaten by meat eaters (carnivores). compare *Detritus food chain.*

Green revolution popular term for the introduction of scientifically bred or selected varieties of a grain (rice, wheat, maize) that with high enough inputs of fertilizer and water can give greatly increased yields per area of land planted.

Greenhouse effect trapping of heat in the atmosphere. Incoming short-wavelength solar radiation penetrates the atmosphere, but the longer wavelength outgoing radiation is absorbed by water vapor, carbon dioxide, and ozone in the atmosphere and is reradiated to earth, causing a rise in atmospheric temperature.

Ground fire low-level fire that typically burns only undergrowth and occa-sionally damages fire-sensitive trees. compare *Crown fire.*

Groundwater water that sinks into the soil, where it may be stored for long times in slowly flowing and slowly renewed underground reservoirs.

Growth rate (population) percentage of increase or decrease of a population. It is the number of births minus the number of deaths per 1,000 population, plus net migration, expressed as a percentage.

Gully reclamation using small dams of manure and straw, earth, stone, or concrete to collect silt and gradually fill in channels of eroded soil.

Habitat place where an organism or community of organisms naturally lives or grows.

Half-life length of time taken for half the atoms in a given amount of a radioactive substance to decay into another isotope. The definition has been extended to refer to biological half-life, or the length of time it takes for half of any substance (such as mercury) in a biological system (such as the brain) to be broken down or excreted.

Hardness (water) condition caused by dissolved salts of calcium, magnesium, and iron, such as bicarbonates, carbonates, sulfates, chlorides, and nitrates.

Hazardous waste discarded solid, liquid, or gaseous material that may pose a substantial threat or potential hazard to human health or the environment when improperly handled.

Heat form of kinetic energy that flows from one body to another as a result of a temperature difference between the two bodies.

Heavy oil black, high-sulfur, tarlike oil found in deposits of crude oil, tar sands, and oil shale.

Heavy water water (D_2O) in which all the hydrogen atoms have been replaced by deuterium (D).

Herbicide chemical that injures or kills plant life by interfering with normal growth.

Herbivore plant-eating organism.

Heterotrophic organism organism that cannot manufacture its own food and must consume organic food compounds found in other plants and animals. compare *Autotrophic organism.*

Homeostasis tendency for a biological system to resist drastic change by maintaining fairly constant internal conditions.

Humus complex mixture of decaying organic matter and inorganic compounds in the soil that serves as a major source of plant nutrients and increases the soil's capacity to absorb water.

Hydrocarbons class of organic compounds containing carbon (C) and hydrogen (H). Hydrocarbons often occur as air pollutants from unburned or partially burned gasoline and from evaporation of industrial solvents, especially from refineries. In the presence of sunlight and oxides of nitrogen, they can form photochemical smog.

Hydroelectric plant electric power plant in which the energy of falling water is used to spin a turbine generator to produce electricity.

Hydrologic cycle biogeochemical cycle that moves and recycles water in various forms through the ecosphere.

Hydropower electrical energy produced by falling water.

Hydrosphere region that includes all the earth's liquid water (oceans, smaller bodies of fresh water, and underground aquifers), frozen water (polar ice caps, floating ice, and frozen upper layer of soil known as permafrost), and the small amounts of water vapor in the earth's atmosphere.

Immigration process of entering one country from another to take up permanent residence.

Incineration controlled process by which combustible wastes are burned and changed into gases.

Incomplete protein protein lacking one or more of the eight essential amino acids.

Industrial smog air pollution, primarily from sulfur oxides and particulates, produced by the burning of coal and oil in households, industries, and power plants.

Infant mortality rate number of deaths of infants under one year of age in a given year per 1,000 births in the same year.

Infectious disease disease resulting from presence of disease-causing organisms or agents, such as bacteria, viruses, and parasitic worms. see also *Non-vector-transmitted infectious disease, Vector-transmitted infectious disease.*

Information feedback information sent back into a homeostatic system so that the system can respond to a new environmental condition.

Inorganic compounds substances that consist of chemical combinations of two or more elements other than those used to form organic compounds.

Inorganic fertilizer synthetic plant nutrients; examples are ammonium sulfate and calcium nitrate. compare *Organic fertilizer*.

Insecticide substance or mixture of substances intended to prevent, destroy, or repel insects.

Integrated pest management (IPM) combination of natural, biological, chemical, and cultural controls designed for a specific pest problem.

Internal costs costs of production that are directly paid by the user or producer.

Interspecific competition see *Competition*.

Inversion see *Thermal inversion*.

Ionizing radiation high-energy radiation that can dislodge one or more electrons from atoms it hits to form charged particles called ions.

Ions species of atoms with either negative or positive electrical charges.

Isotopes two or more forms of a chemical element that have the same number of protons but different numbers of neutrons in their nuclei.

J-shaped curve curve with the shape of the letter J that depicts exponential or geometric growth (1, 2, 4, 8, 16, 32, . . .).

Kerogen rubbery, solid mixture of hydrocarbons that is intimately mixed with a limestonelike sedimentary rock. When the rock is heated to high temperatures, the kerogen is vaporized and much of the vapor can be condensed to yield shale oil, which can be refined to give petroleumlike products. see also *Oil shale, Shale oil*.

Kilocalorie (kcal) unit of energy equal to 1,000 calories. see *Calorie*.

Kilowatt (kW) unit of electrical power equal to 1,000 watts. See *Watt*.

Kinetic energy energy that matter has because of its motion and mass.

Kwashiorkor nutritional deficiency (malnutrition) disease that occurs in infants and very young children when they are weaned from mother's milk to a starchy diet that is relatively high in calories but low in protein.

Landfill land waste disposal site that is located without regard to possible pollution of groundwater and surface water due to runoff and leaching; waste is covered intermittently with a layer of earth to reduce scavenger, aesthetic, disease, and air pollution problems. compare *Open dump, Sanitary landfill, Secured landfill*.

Land-use planning process for deciding the best use of each parcel of land in an area.

Laterite soil found in some tropical areas in which an insoluble concentration of such metals as iron and aluminum is present; soil fertility is generally poor.

Law of conservation of energy see *First law of thermodynamics*.

Law of conservation of matter in any ordinary physical or chemical change, matter is neither created nor destroyed but merely changed from one form to another.

LD-50 (lethal dosage, 50 percent) amount of exposure to a toxic chemical that results in the death of half the exposed population.

Leaching extraction or flushing out of dissolved or suspended materials from the soil, solid waste, or another medium by water or other liquids as they percolate down through the medium to groundwater.

Less developed country (LDC) nation that, compared with more developed countries, typically has: **(1)** a low average per capita income, **(2)** a high rate of population growth, **(3)** a large fraction of its labor force employed in agriculture, **(4)** a high level of adult illiteracy, and **(5)** a weak economy and financial base (because only a few items are available for export).

Lifetime cost initial cost plus lifetime operating costs.

Limiting factor factor such as temperature, light, water, or a chemical that limits the existence, growth, abundance, or distribution of an organism.

Limiting factor principle the existence, growth, abundance, or distribution of an organism can be determined by whether the levels of one or more limiting factors go above or below the levels required by the organism.

Limnetic zone open-water surface layer of a lake through which sunlight can penetrate.

Lithosphere region of soil and rock consisting of the earth's crust, a mantle of partially molten rock beneath this crust, and the earth's inner core of molten rock called magma.

Littoral zone area on or near the shore of a body of water.

Magma molten rock material within the earth's core.

Malnutrition condition in which quality of diet is inadequate and an individual's minimum daily requirements (for proteins, fats, vitamins, minerals, and other specific nutrients necessary for good health) are not met. compare *Overnutrition, Undernutrition*.

Malthusian theory of population the theory of economist Thomas Malthus that population tends to increase as a geometric progression while food tends to increase as an arithmetic progression. The conclusion is that human beings are destined to misery and poverty unless population growth is controlled.

Marasmus nutritional deficiency disease that results from a diet that is low in both calories and protein.

Mariculture deliberate cultivation of fish and shellfish in estuarine and coastal areas. compare *Aquaculture*.

Mass number sum of the number of neutrons and the number of protons in the nucleus of an atomic isotope, giving the approximate mass of that isotope.

Mass transit transportation systems (such as buses, trains, and trolleys) that use vehicles that carry large numbers of people.

Matter anything that has mass and occupies space.

Maximum sustained yield maximum rate at which a renewable resource can be used without impairing or damaging its ability to be renewed.

Megawatt (MW) unit of electrical power equal to 1,000 kilowatts, or 1 million watts. see *Watt*.

Methyl mercury (CH_3Hg^+) deadly form of mercury that apparently can be formed by microscopic organisms from less harmful elemental mercury and inorganic mercury salts.

Metropolitan area a central city and surrounding city in the United States containing 50,000 or more residents and with an average population density of at least 1,000 persons per square mile.

Microconsumer see *Decomposer*.

Microorganism generally, any living thing of microscopic size; examples include bacteria, yeasts, simple fungi, some algae, slime molds, and protozoans.

Migration rate difference between the numbers of people leaving and entering a given country or area per 1,000 persons in its population at midyear.

Mineral either a chemical element or

chemical compound (combination of elements) in solid form.

Mineral deposit   any natural occurrence in the lithosphere of an element or compound in solid form.

Mineral resource   nonrenewable chemical element or compound in solid form that is used by humans. Mineral resources are classified as metallic (such as iron and tin) or nonmetallic (such as fossil fuels, sand, and salt).

Minimum tillage farming   planting crops by disturbing the soil as little as possible and keeping crop residues and litter on the ground instead of turning them under by plowing.

Molecule   chemical combination of two or more atoms of the same chemical element (such as O_2) or different chemical elements (such as H_2O).

Monoculture   cultivation of a single crop (such as maize or cotton) to the exclusion of other crops on a piece of land.

More developed country (MDC)   nation that, compared with less developed countries, typically has: **(1)** a high average per capita income, **(2)** a low rate of population growth, **(3)** a small fraction of its labor force employed in agriculture, **(4)** a low level of adult illiteracy, and **(5)** a strong economy.

Mortality   the death rate.

Multiple use   principle for managing a forest so that it is used for a variety of purposes, including timbering, mining, recreation, grazing, wildlife preservation, and soil and water conservation.

Municipal waste   combined residential and commercial waste materials generated in a given municipal area.

Mutagen   any substance capable of increasing the rate of genetic mutation of living organisms.

Mutation   process of change in the genetic material that determines the characteristics of a species. Mutations caused by chemical compounds are generally regressive; that is, they produce bizarre, grotesque, or nonviable forms of the parent organism.

Mutualism   a symbiotic relationship between two different species that benefits both species. see *Commensalism, Parasitism, Symbiosis*.

Natural community   see *Community*.

Natural eutrophication
see *Eutrophication*.

Natural gas   natural deposits of gases consisting of 50 to 90 percent methane (a hydrocarbon with the chemical formula CH_4) and small amounts of other, more complex hydrocarbons such as propane (C_3H_8) and butane (C_4H_{10}).

Natural increase (or decrease)   difference between the birth rate and the death rate in a given population during a given period.

Natural resource   see *Resources*.

Natural selection   mechanism for evolutionary change in which individual organisms in a single population die off over time because they cannot tolerate a new stress and are replaced by individuals whose genetic traits allow them to cope with the stress and to pass these adaptive traits on to their offspring.

Negative feedback   flow of information into a system that causes the system to counteract the effects of an input or change in external conditions. compare *Positive feedback*.

Neritic zone   portion of the ocean that includes the estuarine zone and the continental shelf.

Net energy   see *Net useful energy*.

Net migration   in a given population, the difference between the numbers of persons immigrating and emigrating during a given period.

Net population change   difference between the total number of live births and the total number of deaths throughout the world or a given part of the world during a specified period (usually a year).

Net useful energy   total useful energy produced during the lifetime of an entire energy system minus the useful energy used, lost, and wasted in making the useful energy available.

Neutron   elementary particle present in all atomic nuclei (except hydrogen-l). It has a relative mass of 1 and no electric charge.

Niche   see *Ecological niche*.

Nitrogen fixation   process in which bacteria and other soil microorganisms convert atmospheric nitrogen into nitrates, which become available to growing plants.

Nondegradable pollutant   pollutant that is not broken down by natural processes; examples are inorganic substances, salts of heavy metals, sediments, some bacteria and viruses, and some synthetic organic chemicals. compare *Degradable pollutant*.

Noninfectious disease   illness not caused by a disease-causing organism and except for genetic diseases, not transmitted from one person to another. Examples include heart disease, bronchitis, cancer, diabetes, asthma, multiple sclerosis, and hemophilia.

Nonmetropolitan area   a combination of rural areas and other urban places.

Nonpoint source   source of pollution in which wastes are not released at one specific, identifiable point but from a number of points that are spread out and difficult to identify and control.

Nonrenewable resource   resource that is not replaced by natural processes or for which the rate of replacement is slower than its rate of use. compare *Renewable resource*.

Nonspontaneous process   process that requires an outside input of energy to occur. compare *Spontaneous process*.

Nonthreshold pollutant   substance or condition harmful to a particular organism at any level or concentration.

Non-vector-transmitted infectious disease   disease that is transmitted from person to person without an intermediate nonhuman live carrier. The transmission usually takes place by: **(1)** close physical contact with infected persons (syphilis, gonorrhea, mononucleosis), **(2)** contact with water, food, soil, clothing, bedding, or other substance contaminated by fecal material or saliva from infected persons (cholera, typhoid fever), or **(3)** inhalation of air containing tiny droplets of contaminated fluid, which are expelled when infected persons cough, sneeze, or talk (common cold, influenza).

Nuclear energy   energy released when atomic nuclei undergo fission or fusion.

Nuclear fission   process in which the nucleus of a heavy isotope splits into two or more nuclei of lighter elements, with the release of neutrons and substantial amounts of energy. The most important fissionable materials are uranium-235 and plutonium-239.

Nuclear fusion   process in which the nuclei of two light, nonradioactive isotopes (such as hydrogen isotopes) are forced together at ultrahigh temperatures to form the nucleus of a slightly heavier element (such as helium); the release of substantial amounts of energy accompanies such reactions.

Nucleus   extremely tiny center of an atom, which contains one or more positively charged protons and in most cases one or more neutrons with no electrical charge. The nucleus contains most of an atom's mass.

Nutrient   element or compound that is an essential raw material for organism growth and development. Examples are carbon, oxygen, nitrogen, phosphorus, and the dissolved solids and gases in water.

Ocean thermal gradients   temperature difference between warm surface waters and cold bottom waters in an ocean. If the difference is large enough, this storage of solar heat could be tapped as a source of energy.

Oil   see *Petroleum*.

Oil shale   underground formation of limestonelike sedimentary rock (marlstone) that contains a rubbery, solid mixture of hydrocarbons known as kerogen. see also *Kerogen, Shale oil*.

Oligotrophic lake   a lake with a low supply of plant nutrients.

Omnivore   creature such as a human that can use both plants and other animals as food sources.

Open dump   land disposal site where wastes are deposited and left uncovered with little or no regard for control of scavenger, aesthetic, disease, air pollution, or water pollution problems. compare *Landfill, Sanitary landfill, Secured landfill*.

Open pit mining   surface mining of materials (primarily stone, sand, gravel, iron, and copper) that creates a large pit.

Open sea (oceanic zone)   the part of an ocean that is beyond the continental shelf.

Ore   mineral deposit containing a high enough proportion of an element to permit it to be mined and sold at a profit.

Organic compounds   molecules that typically contain atoms of the elements carbon and hydrogen; carbon, hydrogen, and oxygen; or carbon, hydrogen, oxygen, and nitrogen.

Organic farming   method of producing crops and livestock naturally by using organic fertilizer (manure, legumes, composting, crop residues), crop rotation, and natural pest control (good bugs that eat bad bugs, plants that repel bugs, and environmental controls such as crop rotation) instead of using commercial fertilizer and synthetic pesticides and herbicides.

Organic fertilizer   animal manure or other organic material used as a plant nutrient. compare *Inorganic fertilizer*.

Organophosphates   diverse group of nonpersistent synthetic chemical insecticides that act chiefly by breaking down nerve and muscle responses; examples are parathion and malathion.

Overfishing   harvesting so many fish of a species that there is not enough

breeding stock left to repopulate the species for the next year's catch.

Overnutrition   diet so high in calories, saturated (animal) fats, salt, sugar, and processed foods, and so low in vegetables and fruits that the consumer runs high risks of diabetes, hypertension, heart disease, and other health hazards. compare *Malnutrition, Undernutrition*.

Oxygen-demanding waste   organic water pollutants that are usually degraded by bacteria if there is sufficient dissolved oxygen (DO) in the water. see also *Biological oxygen demand*.

Ozone layer   layer of gaseous ozone (O_3) in the upper atmosphere that protects life on earth by filtering out harmful ultraviolet radiation from the sun.

PANs   group of chemicals (photochemical oxidants) known as peroxyacyl nitrates, found in photochemical smog.

Parasitism   symbiotic relationship between two different species in which one species (the parasite) benefits and the other species (the host) is harmed. see *Commensalism, Mutualism, Symbiosis*.

Paratransit   transit system such as carpools, vanpools, jitneys, and dial-a-ride systems that carry a relatively small number of passengers per vehicular unit.

Particulate matter   solid particles or liquid droplets suspended or carried in the air.

Parts per billion (ppb)   number of parts of a chemical found in one billion parts of a solid, liquid, or gaseous mixture.

Parts per million (ppm)   number of parts of a chemical found in one million parts of a solid, liquid, or gaseous mixture.

Pathogen   organism that produces disease.

PCBs   (polychlorinated   biphenyls) mixture of at least 50 widely used compounds containing chlorine that can be biologically magnified in the food chain with unknown effects.

Pest   unwanted organism that directly or indirectly interferes with human activities.

Pesticide   any chemical designed to kill weeds, insects, fungi, rodents, and other organisms that humans consider to be undesirable; examples are chlorinated hydrocarbons, carbamates, and organophosphates.

Petrochemicals   chemicals made from natural gas or petroleum.

Petroleum (crude oil)   gooey, dark greenish-brown, foul-smelling liquid found in natural underground reservoirs; it contains a complex mixture of hydrocarbon compounds plus small amounts of oxygen, sulfur, and nitrogen compounds.

pH   numeric value that indicates the relative acidity or alkalinity of a substance on a scale of 0 to 14, with the neutral point at 7.0. Values lower than 7.0 indicate the presence of acids and greater than 7.0 the presence of alkalis (bases).

Photochemical reaction   chemical reaction activated by light.

Photochemical smog   complex mixture of air pollutants (oxidants) produced in the atmosphere by the reaction of hydrocarbons and nitrogen oxides under the influence of sunlight. Especially harmful photochemical oxidants include ozone, peroxyacyl nitrates (PANs), and various aldehydes.

Photosynthesis   complex process that occurs in the cells of green plants whereby sunlight is used to combine carbon dioxide (CO_2) and water (H_2O) to produce oxygen (O_2) and simple sugar or food molecules, such as glucose ($C_6H_{12}O_6$).

Photovoltaic cell (solar cell)   device in which radiant (solar) energy is converted directly into electrical energy.

Physical weathering   breaking down of parent rock into bits and pieces by exposure to temperature changes and the physical action of moving ice and water, growing roots, and human activities such as farming and construction.

Phytoplankton   free-floating, mostly microscopic aquatic plants.

Plankton   microscopic floating plant and animal organisms of lakes, rivers, and oceans.

Plasma   "gas" of charged particles (ions) of elements that exists only at such high temperatures (40 million to several billion degrees Celsius) that all electrons are stripped from the atomic nuclei.

Point source   source of pollution that involves discharge of wastes from an identifiable point, such as a smokestack or sewage treatment plant.

Pollution   undesirable change in the physical, chemical, or biological characteristics of the air, water, or land that can harmfully affect the health, survival, or activities of humans or other living organisms.

Population   group of individual organisms of the same kind (species) that interbreed and occupy a given area at a given time.

Population crash (dieback) extensive deaths resulting when a population exceeds the ability of the environment to support it.

Population density number of organisms in a particular population per square kilometer or other unit of area.

Population distribution variation of population density over a given country, region, or other area.

Positive feedback information sent back into a homeostatic system that causes the system to change continuously in the same direction; as a result, the system can go out of control. compare *Negative feedback*.

Potential energy energy stored in an object as a result of its position or the position of its parts.

Power tower see *Solar furnace*.

Predation situation in which an organism of one species (the predator) captures and feeds on an organism of another species (the prey).

Predator organism that lives by killing and eating other organisms.

Primary air pollutant chemical that has been added directly to the air and occurs in a harmful concentration. compare *secondary air pollutant*.

Primary energy source renewable or nonrenewable energy resource used to supplement the direct input of energy from the sun.

Primary succession ecological succession that begins on an area (such as bare rock, lava, or sand) that has never been occupied by a community of organisms.

Primary treatment (of sewage) mechanical treatment in which large solids, like old shoes and sticks of wood, are screened out, and suspended solids settle out as sludge. compare *Secondary treatment*, *Tertiary treatment*.

Prime reproductive age years between ages 20 and 29 during which most women have most of their children. compare *Reproductive age*.

Producer organism that synthesizes its own organic substances from inorganic substances, such as a plant.

Profundal zone deep-water region of a lake, which is not penetrated by sunlight.

Protein-calorie malnutrition combination of undernutrition and malnutrition.

Pulmonary emphysema lung disease in which the alveoli enlarge and lose their elasticity, thus impairing the transfer of oxygen to the blood.

Pyramid of biomass diagram representing the biomass, or total dry weight of all living organisms, that can be supported at each trophic level in a food chain.

Pyramid of numbers diagram representing the number of organisms of a particular type that can be supported at each trophic level from a given input of solar energy at the producer trophic level in a food chain.

Pyrolysis high-temperature decomposition of material in the absence of oxygen.

Radiation propagation of energy through matter and space in the form of fast-moving particles (particulate radiation) or waves (electromagnetic radiation).

Radioactive waste products of nuclear power plants, research, medicine, weapons production, or other processes involving nuclear reactions.

Radioactivity property of certain chemical elements of spontaneously emitting radiation from unstable atomic nuclei. The emitted radiation may damage organisms. see *Radioisotope*.

Radioisotopes isotopes whose nuclei are capable of spontaneously emitting at a certain rate radiation in the form of alpha particles, beta particles, or gamma rays, to form nonradioactive or radioactive isotopes of a different kind.

Range of tolerance range or span of conditions that must be maintained for an organism to stay alive and grow, develop, and function normally.

Rangeland land on which the vegetation is predominantly grasses, grasslike plants, or shrubs such as sagebrush.

Rate of natural change measure of population change obtained by finding the difference between the birth rate and the death rate.

Rate of population change difference between the birth rate and the death rate plus net migration rate for a particular country or area.

Recycle to collect and treat a resource so it can be used again, as when used glass bottles are collected, melted down, and made into new glass bottles. compare *Reuse*.

Renewable resource resource that potentially cannot be used up because it is constantly or cyclically replenished. Either it comes from an essentially inexhaustible source (such as solar energy), or it can be renewed by natural or human-devised cyclical processes if it is not used

faster than it is renewed. compare *Nonrenewable resource*.

Replacement level of fertility fertility rate of 2.11 children per woman (in the United States), which will supply just enough births to replace the parents and compensate for premature deaths, assuming no net effect of migration.

Reproductive age ages 15 to 44, when most women have all their children. compare *Prime reproductive age*.

Reproductive potential see *Biotic potential*.

Reserves amount of a particular resource in known locations that can be extracted at a profit with present technology and prices.

Resource (natural) any form of matter or energy obtained from the environment that meets human needs. see also *Mineral resource, Nonrenewable resource, Renewable resource*.

Resource recovery extraction of useful materials or energy from waste materials. This may involve recycling or conversion into different and sometimes unrelated products or uses. compare *Recycle, Reuse*.

Resources total amount of a particular resource material that exists on earth.

Respiration complex process that occurs in the cells of plants and animals in which food molecules such as glucose ($C_6H_{12}O_6$) combine with oxygen (O_2) and break down into carbon dioxide (CO_2) and water (H_2O), releasing usable energy.

Reuse to use a product over and over again in the same form, as when returnable glass bottles are washed and refilled. compare *Recycle*.

Runoff surface water entering rivers, freshwater lakes, or reservoirs from land surfaces.

Rural area area in the United States with a population of fewer than 2,500 people.

S-shaped curve leveling off of an exponential or J-shaped curve.

Salinity amount of dissolved salts in a given volume of water.

Sanitary landfill land waste disposal site that is located to minimize water pollution from runoff and leaching; waste is spread in thin layers, compacted, and covered with a fresh layer of soil each day. compare *Landfill, Open dump, Secured landfill*.

Saprotrophic organisms tiny organisms—such as bacteria, fungi, termites, and maggots—that break down complex chemicals in the bodies of dead animals and plants into simpler chemicals.

Scrubber   common antipollution device that uses a liquid spray to remove pollutants from a stream of air.

Second energy law   see *Second law of thermodynamics*.

Second-law energy efficiency   ratio of the minimum amount of useful energy needed to perform a task in the most efficient way that is theoretically possible to the actual amount of useful energy used to perform the task.

Second law of thermodynamics (law of energy degradation)   **(1)** In all conversions of heat energy to work, some of the energy is degraded to a more dispersed and less useful form, usually heat energy given off at a low temperature to the surroundings, or environment; *or*, you can't break even in terms of energy quality. **(2)** Any system and its surrounding (environment) as a whole spontaneously tends toward increasing randomness, disorder, or entropy; *or*, if you think things are mixed up now, just wait.

Secondary air pollutant   harmful chemical formed in the atmosphere through a chemical reaction among air components. compare *Primary air pollutant*.

Secondary consumer   see *Carnivore*.

Secondary succession   ecological succession that begins on an area (such as abandoned farmland, a new pond, or land disrupted by fire) that had been occupied by a community of organisms.

Secondary treatment (of sewage)   second step in most waste treatment systems, in which bacteria break down the organic parts of sewage wastes; usually accomplished by bringing the sewage and bacteria together in trickling filters or in the activated sludge process. compare *Primary treatment, Tertiary treatment*.

Secured landfill   a land site for the storage of hazardous solid and liquid wastes, which are normally placed in containers and buried in a restricted-access area that is continually monitored. Such landfills are located above geologic strata that are supposed to prevent the leaching of wastes into groundwater. compare *Landfill, Open dump, Sanitary landfill*.

Sediment   soil particles, sand, and minerals washed from the land into aquatic systems as a result of natural and human activities.

Sedimentary cycle   biogeochemical cycle in which materials primarily are moved from land to sea and back again.

Examples include the phosphorus and sulfur cycles.

Selection cutting   cutting of mature or diseased trees singly or in small groups to encourage younger trees to grow and to produce an uneven-aged stand with trees of different species, ages, and size.

Septic tank   underground receptacle for wastewater from a home. The bacteria in the sewage decompose the organic wastes, and the sludge settles to the bottom of the tank. The effluent flows out of the tank into the ground through drains.

Shale oil   low-sulfur, very viscous, petroleumlike liquid, obtained when kerogen in shale oil rock is vaporized at high temperatures and then condensed. Shale oil can be refined to yield petroleum products. see also *Kerogen, Oil shale*.

Shelterwood cutting   removal of all mature trees in an area in a series of cuts over two or three decades.

Silviculture   cultivation of forests.

Single-cell protein (SCP)   form of protein made from oil by the action of microorganisms.

Slash-and-burn agriculture   in many tropical areas, the practice of clearing a patch of forest overgrowth, burning the residue, and planting crops. The patch is abandoned after 2 or 3 years to prevent depletion of soil fertility.

Sludge   solid matter that settles to the bottom of sedimentation tanks in a sewage treatment plant and must be disposed of by digestion or other methods or recycled to the land.

Smog   originally a combination of *smoke* and *fog*; now applied also to the photochemical haze produced by the action of sun and atmosphere on automobile and industrial exhausts. compare *Industrial smog, Photochemical smog*.

Soil   complex mixture of small pieces of rock, minerals (inorganic compounds), organic compounds, living organisms, air, and water. It is a dynamic body that is always changing in response to climate, vegetation, local topography, parent rock material, age, and human use and abuse.

Soil erosion   processes by which soil is removed from one place by forces such as wind, water, waves, glaciers, and construction activity and eventually deposited at some new place.

Soil horizons   horizontal layers that make up a particular type of soil.

Soil profile   cross-sectional view of the horizons in a soil.

Soil structure   the way soil particles clump together in larger lumps and clods.

Soil texture   size of a soil's individual mineral particles and the proportion in which particles of different sizes are found in a soil.

Solar energy   direct radiant energy from the sun plus indirect forms of energy—such as wind, falling or flowing water (hydropower), ocean thermal gradients, and biomass—that are produced when solar energy interacts with the earth.

Solar furnace   system for concentrating direct solar energy to produce electricity or high-temperature heat for direct use. Also called a power tower.

Solar pond   saline body of water in which stored solar energy can be extracted as a result of the temperature difference (thermal gradient) between a layer of saline water on the bottom and a layer of less saline water on top.

Solid waste   any unwanted or discarded material that is not a liquid or a gas.

Species   all organisms of a given kind; a group of plants or animals that breed or are bred together but are not bred successfully with organisms outside their group.

Species diversity   ratio between the number of species in a community and the number of individuals in each species. (For example, low diversity occurs when there are few species but many individuals per species.)

Spontaneous process   any process that can occur naturally without an outside input of energy (for example, water flowing downhill or heat energy flowing from hot to cold). compare *Nonspontaneous process*.

Static reserve index   estimated number of years until known world reserves of a nonrenewable resource will be 80 percent depleted if depletion proceeds at the present annual rate. compare *Exponential reserve index*.

Strip cropping   planting regular crops and close-growing plants such as hay or nitrogen-fixing legumes in alternating rows or bands.

Strip (surface) mining   mining in which the earth's surface is removed and the excavated area left bare.

Subsidence   sinking down of part of the earth's crust due to underground excavation, such as a coal mine.

Subsurface mining   underground extraction of a metal ore or fuel resource such as coal.

Succession   see *Ecological succession*.

Superinsulated house   house that contains massive amounts of insulation, is extremely airtight, typically uses active or passive solar collectors to heat water, and has an air-to-air heat exchanger to prevent buildup of excessive moisture and indoor air pollutants.

Surface mining   the process of removing the overburden of topsoil, subsoil, and other strata to permit the extraction of underlying mineral deposits. see *Area surface mining, Contour surface mining, Dredging, Open pit mining.*

Surface water   water that flows in streams and rivers and in natural lakes, in wetlands, and in reservoirs constructed by humans.

Surroundings (environment)   everything outside a specified system or collection of matter.

Sustained yield   principle for managing a forest in which depletion is avoided by striking a balance between new planting and growing and the amount of wood removed by cutting, pests, disease, and fire.

Symbiosis   interaction in which two different species exist in close physical contact, with one living on or in the other so that both species benefit from the association. see *Commensalism, Mutualism, Parasitism.*

Synergistic effect   result of the interaction of two or more substances or factors that could not have been produced by these factors acting separately. compare *Antagonistic effect.*

Synergy   interaction in which the total effect is greater than the sum of two effects taken independently.

Synfuels   fuels such as synthetic natural gas and synthetic fuel oil produced from coal or sources other than natural gas or crude oil.

System   collection of matter under study.

Tar sands (oil sands)   enormous swamps that contain fine clay and sand mixed with water and highly variable amounts of a black, high-sulfur, molasseslike tar known as heavy oil, or bitumen, which is about 83 percent carbon. The heavy oil can be extracted from the tar sand by heating and flotation, and purified and upgraded to synthetic crude oil.

Terracing   planting crops on a long steep slope that has been converted into a series of broad, level terraces at right angles to the slope of the land.

Tertiary treatment (of sewage)   removal from wastewater of traces of organic materials and dissolved solids that remain after *primary* and *secondary treatment.*

Thermal gradient   temperature difference between two areas.

Thermal inversion   layer of cool air trapped under a layer of less dense warm air, thus reversing the normal situation. In a prolonged inversion, air pollution may rise to harmful levels.

Thermal pollution   increase in air or water temperature that disturbs the climate or ecology of an area.

Threshold effect   effect that is not observed until a certain level or concentration is attained. see *Threshold pollutant.*

Threshold pollutant   substance that is harmful to a particular organism only above a certain concentration, or threshold level.

Time delay   lag between the receipt of an information signal or stimulus and the system's making of corrective action by negative feedback.

Tolerance limit   point at and beyond which a chemical or physical condition (such as heat) becomes harmful to a living organism.

Total fertility rate (TFR)   projection of the average number of children that would be born to each woman if throughout her reproductive years (ages 15 to 44) she were to bear children at the same rate as other women did in each of these years.

Transpiration   direct transfer of water from the leaves of living plants to the atmosphere.

Tritium (T: hydrogen-3)   isotope of hydrogen with a nucleus containing one proton and two neutrons, thus having a mass number of 3. compare *Deuterium.*

Trophic level   level at which energy in the form of food is transferred from one organism to another in a food chain or food web.

Troposphere   the lower layer of the atmosphere, which contains about 95 percent of the earth's air, and extends about 8 to 12 kilometers (5 to 7 miles) above the earth's surface.

Undernutrition   condition characterized by an insufficient quantity or caloric intake of food to meet an individual's minimum daily energy requirement. compare *Malnutrition, Overnutrition.*

Upwelling region   area adjacent to a continent where ocean bottom waters rich in nutrients are brought to the surface.

Urban area   place with a population of 20,000 or more.

Vector   living organism (usually an insect) that carries an infectious disease from one host (person or animal) to another.

Vector-transmitted infectious disease   disease carried from one host to another by a living organism (usually an insect), called a vector. Examples include malaria, schistosomiasis, and African sleeping sickness.

Water cycle   see *Hydrologic cycle.*

Water pollution   degradation of a body of water by a substance or condition to such a degree that the water fails to meet specified standards or cannot be used for a specific purpose.

Water table   level below the earth's surface at which the ground becomes saturated with water.

Waterlogging   saturation of soil with irrigation water so that the water table rises close to the surface.

Watershed   land area from which water drains toward a common watercourse in a natural basin.

Wavelength   distance between the crest (or trough) of one wave and that of the next.

Wetland   area that is regularly wet or flooded and has a water table that stands at or above the land surface for at least part of the year. Coastal wetlands extend back from estuaries and include salt marshes, tidal basins, marshes, and mangrove swamps. Inland freshwater wetlands consist of swamps, marshes, and bogs.

Whole-tree harvesting   use of machines to pull entire trees from the ground and reduce them to small chips.

Wilderness   area where the earth and its community of life have not been seriously disturbed by humans and where humans are only temporary visitors.

Windbreaks   rows of trees or hedges planted to reduce soil erosion on cultivated land that is exposed to high winds.

Zero population growth (ZPG)   state in which the birth rate (plus immigration) equals the death rate (plus emigration) so that population is no longer increasing.

Index

Note: Numbers in **boldface** type indicate pages where term is defined. Alphabetization is in letter-by-letter mode. (f) = figure; (t) = table; (n) = footnote.

A

Abiotic components of ecosystem, **31**, 31–34
Abortion, 86–87(t)
Abyssal zone, 318(f), **319**
Acid deposition, 145, 287–288, **294**–296
Acid fog, **294**–296
Acidity, relative (pH), 294, 295(f)
Acid precipitation. *See* Acid deposition
Acid rain. *See* Acid deposition
Active solar energy systems, 239, 240, 241(f)
Acute diseases, **265**, 266(t)
Acute effect of pollutants, 8
Additives, food, **275**–280
   consumer protection from, 277–280
   natural versus synthetic foods, 277
   uses and types, 275–277, 276(t)
Affluent diets, simplifying, 128–130
Agency for International Development (AID), 85
Agent Orange, 336–337, 352
Age structure, effect on population, 70, 76–79, 77(f,t)
   diagrams of, 77(f)–78(f)
Agriculture:
   beginnings of, 13
   cultivating more, 132–135
   industrialized systems, 126–128
   societies, 13–14
   soil erosion, 96–100(f,t)
   systems of world, 126–128
   water pollution from, 323–325
   See also Animals; Crops; Farming; Fertilizers; Manure
Agriculture, Department of, 15, 16, 99, 100, 156, 278, 332, 340, 344
Air:
   global circulation patterns of, 36–37(f)
   in soil, 92
Air conditioning, energy demands of, 194, 202(f)
   and solar system, 240–241
Air masses, 36
Air pollution, **283**–305
   and cancer, 275
   control of photochemical smog, 303–305
   control of sulfur dioxide and particulates in smog, 301–303
   effects on ecosystems, 294–296
   effects on ozone layer and global climate, 296–299

effects on property, plants, animals, human health, 290–294
industrial and photochemical smog, 285–290
principles of control, 299–301
as threat to forests, 145
types and sources of, 283–285
Air quality standards, U.S., 370
Albedo, **38**(f)
Alcohol:
   as auto fuel, 254
   and cancer risk, 273
Aldrin, 337
Alfvén, Hannes, 222
Alkalinity, relative (pH), 294, 295(f)
Allied Chemical Co., 336
Alligator, American, 63
Alpha particles, 224
Alpine grasslands, 42–43
Alveoli, 292
American Indians, Grand Council Fire of, 43
American Petroleum Institute, 196
Amoco Cadiz accident, 321–322
Anadromous species, 162
Animals:
   agricultural, 126. *See also* Agriculture
   domestication of, 13
   endangered. *See* Endangered species
   game, 152
   uses of wastes from, 101–102, 253
   wild. *See* Wildlife
Anthracite coal, 213, 214, 215(f)
Aquaculture, 132
Aquariums, to preserve species, 159–160
Aquatic ecosystems, 30
Aquatic plants for biomass, 253
Aquifers, **108**
   contamination of, 316
   depletion of, 116
Arboretums, 159–160
Arctic grasslands, 42–43
Area strip mining, 179, 180(f)
Army Corps of Engineers, 114, 156, 247
Arsenic, cancer and, 274
Asbestos:
   and cancer, 274
   as indoor pollutant, 285
Ashworth, William, 119
Atmosphere, **27**, 28(f)
   in biogeochemical cycle, 48–49
   and emissivity, 38

greenhouse effect, 38, 297–298
Atomic Energy Commission, 233
Atomic Industrial Forum, 222
Atoms, **29**, 30(f), 31, 223
Audubon Society, 17, 154
Automobile:
   controlling emissions from, 303, 305
   discouraging use of, 172
   engine, energy efficiency of, 22–23, 202(t)
   use of, 170–171
Autotrophs, **34**

B

Baby boom. *See* Birth rate; Population, human
Background radiation, 224, 225(t)
Bacon, Sir Francis, 56
Bacteria, **36**
Balanced multiple-use land-use ethic, 140
Baron, Robert Alex, 280
Barrier islands, 318
Basic solution (pH), 294, 295(f)
Bathyal zone, 318(f), **319**
Bay Area Rapid Transit (BART) system, 171
Beck, A. J., 63
Bedrock, 95(f), 96
Bennett, Hugh H., 16, 99, 100
Benzine, cancer and, 274
Beta particles, 224
Bicycles, use of, 171
Big Muskie, 179(f)
Binary-cycle system, geothermal energy, 219(f)
Biofuels, 254–255
Biogas, 254
Biogeochemical cycles, 48–49
Biological control of insects, 339–340(f)
Biological diversity. *See* Diversity
Biological magnification, **60**–61(f), 334, 335(f)
Biological oxygen demand (BOD), 307
Biomass energy, 252–255
   plantations, 152
Biomes, 31, 40–43
Biosphere, **27**, 28(f)
Biotic components of ecosystem, **31**, 34(f)–36, 35(f)
Biotic potential of species, 81
Birth rate, effect on population, 70–72, 73(f)

Birth rate (*continued*)
net population change, 70–71
percentage growth rates and doubling times, 71–72
rate of natural change, 71(t)
Birth rates, controlling, 84–88
Bitumen, 211
Bituminous coal, 214, 215(f)
Boiling-water reactor (BWR), 227
Boreal forests, 42
Borgstrom, Georg, 115, 164
Botanical gardens, 159–160
Bowel cancer, 271, 272
Bray, Phil, 222
Breast cancer, 271, 272
Breast-feeding, importance of, 126
Breeder nuclear fission, reactors, 235–236
British Water Act of 1973, 118
Bronchitis, chronic, **293**
Brookhaven National Laboratory, 295
The Brothers Karamazov, 163
Brower, David, 15
Brown, Lester R., 104, 128, 134, 208, 238
Brown's Ferry nuclear reactor accident, 229
Bureau of Land Management (BLM), 16, 17, 148
Bureau of Mines, 189
Bureau of the Census, 75, 165
Buses, 171–172

C

Cadmium, 353
Cahn, Robert, 361
Caldicott, Helen, 226
California Department of Industrial Relations, 351
California Energy Commission, 251
Calorific Recovery by Anaerobic Processes (CRAP), 254
Canadian aurora trout, 156
Canadian deuterium uranium (CANDU) reactor, 227(n)
Cancer, **270**–275
diagnosis and treatment, 272(t)
and diet, 273–274
incidence and geography, 270–272, 271(f)
nature and effects, 270
and pollution, 275
risk factors, 272–273
and smoking, 273
and the workplace, 274–275
Cans, aluminum, versus bottles, 349
Carbamates, 330, 331(t)
Carbon cycle, 49–50(f)
Carbon dioxide:
in carbon and oxygen cycles, 49–50(f)
greenhouse effect, 297–298
Carbon monoxide, 285, 292–293
Carcinogens, 270
Carnivores, 31, **34**(f)
Carolina parakeet, 155
Carr, Donald E., 191
Carrying capacity, **54**
of humans, 82, 83(f)
of rangeland, 148–149

Carson, Rachel, 16, 328, 335
Carter, Jimmy, 64
Catlin, George, 14
Cells, 29, 30(f)
Cellular respiration, 49
Center for Science in the Public Interest, 279
Centers for Disease Control, 268
Central city-to-suburbs population shift, 165
Chandler, William U., 264
Channelization, 112–114
Chaparral, 42
Chase, Stuart, 9
Chemical Control Corporation dump fire, 351
Chemical cycling:
in ecosystem, 31, 32(f)
and energy flow, 44(f)–45(f)
Chemical energy, **21**(f)
Chemicals:
disrupting food webs with synthetic, 63
in ecosystems, 31–34, 48–52
equations, formulas, reactions, **33**
fertilizer, 102–104
hazardous dumps, 64
in soils, 92, 102–104, 135–137
Chesapeake Bay, pollution of, 320
Childbirth. *See* Birthrate
China, population control efforts of, 87–88, 89–90
Chlordane, 337
Cholera, 269–270
Chronic diseases, **265**, 266(t)
Chronic effect of pollutants, 8
Cigarette smoking, cancer and, 273
Cities:
building new, 176
daily inputs and outputs of typical U.S., 169(f)
parks serving, 141, 174
pollution, 168–170
repairing, revitalizing, 175–176
See also Urban areas
Citizens for Clean Air, Inc., 283
Civilian Conservation Corps (CCC), 16
Clark, Colin, 82
Clarke, Arthur C., 190
Clay, 94
Clean Air Act of 1970, 10, 283, 300
Clean Air Act of 1977, 300
Clean Water Act of 1977, 325–326
Clearcutting (tree), 143
Cleveland, Grover, 15
Climate, global patterns of, **36**(f)–38(f), 37(f)
air circulation patterns, 36–37(f)
and air pollution, 289–290
and albedo and emissivity, 37–38(f)
and weather, 36
Climax ecosystem, **57**
Clinch River breeder reactor, 235–236
Clouds, 37
seeding of, 118
Cluster development, 174, 175(f)
Coal:
fluidized-bed combustion, 216(f)–217
formation of, 213(f)

gasification and liquefaction, 217
mining, 178–182, 212–214
natural gas from seams, 212
supplies, 197, 214–215(f)
synfuels from, 217
uses, 214
Coal tar, 214
dyes, 278
pitch, and cancer, 274
Coastal wetlands, 318(f)
Coastal zone management, 320
Cogeneration, 203
Cohen, Bernard L., 222, 226
Coke oven emissions and cancer, 274
Cold desert, 41
Colinvaux, Paul, 54
Colon-rectal cancer, 271
Commensalism, **55**
Committee for Energy Awareness, 230
Commoner, Barry, 16, 22
Competition, interspecific, 54–55
Competitive exclusion principle, 53
Complete proteins, 122
Compost, 102, 347
Compounds:
chemical, **32**–34
molecules, **29**
Comprehensive Environmental Response, Compensation and Liability Act, 358
Condor, California, 154(f), 157–158
Congressional Research Service, 196
Coniferous trees, 41
Conservation:
mentality, versus frontier mentality, 14, 361
mineral resources, 185–186
natural resources (U.S.), 14–17
soil, 100–104
water, 118–119
Conservation of energy, law of, 19, **21**–22
energy efficiency, 199–202(t), 200(f)
energy quality, 197–198(f)
net useful energy, 203–205
Conservation of matter, law of, 19, **20**, 44
Consumers, 31, **34**(f)–35(f)
Contact herbicides, 331
Contour cultivation, 100–101, 119
Contour strip mining, 179, 180
Control Data Corporation (CDC), 176
Convention on International Trade in Endangered Species of Wild Flora and Fauna treaty, 158
Cool desert, 39(f)
Cooper, William S., 57
Copaifera langdorfii, 253
Cornucopians, 5–6, 182, 196
Cost-benefit analysis, **299**–300
Cost-effectiveness analysis, **299**(f)–300
Costs. *See* Economics
Council of Agricultural Science and Technology (CAST), 100
Council on Environmental Quality, 149
Cousins, Norman, 1
Cox, Harvey, 80
Critical habitats, 159
Cropland, 134

Crops:
    increasing yields of, 135–137
    residues, burning or converting to biofuel, 253
    rotating, 100, 101
Crown fires, 144
Crude oil. *See* Oil
Crutzen, Paul, 64
Cultural eutrophication, **313**–315, 314(f)
Cycles:
    biogeochemical, 48–49
    carbon and oxygen, 49–50(f)
    nitrogen, 50–51(f)
    phosphorus, 51–52(f)
    water, 52, 53(f)
Cyclone separators, 303

D

Dams, 112–114, 113(t)
Dasmann, Raymond, 110
Day, Lincoln, 81
D-D fusion reaction, 236
DDT (dichlorodiphenyltrichloroethan), 60–61(f), 156, 268, 331–332, 334, 335(f), 337–339
Death rate, effect on population, 70, 70–72, 73(f)
    net population change, 70–71
    percentage growth rates and doubling times, 71–72
    rate of natural change, 71(t)
Decibels (db), 280
Deciduous forest, 39(f), 41
Decomposers, 31, **36**
Deep ecology, 362
Degradable pollutant, **7**
Delaney, James J., 279(n)
Delaney Clause, 279
Delaware Bay, cleanup of, 320
Demographic transition, 84(f)–85
Department of Commerce, 159
Department of Defense, 350, 358
Department of Energy, 196, 204, 220, 231, 232, 233, 246
Department of Housing and Urban Development (HUD), 176
Department of the Interior, 15, 16, 17, 159, 214, 321
Department of Transportation, 351
Depletion:
    of crude oil, 196–197
    curves and rate estimates, 186(f)–187(f)
Desalination, 117
Desert biomes, 41
Desertification, 98, 149
Detergent controversy, 315–316
Detritus, **35**
Detritus feeders, 35–36
Detritus food chain, **45**
Deudney, Daniel, 220, 262
Diablo Canyon power plant, 233
Dial-a-ride systems, 172
Dieldrin, 63, 337
Diesohol, 254–255
Diet:
    and cancer, 273–274
    simplifying affluent, 128–130
Dioxin, 352

Diseases, forest tree, 144–145
Diseases, human, 264–275(f,t)
    acute and chronic, 265
    cancer, 270–275(f,t)
    cholera, 269–270
    malaria, 267(f)–268
    schistosomiasis, 268–269(f)
    social ecology of, 265–267
    types, 264–265(t), 266(t)
Disorder, tendency of system to increasing, 23–24
Disposal, waste. *See* Wastes, disposal
Dissolved oxygen (DO) content of water, 307
District heating system, 203
Diversity:
    species, and ecosystem stability, 62
    versus simplicity, balancing, in ecosystems, 66(t)
Dodo bird, 156
Donora, Pennsylvania, air pollution disaster, 284
Dostoyevsky, Feodor, 163
Doubling times, population, 71–72(f), 73(f)
Drainage basins, 107
Dredging, 179, 180
Drinking water standards, U.S., 371
Dry deposition, 288, 294
D-T fusion reaction, 236(f)
Dubos, René, 9, 263
Ducks Unlimited, 159
Dumps:
    hazardous chemical, 64
    in ocean, 320–321
Dunes, 319(f)–320
Dust Bowl, U.S., 99–100
Dyes, food, as controversial food additives, 278

E

Early-successional species, 161
Earth Day, 9
Earth-sheltered home, 240
Ecclesiastes, 281
Eckholm, Erik P., 305
Ecological argument for preserving wild species, 153
Ecological equivalents, **53**
Ecological land-use ethic and planning, 140, 172–173
Ecological niche, **52**–53
Ecological succession, 57–59(f), 58(f)
Ecological waste management, 324–325
Ecology, **27**, 29–31, 30(f)
Economics:
    development, demographic transition, and, 84(f)–85
    energy costs, resources, alternatives, 203, 234(t), 260(t)–261(t)
    family planning, 87–88
    fish extinction and, 162
    inequities, 2, 134–135
    land-use ethic, 140
    minerals (resources), 188–189(t)
    new international order, 188
    resource supply (reserves), 182–183
    urban succession, 165, 168

wild species, 152
Ecosphere, **27**, 28(f)
Ecosystems, **27**–67(f,t)
    changes in, 57–67
    diversity, 62, 63, 66(t), 136–137
    energy flow in, 31, 45–47(f)
    functioning, 44–56(f)
    humans and, 66(t)–67
    marine, 30, 318(f)–319
    potential problems in, 62–66, 63(t)
    preserving large reserve, 160–161
    reducing variety in, 63
    simplifying, 66(t)–67
    stability, 59–62(f)
    structure, 27–43(f,t)
Ectoparasitism, 35, 56
Edberg, Rolf, 67
Eddington, Arthur S., 26
Edward I, 283
Efficiency, energy:
    first-law, 199–202(t), 200(f), 201(t)
    net useful energy, 203–204(t)
    second-law, 202(t)–203(f)
    See also Conservation
Egret, snowy, 154
Ehrlich, Paul, 16, 136
Ehrlich, Thomas, 188
Eisley, Loren, 106
Electricity, 197–203(f,t)
    cost estimates, 234(t)
    direct solar energy, 239–246(f)
    hydropower, 246–248
    indirect solar energy, 246–250(f)
    ocean thermal energy, 248–250(f)
    ocean tidal power, 255
    solar ponds, 250
    wastes, 252–255
    wind power, 250–252
Electromagnetic fields (nuclear fusion), 237–238
Electromagnetic pulse (EMPs), 206
Electromagnetic radiation, **28**, 29(f)
Electromagnetic spectrum, **28**, 29(f)
Electrons (e), **32**, 223
Elements, 29, **31**–32
    in earth's crust and seawater, 178(f)
Eliot, Charles W., 14
Emerson, Ralph Waldo, 14, 328
Emigration, 72–73
Emissivity, climate and, 38(f)
Endangered species, 152–158(f,t), **153**
Endangered Species Act of 1973, 154, 158–159
Endangered Species Review Committee, 159
Endemic disease, 264
Endoparasitism, 35, 56
Energy, **20**
    concepts, 197–205(f,t)
    conservation (efficiency), 199–204(f,t), 239–243(f), 256–258(t), 368–369
    diet simplification, 128–130
    laws of, 19, 20–26, 44, 46(f)–47(f), 49(f), 198–199
    net useful, 203–204(t)
    plantations, 253
    See also related topics: Conservation; Energy flow; Energy resources; *and specific types*: Coal; Geothermal energy; Natural gas; Nuclear

Energy (*continued*)
    energy; Oil; Solar energy; Water;
    Wind energy
Energy flow:
    chart, 191, 192(f)
    in ecosystem, 31, 32(f)
    to and from earth, 27–28
    rates and energy resource quality,
    197–198
Energy resources:
    average daily per capita use, 193–
    194(f)
    costs, alternatives, 203, 234(t), 258–
    262(t)
    efficiency of different, 199–204(f,t)
    evaluating, 205–207
    hard versus soft paths, 205–207
    quality of different, 197–198(f)
    types and strategies, 191–192(f)
    U.S. alternatives, strategies, 258–
    262(t)
    See also Energy; Energy use; *and spe-*
    cific resources: Coal; Geothermal
    energy; Natural gas; Nuclear en-
    ergy; Oil; Solar energy; Water;
    Wind energy
Energy use:
    history of, 6(f)
    industrialized agriculture and, 126–
    128
    U.S. and Sweden, 259(t)
Entropy, **23**–24, 25, 26
Environmental awareness, four levels
    of, 362
Environmental decade, 16–17
Environmental Defense Fund (EDF),
    16, 75, 88
Environmental impact of resource
    use, 181(f)–182
Environmental law (legislation), 14–
    17, 366
Environmental Program, U.N., 311
Environmental Protection Agency
    (EPA), 16, 17, 64, 117, 280, 281,
    283, 300, 309, 316, 317, 320, 326,
    337, 338, 339, 344, 349, 351, 352,
    353, 354, 357, 358, 359
Environmental resistance, 81
Environmental stress, 62–63(t)
Enzymes, **36**
Epidemic, 264
Epiphytes, 40, 55
Epstein, Samuel, 275
Equilibrium species, 57
Erosion, soil, 96–100
    control of, 100–101
Essential amino acids, 122
Estuarine ecosystems, 30
Estuarine zone, **318**(f)–320
Ethanol, 254–255
Ethics:
    land-use, 138, 140
    wild species, 153
Euphorbia, 253
Euphotic zone, **318**(f)–319
Eutrophication, **313**–315, 314(f)
Eutrophic lake, 313
Exponential growth of population,
    2–3
Exponential reserve index, 187

Extinction of species, and causes,
    152–158(f,t)

F

Fabricated foods, 130–131
Falling water. *See* Hydroelectric power
Family planning, 85–86(t)
Farming:
    contour, 100–101, 119
    fish and shellfish, 132
    organic, 128
    See also Agriculture
Fecal coliform bacteria count, in
    water, 307
Federal Insecticide, Fungicide, and
    Rodenticide Act, 337, 338
Federal Land Policy and Management
    Act of 1976, 17
Federal legislation. *See* Legislation *and*
    titles of specific acts
Federal Water Pollution Act of 1972,
    325–326
Fertility rate, effect on population, 70,
    74–76
    total fertility rate, 74
    U.S. population stabilization, 75–76
    world population stabilization, 74–
    75
Fertilizers:
    animal (manure), 101–102, 253, 254
    for green revolution, 135–137
    and reducing food and energy
    wastes, 128–130
    See also Soils
Fire, as management tool, 142, 144
First-law energy efficiency, **199**–202(t),
    200(f), 201(t)
Fish:
    catching more, 131(f)–132
    farming, 132
Fish and Wildlife Service (FWS), 152,
    159
Fisheries management, 162–163
Fission. *See* Nuclear fission
Fixed-rail rapid transit systems, 171
Flavin, Christopher, 220, 262
Fluidized-bed combustion (FBC) of
    coal, 216(f)–217, 302
Fluorescent light bulbs, 199
Flyways (bird, waterfowl), 161
Food, Drug, and Cosmetic Act of
    1938, 277
Food additives. *See* Additives, food
Food and Agriculture Organization,
    U.N. (FAO), 125, 338
Food and Drug Administration
    (FDA), 277–278, 279, 310, 334,
    342, 354
Food chains, **45**–47(f), 46(f)
    ocean, 131–132
    and second law of thermodynam-
    ics, 46–47, 49(f)
Food resources:
    and additives, **275**–280
    hunger and, 121–137(f,t)
    land reform and, 123, 132–135
    natural versus synthetic, 277
    plant and animal species for hu-
    man, 124
    reducing waste, 128–130

    supplemented and fabricated, 130–
    131
    surplus, 13
    unconventional, 130
    See also Agriculture; Diet; Fish; Soils
Food webs, **46**–47, 48(f), 49(f), 63
Forage, 147
Ford Foundation, 85
Forests, 142–147
    harvesting and regeneration meth-
    ods, 142–143
    importance of, 142
    status of U.S., 146–147(f)
    status of world, 145–146
Forests, National, 15
    clearing of, 13
    management of, 142
    protecting, 142, 144–145
    reserves, establishment of, U.S., 15
Forest Service, U.S., 15, 16, 142, 146,
    147, 148, 334, 337
Formaldehyde, as indoor pollutant,
    285
Fossil fuel era, 5
Fossil fuels. *See* Coal; Natural gas; Oil
Freshwater ecosystems, 30
Freshwater fisheries management,
    162
Freshwater pond ecosystem, 35(f)
Friends of the Earth, 15, 16
Frontier mentality, 14, 361
Fuel cell, 256
Fuels. *See* Energy resources
Fuelwood crisis, 146–147
Fungi, **36**
Fungicides, use, 330

G

Game animals, 154–155, 161
Game fish, 162
Game species, **152**
Gamma rays, 224
Gandhi, Indira, 89
Gardening, home, 128
Gaseous cycles, 48
Gasohol, 254–255
Gene banks, agricultural, 159
Gene pool, **29**
General Accounting Office, 190, 326,
    338, 349
Generalist niche, 54
Generally recognized as safe (GRAS)
    list, 277–278
Genes, 62
Genetic diversity, 62, 136–137
Genetic resistance to pesticides, 332–
    333
Geography of hunger, 121–122
Geography of population, 79(t)–80
Geological Survey, U.S., 182, 189, 220
Geometric growth of population, **2**–3
Georgia Power, 234, 257
Geothermal energy, 217–220(f)
Gilland, Bernard, 82
Global mobility of pesticides, 334–
    335(f)
Gofman, John W., 226
Golden Gate Park, 174

Grains, food. *See* Cropland; Food resources; Gasohol
Grand Canal concept, 114
Grand Council Fire of American Indians, 43
Grant, Ulysses S., 15
Grazing:
    districts, 16
    food chain, **45**
    lands (rangelands), 17, 147–150(f)
Grazing Service, 16
Great Lakes, pollution of, 314
Great Plains, U.S., soil erosion and, 42, 99–100(f)
Greenhouse effect, 38, 297–298
Green manure, as fertilizer, 102
Greenpeace movement, 162–163
Green revolution, 134, 135–137
Groundwater, **107**, 108(f)
    depletion of, 116–117
    pollution of, 116, 117, 316–318(f)
    use, 116(f)–117
Guano, 51
Gully reclamation, 100, 101

H

Haber, Fritz, 50
Habitats, **29**, 53, 153–154, 161
Haeckel, Ernst, 27
Half-life, **224**
Hannon, Bruce, 207
Hardin, Garrett, 16
Hard-path energy strategy, 205
Harrison, Benjamin, 15
Hayes, Denis, 185, 359
Hazardous wastes, **350**–359
Health:
    diseases, 264–275(f,t)
    hazardous wastes threat to, 351–355(f)
    insecticide and herbicide threats to, 336–337
    nuclear war's effects on, 65–66
Heat:
    space, 200(f), 201(t), 204(t)
    water, 201(t)–202(t), 239–243(f), 368–369
Heat islands:
    regional, 170
    urban, 169–170(f)
Heat pollution. *See* Thermal pollution
Heat radiation. *See* Radiation
Heavy oils. *See* Oil
Hegel, Georg, 7
Hemingway, Ernest, 17
Heptachlor, 337
Herbicides:
    bans, 337–339
    case against use, 332–337(f)
    case for use, 331–332
    types and properties, 331
Herbivores, 31, **34**
Heronemus, William E., 251, 252
Hessian fly, 341
Heterotrophs, **34**
Heyerdahl, Thor, 178
High-temperature, gas-cooled reactor (HTGR), 227(n)
Home gardening, 128

Hooker Chemicals and Plastics Corporation, 64
Horizons, soil, 95(f)
Hormones, for insect control, 341–342(f), 343(f)
Horticulture, **13**
Host, **56**
    and parasite relationship, 35
Hot water heating:
    direct solar energy for, 239–243(f)
    efficiency of different systems, 201(t), 202(t), 204(t)
    ways to save energy in, 368–369
Hough, Franklin, 15
Howard, Ebenezer, 176
Hulett, H. R., 82
Humans:
    air pollution's effects on, 292–294(f)
    ecological niche of, 54
    in food chains, 46
    impact on earth, 12–17
Humidity, **37**
Humus, 92
Hunger, world. *See* Food resources
Hunter-gatherer societies, 12–13
Hunting:
    as cause of extinction, 154–155
    wildlife population regulation by controlled, 161
Hydroelectric power, 246–248
Hydrogen gas (H_2) as fuel, 255–256
Hydrologic cycle, 52, 53(f), 107–108
Hydrosphere, **27**, 28(f)

I

Icebergs, towing, 117–118
Illinois Prairie Path, 174
Iltis, Hugh H., 27
Immigration:
    effect on population, 72–73
    restricting, to control population, 88
Immunotherapy, 272
Imports, mineral resources, U.S. dependence on, 188–189(t)
Incandescent light bulbs, 23, 199
Incineration of hazardous wastes, 357
Incomplete proteins, 122
India, population control efforts of, 87–88, 88–89
Indoor air pollution, 285
Industrialized agriculture, 126–128, 127(f)
Industrial smog, 285–290
Industrial societies, 14, 84(f)–85
Infant mortality rate, 266. *See also* Death rate
Infectious disease, **264**, 266(t)
    major types of, 265(t)
Inorganic compounds, **33**
Inorganic fertilizer, commercial, 102–104
Inorganic minerals, in soil, 92
Insecticides:
    bans, 337–339
    case against use, 332–337, 333(f), 335(f)
    case for use, 331–332
    types and properties, 330–331(t)

Insects:
    alternative control methods, 339–344
    damage to forests, 144–145
    infestation effects on land cultivation, 133
    as sources of protein, 130
Institute of Nuclear Power Operations, 230
Integrated pest management (IPM), 343–344
Intercropping, 339
International Atomic Energy Agency (IAEA), 233
International Council for Bird Preservation (ICBP), 158
International Development Corporation Agency, 188
International economic order, new, 188
International Energy Agency, 196
International Monetary Fund, 188
International Planned Parenthood Federation, 85
International Union for the Conservation of Nature (IUCN), 158
International Whaling Commission (IWC), 162
Interspecific competition, 54–55. *See also* Species
Ionizing radiation, **223**–224
Ions, **32**
Iranian oil cutoff, 195–196
Irradiation of foods, for insect control, 342
Irrigation, 133–134
    losses, reducing, 118–119
    problems, 114–116
Isotopes and radiation, 223–226
Itai-itai byo, 353

J

Jacks, G. Y., 92
James River, contamination of, 336
Jitneys, 172
Johns-Manville Corporation, 274(n)
Johnson, Warren, 239
Joint Economic Committee of the House and Senate, 175
J-shaped curves of population growth, **2**–3(f), 81(f)

K

Kepone, 336, 337–338
Kerogen, 209
Kinetic energy, **20**, 21(f)
Kistiakowsky, George, 9
Komanoff, Charles, 222
Krill, harvesting, 132
Kwashiorkor, **125**

L

Lacey Act, 154
Lakes, pollution of, 311–313, 313–316
    detergent controversy, 315–316
    eutrophication, 313–315, 314(f)
    lake characteristics, 313(f)

Landfarming, for hazardous wastes, 357
Landfill, **347**(t)
  for hazardous wastes, 355–357, 356(f)
Land resources:
  classifications of, 101, 103(t), 104(f), 133(f)
  conservation and protection of, 138–150
  cropland (cultivated), 132–135
  rangeland, 147–150
  urban, 164–176(f,t)
  See also Agriculture; Farming; Forests; Land use; Mining; Soils; Wastes, disposal
Landsberg, Hans, 189
Land use:
  economic and ethological ethics of, 138, 140
  planning and control, 172–175(f)
  U.S., 138(f)–140, 139(f)
  urban, 164–176(f,t)
Laterites, 96, 133
Late-successional species, 161
Law of the Sea Treaty, 184
Laws. *See* Conservation of energy, law of; Conservation of matter, law of; Legislation, environmental; Limiting factor principle
LD-50 (lethal dose-50 percent), 8
Lead, 60, 353–354
Legislation, environmental, 14–17, 366
  air pollution control, 300
  bottles and cans, 349
  and control of hazardous wastes, 358–359
  pesticide bans, 337–339
  water pollution control, 325–326
  wildlife protection, 158–159
Lehrer, Tom, 307
Leontief, Wassily, 5
Leopold, Aldo, 15, 137, 138
Leucenda, 130
Levels of organization of matter, 28–29, 30(f)
Lianas, 40
Life expectancy, average, U.S., 267
Light, as limiting factor, 40
Light-rail trolley systems, 171
Light-water reactors (LWR), 227
Limiting factor principle, 38–40, 39(f), **60**, 81, 133–135
Limnetic zone, **313**(f)
Lindane, 337
Lithosphere, **27**, 28(f)
Littoral zone, **313**(f)
Loams, 95
Los Alamos National Laboratory, 232
Love Canal, 64
Lovins, Amory, 199, 205–207
Lung cancer, 271, 293–294

M

McHale, John, 1
McNamara, Robert S., 90
Macronutrients, **48**
Magma, as geothermal energy source, 220
Magnetic confinement, 237(f)

Magnetohydrodynamic (MHD) generation, of coal, 217
Magnification, biological, **60**–61(f), 334, 335(f)
Malaria, 63, 267(f)–268
Malnutrition, nutrition, and hunger, 122, 124–126
Malthus, Thomas Robert, 5
Manure, 101–102, 253, 254
Marasmus, **125**
Mariculture, 132
Marine ecosystems, 30
Marine fisheries management, 162
Markandaya, Kamala, 121
Marriage age, average, effect on population, 70, 79. *See also* Birth rate
Marsh, 16
Marsh, George Perkins, 14
Marshall, Robert, 15
Mass number, 223
Mass transit, 171–172
Mather, Stephen, 14, 15
Matter, **20**
  conservation of, law, 19, **20**, 44
  and energy laws, 19, 20–26
  levels of organization of, 28–29, 30(f)
Maximum sustained yield, of resource, **3**–4
Max Planck Institute of Chemistry, 64
Measurement, units of, 365
Medfly, 340
Meltdown, nuclear reactor, 229
Mencken, H. L., 6
Mercury, 60, 354–355(f)
Metabolic reserve of plant, 148
Metallurgical processes, 180–181
  environmental impact, 182
Methanol, 254–255
Metropolitan Edison, 230
METRO system (Washington, D.C.), 171
Microconsumers, 31
Micronutrients, **48**
Midsuccessional species, 161
Migration, effect on population, 70, 72–74
Migratory Bird Hunting Stamp Act, 161
Minamata, Japan, methyl mercury poisoning, 355
Mineral resources (nonrenewable), 4–5, 178–190(f,t)
  depletion time, 186–187(f)
  oceanic, 183(f)–184
Mining, 178–182, 184, 212–214. *See also* Coal
Minnesota Mining and Manufacturing Company, 358
Mirex, 337
Molecules, **29**, 30(f), 31, 33
Money, as limiting factor, 134–135
Monuments, national, 15
Motor vehicles. *See* Automobile
Muir, John, 14, 15, 57
Multiple use (forests), **15**, 142, 146
Multiple-Use-Sustained-Yield Act, 16
Municipal parks, 174
Mutations, 62
Mutualism, **55**
Mycelia, **36**
Myers, Norman, 153

N

Nash, Roderick, 140
National Academy of Sciences, 137, 225, 226, 230, 274, 279, 295, 297, 298, 337, 351, 356
National Aeronautic and Space Administration (NASA), 157
National ambient air quality standards, 370
National Audubon Society, 159, 262
National Cancer Institute, 336
National Center for Health Statistics, 354
National Coastal Zone Management Act of 1972, 320
National Eutrophication Survey, 314
National Forest Service, 141, 143
National Marine Fisheries Service, 159
National Park Service, 15, 141
National Restaurant Association, 279
National Wilderness Preservation System, 140
National Wilderness System, 16
National Wildlife Federation, 16
National Wildlife Refuge System, 159
Natural community, **29**
  pioneer, 58
Natural eutrophication, **313**
Natural gas, 211–212
  advantages and disadvantages of, 212
  conventional, 211–212
  increase in use of, 194(f), 195(f)
  supplies, 197
  unconventional, 212
Natural radiation, 224, 225(t)
Natural resources, **3**–6
  history of conservation, 14–17
  major types, 5(f)
Natural Resources Defense Council (NRDC), 16
Natural selection, **62**
Nature Conservancy, 17, 159
Neo-Malthusians, 5, 182, 362
Neritic zone, **318**(f)
Net migration rate, 72–**73**
Net population change, **70**–71
Net useful energy, 22, **203**–204(t)
Neutrons (n), **32**, 223, 224
New Alchemy Institute, 128
New international economic order, 188
New towns, 174, 176
New York Bight, 320
Niche concept, 52–54
  competitive exclusion principle, 53
  ecological niches, 52–53
  niche and population size, 53–54
Nitrates, nitrites, 278
Nitrogen cycle, 50–51(f)
Nitrogen dioxide, as indoor pollutant, 285
Nixon, Richard, 16
Noise Control Act of 1978, 281
Noise pollution, 280(t)–281
Nondegradable pollutants, 7
Noninfectious diseases, **265**, 266(t)
Nonmetropolitan areas, 165
Nonpersistent pollutant, 7
Nonpoint sources, 309

Nonrenewable energy resources, 4, 5(f), **191**, 192(f), 193(t)
  changes in consumption, U.S., 195(f)
  depletion curves and rate estimates, 186(f)–187(f), 196–197
  world consumption, 194(f), 196(f)
Nonthreshold pollutants, **7**, 8(f)
Non-vector-transmitted infectious diseases, **264**
Northern coniferous forests, 39(f), 40
No-tillage farming, 101
Nuclear energy, 21(f), 222–238(f,t). *See also* Nuclear fission; Nuclear fusion
Nuclear fission, 226–235(f,t)
Nuclear fuel cycle, 228(f)
Nuclear fusion, 236–238, 237(f)
  achieving controlled, 236–237(f)
  building a reactor, 238
Nuclear Regulatory Commission, 229, 230
Nuclear Safety Analysis Center, 230
Nuclear war:
  as threat to environment, 9
  as ultimate pollutant, 64–66
Nuclear Waste Policy Act, 231–232
Nuclear weapons, proliferation of, 232–233
Nuclear winter, 64–65
Nucleus, 32, **223**
Nutrition, malnutrition, and hunger, 122, 124–126

O

Occupation, cancer and, 274–275
Occupational Safety and Health Act of 1970, 274
Occupational Safety and Health Administration (OSHA), 275
Ocean Dumping Act of 1972, 321
Oceanic zone, 318
Oceans:
  elements in waters, 178(f), 183
  fish and food chains, 131–132
  mineral resources, 183(f)–184
  pollution, 318–322(f,t)
  tidal energy, 255
  wave energy, 248
Ocean thermal energy conversion (OTEC) power plant, 248–250, 249(f)
Odeillo Furnace, 244
Office of Technology Assessment, 155, 234, 292, 316, 332, 356, 359
Oil, 208–211(f)
  advantages and disadvantages of use, 209
  conventional crude, 208–209, 210(f)
  cutoff, Iranian, 195–196
  deposits, U.S., 209(f)
  embargo, OPEC, 194–196
  increase in use, 194(f), 195(f)
  from oil shale and tar sands, 209, 211
  pollution in ocean, 321(f)–322(t)
  supplies, 196–197
Oil shale, extracting oil from, 197, 209, 211
Oligotrophic lake, 313
Olmstead, Frederick Law, 14

Omnivores, **35**
One-way flow of matter and energy, 44
One-way or throwaway society, 24(f)
Open dump, **347**(t)
Open pit mining, 179–180
Open sea, **318**(f)
Open space, urban, preserving, 173–175(f), 174(f)
Opportunist species, 57
Ore, 178
Organic chemicals, in drinking water, 309
Organic compounds, **33**
Organic farming, 128
Organic matter, in soil, 92
Organisms, 29, 30(f)
Organization of matter, levels of, 28–29, 30(f)
Organization of Petroleum Exporting Countries (OPEC), oil embargo, 194–196
Organophosphates, 330, 331(t)
Oryx, Arabian, 160(f)
Osborn, Fairfield, 14–15, 16
Ouch-ouch disease, 353
Overgrazing of rangeland, 148–149
Overnutrition, 124
Overpackaging, 185
Oxygen cycle, 49–50(f)
Oxygen sag curve, 310(f)
Ozone layer, 32, 66, **296**–297

P

Paley Park, 174
Pandemic, 264
Paper, recycling, 349–350
Parasite, **35**
Parasitism, **56**
Paratransit, 172
Parks:
  municipal, 174
  national, 15, 141–142
Particulate matter, 285, 288(f)
  emissions, control, 301, 303, 304(f)
  global cooling from, 298–299
Parts per billion (ppb), **7**
Parts per million (ppm), **7**
Passenger pigeon, 151
Passive solar energy systems, 239–243(f)
PCBs (polychlorinated biphenyls), 7
Pelican Island, 159
Pemex oil well blowout, 321
People's Republic of China. *See* China
Permaculture, **128**
Permafrost, 27, **43**
Persistent pollutant, 7
Pesticides, **328**–344
  export of, 338
  food web disruption caused by, 63
  treadmill, and getting off, 333, 344
  use, U.S., 330
Pests, **328**
  enemies, killing of, 333
  integrated management of, 343–344
  new, creation of, 333–334
  species extinction and control of, 155
Pet foods, 129

Petrochemicals, 208–209, 210(f)
Petroleum, 208–209, 210(f)
  deposits, U.S., 209(f)
  plantations, 253
Pets, threatened or endangered species as, 155
pH, **294**, 295(f)
Phenoxy compounds, 331
Pheromones, for insect control, 341–342(f), 343(f)
Phosphorus cycle, 51–52(f)
Photochemical reaction, 289
Photochemical smog, **286**
  control of, 303, 305
  formation of, 288–289, 290(f)
Photoelectrochemical solar cells, 256
Photosynthesis, 48–**49**
Photovoltaic cells, 245(f)–246
Physical weathering, 93
Phytoplankton, harvesting, 132
Pica, 354
Pimentel, David, 332
Pinchot, Gifford, 15, 143
Pioneer natural community, 58
Pioneer species, 57
Planned Parenthood Federation of America, 85
Planned unit development (PUD), 174, 175(f)
Plants:
  air pollution damage to, 290–292
  endangered, threatened species of, 156, 157(t)
  species for food, 124
Point sources, 307, 309
Polar grasslands, 39(f), 40
Pollution, 2, **6**–9(f,t), 263–359(f,t)
  awareness of, 362
  cancer and, 275
  in cities, 168–170(f)
  degradable and nondegradable, 7
  economics, costs, and controlling, 299(f)–300
  estuaries and coastal wetlands removal of, 320
  of food, 275–280
  in forests, 145
  harmful levels of, 7–8
  laws of conservation of matter and, 20, 24–25
  relationships among population, resources, technology, and, 9
  sources of, 8(t)–9
  types, 7
  wildlife threatened by, 156
  See also specific kinds
Polychlorinated biphenyls (PCBs), 352–353
Population, human:
  controlling, 81–90(f,t)
  crash, 81–82(f)
  dynamics, 70–80(f,t)
  geographic distribution, 79(t)–80
  growth (net change), 2–3(f), 70–72(f,t), 73(f), 82–83, 121, 164–165(f,t)
  overpopulation, 2–3(f)
  resources, technology, economics, and, 9. *See also* Food resources
  stabilization, 74–76(f,t)
  U.S., 75–76(f), 165–168(f)
  urban areas, 164–168(f,t)

Populations, species, **29**, 30(f), 53–54, 60, 61–62, 161. *See also* Diversity; Niche concept
Potential energy, **21**(f)
Poverty:
in cities, 168–169
hunger, malnutrition, and, 121–123, 125(f)
Powell, John Wesley, 15
Power tower, 244(f)
Precipitation, 37, 39(f)–40
waste treatment method, 323
Predators/predation, **55**
control programs, 150, 155
Preservationist land-use ethic, 15, 140
Pressurized-water reactor (PWR), 227(f)
Prevailing ground winds, **37**
Prey, **55**
Price-Anderson Act, 233
Priestly, J. B., 170
Primary air pollutant, **284**, 286(t)
Primary consumers, **31**, 34(f)–35(f)
Primary energy resources, 191, 192(f), 193(t)
changes in consumption, U.S., 195(f)
world consumption, 194(f), 196(f)
Primary succession, 57–58(f)
Primary waste treatment, 323, 324(f)–325(f)
Processed foods, 275
Producers, **31**, 34(f)–35(f)
Products, chemical, **33**
Profiles, soil, 95–96, 97(f)
Profundal zone, of lake, **313**(f)
Protection of wildlife, 158–161
Protein-calorie malnutrition, 124
Proteins, human need for, 122
Protein supplements, 130–131
Protons (p), **32**, 223
Proverbs, 361
Public Utilities Regulatory Policies Act, 250
Pulmonary emphysema, **293**
Pure-Cycle Company, 119
Pure Food and Drug Act of 1906, 277
Pyramid of energy, 46, 49(f)
Pyramid of numbers, 46, 49(f)
Pyrethroids, 330
Pyrethrum, 329

Q
Quality, energy, 197–198(f)
matching to end uses, 198–199
Quality, water:
drinking, 309–310
indicators of, 307
Quarries, 179

R
Radiant energy, 27–28
Radiation, **223**–226
background or natural, 224–225(t)
danger of, 225–226
effects on body, 224(t)
exposure to, 224–225(t)
ionizing, **223**–226
Radioactive isotopes, 223–224

Radioactive wastes, disposal and storage of, 230–232, 231(t)
Radioisotopes, **223**
Rain, acid. *See* Acid deposition
Rain, producing, 118
Rain forests. *See* Temperate rain forests; Tropical rain forests
Rangelands, 133(f), 147–150(f,t). *See also* Grazing
Range of tolerance, 60(f)
Rapidly degradable pollutant, 7
Rasmussen report, 229
Rate of natural (population) change, 71(t), 73(f)
Reactants, chemical, **33**
Reagan, Ronald, 17
Reclamation of mined land, 180
Recycling, 4–5, 24, **185**, 186(t)
hazardous wastes, 357–358
paper, 349–350
plant nutrient wastes, 324(f)–325(f)
Refuges, wildlife, 159
Regional heat islands, 170
Relative humidity, **37**
Renewable energy resources, 3, 191, 192(f), 193(t), 239–262(f,t)
biomass energy, 252–255
changes in consumption, U.S., 195(f)
developing an energy strategy for U.S., 258–262
energy conservation, 256–258(t)
hydrogen as fuel, 255–256
solar energy, direct, 207, 239–246(f,t)
thermal gradient energy, 248–250, 249(f)
tidal power, 255
water and wave energy, 246–248
wind energy, 250–252
world consumption, 194(f), 196(f)
Reserves:
mineral resources, 182
wildlife, 161
Resistant crops and animals, breeding, for insect control, 341
Resource Conservation and Recovery Act of 1976, 348, 350, 355, 358
Resources, 2, **3**–5, 91–262(f,t)
average use (actual and potential) of, 5–6, 182–184
population, technology, economics, and, 9, 82
projection of future, 5–6, 182
total, 182
See also Energy resources; Food resources; Land resources; Mineral resources; Natural resources; Nonrenewable energy resources; Renewable energy resources; Soils; Water; Wildlife
Reuse, 4–5, 185–186(t), 357–358. *See also* Recycling
Rickover, Admiral Hyman G., 222
Risk-benefit analysis, 7
Rivers:
as energy source, 114, 246–248
pollution of, 310–313(f,t)
Roosevelt, Theodore, 14, 15, 159
Roughage, 122
Ruckelshaus, William D., 16

Rudd, Robert L., 344
Runoff, 107
Rural-to-urban population shift, 165

S
Safe Drinking Water Act of 1974, 309
Sakharov, Andrei, 237
Salinity, from irrigation, 114–116, 134. *See also* Desalination
Salk, Jonas, 70
Salk, Jonathan, 70
Saltwater intrusion into groundwater, 116. *See also* Seawater
Sand, 94
Sandia National Laboratory, 220
Sandvig, Earl, 149
Sanitary landfills, **347**(t), 348(f)
Saprophytes, **35**
Savanna, 39(f), 40
Scherer, Robert, 234
Schistosomiasis, 268–269(f)
School of Public Health, Harvard University, 316
Scientific Committee on Effects of Atomic Radiation, U.N., 226
Scientific conservation, land-ethic use, 15, 140
Screwworm fly, 340–341
Scrub, tropical thorn, 41
Sea, Law of the (treaty), 184
Seabed mining, 184
Seabrook nuclear power plant, 233
Sea lamprey, 35(f)
Sears, Paul B., 14, 16
Seasonal forests, 40–41
Seawater, elements in, 178(f), 183. *See also* Oceans; Saltwater intrusion
Secondary air pollutant, **284**–285
Secondary consumer, **31**, 34(f)–35(f)
Secondary succession, 59(f)
Secondary waste treatment, 323, 324(f)–325(f)
Second-law energy efficiency, **202**(t)–203(f)
Secured landfill, **347**–348, 356(f)
Sedimentary cycles, 48
Seed-tree cutting (forests), 143
Selective cutting (forests), 142
Self, Peter, 176
Sewage treatment, 323–324, 324(f)–325(f). *See also* Wastes, disposal
Shale oil, 209, 211
Shallow ecology, 362
Shelterwood cutting (forests), 142–143
Shifting cultivation, 13, 98
Sierra Club, 15, 16
Silent Spring, 16, 328
Silt, 94
Silvex, 338, 352
Silviculture, **142**
Simon, Julian, 82
Simplicity, balancing diversity and, 66(t)
Simplification, as effect of environmental stress, 63(t)
Single-cell protein (SCP), 130
Skin cancer, 271
Slash-and-burn cultivation, 13
Slowly degradable pollutant, 7
Smog, 283, 285–289(f,t)

Smokey the Bear, 143
Smoking, cancer and, 271, 273
Snail darter, 159
Snowy egret, 154
Soap versus detergents, 315–316
Soddy, Frederick, 20
Soft-path energy strategy, 205–207
Soil Conservation Service (SCS), 16, 98, 100
Soils, 40, **92**–104(f,t)
   classifying, 101–104(f,t)
   components, 92–93(f)
   conservation, 100–101, 323
   erosion, 96–100(f), 102(f), 148–149
   fertility of, 101–104
   formation and development, 93–94
   horizons, 95(f)–96
   as limiting factor, 133(f)
   profiles, 95(f)–96
   properties, 94(f)–95
   structure, 94–95
   texture, 94–95
   See also Fertilizers
Soil sterilant herbicides, 331
Solar energy, 24–25, 27–28
   direct, 239–246(f,t)
   effect on earth, 37–38(f)
   evaluation of, 261(t)
   indirect, 246–255(f)
   passive and active systems, 239–243(f)
   photosynthesis, 48–**49**
   from photovoltaic cells, 25, 245(f)–246
Solar Energy Research Institute, 251
Solar furnace, 244
Solar I, 244
Solar photovoltaic cells, 25, 245(f)–246
Solar ponds, **250**
Solid wastes, **346**–350
Space heating:
   energy efficiency, costs, 200(f), 201(t)
   net useful energy of systems, 204(t)
Spaceship earth, 362
Species, **29**
   competition, predation, and symbiosis, 54–56
   diversity, and ecosystem stability, 62
   ecological succession, 57–59(f), 58(f)
   endangered, extinct, threatened, 152–161(f,t)
   extinction prone, 157–158(t)
   introduction (importation), 63, 156–157(t)
   preserving wild, 140, 152–153
S-shaped curve of population growth, 81(f)
Stability, in living systems, 59–62
Steinwill, Gerald D., 109
Sterilization, insect control by, 340–341
Sternglass, E. J., 226
Stevenson, Adlai E., 2
Stomach cancer, 272
Stress, effects/responses, 60–63(t)
Strip cropping, 100, 101
Strip mining, 180(f)
Subatomic particles, **29**, 30(f)

Subbituminous coal, 214, 215(f)
Subsidence, land, 116, 213
Subsistence agriculture, 126
Subsoil, 95(f), 96
Substituted ureas, 331
Substitute resource, 4
Substitution, for scarce resources, 184–185
Subsurface mining, 178–179, 181
Succession:
   ecological, 57–59
   urban economic and social, 165, 168
Sulfites, 278–279
Sulfur cycle, and industrial smog, 286–288, 287(f)
Sulfur dioxide, as dangerous food additive, 278
Sulfur oxides, 285, 301–302
Supercity, 165
Superfund program, 358–359
Superinsulated house, 257
Superphénix, 235
Surface mining, 179(f)–180(f), 181, 213
Surface Mining Control and Reclamation Act of 1977, 213–214
Surface water, 107
   pollution, 310–316(f,t)
Sustainable earth, 25(f)–26, 140, 186, 361, 362
Sustained yield, 15, 142, 146
Swift, Ernest, 15
Symbiosis, 55–56
Synergism, 8
Synfuels, 217
Systemic herbicides, 331

T

Taiga, 42
Tansley, A. G., 27
Tapiola, Finland, 176
Tar sands, extracting oil from, 197, 209, 211
Taylor Grazing Act, 16
TCCD, 352
Technology, relationship with pollution, population, and resource use, 9
Tellico Dam, 159
Tempeh, 131
Temperate biomes, 41–42
Temperate deciduous forests, 41–42
Temperate desert, 39(f)
Temperate forests, 39(f), 40, 41
Temperate grasslands, 39(f), 40, 42
Temperate rain forests, 41
Temperate shrubland, 42
Temperate woodlands, 42
Temperature, as limiting factor, 39(f)–40
Tennessee Valley Authority, 159
Ten percent rule, **46**
Terracing, 100, 101, 119
Tertiary consumer, 35(f)
Tertiary waste treatment, 323, 324(f)–325(f)
Thames River, cleanup, 10, 311
Thermal enrichment, 311
Thermal gradients, energy from, 248–250, 249(f)

Thermal inversion, **289**, 291(f)
Thermal pollution, of rivers and lakes, 311–313(f,t)
Thermodynamics, laws of, 20, **21**–22, **22**(f)–24, 26, 44, 46–47, 49(f)
   end uses, energy quality and, 197–203(f,t)
   See also Energy
Third-generation pesticides, 341
Thoreau, Henry David, 10, 14
Thorn woodland biome, 41
Threatened species, **153**, 158–159
Three Mile Island (TMI) accident, 222, 229–230, 232, 234
Threshold effect, 60
Threshold pollutants, **7**, 8(f)
Throwaway society, 24(f), 186(t), 361
Tidal power, 255
Tight sands, natural gas from, 197, 212
TMI Public Health Fund, 229
Tokamak, fusion test reactor, 237(f)
Tolerance:
   law of. *See* Limiting factor principle
   range of, **60**(f)
Topography, air pollution and, 289–290
Topsoil, 95
Total fertility rate (TFR), 74
Toxaphene, 337
Toxic Substances Control Act of 1975, 274
Toxic Substances Control Act of 1976, 358
Tragedy of the Commons, 162
Transpiration, 92
Transportation:
   net useful energy of systems, 204(t)
   urban, 170–172
   ways to save energy in, 369
Treaties:
   Law of Sea, 184
   on Nonproliferation of Nuclear Weapons, 233
   wildlife protection, 158–159
Tree harvesting and regeneration methods, 142–143
Triage, environmental, 160
Triazines, 331
Trombe, Felix, 239(n)
Trombe wall, 239(n)
Trophic levels, **45**, 46(f)
Tropical biomes, 40–41
Tropical deciduous forests, 40–41
Tropical desert, 39(f)
Tropical grasslands, 39(f), 40
Tropical rain forests, 39(f), 40
Tropical savanna, 41
Tropical thorn scrub biome, 41
Troposphere, **283**, 284(f)
Tundra, 39(f), 40, 42–43
2,4,5-T, 338, 352

U

Udall, Stewart L., 15, 44, 326
Ultimate sink, ocean as, 319
Unconventional foods, 130
Underground aquifers, 108, 116, 316
Underground mining, 213
Undernutrition, 124

Undiscovered resources, 182
Union Carbide pesticide plant leakage, 336
United Nations, 164, 165, 252, 338
United Nations Fund for Population Activities (UNFPA), 85
United Nations International Conference on Population, 83
Urban areas, **164**
  benefits and stresses, coping with, 168–169(f), 175–176
  growth, 164–165(f,t)
  land use, 164–176(f,t)
  parks, open space in, 141–142, 173–175(f)
  transportation within and between, 170–172
  U.S. regions, 165–168(f)
  wastes of, 169(f), 346–350
  See also Cities
Urban heat islands, 169–170(f)
Urban homesteading programs, 175
Urban succession, 165, 168
Uterine cancer, 271

V

Valley of the Drums, 351
Van Hise, Charles, 15
Variety, reducing in ecosystems, 63
Vector-transmitted infectious diseases, **264**, 265(t)
Veterans Administration, 337
Vinyl chloride, cancer and, 274
Vitamin supplements, 130–131

W

Warm desert, 41
WASH-1400, 229
Waste clearinghouses, 357–358
Waste exchanges, 357–358
Waste heat, using, 203
Waste Isolation Pilot-Plant Project, 231
Wastes, disposal:
  animal, 101–102, 253
  crop, 253
  as energy sources, 252–253
  hazardous, **350**–359
  nuclear, radioactive, 230–232, 231(t)

point sources, 307, 309
reducing food, 128–130
sewage treatment, 323–324, 324(f)–325(f)
solid, **346**–350(f,t)
typical daily U.S. city, 169(f)
urban, 169(f), 346–350
Water:
  conservation, 119, 367
  drinking, standards, quality, 309–310, 371
  elements in seawater, 178(f), 183
  for green revolution, 135–137
  groundwater, **107**, 108(f), 116–117, 316–318(f)
  heating hot, 201(t)–202(t), 239–243(f), 368–369
  in human nutrition, 122
  importance and properties of, 106–107
  as limiting factor, 133–134
  in soil, 92
  use, U.S. and world, 108–109(f), 110(t)
  See also Hydroelectric power; Oceans; Water pollution; Water resources
Water diversion projects, 114
Waterfowl, migratory, 161
Water hyacinths, 156–157(t)
Water (hydrologic) cycle, 52, 53(f)
Waterlogging, from irrigation, 114–116, 134
Water pollution, 114–116, 307–326(f,t)
  agricultural, 323–325(f)
  and cancer, 275
  groundwater, 116, 117, 316–318(f)
  surface, 310–316(f,t)
  underground aquifers, 316
Water resources, 106–119(f,t)
  methods for managing, 112–119(f,t)
  worldwide, 107–112, 113(f)
Water Resources Council, 109, 110
Watersheds, 107
Water table, 107
Wave energy, 248
Wavelength, energy, 28
Weather:
  and climate, **36**
  controlling, 118
Weathering, 93

Weinberg, Alvin M., 222
West Falmouth, Mass., oil spill, 322
Whaling industry, 162–163
Whitehead, Alfred North, 6
Whole-tree harvesting, 143
Whyte, R. O., 92
Wild and Scenic Rivers Act of 1968, 140, 247
Wilderness, **16**, 140–141
  species, 161
Wilderness Act of 1964, 140
Wilderness Society, 15, 16
Wildlife, **152**
  pesticides as threat to, 334–336
  protection, management, refuges, 158–161
  and wildlife resources, 151–163, **152**
Wilson, Alexander, 151
Wilson, Edward O., 150
Windbreaks, 100, 101
Wind energy, 250–252
Winds. *See* Air
Women's roles, changes in, effects on birth rate, 87
Workplace, cancer and, 274–275
World Bank, 85, 121, 146, 188
World Health Organization (WHO), 3, 63, 65, 109, 125, 265, 268, 331, 332
World Trade Center, 257
Worldwatch Institute, 96, 119, 146, 247
World Wildlife Charter, 151
World Wildlife Fund (WWF), 158

X

X rays, 224, 225(t), 226

Y

Ye-ed, 130
Yellowstone National Park, 15
Yellowstone Timberland Reserve, 15
Yosemite Forest, 15

Z

Zero population growth (ZPG), 10, 75
Zoos, 156, 159–160

Tropical savanna biome

Shortgrass grassland biome

Lynn Erckmann, University of Washington / BPS

Arctic tundra biome

Dennis Brokaw

Chaparral biome

Dennis Brokaw

Taiga biome